# Problem Books in Mathematics

Edited by P.R. Halmos

# Problem Books in Mathematics

Series Editor: P.R. Halmos

**Polynomials**
by *E.J. Barbeau*

**Problems in Geometry**
by *Marcel Berger, Pierre Pansu, Jean-Pic Berry and Xavier Saint-Raymond*

**Problem Book for First Year Calculus**
by *George W. Bluman*

**Exercises in Probability**
by *T. Cacoullos*

**An Introduction to Hilbert Space and Quantum Logic**
by *D. Cohen*

**Problems in Analysis**
by *Bernard Gelbaum*

**Exercises in Integration**
by *Claude George*

**Algebraic Logic**
by *S.G. Gindikin*

**An Outline of Set Theory**
by *James M. Henle*

**Demography Through Problems**
by *N. Keyfitz and J.A. Beekman*

**Theorems and Problems in Functional Analysis**
by *A.A. Kirillov and A.D. Gvishiani*

**Problem-Solving Through Problems**
by *Loren C. Larson*

**A Problem Seminar**
by *Donald J. Newman*

**Exercises in Number Theory**
by *D.P. Parent*

E.J. Barbeau

# Polynomials

With 36 Illustrations

Springer-Verlag
New York Berlin Heidelberg
London Paris Tokyo

✓

E.J. Barbeau
Department of Mathematics
University of Toronto
Toronto, Ontario M5S 1A1
Canada

*Series Editor*
P.R. Halmos
Department of Mathematics
University of Santa Clara
Santa Clara, CA 95053
USA

Mathematics Subject Classification (1980): 12-01, 26-01, 12D05, 12E05, 12E12, 13M10, 26C05, 26C10, 30C10, 65H05, 65H10

Library of Congress Cataloging-in-Publication Data
Barbeau, E.J.
    Polynomials/E.J. Barbeau.
        p.    cm.—(Problem books in mathematics)
    1. Polynomials.  I. Title.  II. Series.
QA281.B37   1989
512.9′42—dc19                          88-39062

Printed on acid-free paper.

Camera-ready copy prepared using LaTeX.
Printed and bound by R.R. Donnelley & Sons, Harrisonburg, Virginia.
Printed in the United States of America.

9 8 7 6 5 4 3 2 1

ISBN 0-387-96919-5 Springer-Verlag New York Berlin Heidelberg
ISBN 3-540-96919-5 Springer-Verlag Berlin Heidelberg New York

873796

To Eileen

# Preface

Particularly during the last thirty years, many criticisms have been directed at the school mathematics curriculum. In response, a number of movements have left their trace—New Mathematics, Real-world Applications, Problem Solving and now Back to the Basics. Moreover, with so many students encouraged to take mathematics for the sake of their careers, educators have tried to respond in a practical way to the difficulties they find in the subject.

The result is that mathematics in school is suffering from ecological overload. The attempt to respond in a piecemeal way to often conflicting advice has threatened the enterprise with being swamped. Whatever the merits of the criticisms of the traditional mathematics program and however compelling the psychological and political consequences of high failure rates, the attempt at a resolution seems often to have resulted in a denatured curriculum, one from which any depth, sophistication or joy has been rigorously expunged. Students nibble at topics, abandoning them before they discover any reason to master them. The mathematics taught is quickly lost to memory and must be reviewed at a later stage (often in a remedial class).

Rather than fragment mathematics, it may be more productive to take an integrated approach, in which students are encouraged to dwell on a mathematical topic long enough to sense how it is put together and what its proper context is. Formerly, students might spend a whole year in a single area of mathematics—Euclidean geometry, the analytic geometry of conic sections, trigonometry and statics, theory of equations. They had the chance to learn many techniques and experience through astute reasoning and manipulation the power of mathematics. Better students would develop a sensitivity to pattern and elegance, and find mathematics both substantial and satisfying.

This book is not a textbook. Nor is its topic being particularly recommended for inclusion, indiscriminately, into the school curriculum. However, it should convey some of the breadth and depth found close to the traditional school and college curricula, and encourage the reader not only to follow up on some of the historical and technical references, but to pull out pen and paper to tackle some problems of special interest. Some of the mathematics will be difficult, but I believe that it will all be accessible.

The intended audience consists of students at both high school and college who wish to go beyond the usual curriculum, as well as teachers who wish to broaden their mathematical experience and discover possible material for use with their regular or enriched students. In particular, I am concerned about two groups of students.

There are those who romp through the school curriculum in mathematics while they have yet to complete other subjects. A standard response to this situation is to accelerate them, either into calculus or into college prematurely. While this is undoubtedly appropriate for some, my experience is that very often such acceleration is counterproductive and leads to an unsettled academic experience.

Then there are those who get caught up in contest activity. It is now possible to spend much of the spring semester preparing for and writing contests, and this may have some value. However, there are some for whom contests are not congenial and others who emphasize the short-term goal of solving problems and winning contests at the expense of proper mathematical growth.

What seems to be needed is a mathematical enrichment which starts with school mathematics, broadens it and yet is sufficiently down-to-earth that the student can explore it in an elementary way with pencil and paper or calculator.

The theory of equations seems to fill the bill. There is a large algorithmic component, so that students can enjoy technical mastery. At the same time, they are led through their experiences into an appreciation of structure and a sense of historical and mathematical context. Beginning with topics of high school—factoring, theory of the quadratic, solving simple equations—polynomial theory looks forward to central areas of the university curriculum. Having seen the derivative and the Taylor expansion in an algebraic setting, and having graphed polynomials and appreciated the role of continuity of polynomials in root approximation, students will then see in calculus how these ideas can be adapted to a wider class of functions. The algorithms of evaluation, factoring and root approximation will provide a base of experience upon which a college numerical analysis course can be built. The ring of polynomials provides a concrete model of an abstract structure encountered in a modern algebra course. Having studied the role of the complex plane in the analysis of polynomials, students will better be able to appreciate the richness of a complex variable course and see many of the results there as extensions from polynomials to a wider class of functions. Other areas, such as combinatorics, geometry and number theory, also make a brief appearance.

I offered a course on polynomials for four successive years to high school students in the Toronto area. They were given a set of notes, a monthly problem set for which solutions were submitted for grading, a monthly lecture at the university and a set of videotaped lectures. It was advertised for those who had completed school mathematics, but were still in high school. Many students enrolled in the course, some stuck with it and only a few wrote the optional examination at the end of the year. However, the profile of the students who did well is interesting. They were not always the final-year senior students, who were "busy" making sure they got grades high enough to get into college, nor were they, on the whole, local contest winners. Several were students who still had another year to spend at high school (with some mathematics left to take); they struggled with the problem sets, but their work improved steadily during the year. One participant gave the following assessment of her experience:

> After innumerable years of "math enrichment" consisting of pointless number games, I was prepared for another similar course. Great was my surprise when I found this course to be extremely challenging. Its difficulty was somewhat dismaying at the start, but now I find that many doors have been opened and that I have the confidence to tackle more complex ideas in math. ... I have gained a great deal of insight into a subject I trivially used to discard as an easy school course. But most important for me is that I have gained a vast amount of faith in my ability to solve challenging problems.

It is assumed that the reader can manipulate simple algebraic expressions and solve linear and quadratic equations as well as simple systems in two variables. Some knowledge of trigonometry, exponentials and logarithms is required, but a background in calculus is not generally needed. The few places in which calculus intervenes can be passed over. While many of the topics of this book will not appear in regular courses, they should be of value through their historic importance, application or intrinsic interest and as a backdrop to other college-level material.

Since this is not intended to be a comprehensive treatment, readers are encouraged to delve into the often excellent publications that are recommended. They will find that the boundary between elementary and deep mathematics is often very thin, and that close to results known for centuries one finds frontiers of modern research.

The book is organized along the following lines:

(a) Exercises: These introduce the basic ideas and advance the required theory. Through examples, students should grasp the principal results and techniques. The emphasis is on familiarity rather than proof; while readers should get some sense of why a given result is true, it is expected that they will have recourse to some other text for a formal treatment. Students should work through the exercises in order, consulting the hints and answers where necessary. However, if they feel that they have a general understanding, they might skim through and work ahead, backtracking if necessary to pick up a lost idea. Readers who find the last three sections of Chapter 1 difficult may wish to proceed to Chapter 2 and 3 and return to these sections later.

(b) Explorations: While these are inserted near related material in the exercises, readers should not feel obliged to work at them right away. In general, they are not needed to follow the main thread. Their purpose is to raise questions and encourage investigation; some explorations involve new theory, some are straightforward problems and others involve questions which have deep ramifications. All are intended only as starting points. The investigations should be revisited as more experience is gained.

(c) Problems: Each chapter concludes with problems drawn from a variety of sources: journals such as the *American Mathematical Monthly* and *Crux Mathematicorum*, contests and Olympiads, examination and scholarship papers. The first ten or so of each set are moderately difficult, but after that they are not arranged in any particular order. Some are tough. Students who get blocked should return to the problem intermittently. Hints are provided.

In referring to exercises and problems, I will use a single number to refer to a question in the same section, and the section number with the question number separated by a dot to refer to a question in a different section of the same chapter. A triple designation will refer to a question in a different chapter; for example, 2.3.4 refers to Exercise 4 of Section 3 of Chapter 2.

One source of problems is worth special mention. Until the mid 1960s, students in Ontario wrote Grade 13 examinations set by the provincial Department of Education. Besides the regular papers (Algebra, Analytic Geometry, Trigonometry and Statistics), students vying for a university scholarship had the opportunity to write a Mathematical Problems Paper. Through Jeff Martin of the Etobicoke Board of Education, I have acquired copies of these papers. In many of the problems, I have been struck by the emphasis on mathematical competence; they could be done, not by a leap of ingenuity, but rather through a thorough grasp of standard but somewhat sophisticated techniques. These problem papers (and I am sure they had their counterpart in other jurisdictions) should not be lost to our collective memories; they are indicative of the skills which were expected of a previous generation of students who planned to do university level mathematics. I believe that students still need to be skillful, and indeed should not be denied the pleasure of feeling competent in what they do.

I would like to acknowledge the assistance and advice of various organizations and individuals. In particular, I am indebted to the Ontario Ministry of Education and the Queen's Printer of Ontario for permission to use problems from the Ontario Problems Papers, the Canadian Mathematical Society for permission to use problems appearing in *Crux Mathematicorum*, the *Canadian Mathematical Olympiad* and its other publications, and the Mathematical Association of America for permission to use problems from the *Putnam Competition*, the *Monthly* and the *Magazine*.

I am grateful to the Samuel Beatty Fund, administered by a board representing the graduates of Mathematics and Physics at the University of

Toronto, for a grant to hire a student to check over the manuscript. Miss Azita Bassiji, a Toronto undergraduate, has been helpful with her suggestions.

Various colleagues have looked through the material and offered useful advice, in particular Peter Borwein of Dalhousie University in Halifax, Nova Scotia, Tony Gardiner of the University of Birmingham in England, Abe Shenitzer of York University in Toronto, Ontario, and John Wilker of the University of Toronto. I am also thankful to the many students and teachers who functioned as guinea pigs, especially, to Jim Farintosh, a teacher at the George S. Henry Academy in North York, Ontario, for his enthusiasm and insights and to Ravi Vakil, currently an undergraduate at the University of Toronto, for his comments on an early draft.

I heartily praise Paul Halmos, the general editor of the series to which this book belongs, for his open-hearted acceptance of the concept of the book and his encouragement in bringing it to fruition. While I was preparing the manuscript, University College of the University of Toronto provided access to a word processor and printer. With pleasure, I acknowledge the understanding and efficiency of Springer-Verlag in preparing the book for publication, notably those in the Editorial and Production departments. Finally, I wish to express my deep appreciation to my wife, Eileen, and children, Judy and Paul, for their support and encouragement over the many years that this project was maturing.

E.J. Barbeau

# Acknowledgment of Problem Sources

Virtually all of the problems in this book come from elsewhere. Some of them I have gotten from colleagues and students, or from competitions and examinations long since forgotten, so that I cannot now identify from where they came. However, others have been deliberately drawn from specific sources, which I would like to acknowledge.

1. *Ontario Problems Papers.* With the permission of the Ministry of Education and the Queen's Printer of Ontario, the following are taken from the Grade 13 Problems Examinations set annually by the Department of Education prior to 1967:

    Section 1.8: 1, 9, 10, 11, 19, 20, 21, 23
    Section 1.9: 5, 6, 7, 9
    Section 2.5: 3, 4, 6, 12, 13
    Section 3.6: 7
    Section 3.7: 1(j), 3, 4, 5, 6
    Section 3.8: 5, 28, 29
    Section 4.1: 9
    Section 4.7: 8
    Section 4.8: 4, 5, 6, 7, 8, 9
    Section 6.1: 1
    Section 6.4: 1, 2, 4

2. *Canadian Mathematical Society.* With the permission of the Canadian Mathematical Society, the following are included:

(a) from *Crux Mathematicorum:*

    Section 1.8: 3, 8, 14, 16
    Section 1.9: 4, 14
    Section 2.5: 17
    Section 3.7: 2, 17, 22
    Section 3.8: 7, 11, 12, 13
    Section 4.7: 3, 7
    Section 4.8: 20, 21, 22, 29, 33
    Section 4.9: 3, 4
    Section 5.4: 8, 9, 17, 30
    Section 7.4: 1
    Section 7.5: 1

Chapter 8: 1, 8, 11, 12, 15, 22, 28, 48
Problems from Olympiads and other competitions listed in Section 4

(b) from the *Canadian Mathematical Olympiad*
Section 6.4: 3
Chapter 8: 19

(c) from the *Canadian Mathematical Bulletin*
Chapter 8: 23

3. *Mathematical Association of America*. With the permission of the Mathematical Association of America, the following are included:

(a) from the *Putnam Competition*
Section 1.9: 2, 13, 19
Section 2.5: 10, 14, 15
Section 3.7: 8
Section 3.8: 17, 18, 19
Section 4.7: 9
Section 4.9: 8, 9
Section 5.2: 15
Section 5.4: 13, 14, 16, 19, 20, 25
Section 6.4: 14

(b) from the *American Mathematical Monthly*
Section 1.8: 2, 15, 18
Section 1.9: 15
Section 2.5: 1, 16
Section 3.7: 12, 14, 16
Section 3.8: 20, 21, 22, 23, 24, 27, 30
Section 4.7: 4, 11
Section 4.8: 11, 17, 27
Section 4.9: 2, 5
Section 5.4: 3, 11, 18, 22, 23, 24, 27, 28, 29, 31, 33, 34, 35
Section 6.4: 11, 13, 15, 16
Section 7.4: 4, 7
Section 7.5: 7, 10
Chapter 8: 2, 3, 7, 29, 31, 32, 49

(c) *USA Mathematical Olympiad*
5.4.12; 8.5

(d) *Mathematics Magazine*
5.4.4

(e) *College Mathematics Journal*
4.7.13; 4.8.33

4. *Other Sources.* Problems from *Elemente der Mathematik* are used with the permission of the editor. Problems from *Normat* are used with the permission of the editor and the publisher. Problems from the Olympiads

and other competitions were taken from the journal *Crux Mathematicorum.*
I am grateful to these sources and to organizers of the contests for allowing
the use of these problems.

University of Toronto examinations (1859–1865):
  3.7.1 (t, u); 4.8.23, 30, 31, 32; 8.21, 47, 56
Cambridge Tripos (1870s): 1.7.17; 2.5.5; 3.7.9; 3.8.15; 4.8.15, 16
Elemente der Mathematik: 3.8.2; 4.8.28; 5.4.21, 26; 7.4.10
Normat: 8.30, 8.39, 8.50
Olympiads: International 4.8.25; 8.35
  Austrian 1.8.12; 6.4.8
  Leningrad 2.5.2
  Moscow 3.7.1(h)
  Australian 6.4.7
  Hungarian 7.5.8
  Bulgarian 7.5.9; 8.25
Other competitions: Albertan (Canada) 3.7.2
  Greek High School (1984) 4.7.2
  Austrian-Polish 4.8.1
  Bulgarian 4.8.26
  Romanian 8.16; 8.20

# Contents

# 1

# Fundamentals

## 1.1 The Anatomy of a Polynomial of a Single Variable

$3t^3 - 7t^2 + 4t + 1$ and $8t^6 - t^5 + \sqrt{2}t^2 + \sqrt{-1}t - 1$ are polynomials. So are $(t^2 - 1)/(t - 1)$ for $t \neq 1$ and $\cos 2(\arccos t)$ for $-1 \leq t \leq 1$. But $t^{1/3}$ and $\sin t$ are not polynomials. What do we mean by a polynomial?

> *A function of a single variable t is a polynomial on its domain if we can put it in the form*
>
> $$a_n t^n + a_{n-1}t^{n-1} + \cdots + a_1 t + a_0$$
>
> *where $a_n, a_{n-1}, \ldots, a_1, a_0$ are constants.*

This definition says that every polynomial can be expressed as a *finite* sum of monomial terms of the form $a_k t^k$ in which the variable is raised to a nonnegative integer power. We use the convention that $t^0 = 1$, so that $a_0 t^0 = a_0$. To begin with, we will look at polynomials for which the constants $a_i$ are real or complex numbers.

With this definition in hand, we can immediately agree that the first two functions are polynomials. For the next two, we have to remove a disguise:

$$(t^2 - 1)/(t - 1) = t + 1$$

$$\cos 2(\arccos t) = \cos 2\theta = 2\cos^2 \theta - 1 = 2t^2 - 1$$

where $t = \cos \theta$, $0 \leq \theta \leq \pi$. The last two, $t^{1/3}$ and $\sin t$, do not look like polynomials, but how can we decide for sure? One way is to look for properties which distinguish these functions from polynomials. One of the tasks of this book will be to provide a number of such characteristics to assist in this sort of classification question.

In the title of this section, we promised you some anatomy. Here it is. For the polynomial, $a_n t^n + a_{n-1}t^{n-1} + \cdots + a_1 t + a_0$, with $a_n \neq 0$, the numbers $a_i$ ($0 \leq i \leq n$) are called *coefficients*. $a_n$ is the *leading coefficient*, and $a_n t^n$ the *leading term*. $a_0$ is the *constant term* or the *constant coefficient*. $a_1$ is the *linear coefficient* and $a_1 t$ the *linear term*. When the leading coefficient $a_n$ is 1, the polynomial is said to be *monic*.

The nonnegative integer $n$ is the *degree* of the polynomial; we write $\deg p = n$. A *constant* polynomial has but a single term, $a_0$. A nonzero constant polynomial has degree 0, but, by convention, the zero polynomial (all coefficients vanishing) has degree $-\infty$. Special names are given to polynomials of low degree:

degree of polynomial    type of polynomial

|  |  |
|:---:|:---:|
| 1 | linear |
| 2 | quadratic |
| 3 | cubic |
| 4 | quartic |
| 5 | quintic |

We can *evaluate* a polynomial by replacing its variable by any number and carrying out the computation. The value of a polynomial $p(t)$ at $t = r$ is denoted by $p(r)$. For example, if $p(t) = 3t^3 - 2t^2 - t + 4$, its value when $t = 2$ is $p(2) = 3.2^3 - 2.2^2 - 2 + 4 = 24 - 8 - 2 + 4 = 18$. Since polynomials are a simple type of function easy to evaluate, they are very useful in approximating other more complex functions.

A *zero* of a polynomial $p(t)$ is any number $r$ for which $p(r)$ takes the value 0. When $p(r) = 0$, we say that $r$ is a *root* or a *solution* of the equation $p(t) = 0$. There are many situations in which we need to have information about the zeros of a polynomial, and considerable amount of attention is devoted to methods of solving equations $p(t) = 0$ either exactly or approximately. In particular, knowing the zeros of polynomials is often helpful in graphing a wide variety of functions and obtaining inequalities.

In operating with polynomials, we treat the variables as though they were numbers. Let

$$p(t) = a_0 + a_1 t + a_2 t^2 + \cdots + a_n t^n$$

$$q(t) = b_0 + b_1 t + b_2 t^2 + \cdots + b_m t^m$$

be any two polynomials.

*Sum:* $(p + q)(t) = (a_0 + b_0) + (a_1 + b_1)t + (a_2 + b_2)t^2 + \cdots$.

*Difference:* $(p - q)(t) = (a_0 - b_0) + (a_1 - b_1)t + \cdots$.

*Product of a constant and a polynomial:* $(cp)(t) = ca_0 + ca_1 t + ca_2 t^2 + \cdots$.

*Product of two polynomials:* $(pq)(t) = a_0 b_0 + (a_0 b_1 + a_1 b_0)t +$
$(a_0 b_2 + a_1 b_1 + a_2 b_0)t^2 + \cdots + (a_0 b_r + a_1 b_{r-1} + \cdots + a_i b_{r-i} + \cdots + a_r b_0)t^r +$
$\cdots + (a_n b_m)t^{m+n}$.

*Composition of two polynomials:* $(p \circ q)(t) = p(q(t))$. This definition instructs us to replace each occurrence of $t$ in the expression for $p(t)$ by $q(t)$.

## Exercises

1. State the degree, and the constant, linear and leading coefficients of the following polynomials:

   (a) $7t^5 - 6t^4 + 3t^2 + 1$
   (b) $8t^5 + 2t + 3$
   (c) $4t^3$
   (d) $(3t - 1)(2t + 1)$.

2. Give examples of

   (a) a monic polynomial of degree 7
   (b) a non-monic polynomial of degree 3
   (c) a polynomial of degree $-\infty$.

3. Decide which of the following functions are polynomials. For each polynomial in the list, specify its degree, its constant coefficient, its linear coefficient, its leading coefficient, and its values at $t = 0$ and $t = -(1/2)$.

   For some of the functions, you may not be able to make a firm decision at this point. As you master more of the theory of polynomials, you should return to them.

   (a) $0$
   (b) $3t^4$
   (c) $3 + t^2$
   (d) $8t^2 - 3t$
   (e) $t + t^{-1}$
   (f) $8t^2 + t^{3/4} + 2t^{3/2} - 3t^{9/4} + 8$ $(0 < t)$
   (g) $\sin 2(\arc \sin t)$ $(-1 \le t \le 1)$
   (h) $\sin 3(\arc \sin t)$ $(-1 \le t \le 1)$
   (i) $\cos 4(\arc \cos t)$ $(-1 \le t \le 1)$
   (j) $2^t$
   (k) $3t^3 - 2t^4 + 5t^2 + 6t^5$
   (l) $9^t + 4^t - 2^t + 6$
   (m) $\log t$
   (n) $t^{1/2}$ $(0 \le t)$
   (o) $t3^{-t}$
   (p) $\tan t$

(q) $(1+t^2)^{-1}$

(r) $\frac{3+t-2t^2}{t+7}$.

4. Let $p(t) = 3t - 4$ and $q(t) = 2t^2 - 5t + 8$. Verify that

   (a) $(p+q)(t) = 2t^2 - 2t + 4$
   (b) $(7p - 6q)(t) = -12t^2 + 51t - 76$
   (c) $(pq)(t) = (qp)(t) = 6t^3 - 23t^2 + 44t - 32$
   (d) $(p \circ q)(t) = 6t^2 - 15t + 20$
   (e) $(q \circ p)(t) = 18t^2 - 63t + 60$.

5. In multiplying two polynomials together, we can use the method of detached coefficients. In finding the product of the polynomials $t^3 + 3t^2 - 2t + 4$ and $2t^2 + t + 6$, the paper-and-pencil computation looks like this:

|     |     |     | 1   | 3   | -2  | 4   |
|-----|-----|-----|-----|-----|-----|-----|
|     |     |     |     | 2   | 1   | 6   |
|     |     |     | 6   | 18  | -12 | 24  |
|     |     | 1   | 3   | -2  | 4   |     |
|     | 2   | 6   | -4  | 8   |     |     |
|     | 2   | 7   | 5   | 24  | -8  | 24. |

Justify this algorithm and use it to read off the product of the two polynomials.

6. (a) Multiply the polynomials $4t^3 + 2t^2 + 7t + 1$ and $2t^2 + t + 6$ by using the method of detached coefficients.

   (b) Evaluate each of the two polynomials and their product at $t = 10$.

   (c) Compare the paper-and-pencil long multiplication for the product of the numbers 4271 and 216 with the table given in (a).

7. Using a pocket calculator, multiply 11254361 by 57762343 by each of the following methods:

   (a) Multiply the polynomials $1125t + 4361$ and $5776t + 2343$, and evaluate the product at $t = 10^4$.

   (b) Multiply the polynomials $11t^2 + 254t + 361$ and $57t^2 + 762t + 343$, and evaluate the product at $t = 10^3$.

8. Find the product of 26543645132 and 27568374445.

9. Let $p$ and $q$ be nonzero polynomials. Show that

(a) $\deg(p + q) \leq \max(\deg p, \deg q)$. $(\max(a, b)$ is the larger of the two numbers $a$ and $b)$

(b) $\deg(pq) = \deg p + \deg q$.

Give examples when equality and strict inequality hold in (a). Observe that, because of the convention that the sum of $-\infty$ and any nonnegative number is $-\infty$, the degree of the zero polynomial is defined in such a way as to make (b) valid when one of the polynomials is zero.

10. Is $\deg(p \circ q)$ related in any way to $\deg(q \circ p)$?

11. Find a pair $p$, $q$ of polynomials for which $p \circ q = q \circ p$.

12. (a) Is it possible to find a polynomial, apart from the constant 0 itself, which is identically equal to 0 (i.e. a polynomial $p(t)$ with some nonzero coefficient such that $p(c) = 0$ for each number $c$)? Try to justify your answer. [This is not an easy question, although the answer is not surprising. Examine your justification carefully to see what you are assuming about polynomials; can you explain why it is valid to use the properties you think you need?]

(b) Use your answer to (a) to deduce that, if two polynomials assume exactly the same values for all values of the variable, then their respective coefficients are equal. [Thus, there is only one way, up to order of writing down the terms, of presenting a polynomial as a sum of monomials.]

13. (a) Find all polynomials $f$ such that $f(2t)$ can be written as a polynomial in $f(t)$, i.e. for which there exists a polynomial $h$ such that

$$f(2t) = h(f(t)).$$

(b) Use the identity $\sin^2 2t = 4\sin^2 t(1 - \sin^2 t)$ to show that $\sin t$ is not a polynomial.

14. Show that, for $t > 0$, $\log t$ is not a polynomial.

15. Find all periodic polynomials, i.e. polynomials $g(t)$ which satisfy an identity of the type $g(t + k) = g(t)$ for some $k$ and all $t$. Deduce that the trigonometric functions $\sin t$, $\cos t$ and $\tan t$ are not polynomials.

16. Prove that $t^{1/3}$ is not a polynomial.

17. Show that, if $p$, $f$, $g$ are nonzero polynomials for which $pf = pg$, then $f = g$.

18. Show that, for any positive integer $k$,

$$(1+t)(1+t^2)(1+t^4)\dots(1+t^{2^{k-1}}) = 1+t+t^2+t^3+\cdots+t^{2^{k}-l}.$$

19. Characterize those polynomials $p(t)$ for which

    (a) $p(t) = p(-t)$
    (b) $p(t) = -p(-t)$.

20. As seen in Question 4, it is not always the case that $p \circ q = q \circ p$ for polynomials $p$ and $q$. If $p \circ q = q \circ p$, then $q$ is said to commute with $p$ under composition. Determine all polynomials $p(t)$ which commute under composition with $t^2$, i.e. for which $p(t^2) = [p(t)]^2$.

# Explorations

**E.1. Square of a Polynomial.** The square of a polynomial is the product of a polynomial with itself. Normally, the square has more nonzero terms than the polynomial itself. Show that this always occurs for polynomials of degrees 1, 2 and 3 having more than one term. Find a polynomial with more than one term whose square has exactly the same number of terms as the polynomial. Is it possible to find a polynomial whose square actually has fewer terms?

**E.2.** (a) Let $p(t) = at^2 + bt + c$ be any quadratic polynomial. Verify that $p(1) + p(4) + p(6) + p(7) = p(2) + p(3) + p(5) + p(8)$.

(b) Partition the set of numbers $\{1, 2, 3, \dots, 14, 15, 16\}$ into two sets such that, given any cubic polynomial $p(t)$ with integer coefficients, the sum of the numbers $p(k)$ where $k$ ranges over one of the two sets is the same as the sum where $k$ ranges over the other.

(c) Let $m$ be a positive integer. It is a remarkable fact that the numbers from 1 to $2^{m+1}$ inclusive can be subdivided into two subsets $A$ and $B$ such that, for any polynomial $p(t)$ of degree not exceeding $m$, the sum of the values of the polynomials over the numbers in $A$ is equal to the sum of the values over the numbers in $B$. Show that we can reduce the problem to finding sets $A$ and $B$ for which the sum of the $k$th powers of the numbers in one set is equal to the sum of the $k$th powers for the other, for $k = 0, 1, 2, \dots, m$.

(d) This situation can be generalized. If $d$ and $m$ are any integers with $d \geq 2$, the set of numbers from 1 to $d^{m+1}$ can be subdivided into $d$ disjoint subsets such that, for any polynomial of degree not exceeding $m$, the sum of its values over any of the subsets is the same.

(e) Problem (a) can be generalized in another way. Consider the question of looking for disjoint sets $(a_1, a_2, \dots, a_n)$ and $(b_1, b_2, \dots, b_n)$ of integers for which the sum of the $k$th powers of the numbers of the two subsets are equal for $k = 1, 2, 3, \ldots, m$.

For the case $m = 2$, (1, 5, 6) and (2, 3, 7) have equal sums (12) and square sums (62). Show that it is not possible to find two sets with only two numbers in each whose sums and square sums are equal.

For each fixed $m$, we ask for sets $(a_i)$ and $(b_i)$ for which the number $n$ of elements is as small as possible. From (a), we can see that $n$ can be made equal to $2^m$. Can it be made significantly smaller? Examine the cases $m = 3, 4, 5, 6$.

**E.3. Polynomials as Generating Functions.** We do not always want to think of the variable as a placeholder for a number. In combinatorial problems, we focus on the coefficients of polynomials as carriers of information. Consider this problem: A furniture company has warehouses at Albany, Buffalo, Montreal and Toronto. Deliveries have to be made to Kingston, Rochester and Syracuse. Each warehouse has one truck which can visit at most one city in a day. There are other constraints:

(i) the Albany warehouse does not serve Kingston and Rochester;

(ii) the Buffalo warehouse does not serve Kingston;

(iii) the Toronto warehouse does not serve Syracuse;

(iv) the Montreal warehouse does not serve American cities.

In how many ways can the dispatcher arrange the deliveries for today?

The dispatcher is free to defer any delivery until a later day. However, there is no point sending trucks from two different depots to the same destination. With these points in mind, we can reformulate the problem. Make a chart to show the origins, destinations and forbidden links:

|   | A | B | M | T |
|---|---|---|---|---|
| K | X | X |   |   |
| R | X |   | X |   |
| S |   |   | X | X |

Each city is represented by its initial letter. We can regard the chart as a $3 \times 4$ chessboard with the $X$-squares not available for the placement of a chessman. The problem is to find the number of ways of choosing no more than three of the available squares so that no two are in the same row or column. Choice of the $(R, B)$ square, for instance, indicates that the Buffalo truck is to make the delivery to Rochester. Equivalently, we have to find the number of ways of placing at most three rooks (castles) in the available squares of the chessboard so that no one threatens any other.

This particular problem is sufficiently simple that the possibilities can be enumerated without difficulty. However, for more complicated problems, we can enlist the aid of rook polynomials to help us avoid omissions or repeats in our counting.

Let an $m \times n$ ($m$ rows, $n$ columns) chessboard $C$ be given, with some of its squares forbidden. For each nonnegative integer $k$, let $r_k$ be the number of ways of placing $k$ rooks so that none is on a forbidden square and no two are in the same row or column. By convention, $r_0 = 1$. Also, $r_k = 0$ if $k > c = \min(m, n)$. The *rook polynomial* $R(C; t)$ is defined to be

$$r_0 + r_1 t + r_2 t^2 + \cdots + r_c t^c.$$

Since smaller chessboards have shorter and more easily defined rook polynomials, we look for ways of building more complex polynomials from simple ones.

Suppose $S$ is one of the available squares on the chessboard $C$. Form chessboards $C_1$ and $C_2$ as follows:

(1) $C_1$ is the same as $C$ except that the square $S$ is forbidden;

(2) $C_2$ is the $(m - 1) \times (n - 1)$ board obtained from $C$ by deleting the entire row and column containing $S$.

Let $R(C_1; t)$ and $R(C_2; t)$ be the corresponding rook polynomials.

Show that $R(C; t) = R(C_1; t) + t R(C_2; t)$.

Apply this process to obtain the rook polynomial for the problem of the delivery trucks. Read off from its coefficients the number of ways in which the dispatcher can send out one, two or three trucks, and give the answer to the problem.

Here are some further questions to consider:

(a) What is the rook polynomial of an $m \times n$ board with no forbidden squares?

(b) Other sorts of generating functions can be found. Consider the problem of choosing 6 coins from among 3 coppers, 2 nickels, 2 dimes, 1 quarter and 1 half-dollar. You are permitted more than one coin of any denomination; coins of the same value are regarded as indistinguishable. To tackle the problem systematically, introduce a variable $t$ to act as a counter for the number of coins, and the symbols $c$, $n$, $d$, $q$, $h$ for the coins. We can think of the product $nd^2q$ as standing for the choice of four coins consisting of one nickel, two dimes and one quarter. Using the counter $t$, we form the term $nd^2qt^4$ with the exponent of $t$ indicating the number of coins chosen. Let

$$P(t) = (1 + ct + c^2 t^2 + c^3 t^3)(1 + nt + n^2 t^2)(1 + dt + d^2 t^2)(1 + qt)(1 + ht).$$

Expand $P(t)$ in ascending powers of $t$ and interpret the coefficients. In particular, what is the relevance of the coefficient of $t^6$ to our problem? Now set $c = n = d = q = h = 1$. How do you interpret the coefficients now? What is the solution to the coin problem?

(c) Let $n$ be a positive integer and let $a_1, \ldots, a_n$ be $n$ symbols. What is the coefficient of $t^r$ in the expansion of the product $(1 + a_1 t)(1 + a_2 t)$ $(1 + a_3 t) \ldots (1 + a_n t)$? Argue that the coefficient of $t^r$ in the expansion of $(1 + t)^n$ is $\binom{n}{r}$ (read: "$n$ choose $r$"), the number of distinct ways of choosing $r$ objects from $n$ distinct objects.

## 1.2 Quadratic Polynomials

Many of the issues which arise for polynomials in general can be illustrated in the special case for which the degree is 2. We review results about quadratics.

## Exercises

1. Let $p(t) = at^2 + bt + c$ be a quadratic polynomial.

   (a) Show that $p(t)$ can be written in the form
   $$a\left(t + \frac{b}{2a}\right)^2 - \frac{1}{4a}(b^2 - 4ac).$$

   (b) Use (a) to determine all the roots of the equation
   $$t^2 - 7t + 12 = 0.$$

   (c) Give a general formula for the roots of a quadratic.

2. (a) Verify that $t^2 - r^2 = (t - r)(t + r)$.

   (b) Let $r$ be a zero of the polynomial $p(t) = at^2 + bt + c$. Verify that $p(t) = p(t) - p(r) = (t - r)(at + ar + b)$.

   (c) Show that $r$ is a zero of a quadratic polynomial $p(t)$ if and only if $p(t)$ can be written in the form $(t - r)q(t)$ for some linear polynomial $q(t)$.

3. Solve for $x$ the equation
   $$2m(1 + x^2) - (1 + m^2)(x + m) = 0.$$

4. *Theory of the quadratic.* Let $at^2 + bt + c$ be a polynomial whose coefficients are complex numbers.

   (a) Deduce from Exercise 1(a) that $a^2 + bt + c$ can be written as a constant times the square of a linear polynomial if and only if its *discriminant* $b^2 - 4ac$ vanishes. In this case, show that there is only one zero of the polynomial.

(b) Show that if the discriminant of the polynomial does not vanish, then it has two zeros.

(c) Let $m$ and $n$ denote the zeros of $at^2 + bt + c$. (When the discriminant vanishes, there is only one zero so in this case we set both $m$ and $n$ equal to that zero.) Show that $at^2 + bt + c = a(t-m)(t-n)$.

(d) Show that the sum of the zeros of the quadratic $at^2 + bt + c$ is $-b/a$, and that the product of the zeros is $c/a$.

5. For which values of $m$ will the polynomial

$$m^2 t^2 + 2(m+1)t + 4$$

have exactly one zero?

6. Let $s$ and $p$ be numbers. Show that the solutions $(x, y)$ of the system

$$x + y = s$$
$$xy = p$$

are the zeros in some order of the quadratic $t^2 - st + p$.

7. Determine the values of $x$ for which $6x^2 - 5x - 4$ is negative.

8. Determine those values of $k$ for which the equation

$$\frac{x^2 + x + 2}{3x + 1} = k$$

is solvable for real $x$.

9. If the domain of the function

$$\frac{x^2 + x - 1}{x^2 + 3x + 2}$$

is the set of all real numbers, show that it assumes all real values.

10. Given that $m$ and $n$ are the roots of the quadratic $6t^2 - 5t - 3$, find a quadratic whose roots are $m - n^2$ and $n - m^2$, without actually finding the values of $m$ and $n$ individually.

11. Let $m$ and $n$ be the roots of the equation $t^2 + bt + c = 0$. Show that $b$ and $c$ are the roots of the equation

$$t^2 + (m + n - mn)t - mn(m + n) = 0.$$

12. (a) Let $p(t)$ and $q(t)$ be two quadratic polynomials with integer coefficients. Prove that, if they have a nonrational zero in common, then one must be a constant multiple of the other.

   (b) Find a counterexample to (a) if the word "nonrational" is re-
       placed by "rational."

13. Show that, if $x = (b - d)/(a - c)$ satisfies one of the equations

$$x^2 - ax + b = 0$$
$$x^2 - cx + d = 0,$$

   then it satisfies the other as well.

14. (a) Let $a_i$, $b_i$ be nonnegative reals $(1 \le i \le n)$. The function

$$\sum_{i=1}^{n}(a_i t + b_i)^2$$

   is a polynomial in $t$. Explain why its discriminant is nonpositive.

   (b) Use (a) to establish the *Cauchy–Schwarz–Bunjakovsky Inequal-
       ity*:

$$\sum_{i=1}^{n} a_i b_i \le \sqrt{\sum_{i=1}^{n} a_i^2} \sqrt{\sum_{i=1}^{n} b_i^2}.$$

   When does equality occur?

15. (a) Verify the *Lagrange identity*:

$$\left(\sum_{i=1}^{n} a_i^2\right)\left(\sum_{i=1}^{n} b_i^2\right) - \left(\sum_{i=1}^{n} a_i b_i\right)^2 = \sum_{1 \le i < j \le n} (a_i b_j - a_j b_i)^2.$$

   (b) Use (a) to establish the Cauchy–Schwarz–Bunjakovsky Inequal-
       ity.

16. *Diameters of an ellipse.* The equation of an ellipse whose axes lie
   along the axes of coordinates and whose center is at the origin can
   be written $b^2 x^2 + a^2 y^2 = b^2 a^2$, where $a$ and $b$ are the lengths of the
   semi-axes. Find the locus of the midpoints of chords of the ellipse
   with fixed slope $m$; such a locus is called a *diameter* of the ellipse.

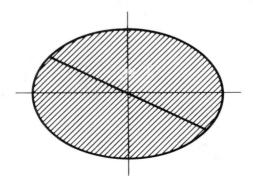

(Let the equation of a typical chord be $y = mx + k$, where $k$ is a parameter. The midpoint of the chord is given by $((x_1 + x_2)/2,$ $(y_1 + y_2)/2)$ where $(x_i, y_i)$ $(i = 1, 2)$ are the endpoints of the chord. The $x_i$ are found by solving the system consisting of the equations of the chord and the ellipse; eliminating $y$ yields a quadratic in $x$. However, it is not necessary to actually determine the $x_i$ individually.)

17. Let $a$, $b$ be two nonnegative real numbers. Use the fact that the zeros of the quadratic $(t - \sqrt{a})(t - \sqrt{b})$ are real to establish the arithmetic-geometric mean inequality $(ab)^{1/2} \leq \frac{1}{2}(a + b)$ with equality if and only if $a = b$.

18. An interesting question in numerical approximation is how closely a nonrational root of an equation can be approximated by a rational. In this exercise, we see that if a quadratic equation with integer coefficients has a nonrational root $r$, then no rational number can be any closer to it than the reciprocal of the square of its denominator multiplied by a constant.

    Suppose that $a$, $b$, $c$ are integers and that $r$ is a nonrational root of the quadratic equation $at^2 + bt + c = 0$. Let $u = p/q$ be any rational number, and suppose that $|u - r| < 1$.

    Prove that
    $$1/q^2 \leq |p(u)| \leq |u - r|K$$
    where $K = 2|ar| + |a| + |b|$.

    Deduce that there is a constant $M$ such that
    $$|r - p/q| \geq M/q^2 \text{ for any rational } p/q.$$

## Explorations

**E.4. Graphical Solution of the Quadratic.** Suppose a quadratic equation $x^2 - ux + v = 0$, with real coefficients and real roots is given. How can this equation be solved graphically? In other words, segments of length $u$ and $v$ are given and it is required to use them in determining points in the plane from which the roots might be found using the ancient Greek tools, ruler and compasses.

One such method is attributed to Thomas Carlyle. Assume for convenience, that $u$ and $v$ are positive. Construct the circle with the segment joining $(0, 1)$ and $(u, v)$ as diameter. Verify that the abscissae of its points of intersection with the $x$-axis are the required roots. Relate the condition that the circle intersects the $x$-axis to the discriminant condition for real roots.

Can you find other methods?

**E.5. Polynomials, some of whose values are squares.** The square integers $1 = 1^2$, $25 = 5^2$ and $49 = 7^2$ are in arithmetic progression. This means that there is a linear polynomial, for example $1 + 24t$, whose values are squares for three consecutive integer values of $t$. Is it possible to find a linear polynomial which takes a square value at four consecutive integer values of the variable $t$?

If a quadratic polynomial $p(t)$ is the square, $[q(t)]^2$ of a linear polynomial with integer coefficients, then it will always assume square values for integer values of $t$. Is the converse true? If not, what is the maximum number of square values which a quadratic polynomial (not equal to the square of a linear polynomial) might assume at consecutive integer values of its variable. In particular, determine a quadratic polynomial which assumes six consecutive square values.

Somewhat related to these questions is that of taking two finite disjoint subsets $U$ and $V$ of the integers and seeing whether there exists a polynomial $p(t)$ with integer coefficients which is square when $t$ belongs to $U$ and nonsquare when $t$ belongs to $V$. For example, one can find a quadratic polynomial $f(t)$ for which $f(1)$, $f(9)$, $f(8)$, $f(6)$ are squares but $f(1986)$ is a nonsquare.

## 1.3   Complex Numbers

The roots of the quadratic equation $t^2 + 2t + 8 = 0$ are $-1 + \sqrt{-7}$ and $-1 - \sqrt{-7}$. Thus, even simple polynomial equations lead us beyond the real number system. If we expand the system to include such "imaginaries" as $\sqrt{-7}$, we shall see that the theory of polynomials can be placed in a very natural setting indeed. A *complex number* is one which can be written in the form $x + yi$ where $x$ and $y$ are real and $i^2 = -1$. Do not worry about what $i$ "means"; all we need to know is that its square is $-1$. The set **C** of all complex numbers $x + yi$ can be represented by points $(x, y)$ in the Cartesian plane; such a representation is called the *Argand diagram* and we refer to the *complex plane*. The $x$-axis is called the *real axis* and the $y$-axis the *imaginary axis*.

In discussing complex numbers, it is useful to have some more terminology:

Let $z = x + yi$ denote a complex number, with $x$ and $y$ real.

The *real part*, Re $z$, of $z$ is the number $x$.

The *imaginary part*, Im $z$, of $z$ is the number $y$.

The *complex conjugate*, $\overline{z}$, of $z$ is $x - yi$.

The *modulus* or *absolute value* of $z$, denoted by $|z|$, is $\sqrt{x^2 + y^2}$. This is the distance from the origin to the point representing $z$.

The *argument* of $z$ is the angle between the real axis and the line joining $0$ and $z$, measured in the counterclockwise direction. It is denoted by arg $z$. Generally, we assign to arg $z$ a value between $0$ and $2\pi$.

*The polar decomposition.* Let $r = |z|$, $\theta = \arg z$, then $z = r(\cos\theta + i\sin\theta)$.

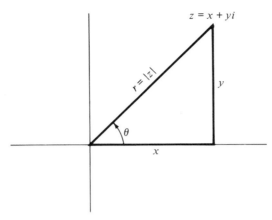

Through the introduction of complex numbers, we can find an expression for the solution of any quadratic equation with real coefficients. Suppose we try to solve quadratic equations with complex coefficients or polynomial equations of higher degree. Would it be necessary to extend our number system still further to accommodate the situation? For example, Leibniz recognized that a root of $t^4 + a^4 = 0$ is given by $a\sqrt[4]{-1}$, but apparently did not realize that $a\sqrt[4]{-1}$ could be expressed in the form $a + bi$. It is a remarkable fact that no further extension of the number system is required in order to solve any polynomial equation. In Exercises 10 and 12, this will be shown insofar as quadratic equations with complex coefficients are concerned; the more general result will be discussed in Chapter 4.

## Exercises

1. Given that the square of every real number is nonnegative, show that a complex number can be written in exactly one way in the form $x + yi$ with $x$ and $y$ real.

2. (a) Show that the transformation $z \longrightarrow iz$ corresponds to a rotation of the complex plane counterclockwise through an angle of $\pi/2$.

   (b) Describe the result of applying the transformation in (a) twice.

   (c) Let $w$ be an arbitrary fixed complex number. Give a geometric description of the transformation $z \longrightarrow wz$ on the complex plane.

3. Let $z = x + yi = r(\cos\theta + i\sin\theta)$, $w = u + vi$ be two complex numbers. Show that

   (a) $z + w = (x + u) + (y + v)i$
   (b) $zw = (xu - yv) + (xv + yu)i$

(c) $\overline{z+w} = \overline{z} + \overline{w}$

(d) $\overline{zw} = \overline{z}\,\overline{w}$

(e) $\overline{\overline{z}} = z$

(f) Re $z = r\cos\theta = \frac{1}{2}(z+\overline{z}) \le |z|$

(g) Im $z = r\sin\theta = \frac{1}{2i}(z-\overline{z}) \le |z|$

(h) $|z|^2 = z\overline{z}$

(i) $|zw| = |z|\,|w|$

(j) $\arg(zw) = \arg z + \arg w$ (up to an integer multiple of $2\pi$)

(k) $|z+w| \le |z| + |w|$

(l) $2^{-1/2}(|x|+|y|) \le |z| \le |x| + |y|$

(m) for $z \ne 0$, $1/z = \overline{z}/|z|^2$.

4. The Greek geometers were interested in discovering which geometric entities could be constructed from given data using only ruler (straightedge) and compasses. Given the points representing $0$, $z$ and $w$ in the Argand diagram, determine ruler and compasses constructions for $\overline{z}$, $z+w$, $zw$ and $1/z$.

5. Let $c$ be a fixed real number and $w$ a fixed complex number. Find the locus of points $z$ in the Argand diagram which satisfy the equation Re $(zw) = c$.

6. Let $k$ be a fixed positive constant. Describe the locus of the equation $|z| = k|z+1|$.

7. Use complex numbers and an Argand diagram to solve the following problem: Some pirates wish to bury their treasure on an island. They find a tree $T$ and two rocks $U$ and $V$. Starting at $T$, they pace off the distance from $T$ to $U$, then turn right and pace off an equal distance from $U$ to a point $P$, which they mark. Returning to $T$, they pace off the distance from $T$ to $V$, then turn left and pace off an equal distance (to $TV$) to a point $Q$, which they mark. The treasure is buried at the midpoint of the line segment $PQ$.

   Years later, they return to the island and discover to their dismay at the tree $T$ is missing. One of them decides to just assume any position for the tree and then carry out the procedure. Is this strategy likely to succeed?

8. Prove *De Moivre's Theorem:* For any integer $n$,

$$(r(\cos\theta + i\sin\theta))^n = r^n(\cos n\theta + i\sin n\theta).$$

9. Determine all those complex numbers $z$ for which

(a) $z^3 = 1$

(b) $z^4 = 1$

(c) $z^6 = 1$

(d) $z^8 = 1$.

Indicate the solutions of each equation on an Argand diagram.

10. (a) Let $a, b$ be real numbers. Find real numbers $x$ and $y$ for which $(x + yi)^2 = a + bi$.

(b) Determine the square roots of $-7 - 24i$.

11. Solve the equations

(a) $t^2 + 3t + 3 - i = 0$

(b) $t^2 + (2i - 1)t + (5i + 1) = 0$.

12. Show that every quadratic equation with complex coefficients has at least one complex root, and therefore can be written as the product of two linear factors with complex coefficients.

13. Prove that $|1 + iz| = |1 - iz|$ if and only if $z$ is real.

14. (a) Let $p(t)$ be a polynomial with real coefficients. Show that, for any complex number $w$, $p(\overline{w}) = \overline{p(w)}$. Deduce that, if $w$ is a zero of $p(t)$, then so is $\overline{w}$.

(b) Give a counterexample to show that (a) is not true in general if $p(t)$ has some nonreal coefficients.

15. Let $n$ be a nonnegative integer. The Tchebychef Polynomial $T_n(x)$ is defined, for $-1 \le x \le 1$, by

$$T_0(x) = 1$$

$$T_n(x) = \cos n(\text{arc } \cos x) \quad (n \ge 1).$$

(a) Show that $T_{n+1}(x) - 2xT_n(x) + T_{n-1}(x) = 0$ $(n \ge 1)$.

(b) Find $T_1(x)$, $T_2(x)$, $T_3(x)$ and $T_4(x)$. Sketch the graphs of these functions.

(c) For each $n$, establish that $T_n(x)$ is a polynomial and determine its degree. (Do this in two ways, by (a) and by de Moivre's Theorem.)

(d) Show that

$$T_n(x) = \sum_{r=0}^{[n/2]} (-1)^r \binom{n}{2r} x^{n-2r}(1 - x^2)^r.$$

## Exploration

**E.6. Commuting Polynomials.** Two polynomials are said to commute under composition if and only if $(p \circ q)(t) = (q \circ p)(t)$ (i.e. $p(q(t)) = q(p(t))$. We define the composition powers of a polynomial as follows

$$p^{[2]}(t) = p(p(t))$$

$$p^{[3]}(t) = p(p(p(t)))$$

and, in general, $p^{[k]}(t) = p(p^{[k-1]}(t))$ for $k = 2, 3, \ldots$.

Show that any two composition powers of the same polynomial commute with each other.

One might ask whether two commuting polynomials must be composition powers of the same polynomial. The answer is no. Show that any pair of polynomials in the following two sets commute

I. $\{t^n : n = 1, 2, \ldots\}$

II. $\{T_n(t) : n = 1, 2, \ldots\}$.

Let $a$ and $b$ be any constants with $a \neq 0$. Show that, if $p$ and $q$ are two polynomials which commute under composition, then the polynomials $(t/a - b/a) \circ p \circ (at + b)$ and $(t/a - b/a) \circ q \circ (at + b)$ also commute under composition. Use this fact to find from sets I and II other families which commute under composition.

Can you find pairs of polynomials not comprised in the foregoing discussion which commute under composition? Find families of polynomials which commute under composition and within which there is exactly one polynomial of each positive degree.

## 1.4   Equations of Low Degree

With access to complex numbers, we are able to determine the solutions to any quadratic equation whose coefficients are real, or even complex. It is a notable result, realized by mathematicians in the sixteenth century, that one does not have to extend the number system any further in order to solve real cubic or quartic equations. One phenomenon which led to the adoption of complex numbers was the use of nonreal roots of a quadratic equation in determining the real roots of a real cubic equation.

## Exercises

1.  (a) Let $p(t)$ be a cubic polynomial. Show that $r$ is a root of the polynomial equation $p(t) = 0$ if and only if $p(t) = (t - r)q(t)$ for some quadratic polynomial $q(t)$.

(b) Determine by inspection a root of the polynomial equation

$$t^3 - 4t + 3 = 0,$$

and use this information to find a complete set of solutions to the equation.

2. Consider the cubic equation

$$x^3 - 12x^2 + 29x - 18 = 0.$$

As we shall see below, it is possible to solve cubic equations in which the quadratic coefficient vanishes. Fortunately, a simple transformation permits us to reduce any cubic to this form. Verify that the substitution $x = t + 4$ converts the equation to

$$t^3 - 19t - 30 = 0.$$

By inspection, obtain a solution to the equation in $t$, and thence solve the equation in $x$.

3. The solutions of the cubic equations so far have involved finding one solution by guessing. This is too much to expect in general. Argue that, if a general method can be found to solve cubic equations of the form
$$t^3 + pt + q = 0,$$
then it is possible to solve any cubic equation.

4. *The cubic equation: Cardan's Method.* An elegant way to solve the general cubic is due to Cardan. The strategy is to replace an equation in one variable by one in two variables. This provides an extra degree of freedom by which we can impose a convenient second constraint, allowing us to reduce the problem to that of solving a quadratic.

(a) Suppose the given equation is

$$t^3 + pt + q = 0.$$

Set $t = u + v$ and obtain the equation

$$u^3 + v^3 + (3uv + p)(u + v) + q = 0.$$

Impose the second condition $3uv + p = 0$ (why do we do this?) and argue that we can obtain solutions for the cubic by solving the system
$$u^3 + v^3 = -q$$
$$uv = -p/3.$$

(b) Show that $u^3$ and $v^3$ are roots of the quadratic equation

$$x^2 + qx - p^3/27 = 0.$$

(c) Let $D = 27q^2 + 4p^3$. Suppose that $p$ and $q$ are both real and that $D > 0$. Show that the quadratic in (b) has real solutions, and that if $u_0$ and $v_0$ are the real cubic roots of these solutions, then the system in (a) is satisfied by

$$(u, v) = (u_0, v_0), (u_0\omega, v_0\omega^2), (u_0\omega^2, v_0\omega)$$

where $\omega$ is the imaginary cube root $(-1 + \sqrt{-3})/2$ of unity. Deduce that the cubic polynomial $t^3 + pt + q$ has one real and two nonreal zeros.

(d) Suppose that $p$ and $q$ are both real and that $D = 0$. Let $u_0$ be the real cube root of the solution of the quadratic in (b). Show that, in this case, the cubic has all its zeros real, and in fact can be written in the form

$$(t + u_0)^2(t - 2u_0).$$

(e) Suppose that $p$ and $q$ are both real and that $D < 0$. Show that the solutions of the quadratic equation in (b) are nonreal complex conjugates, and that it is possible to choose cube roots $u$ and $v$ of these solutions which are complex conjugates and satisfy the system in (a). If $u = r(\cos\theta + i\sin\theta)$ and $v = r(\cos\theta - i\sin\theta)$, show that the three roots of the cubic equation are the reals

$$2r\cos\theta, 2r\cos(\theta + 2\pi/3), 2r\cos(\theta + 4\pi/3).$$

(f) Prove that every cubic equation with real coefficients has at least one real root.

5. Use Cardan's Method to solve the cubic equations:

(a) $x^3 - 6x + 9 = 0.$

(b) $x^3 - 7x + 6 = 0.$

[(b) will require the use of a pocket calculator and some trigonometry; remember de Moivre's Theorem (Exercise 3.8); work to an accuracy of three decimal places.]

6. By means of a transformation, convert the equation

$$x^3 - 15x^2 - 33x + 847 = 0$$

to the form $t^3 + pt + q = 0$, and verify that $D = 0$ (in the notation of Exercise 4). Solve the given equation for $x$.

7. On August 7, 1877 (?), Arthur Cayley (1821–1895) wrote to Rudolf
   Lipschitz (1832–1903) a letter containing the following paragraph:

   > As to the cubic, there is a variation of Cardano's solu-
   > tion which I think is theoretically interesting: if instead of
   > assuming $x = a + b$, we assume $x = a^2b + ab^2$, then instead
   > of $ab$ and $a^3 + b^3$, we have only $a^3b^3$ and $a^3 + b^3$ rationally
   > determined: $a$ and $b$ may therefore be any values whatever
   > of the cube roots of $a^3$, $b^3$, but the apparently 9-valued
   > function $a^2b + ab^2$ will be only 3-valued.
   >
   > (Rudolf Lipschitz, Briefwechsel mit Cantor, Dedekind,
   > Helmholtz, Kronecker, Weierstrass. DMV, Vieweg & Sohn,
   > 1986).[1]

   Verify that Cayley's remark is true. (Cayley's $x$, $a$, $b$ correspond to
   our $t$, $u$, $v$ in Exercise 4.)

8. Why is the general method of solving a cubic equation not a part of
   most school curricula?

9. Find the relationship between $p$ and $q$ in order that the equation
   $x^3 + px + q = 0$ may be put into the form

   $$x^4 = (x^2 + ax + b)^2.$$

   Hence, solve the equation

   $$8x^3 - 36x + 27 = 0.$$

10. Vieta [Cajori, History, p. 138] had an alternative method of solving
    a cubic of the form
    $$x^3 - 3a^2x = a^2b$$
    where $a$, $b$ are real numbers which satisfy $|b| \le 2|a|$. Show that, if $\phi$
    is defined by $b = 2a \cos \phi$, then a solution of the equation is given by
    $x = 2a \cos(1/3)\phi$. Use this method to locate a solution of each of the
    following equations:

    (a) $x^3 - 3a^2x = 0$.
    (b) $x^3 - 3x - 2 = 0$.

11. *The quartic equation: Descartes' Method (1637).*

    (a) Argue that any quartic equation can be solved once one has a
        method to handle quartic equations of the form

        $$t^4 + pt^2 + qt + r = 0.$$

---

[1] Used with permission of the editor, Dr. Winfried Scharlau, and the publisher.

(b) Show that the quartic polynomial in (a) can be written as the product of two factors

$$(t^2 + ut + v)(t^2 - ut + w)$$

where $u$, $v$, $w$ satisfy the simultaneous system

$$v + w - u^2 = p$$
$$u(w - v) = q$$
$$vw = r.$$

Eliminate $v$ and $w$ to obtain a cubic equation in $u^2$.

(c) Show how any solution $u$ obtained in (b) can be used to find all the roots of the quartic equation.

(d) Use Descartes' Method to solve

$$t^4 + t^2 + 4t - 3 = 0$$
$$t^4 - 2t^2 + 8t - 3 = 0.$$

12. *The quartic equation: Ferrari's method.*

(a) Let a quartic equation be presented in the form

$$t^4 + 2pt^3 + qt^2 + 2rt + s = 0.$$

The strategy is to complete the square on the left side in such a way as to incorporate the cubic term. Show that the equation can be rewritten in the form

$$(t^2 + pt + u)^2 = (p^2 - q + 2u)t^2 + 2(pu - r)t + (u^2 - s),$$

where $u$ is indeterminate.

(b) Show that the right side of the transformed equation in (a) is the square of a linear polynomial if $u$ satisfies a certain cubic equation. Explain how such a value of $u$ can be used to completely solve the quartic.

(c) Use Ferrari's Method to solve

$$t^4 + t^2 + 4t - 3 = 0$$
$$t^4 - 2t^3 - 5t^2 + 10t - 3 = 0.$$

13. *Reciprocal equations.* A *reciprocal polynomial* has the form

$$ax^n + bx^{n-1} + cx^{n-2} + \cdots + cx^2 + bx + a,$$

in which $a \neq 0$ and the coefficients are symmetric about the middle one. A *reciprocal equation* is of the form $p(t) = 0$ with $p(t)$ a reciprocal polynomial.

(a) Verify that each of the following polynomials is a reciprocal polynomial:
$$x^3 + 4x^2 + 4x + 1$$
$$3x^6 - 7x^5 + 5x^4 + 2x^3 + 5x^2 - 7x + 3.$$

(b) Show that 0 is not a zero of any reciprocal polynomial.

(c) Show that $-1$ is a zero of any reciprocal polynomial of odd degree, and deduce that any reciprocal polynomial of odd degree can be written in the form $(x + 1)q(x)$, with $q(x)$ a reciprocal polynomial of even degree.

(d) Show that, if $r$ is a root of a reciprocal equation, then so also is $1/r$.

14. (a) Let $ax^{2k} + bx^{2k-1} + \ldots + rx^k + \ldots + bx + a = 0$ be a reciprocal equation of even degree $2k$. Show that this equation can be rewritten
$$a(x^k + x^{-k}) + b(x^{k-1} + x^{-k+1}) + \ldots + r = 0.$$

(b) Let $t = x + x^{-1}$. Verify that $x^2 + x^{-2} = t^2 - 2$ and that $x^3 + x^{-3} = t^3 - 3t$. Prove that, in general, $x^m + x^{-m}$ is a polynomial of degree $m$ in $t$.

(c) Use the substitution in (b) to show that the reciprocal equation in (a) can be rewritten as an equation of degree $k$ in the variable $t$. Deduce that the solution of a reciprocal equation of degree $2k$ can in general be reduced to solving one polynomial equation of degree $k$ as well as at most $k$ quadratic equations.

15. (a) Show that the transformation $t = x + x^{-1}$ applied to the equation
$$2x^4 + 5x^3 + x^2 + 5x + 2 = 0$$
leads to the equation
$$2t^2 + 5t - 3 = 0.$$
Solve the latter equation for $t$ and use the result to obtain solutions to the original equation.

(b) As a check, verify that the left side of the equation in $x$ can be written as the product of the two quadratic polynomials which arise in solving for $x$ once the two values of $t$ are found.

16. (a) Show that a product of reciprocal polynomials is a reciprocal polynomial.

(b) Show that, if $f$, $g$, $h$ are polynomials with $f = gh$ and $f$ and $h$ are both reciprocal polynomials, then $g$ is also a reciprocal polynomial.

17. (a) A quartic equation of the form

$$x^4 + px^3 + qx^2 + rx + r^2/p^2 = 0$$

is said to be quasi-reciprocal. Show that the substitution

$$t = x + r/px$$

leads to the equation $t^2 + pt + q - 2r/p = 0$.

(b) A method for solving the general quartic equation can be formulated as follows. Suppose the given equation can be written in the form

$$x^4 - qx^2 - rx - s = 0.$$

Set $x = u + v$ to obtain the equation

$$u^4 + 4vu^3 + (6v^2 - q)u^2 + (4v^3 - 2qv - r)u + (v^4 - qv^2 - rv - s) = 0.$$

Show that this becomes a quasi-reciprocal equation in $u$ if $v$ is chosen so that

$$v^3 + (1/2r)(q^2 + 4s)v^2 + (q/2)v + (r/8) = 0.$$

(c) Use (a) and (b) to obtain a solution to the equation

$$x^4 + 3x^2 - 2x + 2 = 0.$$

# Exploration

**E.7. The Reciprocal Equation Substitution.** The substitution $t = x + x^{-1}$ is used in solving reciprocal equations. The quantity $x^n + x^{-n}$ can be expressed as a polynomial $p_n(t)$ of $t$ (see Exercise 14). Verify that $p_0(t) = 2$, $p_1(t) = t$ and that $p_{n+1}(t) = t\, p_n(t) - p_{n-1}(t)$ for $n \geq 1$. Tabulate these polynomials and look for patterns among their coefficients. Examine the composition $p_m \circ p_n(t)$ for indices $m$ and $n$. Test the conjecture that all coefficients of $p_n(t)$ except the leading coefficient are divisible by $n$ when $n$ is prime. Is this true? Is there any connection between the polynomials $p_n(t)$ and the Tchebychef polynomials (Exercise 3.15)?

## 1.5   Polynomials of Several Variables

If $x$ and $y$ are the roots of a quadratic equation $at^2 + bt + c = 0$, then $-b/a = x + y$ and $c/a = xy$. The expressions $x + y$ and $xy$ are examples of polynomials in the two variables $x$ and $y$. In general, a function $f(x, y)$ is a polynomial in $x$ and $y$ if and only if it can be represented as a finite sum of terms of the form

$$cx^k y^m ,$$

where $c$ is a coefficient and $k$ and $m$ are nonnegative integers. The number $k + m$ is called the degree of the term, and the degree of the polynomial $f(x, y)$ is equal to the highest degree of its terms. Polynomials of several variables can be added, subtracted and multiplied in a way analogous to polynomials of a single variable, in which like terms are collected and the variables are assumed to adhere to all the usual arithmetic laws.

There are two classes of polynomials of two variables which we shall consider:

(i) symmetric polynomials $f(x, y)$ which satisfy $f(x, y) = f(y, x)$;

(ii) homogeneous polynomials in which all the terms are of the same degree.

For example, $s_1 = x + y$ is symmetric and homogeneous of degree 1, while $s_2 = xy$ is symmetric and homogeneous of degree 2. However, $x^2 + x + y + y^2$ is symmetric but not homogeneous, while $x^2 y + 2x^3$ is homogeneous but not symmetric.

Similar definitions can be made for functions of three variables, say $x$, $y$, $z$. A polynomial is any finite sum of the type $cx^k y^m z^n$, with $k$, $m$, $n$ nonnegative integers. The degree of the polynomial is the highest degree $k + m + n$ of any of its terms. If all the terms have the same degree, the polynomial is said to be homogeneous. If the polynomial $f(x, y, z)$ satisfies $f(x, y, z) = f(x, z, y) = f(y, x, z) = f(y, z, x) = f(z, x, y) = f(z, y, x)$, then $f(x, y, z)$ is said to be symmetric.

The *elementary symmetric functions*

$$s_1 = x + y + z$$

$$s_2 = xy + yz + zx$$

$$s_3 = xyz$$

are both homogeneous and symmetric.

The purpose of this section is to introduce some elementary properties of polynomials of several variables.

# Exercises

1. What are the degrees of the following polynomials? Are they homogeneous? symmetric?

   (a) $x - y$
   (b) $3xy + 2x^2$
   (c) $4xy^2 + 3x + 3y + 4x^2y$
   (d) $5x + 7$
   (e) $xy^2 + yz^2 + zx^2$
   (f) $xy^2 + yz^2 + zx^2 - x^2y - y^2z - z^2x$

2. Show that a polynomial $f(x, y, z)$ in the variables $x$, $y$, $z$ is homogeneous of degree $d$ if and only if

$$f(tx, ty, tz) = t^d f(x, y, z).$$

3. One can also define the notion of homogeneity for polynomials of a single variable. What are the homogeneous polynomials of degree $k$ in a single variable $x$?

4. Show that each symmetric polynomial in two or three variables can be written as a sum of homogeneous symmetric polynomials.

5. The elementary symmetric polynomials $s_1 = x + y$ and $s_2 = xy$ of two variables are the building blocks for all symmetric polynomials in the sense that every symmetric polynomial can be expressed as a polynomial in the elementary symmetric polynomials.

   For example,

$$x^2 + y^2 = (x + y)^2 - 2xy = s_1^2 - 2s_2$$

$$x^3 + 2x^2y + 2xy^2 + y^3 = (x + y)^3 - xy(x + y) = s_1^3 - s_2s_1.$$

   Prove that every symmetric polynomial can be written as a polynomial of the elementary symmetric function $s_1$, $s_2$.

   As we have seen in Exercise 2.4, if $x$ and $y$ are the zeros of a quadratic polynomial, then $x + y$ and $xy$ are expressible in terms of the coefficients. The result just established means that we can evaluate any symmetric polynomial function of the roots of a quadratic equation without actually having to solve it. An analogous result for polynomials of higher degree is of great practical and theoretical use, for, as we have seen, the task of obtaining solutions to a polynomial equation becomes heavier with the degree.

6. Let $p(t) = at^3 + bt^2 + ct + d$ be a cubic polynomial whose zeros are $x$, $y$, $z$.

   (a) Show that

   $$p(t) = p(t) - p(x) = (t - x)(at^2 + (ax + b)t + (ax^2 + bx + c)).$$

   (b) Show that $p(t)$ can be written in the form

   $$a(t - x)(t - y)(t - z).$$

   (c) By expanding the product in (b) and comparing coefficients, verify that

   $$x + y + z = -b/a$$
   $$xy + yz + zx = c/a$$
   $$xyz = -d/a.$$

7. Find a necessary and sufficient condition on $p$, $q$, $r$ that the zeros of

   $$t^3 + pt^2 + qt + r$$

   are in arithmetic progression.

8. Express each of the following polynomials as a polynomial in the elementary symmetric functions $s_1 = x + y + z$, $s_2 = xy + yz + zx$, $s_3 = xyz$:

   $$x^3 + y^3 + z^3$$
   $$x^2y^3 + x^3y^2 + x^2z^3 + x^3z^2 + y^2z^3 + y^3z^2$$
   $$(x + y)(y + z)(z + x).$$

9. (a) Verify that

   $$x^3 + y^3 + z^3 - 3xyz = (x + y + z)(x^2 + y^2 + z^2 - xy - xz - yz).$$

   (b) Write $x^2 + y^2 + z^2 - xy - xz - yz$ as the sum of three squares of polynomials and deduce that this quantity is nonnegative whenever $x$, $y$, $z$ are real.

   (c) Prove the arithmetic-geometric mean inequality: if $a, b, c \geq 0$, then

   $$(abc)^{1/3} \leq (a + b + c)/3$$

   with equality if and only if $a = b = c$.

10. A polynomial of several variables $t_1, t_2, \ldots, t_m$ is a finite sum of monomials of the type

$$at_1^{r_1} t_2^{r_2} \cdots t_m^{r_m}$$

where $a$ is a coefficient and the $r_i$ are nonnegative integers. The degree of this term is $r_1 + r_2 + r_3 + \cdots + r_m$, and the *degree of the polynomial* is equal to the highest degree of any of its terms.

A polynomial of several variables is *homogeneous* (of degree $d$) if and only if each term is of the same degree ($d$).

A polynomial of several varibles is *symmetric* if it remains unchanged no matter how we interchange its variables.

Given a set of variables, $t_1, t_2, \ldots, t_m$, there is a special class of symmetric polynomials associated with them. There are the *elementary symmetric functions:*

$$s_1 = s_1(t_1, t_2, \ldots, t_m) = t_1 + t_2 + \cdots + t_m$$

$$s_2 = s_2(t_1, t_2, \ldots, t_m) = t_1 t_2 + t_1 t_3 + \cdots + t_1 t_m$$
$$+ t_2 t_3 + \cdots + t_{m-1} t_m$$

$$\cdots$$

$s_r$ is the sum of all possible products of $r$ of the variables (this sum has $\binom{m}{r}$ terms)

$$\cdots$$

$$s_{m-1} = s_{m-1}(t_1, t_2, \ldots, t_m) = \sum_{i=1}^{m} t_1 t_2 \ldots \hat{t}_i \ldots t_m$$

(A "hat" denotes a deleted term.)

$$s_m = s_m(t_1, t_2, \ldots, t_m) = t_1 t_2 \ldots t_m$$

Give all symmetric homogeneous polynomials of degree 0, 1 and 2 in the variables $t_1, t_2, t_3, \ldots, t_m$, and show how they can be expressed as a polynomial in the functions $s_1$ and $s_2$.

11. Formulate and prove the analogue of Exercise 2 for any number of variables.

12. The polynomial $g(x, y)$ has the property that, for any numerical substitutions of $x$ and $y$, $g(x, y) = g(y, x)$. Must $g(x, y)$ be symmetric in the variables $x$ and $y$?

## Explorations

**E.8.** Suppose that $f(x, y)$ is a function of the two real variables $x$ and $y$. For each fixed value of $x$, $f(x, y)$ is a polynomial in $y$. For each fixed value of $y$, $f(x, y)$ is a polynomial in $x$. Is $f(x, y)$ necessarily a polynomial of the two variables $x$ and $y$?

There is more to this question than might seem initially apparent. The hypothesis says, for example that, for each specific $x$, $f(x, y)$ can be written in the form

$$f(x, y) = a_0 + a_1 y + \ldots + a_n y^n,$$

where not only the coefficients $a_i$ but also the degree $n$ depends on $x$. On the face of it, it might happen that for certain choices of $x$, $n$ could be arbitrarily large. However, if $f(x, y)$ were a polynomial in $x$ and $y$ jointly, the number $n$ would not exceed some fixed number independently of $x$.

**E.9. The Range of a Polynomial.** Any polynomial is a continuous function of its variables. One important consequence is the restriction it imposes on its possible range of values. Let $f(\mathbf{x})$ be a polynomial with real coefficients of $n$ real variables, where $\mathbf{x} = (x_1, x_2, \ldots, x_n)$. For any vectors $\mathbf{a}$ and $\mathbf{b}$, the line segment joining $\mathbf{a}$ and $\mathbf{b}$ consists exactly of the points $(1 - t)\mathbf{a} + t\mathbf{b}$ with $0 \leq t \leq 1$. Then $p(t) = f((1 - t)\mathbf{a} + t\mathbf{b})$ is a polynomial in $t$; as $t$ varies between 0 and 1, $p(t)$ varies continuously between $f(\mathbf{a})$ and $f(\mathbf{b})$ and accordingly assumes every value between $f(\mathbf{a})$ and $f(\mathbf{b})$.

For any polynomial $f$ with real coefficients, define its *range* as the set $R_f = \{f(\mathbf{x}) : \mathbf{x} = (x_1, \ldots, x_n) \text{ with } x_i \text{ real}\}$. Argue that $R_f$ must be a subset of $R$ of one of the following types:

(a) a singleton (i.e. a set with a single element);

(b) a finite interval with or without either endpoint;

(c) a closed halfline $\{r : r \leq c\}$ or $\{r : r \geq c\}$;

(d) an open halfline $\{r : r < c\}$ or $\{r : r > c\}$;

(e) the entire set of real numbers.

Show that (a) occurs if and only if $f$ is a constant polynomial. Give examples in which (c) and (e) occur. Show that (b) can never occur. Can (d) occur for polynomials of one variable? more than one variable?

**E.10. Diophantine Equations.** Who has not seen the pythagorean equation $X^2 + Y^2 = Z^2$? This is a *diophantine equation* with integer coefficients and exponents and for which integer solutions are sought. This one, for example, is satisfied by $(X, Y, Z) = (3, 4, 5)$, $(8, 15, 17)$, $(5, 12, 13)$. Often, diophantine equations have infinitely many solutions and the solver seeks some formula which will give all, or at least a significant portion of, the solutions. These formulae may be in the form of polynomials with integer coefficients. For example, verify that $(X, Y, Z) = (z(x^2 - y^2), 2zxy, z(x^2 + y^2))$ satisfies the pythagorean equation and verify that every numerical solution (up to the order of $X$ and $Y$) can be found by suitable numerical substitutions for $x$, $y$ and $z$.

How can such polynomial solutions to diophantine equations be found? For the pythagorean equation, the usual argument uses some basic number theory. But such an argument is not always readily available. Rather, one might work empirically, using a computer to churn out a large number of numerical solutions, and then examining these for some pattern from which to indicate that they are values of certain polynomials. Here are some examples for you to work on.

(a) $X^3 + Y^3 + Z^3 = W^3$ is satisfied by $(X, Y, Z, W) =$

| | | |
|---|---|---|
| $(3, 4, 5, 6)$, | $(3, 10, 18, 19)$, | $(4, 17, 22, 25)$, |
| $(11, 15, 27, 29)$, | $(7, 14, 17, 20)$, | $(12, 19, 53, 54)$, |
| $(12, 31, 102, 103)$, | $(20, 54, 79, 87)$, | $(23, 94, 105, 126)$, |
| $(27, 46, 197, 198)$, | $(27, 64, 306, 307)$, | $(28, 53, 75, 84)$, |
| $(34, 39, 65, 72)$, | $(38, 48, 79, 87)$, | $(48, 85, 491, 492)$, |
| $(48, 109, 684, 685)$, | | $(65, 127, 248, 260)$, |
| $(107, 230, 277, 326)$, | | $(227, 230, 277, 356)$. |

Look for patterns which may yield solutions which are polynomials and for which some of the numerical solutions above are obtained by evaluation of the polynomials.

Euler generated polynomial solutions for $X^3 + Y^3 + Z^3 = W^3$ in the following way. Let $X = p + q$, $Y = p - q$, $Z = r - s$, $W = r + s$. Show that this leads to the requirement

$$p(p^2 + 3q^2) = s(s^2 + 3r^2).$$

At this point, we introduce parameters $u$, $v$, $x$, $y$, $z$, $w$ in such a way that $u$ and $v$ appear only linearly in an equation; this will enable us to determine their ratio in terms of $x$, $y$, $z$, $w$. Set

$$p = xu + 3yv \qquad s = 3zv - wu$$
$$q = yu - xv \qquad r = wv + zu.$$

Plug these into the equation for $p$, $q$, $r$, $s$ and determine what the ratio of $u$ to $v$ must be. Now substitute back in to obtain expressions for $p$, $q$, $r$, $s$ and ultimately $X$, $Y$, $Z$, $W$ in terms of $x$, $y$, $z$, $w$.

(b) The simultaneous system $2(B^2 + 1) = A^2 + C^2$; $2(C^2 + 1) = B^2 + D^2$

is satisfied by $(A, B, C, D) =$

$$
\begin{array}{ll}
(6, 23, 32, 39), & (16, 87, 122, 149), \\
(39, 70, 91, 108), & (51, 148, 203, 246), \\
(59, 228, 317, 386), & (59, 630, 889, 1088), \\
(79, 242, 333, 404), & (83, 516, 725, 886), \\
(108, 157, 194, 225), & (147, 302, 401, 480), \\
(225, 296, 353, 402), & (324, 557, 718, 849), \\
(402, 499, 580, 651). &
\end{array}
$$

There is a family of solutions in which $A$, $B$, $C$, $D$ are given by evaluating linear polynomials at integer values. However, the numerical data above will suggest polynomial solutions of higher degree.

(c) Let $r$ be a fixed numerical parameter. Find polynomial solutions to

$$
X^2 + rXY + Y^2 = Z^2.
$$

(d) Show that there are infinitely many integers which are equal to the sum of the squares of their digits written to some base.

(e) Show that there are infinitely many integers which are equal to the sum of the cubes of their digits written to some base. For example, 17 written in base 3 has the representation $(122)_3$ and is the sum of the cubes of 1, 2 and 2.

(f) In Exploration E.2, we considered pairs of sets of numbers for which the sum of various powers of the elements of one were equal to the corresponding powers of the elements of the other. Look for pairs of sets of polynomials which have the same property.

# 1.6   Basic Number Theory and Modular Arithmetic

What numbers can be expressed as the difference of two integer squares? Since $(x + 1)^2 - x^2 = 2x + 1$, it is clear that every odd number can be so expressed. How about 98? If $x^2 - y^2 = 98$, then $x$ and $y$ must be either both even or both odd. But in this case, it is straightforward to argue that $x^2 - y^2$ is divisible by 4. Thus, the representation of 98 is not possible.

This type of argument occurs frequently in studying polynomials with rational and integer coefficients. Accordingly, in this section we will review some basic properties of the number system. Another reason for the importance of knowledge about the structure of integers is the fact that the family of polynomials shares much of this structure and the theory is developed in an analogous way.

First, some terminology. $\mathbf{N}$ denotes the set $\{1, 2, 3, \ldots\}$ of natural numbers and $\mathbf{Z}$ the set $\{\ldots, -2, -1, 0, 1, 2, \ldots\}$ of all integers. For any pair $a$, $b$, of integers, we say that $a$ *divides* $b$ (in symbols: $a|b$) if and only if there is an integer $c$ for which $b = ac$. Thus, for example, $37|111$ since $111 = 37 \times 3$. The only integers which divide every other integer are $+1$ and $-1$. Every integer divides 0, since, for each integer $k$, we can write $0 = k0$.

If $a|b$, then $|a| \leq |b|$, so that, if $a|b$ and $b|a$, then either $a = b$ or $a = -b$. For two nonzero integers $a$ and $b$, an integer $d$ such that $d|a$ and $d|b$ is called a *common divisor* of $a$ and $b$. There is a unique largest integer $g$ which is a common divisor of $a$ and $b$; this is the *greatest common divisor* of $a$ and $b$. We denote $g$ by $\gcd(a, b)$.

If $a|c$, then $c$ is a *multiple* of $a$. If $c$ is a multiple of both $a$ and $b$, then $c$ is a *common multiple* of $a$ and $b$. There is a unique smallest positive integer which is a multiple of both $a$ and $b$; this is the *least common multiple* of $a$ and $b$.

The greatest common divisor of two integers is a multiple of every common divisor. The least common multiple divides every common multiple.

An integer $p$ is *prime* if and only if $p$ is positive, $p \neq 1$ and the only positive divisors of $p$ are 1 and $p$. A pair of integers is *coprime* if their greatest common divisor is 1.

A fundamental result is the following.

**Division Theorem.** *Let $a, b$ belong to $\mathbf{Z}$ with $a > 0$. There are integers $q$* (quotient) *and $r$* (remainder) *such that*

$$b = qa + r \quad and \quad 0 \leq r < a.$$

*Furthermore, $q$ and $r$ are uniquely determined. That is, if the foregoing conditions are satisfied with $(q, r)$ replaced by $(q', r')$, then $r' = r$ and $q' = q$.*

## Exercises

1. *The Euclidean algorithm.* There is an ancient algorithm for finding the greatest common divisor of two given numbers which makes repeated use of the Division Theorem. The original context for the algorithm was not whole numbers but what the Greek geometers called magnitudes. Length is an example. One magnitude measures a second if the second is an positive integer multiple of the first; two magnitudes are commensurable if there is a magnitude which measures them both. In Book X, Propositions 1, 2, and 3, of his *Elements,* Euclid presents a practical method for determining whether or not two magnitudes are commensurable, and, in the latter case, of arriving at the greatest common measure. The Greeks might have used these results in geometry, for example, to show that the side and diagonal of a square are incommensurable.

   To see how Euclid's algorithm works in a numerical situation, let us find the greatest common divisor of 418 and 1606. Divide the smaller number into the larger to get

$$1606 = 3.418 + 352.$$

Explain why the greatest common divisor of 418 and 1606 is the same as the greatest common divisor of 352 and 418. Accordingly, we look for the greatest common divisor of 352 and 418.

$$418 = 1.352 + 66$$

Continue on:

$$352 = 5.66 + 22$$
$$66 = 3.22 + 0.$$

What is the greatest common divisor of 418 and 1606? Justify your answer. Carry out the same process to find gcd(20119, 34782).

Explain the following pencil-and-paper rendition of the Euclidean algorithm:

$$
\begin{array}{r}
3 \\
418 \overline{)1606} \\
\underline{1254} \quad 1 \\
352 \quad \overline{)418} \\
\underline{352} \quad 5 \\
66 \quad \overline{)352} \\
\underline{330} \quad 3 \\
22 \quad \overline{)66} \\
\underline{66} \\
0.
\end{array}
$$

2. (a) An important application of the Euclidean algorithm is to obtain a representation for the greatest common divisor of two numbers. Taking the numerical example of the last exercise, we can start with the representation of the greatest common divisor given by the second last equation and work our way back through the equations:

$$
\begin{aligned}
22 &= 1.352 - 5.66 = 1.352 - 5(418 - 1.352) \\
&= 6.352 - 5.418 = 6(1606 - 3.418) - 5.418 \\
&= 6.1606 - 23.418.
\end{aligned}
$$

This can be rendered in a handy paper-and-pencil form. Suppose we have performed the paper-and-pencil calculation to find the greatest common divisor. Now construct the table:

$$
\begin{array}{rrrr}
 & -5 & -1 & -3 \\
1 & -5 & 6 & -23.
\end{array}
$$

Explain where the numbers in the top row come from. Show that the numbers in the bottom row can be obtained successively by following the scheme

$$
\begin{array}{ccc}
\cdot & \cdot & u \\
v & w & uw + v.
\end{array}
$$

Explain the relevance of the scheme. Finally, show how the table can be used to find numbers $x_1$, $y_1$, $x_2$, $y_2$, $x_3$, $y_3$ to satisfy

$$22 = 352x_1 + 66y_1 = 418x_2 + 352y_2 = 1606x_3 + 418y_3.$$

(b) Find integers $x$ and $y$ to satisfy

(i) $3 = 12x + 21y$

(ii) $9 = 12x + 21y$

(iii) $6 = 24x + 66y$

(iv) $\gcd(20119, 34782) = 20119x + 34782y.$

This exercise illustrates the general result: *Let $a$ and $b$ be two integers whose greatest common divisor is $g$. Show that there exists integers $x$ and $y$ such that $g = ax + by$.*

The set $\{ax + by : x, y \in \mathbf{Z}\}$ is precisely the set of all multiples of the greatest common divisor of $a$ and $b$.

(c) Using the general result enunciated in (b), show that every common divisor of a pair of integers divides the greatest common divisor.

3. (a) Let $p$ be a prime and $a$ be any integer. Show that $a$ is a multiple of $p$ if and only if the greatest common divisor of $p$ and $a$ is not 1.

(b) Show that, if $a$ is not a multiple of the prime $p$, then there exist integers $x$ and $y$ for which $1 = ax + py$.

(c) Prove that, if $a$ and $b$ are any integers, and if the prime $p$ is a divisor of the product $ab$, then either $p|a$ or $p|b$.

(d) Let $n \geq 2$ be a positive integer. Show that there are prime numbers $p_1, p_2, \ldots, p_k$ and positive exponents $e_1, e_2, \ldots, e_k$ such that

$$n = p_1^{e_1} p_2^{e_2} \cdots p_k^{e_k}.$$

Show that this representation is unique up to the order of the prime power factors.

(e) Give the representation described in (e) for the numbers 418, 1606, 20119 and 34782.

4.  (a) Let two positive integers $m$ and $n$ be written out as a product of prime powers. Express the greatest common divisor $u$ and least common multiple $v$ of $m$ and $n$ as a product of the powers of the primes involved in representing $m$ and $n$.

    (b) Show that $mn = uv$.

    (c) Show that $v$ divides every common multiple of $m$ and $n$.

5.  Let $m \in \mathbf{N}$ and $a, b \in \mathbf{Z}$. We say that $a \equiv b \pmod{m}$ (read: "$a$ is congruent to $b$ modulo $m$") if and only if $m | a - b$.

    (a) Show that $a \equiv b \pmod{m}$ if and only if $a$ and $b$ have the same remainder upon division by $m$.

    (b) If $a \equiv b$ and $c \equiv d \pmod{m}$, show that $a + c \equiv b + d \pmod{m}$ and $ac \equiv bd \pmod{m}$. Explain how we take account of these facts every time we add up a column of figures or multiply two large numbers using paper and pencil.

    (c) Show that if $p$ is any polynomial with integer coefficients, and if $a \equiv b \pmod{m}$, then $p(a) \equiv p(b) \pmod{m}$.

6.  Let $m \in \mathbf{N}$, and $a, b \in \mathbf{Z}$. Consider the problem of solving the following congruence

$$ax \equiv b \pmod{m}.$$

A solution is any number $k$ for which $ak - b$ is divisible by $m$.

    (a) Show that 7 is a solution of the congruence $4x \equiv 3 \pmod{5}$. Find all other solutions of this congruence.

    (b) Find all solutons of the congruences

       (i) $4x \equiv 3 \pmod{6}$

       (ii) $4x \equiv 2 \pmod{6}$.

    (c) Let $g = \gcd(a, m)$. Show that, if $ax \equiv b \pmod{m}$ has a solution, then $g | b$.

    (d) Conversely, show that if $g = \gcd(a, m)$ and $g | b$, then the congruence $ax \equiv b \pmod{m}$ has a solution.

    (e) We say that the solution of a congruence $ax \equiv b \pmod{m}$ is *unique modulo m*, or, simply, *unique* if the difference between any solutions is divisible by $m$. In other words, the requirement is that $au \equiv av \equiv b \pmod{m}$ implies $u \equiv v \pmod{m}$.

    Show that the solution of the congruence is unique if and only if $\gcd(a, m) = 1$.

    (f) Show that, if $p$ is a prime, and $a$ is not a multiple of $p$, then there is exactly one value of $x$ satisfying

$$ax \equiv 1 \pmod{p} \quad \text{and} \quad 1 \le x \le p - 1.$$

## Explorations

**E.11.** Define the length of the Euclidean algorithm as the number of divisions required to complete it. Find that pair of numbers not exceeding 20 for which the Euclidean algorithm has its greatest length. Answer the same question replacing 20 by higher numbers. In general, enumerate those pairs $(a, b)$ of numbers for which the algorithm for the greatest common divisor is at least as long as the algorithm for any pair $(u, v)$ with $1 \leq u \leq a$, $1 \leq v \leq b$.

**E.12.** How many solutions (modulo $m$) are there to the congruence $ax \equiv b$ (mod $m$) in general?

**E.13.** Is it possible to find a polynomial of a single variable $n$ which assumes a prime value for every integer value of $n$? One famous attempt turned up the example $n^2 - n + 41$ which is prime for $0 \leq n \leq 40$, but composite when $n = 41$. More generally, there are other primes $p$ such as 11 and 17 for which $n^2 - n + p$ is prime for $0 \leq n \leq p - 1$. Checking this is made easier by the result that, if $n^2 - n + p$ is prime for $0 \leq n \leq \sqrt{p/3} + 1$, then it is prime for $0 \leq n \leq p - 1$. This was posed as a problem in the 1987 International Mathematical Olympiad.

Show that, no matter what polynomial $p(n)$ with integer coefficients is given, there are infinitely many values of the integer $n$ for which the polynomial assumes a composite value.

However, it is possible to find a polynomial of several variables with integer coefficients for which all the *positive* values it assumes are primes, although it will also assume nonpositive values. Such a polynomial is very complicated. In the next Exploration, you will see how to construct a polynomial all of whose positive values coincide with another well known set.

**E.14. Polynomials Whose Positive Values Are Fibonacci Numbers.** In the year 1202, the eminent mathematician Leonardo of Pisa (Fibonacci) posed in his book, *Liber abaci,* a famous problem: How many pairs of rabbits can be produced in a year from a single pair provided that it begets a new pair at the end of each month from the second month on and each new pair similarly reproduces? Thus, the original pair survives without issue through the first two months and produces a second pair at the end of the second month. At the end of the third month, the older pair gives rise to a third pair, while at the end of the fourth month, the offspring of the two oldest pairs bring the number of pairs up to five.

Denote by $F_k$ the number of pairs at the beginning of the $k$th month (at the end of the $(k - 1)$th month). Then for each positive integer $n \geq 2$, the number $F_{n+1}$ of pairs extant during the $(n + 1)$th month will include the $F_n$ pairs alive in the previous month plus the offspring of the $F_{n-1}$ pairs who were alive two months before. Thus, $F_n$ satisfies the recursion relation

$$F_1 = F_2 = 1 \quad F_{n+1} = F_n + F_{n-1} \quad (n = 2, 3, 4, \ldots).$$

We can answer Fibonacci's question by computing each $F_n$ in turn:

$$1, 1, 2, 3, 5, 8, 13, 21, 34, 55, 89, 144, 233, 377, 610, \ldots .$$

The number of pairs present at the beginning of the 13th month is $F_{13} = 233$, so that after one year, the original pair is responsible for the production of 232 new ones.

We will look at the question of finding a general formulae for the terms of the sequence in Exploration **E.50**. Our interest here is to construct a polynomial of two variables whose positive values are precisely the numbers $F_n$ where $n$ is a positive integer.

(a) Show that for each value of $n$ exceeding 1

$$F_{n+1}F_{n-1} - F_n^2 = \begin{cases} 1 & \text{if } n \text{ is even} \\ -1 & \text{if } n \text{ is odd.} \end{cases}$$

(b) Let $x$ and $y$ be positive integers such that

$$|(y-x)y - x^2| = 1. \qquad (*)$$

It can be shown that $y - x$, $x$, $y$ are consecutive terms in the Fibonacci sequence. First note that $x \le y$ and $y - x \le x$, and that, if $x = y$, then $(x, y) = (1, 1)$ and, if $y - x = x$, then $(x, y) = (1, 2)$.

The desired result can be proved by induction. It holds for $y \le F_3$. Assume that $n \ge 3$ and it holds for $y \le F_n$. Now let $(*)$ be valid when $x > 0$ and $F_n < y \le F_{n+1}$. Show that $x \le F_n$ and that, if $z = y - x$, then

$$|(x-z)x - z^2| = 1.$$

Use the induction hypothesis to argue that $x - z$, $z$, $x$, and hence $z$, $x$, $y$ are consecutive Fibonacci numbers.

(c) What are the positive values assumed by the polynomial

$$2 - [(y-x)y - x^2]^2$$

when $x$ and $y$ are integers?

(d) Determine a polynomial $f(x, y)$ with integer coefficients such that, whenever $x$ and $y$ are integers for which $f(x, y) > 0$, $f(x, y)$ belongs to the Fibonacci sequence.

# 1.7   Rings and Fields

Problems involving polynomials often require us to distinguish whether the coefficients are rational or nonrational, real or complex. The solution of even real equations require us to draw in nonreal entities. Since there are rules of operation equally valid for the various number systems–rational, real, complex—which we wish to consider, it is convenient to define abstract structures which embody these.

In thinking about these abstract structures, it is usually adequate to imagine you are dealing with some concrete model. Thus, a field is a structure which embraces as particular cases the sets of rationals, reals or complex numbers, so you can think of a field as being very like any one of these sets in the way in which the elements can be combined by addition, subtraction, multiplication and division. However, any result which can be established for fields in general holds for rational, real or complex numbers in particular. Rings and integral domains, in which division is not always possible, are exemplified by the set of integers or the set of polynomials.

Here are the axioms, or ground rules by which we shall operate.

Let $S$ be a system of entities for which there are two operations, $+$ (which we will call *addition*) and $\cdot$ (which we will call *multiplication*). Consider the following axioms:

**A.1.** If $a$ and $b$ belong to $S$, then $a + b$ belongs to $S$.

**A.2.** For $a$ and $b$ in $S$, $a + b = b + a$.

**A.3.** For $a$, $b$, $c$ in $S$, $(a + b) + c = a + (b + c)$.

**A.4.** There is an element in $S$, denoted by 0 and called the *zero* for which $a + 0 = 0 + a = a$ whenever $a$ belongs to $S$.

**A.5.** Given any element $a$ in $S$, there is exactly one element, denoted by $-a$ and called the additive inverse, such that $a + (-a) = (-a) + a = 0$.

**M.1.** If $a, b$ belong to $S$, then $ab$ belongs to $S$.

**M.2.** For $a, b$ in $S$, $ab = ba$.

**M.3.** For $a$, $b$, $c$ in $S$, $(ab)c = a(bc)$.

**M.4.** There is an element in $S$, denoted by 1 and called the identity, for which $a \cdot 1 = 1 \cdot a = a$ whenever $a$ belongs to $S$.

**M.5.** For any $a$ in $S$ with $a \neq 0$, there is an element, denoted by $a^{-1}$ and called the multiplicative inverse, such that

$$a \cdot a^{-1} = a^{-1} \cdot a = 1.$$

**D.** For $a$, $b$, $c$ in $S$, $a(b + c) = ab + ac$ and $(b + c)a = ba + ca$.

Any system of entities which satisfies all of these axioms is called a *field*. Some structures do not quite manage to be fields, such as:

*ring:* a system satisfying **A.1–5**, **M.1**, **M.3**, **D**.
*commutative ring:* a ring which satisfies **M.2**.
*commutative ring with an identity:* a ring which satisfies **M.2** and **M.4**.
*integral domain:* a commutative ring with identity which has no zero divisors (this means that if $ab = 0$, then either $a = 0$ or $b = 0$).

# Exercises

1. Show that the following are fields with the usual definitions of addition and multiplication:

   (a) $\mathbf{R}$ : the set of all real numbers
   (b) $\mathbf{Q}$ : the set of all rational numbers
   (c) $\mathbf{C}$ : the set of all complex numbers $x + yi$ with $x$, $y$ real, and $i^2 = -1$.

2. Show that $\mathbf{Z}$ is not a field, but is an integral domain.

3. Show that $\mathbf{N}$ is not even a ring.

4. Let $\mathbf{F}[t]$ denote the set of all polynomials in the variable $t$ whose coefficients lie in a field $\mathbf{F}$ and for which addition and multiplication are defined as in Section 1.1. Thus, $\mathbf{Q}[t]$, $\mathbf{R}[t]$ and $\mathbf{C}[t]$ are the sets of polynomials whose coefficients are, respectively, rational, real and complex.

   (a) Show that $\mathbf{Q}[t]$, $\mathbf{R}[t]$, $\mathbf{C}[t]$ are integral domains.
   (b) Show that $\mathbf{F}[t]$ is an integral domain.
   (c) Interpret $\mathbf{Z}[t]$ and show that $\mathbf{Z}[t]$ is an integral domain.
   (d) Interpret $\mathbf{F}[t_1, t_2, \ldots, t_m]$. Show that this is an integral domain.

   We say that $\mathbf{F}[t]$ is the set of polynomials *over* $\mathbf{F}$.

5. (a) Show that every field is an integral domain.
   (b) Show that every integral domain satisfies the *cancellation law:* if $ac = bc$ and $c \neq 0$, then $a = b$.

6. Let $m \geq 2$ be a positive integer. The set $\mathbf{Z}_m$ consists of the numbers $\{0, 1, 2, 3, \ldots, m - 1\}$. We define addition and multiplication on this set modulo $m$:

   $$a + b = c \quad \text{means that } 0 \leq c \leq m - 1 \text{ and } a + b \equiv c \ (\mathrm{mod}\, m)$$

   $$ab = c \quad \text{means that } 0 \leq c \leq m - 1 \text{ and } ab \equiv c \ (\mathrm{mod}\, m).$$

   (a) Fill in the addition and multiplication tables for $\mathbf{Z}_7$:

| + | 0 | 1 | 2 | 3 | 4 | 5 | 6 |
|---|---|---|---|---|---|---|---|
| 0 |   |   |   | 3 |   |   |   |
| 1 | 1 |   |   | 4 |   | 6 |   |
| 2 |   | 3 |   |   |   | 0 |   |
| 3 |   |   | 5 |   |   |   | 2 |
| 4 | 4 |   |   |   | 1 |   |   |
| 5 |   |   |   | 1 |   |   |   |
| 6 |   |   |   |   | 3 |   |   |

| · | 0 | 1 | 2 | 3 | 4 | 5 | 6 |
|---|---|---|---|---|---|---|---|
| 0 |   |   | 0 |   |   |   |   |
| 1 |   |   |   |   | 4 |   |   |
| 2 |   |   |   | 6 |   | 3 |   |
| 3 | 0 |   | 6 |   | 5 |   | 4 |
| 4 |   | 4 |   |   | 2 |   | 3 |
| 5 |   |   |   |   |   | 4 |   |
| 6 |   |   | 5 |   |   |   |   |

(b) Show that $\mathbf{Z}_m$ is a commutative ring with identity.

(c) Characterize those values of $m$ for which $\mathbf{Z}_m$ is a field.

(d) Show that, if $\mathbf{Z}_m$ is not a field, then it is not even an integral domain.

7. Write down a complete list of polynomials of degrees 0, 1, 2, 3, 4 in the ring $\mathbf{Z}_2[t]$. Indicate in your list which of the polynomials cannot be obtained by multiplying two polynomials of lower degree. This will include all polynomials of degrees 0 and 1. Will it also include any polynomials of degrees 2, 3 and 4?

8. Show that the polynomial $t^7 - t$ takes the value 0 for every value of $t$ in $\mathbf{Z}_7$. (This shows that, in contrast to the complex field, there are fields in which nonzero polynomials take the value 0 no matter what value is substituted for the variable.)

## Exploration

**E.15.** Let $p$ be a prime. How many different polynomials of degree $n$ over $\mathbf{Z}_p$ are there? Try to find a formula for the number of monic polynomials in $\mathbf{Z}_p[t]$ of degrees 2, 3, 4 which cannot be expressed as a product of polynomials of lower degree.

## 1.8 Problems on Quadratics

1. Given that $\tan A$ and $\tan B$ are the roots of the equation $x^2 + px + q = 0$, find the value of

$$\sin^2(A + B) + p\sin(A + B)\cos(A + B) + q\cos^2(A + B).$$

2. Find the value of the positive integer $n$ for which the quadratic equation

$$\sum_{i=1}^{n}(x + i - 1)(x + i) = 10n$$

has solutions $x = r$ and $x = r + 1$ for some number $r$.

If the coefficient 10 is replaced by an integer $p$, for which values of $p$ does a corresponding value of $n$ exist?

3. Find a necessary and sufficient condition that one root of the quadratic equation $ax^2 + bx + c = 0$ is the square of the other.

4. Let $p(t)$ be a monic quadratic polynomial. Show that, for any integer $n$, there exists an integer $k$ such that

$$p(n)p(n + 1) = p(k).$$

5. Prove that, if the roots of $x^2 + px + q = 0$ are real, then the roots of $x^2 + px + q + (x + a)(2x + p) = 0$ will be real for every real number $a$.

6. A mathematics teacher wrote the quadratic $x^2 + 10x + 20$ on the board. Then each student either increased by 1 or decreased by 1 either the constant or the linear coefficient. Finally $x^2 + 20x + 10$ appeared. Did a quadratic trinomial with integer zeros necessarily appear on the board in the process?

7. Suppose $a < b$ and $c < d$. Solve the system

$$a^2 + b^2 = c^2 + d^2$$

$$a + b + c + d = 0.$$

8. Find necessary and sufficient conditions on the real numbers $a$, $b$, $c$, $d$ for the equation

$$z^2 + (a + bi)z + (c + di) = 0$$

to have exactly one real and one nonreal root.

9. Show that if $x^2 + px + q = 0$ and $px^2 + qx + 1 = 0$ have a common root, then either $p + q + 1 = 0$ or $p^2 + q^2 + 1 = pq + p + q$.

10. If $p$ and $q$ are real numbers which do not take simultaneously the values $p = 0$, $q = 1$, and if the roots of the equation

$$\left(1 - q + \frac{p^2}{2}\right) x^2 + p(1 + q)x + q(q - 1) + \frac{p^2}{2} = 0$$

are equal, show that $p^2 = 4q$.

11. Show that all the real values of $x$ which satisfy the equation $\tan(\pi \cot x) = \cot(\pi \tan x)$ are given by

$$4 \tan x = 2n + 1 \pm \sqrt{4n^2 + 4n - 15},$$

when $n$ is a positive or negative integer different from $-2$, $-1$, $1$.

12. Find all positive integers $n$ for which the quadratic equation

$$a_{n+1}x^2 - 2x\sqrt{a_1^2 + a_2^2 + \cdots + a_{n+1}^2} + (a_1 + a_2 + \cdots + a_n) = 0$$

has real roots for all reals $a_1, a_2, \ldots, a_{n+1}$.

13. Let $p(z) = z^2 + az + b$ have complex coefficients and satisfy $|p(z)| = 1$ whenever $|z| = 1$. Prove that $a = b = 0$.

14.  (a) Find necessary and sufficient conditions on $a$, $b$, $w$ so that the
     roots of $z^2 + 2az + b = 0$ and $z - w = 0$ are collinear in the
     complex plane.

     (b) Find necessary and sufficient conditions on $a$, $b$, $c$, $d$ so that the
     roots of $z^2 + 2az + b = 0$ and $z^2 + 2cz + d = 0$ are collinear in
     the complex plane.

15.  Show that, if $-(b/a)\cos^2\{(1/4)\text{arc }\cos[(b^2 - 8ac)/b^2]\}$ exists, then it
     is a root of the equation $ax^2 + bx + c = 0$.

16.  Let $u$ and $v$ be the roots of the equation

$$z + (1/z) = 2(\cos\phi + i\sin\phi) \quad \text{where} \quad 0 < \phi < \pi.$$

     (a) Show that $u+i$ and $v+i$ have the same argument and that $u-i$
     and $v - i$ have the same modulus.

     (b) Find the locus of the roots $u$, $v$ in the complex plane when $\phi$
     varies from 0 to $\pi$.

17.  Solve

$$x^2 - (2a - b - c)x + (a^2 + b^2 + c^2 - bc - ca - ab) = 0.$$

18.  Let $a$, $b$, $c$ be nonzero integers such that the greatest common divisor
     of $b$ and $ac$ is 1. Prove that $ax^2 + bx + c$ and $ax^2 + bx - c$ can both be
     written as the product of linear polynomials with integer coefficients
     if and only if $ac = rs(r^2 - s^2)$ and $b^2 = (r^2 + s^2)^2$, where $r$ and $s$ are
     relatively prime integers.

19.  If $a$, $b$ and $h$ are constants, prove that the maximum and minimum
     values of $a\cos^2\theta + 2h\sin\theta\cos\theta + b\sin^2\theta$ are the roots of the equation
     $(x - a)(x - b) = h^2$.

20.  What conditions must be satisfied by the constants $a$, $b$, $c$ for the
     quadratic function

$$f(x,y) = x^2 - y^2 + 2ax + 2by - c$$

     to be the product of two linear factors?

21.  (a) If the line $y = mx + c$ is tangent to the curve $b^2x^2 - a^2y^2 = a^2b^2$,
     show that $a^2m^2 = b^2 + c^2$.

     (b) Chords of the circle $x^2 + y^2 = r^2$ touch the hyperbola $b^2x^2 - a^2y^2 = a^2b^2$. Find the equation of the locus of the midpoints.

22.  Determine all those quadratic polynomials whose zeros are symmetric
     about the imaginary axis, i.e. $r + is$ is a zero if and only if $-r + is$ is
     a zero.

23. Find the equations of those conjugate diameters of the ellipse $b^2x^2 + a^2y^2 = a^2b^2$ which are of equal length. (Two diameters are conjugate if each is the locus of midpoints of chords parallel to the other. Refer to Exercise 2.16.)

24. Let $ax^2 + bx + c$ be a quadratic polynomial with real coefficients for which $|ax^2 + bx + c| \le 1$ for $0 \le x \le 1$. Prove that $|a| + |b| + |c| \le 17$. Give an example for which equality holds.

## 1.9   Other Problems

1. Suppose that $t^3 + pt + q = 0$ has a nonreal root $a + bi$, where $a$, $b$, $p$, $q$ are all real and $q \ne 0$. Show that $aq > 0$.

2. Consider a polynomial $f(x)$ with real coefficients having the property $f(g(x)) = g(f(x))$ for every polynomial $g(x)$ with real coefficients. Determine and prove the nature of $f(x)$.

3. If $a$, $b$, $c$, $d$ are real numbers, show that each of the two systems of three equations is equivalent to the other:

$$\begin{array}{llll} \text{I.} & a^2 + b^2 = 2 & c^2 + d^2 = 2 & ac = bd \\ \text{II.} & a^2 + c^2 = 2 & b^2 + d^2 = 2 & ab = cd. \end{array}$$

4. Find a simple expression for the positive root of

$$x^3 - 3x^2 - x - \sqrt{2} = 0.$$

5. Show that any root of

$$(x + a + b)(x^{-1} + a^{-1} + b^{-1}) = 1$$

is a root of

$$(x^n + a^n + b^n)(x^{-n} + a^{-n} + b^{-n}) = 1,$$

where $n$ is any odd integer and where $a$ and $b$ are both different from 0.

6. (a) Given that $x + a + \sqrt{a^2 - b} = 0$, where $x$ is not 0, verify that

$$x + \frac{b}{x} + 2a = 0.$$

(b) Given that $y = px + q$, where $p \ge 0$ and $x + a + \sqrt{a^2 - b} = 0$, verify that

$$y + (ap - q) + \sqrt{(ap - q)^2 - (bp^2 - 2apq + q^2)} = 0.$$

If $y$ is nonzero, deduce that

$$y + (bp^2 - 2apq + q^2)/y + 2(ap - q) = 0.$$

7. Find the square roots of $1 - x + \sqrt{22x - 15 - 8x^2}$.

8. Determine necessary and sufficient conditions that $ax^4 + bx^3 + cx^2 + dx + e \ (a \neq 0)$ is of the form $p(q(x))$, where $p$ and $q$ are both quadratic.

9. From the pair of equations

$$x = 1 - v + (v/u), \quad y = 1 - u + (u/v),$$

deduce the pair of equations

$$u = 1 - y + (y/x), \quad v = 1 - x + (x/y),$$

and conversely.

10. Solve the equation

$$(x - 2)(x - 3)(x - 4)(x - 5) = 360.$$

11. Show that the polynomial $x^4 y^2 + y^4 z^2 + z^4 x^2 - 3x^2 y^2 z^2$ always assumes a nonnegative value when $x$, $y$, $z$ are real, but cannot be written as the sum of squares of polynomials over $\mathbf{R}$ in $x$, $y$, $z$.

12. Express $x^4 + y^4 + x^2 + y^2$ as the sum of the squares of three polynomials over $\mathbf{R}$ in $x$, $y$.

13. Let $P(x, y) = x^2 y + xy^2$ and $Q(x, y) = x^2 + xy + y^2$. For each positive integer $n$, define

$$F_n(x, y) = (x + y)^n - x^n - y^n$$

$$G_n(x, y) = (x + y)^n + x^n + y^n.$$

Observe that $G_2 = 2Q$, $F_3 = 3P$, $G_4 = 2Q^2$, $F_5 = 5PQ$, $G_6 = 2Q^3 + 3P^2$. Prove that, for each positive integer $n$, either $F_n$ or $G_n$ is expressible as a polynomial in $P$ and $Q$ over $\mathbf{Z}$.

14. Define a sequence of polynomials $P_m(x, y, z)$ as follows:

$$P_0(x, y, z) = 1$$

$$P_m(x, y, z) = (x + z)(y + z)P_{m-1}(x, y, z + 1) - z^2 P_{m-1}(x, y, z).$$

Prove that each $P_m(x, y, z)$ is symmetric in $x$, $y$, $z$.

15. How many distinct terms are there in the expansion of

$$x_1(x_1 + x_2)(x_1 + x_2 + x_3) \cdots (x_1 + x_2 + \cdots + x_n)?$$

16. Let $u$ be an integer. Simplify

$$\sqrt[3]{3u - 1 + \sqrt{8u^3 - 3u^2}} + \sqrt[3]{3u - 1 - \sqrt{8u^3 - 3u^2}}.$$

17. Show that there are infinitely many pairs of positive integers $m$ and $n$ for which $4mn - m - n + 1$ is a perfect square.

18. Determine all numbers $u$ for which

    (i) there is a cubic polynomial $p$ with integer coefficients for which $u$, $u^2$, $u^3$ are distinct zeros;

    (ii) $u$ is nonrational.

19. For any polynomial $p(t) = a_m t^m + a_{m-1} t^{m-1} + \cdots + a_1 t + a_0$, let $\Gamma(p(t)) = a_m^2 + a_{m-1}^2 + \cdots + a_1^2 + a_0^2$.

    Let $f(t) = 3t^2 + 7t + 2$. Find, with proof, a polynomial $g(t)$ for which

    (i) $g(0) = 1$;
    (ii) $\Gamma(f(t)^n) = \Gamma(g(t)^n)$ for $n = 1, 2, \ldots$ .

20. Given that $x^2 + y^2 = 6xy$ and $x > y > 0$, determine

$$\frac{x + y}{x - y}.$$

## Hints

### Chapter 1

1.12. (a) The constant term is the value of the polynomial at 0.

    (b) The difference of the two polynomials is identically zero.

1.13. (a) deg $f(2t) = $ deg $f(t)$. What is deg $h(t)$?

1.14. $\log 2t = \log 2 + \log t$.

1.15. $g(t + k) - g(t)$ is identically equal to 0.

1.17. $p(f - g)$ is identically zero.

1.18. Either use induction or multiply both sides by $1 - t$.

1.20. $p(t^2)$ has terms in only even powers of $t$.

2.3. One root is $x = m$.

2.5. The discriminant should vanish.

2.8. The equation leads to a quadratic in $x$. For which values of $k$ is the discriminant nonnegative?

2.10. Express the sum and product of $m - n^2$ and $n - m^2$ in terms of $m + n = 5/6$ and $mn = -1/2$.

2.14. (a) The polynomial does not take negative values.

2.17. The discriminant of the quadratic, i.e. $(\sqrt{a} + \sqrt{b})^2 - 4\sqrt{ab}$, is not less than 0.

3.3. (k) For a clean proof, apply (h), (d) and (f) to $|z + w|^2$.

3.5. The given locus is $1/w$ times the locus of $\text{Re}(z) = c$.

3.7. Let $U$ and $V$ be represented by the points 0 and 1 in the complex plane, and suppose the tree $T$ is at $z$. Locate the points $P$ and $Q$, noting that multiplication by $i$ corresponds to a rotation through a right angle. Show that the midpoint of $PQ$ does not depend on $z$.

3.10. The solution of the equations for $x$ and $y$ in terms of $a$ and $b$ can be facilitated using the theory of the quadratic.

3.14. (a) Make use of Exercise 3(c) and 3(d).

3.15. (d) Use de Moivre's Theorem, Exercise 3.8.

4.4. (b) Use Exercise 2.4.

4.9. To convert the particular equation to the general form, let $y = 2x$.

4.16. (b) Let $u$ be the polynomial $g$ with its coefficients in the opposite order, i.e. $u(t) = t^k g(1/t)$ where $k = \deg g$. Show that $uh = f = gh$ and use Exercise 1.17.

5.5. It suffices to prove the result for polynomials of the form $x^a y^b + x^b y^a$.

5.7. Three numbers are in arithmetic progression if and only if their sum is equal to three times one of the numbers.

5.9. (c) Let $a = x^3$, etc., and apply (a) and (b).

6.3. (c) Multiply the equation in (b) by $b$.

6.5. (b) Note that $ac - bd = a(c - d) + (a - b)d$.

7.7. A complete collection of quadratic polynomials which can be expressed as the product of two linears can be obtained by multiplying together all possible (not necessarily distinct) linear parts; the number of ways of doing this is easily determined.

8.1. Express $\tan(A+B)$ in terms of $\tan A$ and $\tan B$, and thence in terms of $p$ and $q$.

8.2. Write the left side of the equation in the form $ax^2 + bx + c$. Verify that

$$3\sum_{i=1}^{n}(i-1)i = (n+1)n(n-1).$$

8.3. Consider $(r - s^2)(s - r^2)$, where $r$ and $s$ are the roots. Whatever method you use, be sure to show that the condition you obtain implies *and* is implied by one root being the square of the other.

8.4. By considering $q(t) = p(n+t)$, it suffices to prove the result is true for $n = 0$ and any quadratic.

8.5. Express the discriminant of the second quadratic as the sum of a square and a multiple of the discriminant of the first quadratic.

8.6. $x^2 + (b+1)x + b$ has integer zeros for $b \in \mathbf{Z}$.

8.7. $a$ and $b$ have, respectively, the same sum and the same product as $-c$ and $-d$.

8.8. Substitute $z = r$, the real root, into the equation and separate the real and imaginary parts.

8.9. A common root of the two equations is a root of any equation of the form $f(x)(x^2 + px + q) + g(x)(px^2 + qx + 1) = 0$.

8.11. $\tan(\pi \cot x) = \tan(\pi/2 - \pi \tan x)$.

8.12. Form the discriminant and complete some obvious squares.

8.13. $p(1)$, $p(-1)$, $p(i)$, $p(-i)$ all belong to the unit disc; what does this mean in terms of the coefficients $a$, $b$?

8.14. Solve the quadratic equation by completing the square. The line joining the complex numbers $r$ and $s$ consists of the points $(1-t)r + ts$ where $t$ is real.

8.15. Let $\cos 4\theta = (b^2 - 8ac)/b^2$. Determine $\cos^2 2\theta$ then take its square root and find $\cos^2 \theta$.

8.16. Solve the equation for $z$ by completing the square. Note that

(i) $i = \cos(\pi/2) + i\sin(\pi/2)$, so that

$$i(\cos\phi + i\sin\phi) = \cos(\phi + \pi/2) + i\sin(\phi + \pi/2);$$

(ii) $\cos\phi = \sin(\phi + \pi/2)$ and $\sin\phi = -\cos(\phi + \pi/2)$.

Ultimately, obtain expressions for the roots of the equation in terms of $\theta = \phi/2 + \pi/4$.

8.18. Assume that the factorization over $Z$ occurs. Show that the discriminants are perfect squares of the same parity. Find $u$ and $v$ such that $b^2 = u^2 + v^2$. To get $|b|$ as a sum of squares, use the equation $(r + is)^2 = u + iv$ as inspiration.

8.19. Put the expression in the form $A + B\cos 2\theta + C\sin 2\theta = A + D\sin(2\theta + \phi)$.

8.20. If $f(x, y)$ has linear factors, the discriminant of $f(x, y)$ as a quadratic in $x$ must be square as a quadratic in $y$; what can be said about the discriminant of the second quadratic?

8.21. (a) Substituting $y = mx + c$ into the other equation yields a quadratic equation with equal roots.

(b) Use the theory of the quadratic to determine the midpoints of the chords. If $f(x, y) = 0$ is on the locus, find $m$ and $c$ in terms of $x$ and $y$, and substitute into the condition obtained in (a).

9.1. What are the other two roots? Do not solve the equation; just use the fact that the coefficients are real and one of them is 0.

9.2. In particular, $f$ commutes with any constant polynomial.

9.3. $A$, $B$, $C$ all vanish iff $A^2 + B^2 + C^2 = 0$.

9.4. The observation $(\sqrt{2})^3 - \sqrt{2} = \sqrt{2}$ suggests trying $x = \sqrt{2} + u$.

9.6. (a) With the help of a surd conjugate, determine $1/x$.

(b) Simplify the left side of the first equation.

9.7. Can the square root ever be real? pure imaginary (i.e. a real multiplied by $i$)?

9.9. What is $xu$? $yv$?

9.10. Expand the left side as the product of two quadratics whose leading and linear coefficients agree, then put it in the form $(u - 1)(u + 1)$.

9.11. Use the arithmetic-geometric mean inequality (Exercise 5.9).

9.12. Let the given polynomial be equal to $\sum [f_i(x,y)]^2$. Since constant terms of the $f_i$ must vanish, we can write $f_i(x,y) = a_i x^2 + b_i xy + c_i y^2 + u_i x + v_i y$. Use vectors: let $\mathbf{a} = (a_1, a_2, a_3)$, etc. and verify that $\mathbf{a} \cdot \mathbf{b} = \mathbf{a} \cdot \mathbf{u} = \mathbf{b} \cdot \mathbf{c} = \mathbf{c} \cdot \mathbf{v} = \mathbf{u} \cdot \mathbf{v} = 0, \mathbf{b} \cdot \mathbf{u} + \mathbf{a} \cdot \mathbf{v} = 0, \mathbf{b} \cdot \mathbf{v} + \mathbf{u} \cdot \mathbf{c} = 0$. Try $\mathbf{u} = (0,1,0)$, $\mathbf{v} = (0,0,1)$.

9.13. Compare $F_n$ and $(x+y)^2 F_{n-2}$ with a view to setting up an induction process. Note that $(x+y)^2 = Q + xy$.

9.14. The hard part is to show symmetry in $x$ and $z$. Establish this by induction. Look at $P_{n+1}(x,y,z) - P_{n+1}(z,y,x)$.

# 2

# Evaluation, Division, and Expansion

## 2.1 Horner's Method

A Knight wishes to evaluate the polynomial $8t^3 - 5t^2 + 4t + 1$ at $t = 2$. He takes it to the Royal Reckoner, who charges 10 sous for each multiplication and 5 sous for each addition. Since there are three multiplications required for the first term, two for the second and one for the third, the multiplications will cost 60 sous. In addition, there will be a 15 sou charge for adding the terms, for a grand total of 75 sous. The Knight wonders whether the job could be done more cheaply.

After some thought, he makes a suggestion. Write the first two terms in the form $(8t - 5)t^2$, and substitute in $t = 2$. We then have one multiplication and one subtraction inside the bracket, followed by two other multiplications, for a total cost of 35 sous. This compares very favorably with the 55 sous it would have cost using the Royal Reckoner's method. Why not carry this regrouping further? The sum of the first three terms is equal to

$$((8t - 5)t + 4)t.$$

We still have only three multiplications along with two additions or subtractions when we substitute in $t = 2$ and evaluate. All we have to do now to get the value of the polynomial we started with is to make one more addition. The total cost is 45 sous. The nested form is going to save money!

## Exercises

1. Consider the problem of evaluating the polynomial

$$3t^3 - 4t^2 + 7t + 2$$

at $t = 3$.

(a) Show that the polynomial can be written in the form

$$(((3t - 4)t + 7)t + 2)$$

and use this to effect a cheap evaluation at $t = 3$.

(b) A paper-and-pencil algorithm for making the evaluation in (a) has the following table as its final result.

$$
\begin{array}{rrrr}
3 & -4 & 7 & 2 \\
  & 9 & 15 & 66 \\
\hline
3 & 5 & 22 & 68
\end{array}
$$

Explain how the table is to be filled in and where the answer is found.

(c) Show that the computation can be done on a pocket calculator using one of the following two procedures, depending on the calculator:

(i)

| Press buttons | Result |
|---|---|
| $3 \times 3 =$ | 9 |
| $-4 =$ | 5 |
| $\times 3 =$ | 15 |
| $+7 =$ | 22 |
| $\times 3 =$ | 66 |
| $+2 =$ | 68 |

(ii)

| Press button | Operator | Result |
|---|---|---|
| 3 | Enter | 3 |
| 3 | Multiply | 9 |
| 4 | Subtract | 5 |
| 3 | Multiply | 15 |
| 7 | Add | 22 |
| 3 | Multiply | 66 |
| 2 | Add | 68 |

This method of evaluating a polynomial is called *Horner's Method*.

2. Show that the Horner table for evaluating the polynomial $3t^3 - 4t^2 + 7t + 2$ at $t = -2$ is

$$
\begin{array}{rrrr}
3 & -4 & 7 & 2 \\
  & -6 & 20 & -54 \\
\hline
3 & -10 & 27 & -52
\end{array}
$$

What is the required value? Check your answer independently.

3. In applying Horner's Method, you should not fail to record zero coefficients. Check that the table for evaluating at $t = 6$ the polynomial

$$t^5 - 4t^3 + 2t^2 - 7$$

is

| 1 | 0 | −4 | 2 | 0 | −7 |
|---|---|----|---|---|----|
|   | 6 | 36 | 192 | 1164 | 6984 |
| 1 | 6 | 32 | 194 | 1164 | 6977 |

and read off the required value of the polynomial. Check the value independently.

4. Make up several polynomials of various degrees and evaluate them for a number of values of $t$. Compare the number of operations required in Horner's Method to the number that would be required for a term-by-term evaluation.

5. Programme a computer to carry out an efficient calculation of the value of $t = 2.376$ of the polynomial

$$4.82t^5 + 87.2433t^4 - 764.331t^2 + 12.354t + 77.4412.$$

6. A student, evaluating a polynomial, presses the following buttons on his pocket calculator:

$$7 \times 6 = \times 6 = -2 = \times 6 = -3 = \times 6 = +1 = \times 6 = +2 =$$

Find the polynomial being evaluated and the point of evaluation. Determine the value of the polynomial.

7. Find the polynomial, the point of evaluation and the required value of the polynomial from the following table:

| 3 | 5 | 1 | −2 | 6 |
|---|---|---|----|---|
|   | 6 | 22 | 46 | 88 |
| 3 | 11 | 23 | 44 | 94 |

Check your answer using a pocket calculator.

8. (a) Verify that

$$t^4 + t^2 - 3t + 7 = (t^3 + 3t^2 + 10t + 27)(t - 3) + 88.$$

(b) Construct the Horner table for evaluating this polynomial at $t = 3$. The last entry in the bottom row gives the value sought. Interpret the remaining entries in the bottom row. Account for your interpretation.

(c) Construct the Horner table for evaluating at $t = 3$ the polynomial

$$t^3 + 3t^2 + 10t + 27.$$

(d) Use the table in (c) to express the polynomial there in the form $q(t)(t - 3) + r$, for some polynomial $q(t)$ and constant $r$. Check your answer by direct computation.

(e) Write $q(t)$ in the form $u(t)(t - 3) + v$ for some polynomial $u(t)$ and constant $v$.

(f) Combine the results of the previous parts of this problem to write the polynomial $t^4 + t^2 - 3t + 7$ in the form

$$b_0 + b_1(t - 3) + b_2(t - 3)^2 + b_3(t - 3)^3 + b_4(t - 3)^4$$

for some constants $b_i$. Show how the computation can be displayed in a convenient table.

9. Explain the connection between the table

$$
\begin{array}{r r r r}
3 & -2 & 4 & 7 \\
  & 3 & 1 & 5 \\
\hline
3 & 1 & 5 & 12 \\
  & 3 & 4 & \\
\hline
3 & 4 & 9 & \\
  & 3 & & \\
\hline
3 & 7 & & \\
\hline
3 & & &
\end{array}
$$

and the identity

$$3t^3 - 2t^2 + 4t + 7 = 12 + 9(t - 1) + 7(t - 1)^2 + 3(t - 1)^3.$$

10. For each of the following polynomials $p(t)$ and constants $c$, use Horner's Method to write $p(t)$ in the form $(t - c)q(t) + p(c)$ and expand $p(t)$ in terms of powers of $(t - c)$. In each case, check your answer by making the substitution $t = c + s$, expanding out $p(c + s)$ as a polynomial in $s$, and then substituting $t - c$ for each occurrence of $s$.

(a) $p(t) = t^4 + t^2 - 3t + 7$    $c = 3$

(b) $p(t) = t^5 - 4t^3 + 2t^2 - 7$    $c = -5$

(c) $p(t) = t^7 + t^6 - t^4 + t^2 - 5t - 1$    $c = 6$.

11. Expand the polynomial $y^7 - 4y^6 + 2y^4 - y^2 + y + 1$ in terms of $(y+2)$ and evaluate it at $y = -1.95$ to two decimal places.

12. What does the table for expanding $p(t)$ in powers of $t$ look like?

13. Let $p$, $q$, $r$, $s$ be the zeros of the quartic $t^4 - 3t^3 + 2t^2 + 5t - 2$. Use Horner's Method to find a polynomial with integer coefficients whose zeros are $p + 3$, $q + 3$, $r + 3$, $s + 3$.

## Explorations

**E.16.** We can evaluate $6^5$ by means of four multiplications. However, the number of multiplications can be reduced to 3:

$$6 \times 6 = 36, \quad 36 \times 36 = 1296, \quad 6 \times 1296 = 7776.$$

In general, for an arbitrary positive integer $n$ and constant $c$, what is the minimum number of multiplications necessary to compute $c^n$?

(a) Show that, if $n = 2^{k+1}$, then $c^n$ can be obtained with no more than $k$ multiplications. Is it possible to get by with fewer multiplications?

(b) Show that $c^n$ can be computed using a pocket calculator or a computer by some sequence of the following two operations:

(i) multiply the display by $c$ (which can be stored in memory);

(ii) square the display.

(c) Plan the procedure based on (b) which you would use to determine $c^{51}$. How many multiplications are required? The binary representation (i.e. to base 2) of 51 will give a clue as to the order in which operations (i) and (ii) might be taken.

**E.17.** For small positive values of the integer $n$, determine the expansion of $t^n$ in terms of $(t - 1)$ using Horner's table. Look for patterns, depending on $n$ and $k$, which govern the coefficients of $(t - 1)^k$. Try to find general formulae for these coefficients. Rewrite the equation you get by making the substitution $t = 1 + x$.

**E.18. Factorial Powers and Summations.** The formulae

$$1 + 2 + \cdots + n = \frac{1}{2}n(n + 1) \quad n = 1, 2, \ldots$$

$$1^2 + 2^2 + \cdots + n^2 = \frac{1}{6}n(n + 1)(2n + 1) \quad n = 1, 2, \ldots$$

are familiar to many high school students. The task of finding analogous closed formulae for the sums of higher powers such as cubes and fourth powers increases in complexity with the exponent. Is there a systematic way of proceeding in general?

Taking notice of the fact that the $n$th term of any series is equal to the difference between the sums of the first $n$ terms and of the first $n - 1$ terms, one can see that the two formulae given above can be established by verifying the algebraic results

$$n = \frac{1}{2}n(n + 1) - \frac{1}{2}(n - 1)n$$

$$n^2 = \frac{1}{6}n(n + 1)(2n + 1) - \frac{1}{6}(n - 1)n(2n - 1).$$

With these formulae which express the general term of the series as a difference, we can obtain the required sum by "summation by differences." For example,

$$1 + 2 + \cdots + n = \frac{1}{2}1 \cdot 2 + \left( \frac{1}{2}2 \cdot 3 - \frac{1}{2}1 \cdot 2 \right) + \cdots$$

$$+ \left[ \frac{1}{2}n(n + 1) - \frac{1}{2}(n - 1)n \right] = \frac{1}{2}n(n + 1)$$

after a cancellation of terms. This method seems unsatisfactory since we do not usually know in advance how to find a function whose differences are $n$ and $n^2$.

To get around this, we try to find functions which have differences which are simply described, and then try to express the summands of the series in terms of these functions.

Let $g(n)$ be a function of the integer $n$. The first order difference of $g(n)$ is defined by

$$\Delta g(n) = g(n + 1) - g(n).$$

Find $\Delta g(n)$ when $g(n) = n^k$ $(k = 0, 1, 2, 3, 4, 5)$? Verify that $\Delta n^6 = 6n^5 + 15n^4 + 20n^3 + 15n^2 + 6n + 1$. The result is quite complicated. A function which has a difference of a simpler type is the *factorial power* of $n$. This is defined as follows:

let $k$ be a positive integer. Then the *$k$th factorial power of $n$* is given by

$$n^{(k)} = n(n - 1)(n - 2) \cdots (n - k + 1),$$

where there are $k$ factors, each 1 less than its predecessor.

Verify that $\Delta n^{(k)} = kn^{(k-1)}$.

The value of this formula is that we can now conveniently sum by differences. Using the fact that

$$(r + 1)^{(k+1)} - r^{(k+1)} = (k + 1)r^{(k)},$$

show that

$$\sum_{r=1}^{n} r^{(k)} = \frac{1}{k+1}(n+1)^{(k+1)}$$

for $k = 1, 2, 3, \ldots$ .

Since we know how to sum factorial powers, we can now sum ordinary powers by expressing them first in terms of factorial powers. For example, verify that $r^2 = r^{(2)} + r^{(1)}$, and use this fact to derive the summation formula for the first $n$ squares.

Express $r^3$ as a "polynomial" in factorial powers and use the result to derive a formula for the sum of the first $n$ cubes. Try out the process for higher powers.

How can we systematically determine the factorial power expansion of a given polynomial, such as $r^k$? Horner's method can be adapted to this purpose. For example, suppose $r^4$ is to be written in the form

$$r^4 = c_n r^{(n)} + c_{n-1} r^{(n-1)} + \cdots + c_1 r + c_0.$$

Observe that the polynomial $r^4 - c_0$ is divisible by $r$; the polynomial $r^4 - (c_0 + c_1 r)$ is divisible by $r - 1$, and so on. We can use this to design a "Horner's" table whose entries are the desired coefficients.

Thus, a suitable table for $r^4$ would be

$$
\begin{array}{ccccc}
1 & 0 & 0 & 0 & 0 \\
  & 0 & 0 & 0 & 0 \\
\hline
1 & 0 & 0 & 0 & 0 \\
  & 1 & 1 & 1 & \\
\hline
1 & 1 & 1 & 1 & \\
  & 2 & 6 & & \\
\hline
1 & 3 & 7 & & \\
  & 3 & & & \\
\hline
1 & 6 & & & \\
\end{array}
$$

Justify this table and read off the factorial expansion of $r^4$ from it. Check directly that the expansion is correct. Use this to derive a formula for the sum of the first $n$ fourth powers.

For the polynomial $4r^3 + 2r^2 - r - 1$, verify that the Horner's table is

$$
\begin{array}{rrrr}
4 & 2 & -1 & -1 \\
  & 0 & 0  & 0  \\
\hline
4 & 2 & -1 & -1 \\
  & 4 & 6  &    \\
\hline
4 & 6 & 5  &    \\
  & 8 &    &    \\
\hline
4 & 14 &   &
\end{array}
$$

and use this to determine a factorial power expansion of the given cubic. (For convenience, we can simply delete the first two lines of the table.) [See also Exercise 7.1.16–17, Explorations E.54 and E.57.]

## 2.2 Division of Polynomials

In Exercise 1.8, we observed that the bottom line of Horner's table gave us the coefficients of the quotient $q(t)$ when we divided the polynomial $p(t)$ by $(t - c)$ to obtain an identity of the form

$$p(t) = q(t)(t - c) + k.$$

By analogy with numbers, we can look upon this equation as representing a division. The polynomial $p(t)$ is divided by $(t-c)$, yielding a quotient $q(t)$ and a remainder $k$. However, there is no reason to restrict our attention to divisors of degree 1.

The exercises in this section will sketch in the details of a theory of division for polynomials which is similar to that for integers. The extent to which we can discuss division of one polynomial by another depends on the domain from which the coefficients are taken, so let us establish some terminology:

Let $\mathbf{D}$ be an integral domain, and $\mathbf{D}[t]$ be the set of all polynomials in the variable $t$ with coefficients in $\mathbf{D}$. For short, if $f(t)$ belongs to $\mathbf{D}[t]$, we say that $f(t)$ is a polynomial *over* $\mathbf{D}$. For any pair $f(t)$, $g(t)$ of polynomials in $\mathbf{D}[t]$, we say that $g(t)$ *divides* $f(t)$ (in symbols: $g(t) \mid f(t)$) if there is a polynomial $h(t)$ in $\mathbf{D}[t]$ for which $f(t) = g(t)h(t)$. In this situation, $g(t)$ is a *divisor* or *factor* of $f(t)$ and $f(t)$ a *multiple* of $g(t)$.

## Exercises

1. Let $p(t)$ be any polynomial over an integral domain $\mathbf{D}$ and $c$ be any element of $\mathbf{D}$. Consider the equation

$$p(t) = (t - c)q(t) + k$$

where $q(t)$ is a polynomial and $k$ a constant polynomial over $\mathbf{D}$.

(i) Show that $\deg q = (\deg p) - 1$.

(ii) By making the substitution $t = c$, verify that $k = p(c)$.

2. *Factor Theorem.* Let $c$ belong to an integral domain $\mathbf{D}$ and $p(t)$ be any polynomial over $\mathbf{D}$. Show that $(t - c)$ divides $p(t)$ if and only if $p(c) = 0$.

3. Let $r$ be a zero of the polynomial $p(t)$ over an integral domain, so that for some polynomial $q(t)$, $p(t) = (t - r)q(t)$. Prove that, if $s \neq r$, then $s$ is a zero of $p(t)$ if and only if $s$ is a zero of $q(t)$.

4. (a) Suppose $r_1, r_2, \ldots, r_k$ are distinct zeros of a polynomial $p(t)$ over an integral domain. Show that there exists a polynomial $q(t)$ for which $p(t) = (t - r_1)(t - r_2) \cdots (t - r_k)q(t)$.

(b) Prove that the number of distinct zeros of a nonzero polynomial over an integral domain cannot exceed its degree.

(c) Verify that the polynomial $t^2 - 5t + 6$ over $\mathbf{Z}_{12}$ has more than two zeros. This example shows that (b) may fail when the condition that the coefficients belong to an integral domain is dropped.

5. Let $a$ and $b$ be two distinct zeros of a polynomial $f(t)$, so that, for some polynomials $u(t)$ and $v(t)$,

$$f(t) = (t - a)u(t) = (t - b)v(t).$$

Prove that the remaining zeros of $f(t)$ are the solutions of the equation

$$u(t) - v(t) = 0.$$

6. Consider the equation $t^4 - 5t - 6 = 0$.

(a) By inspection, determine two integer solutions.

(b) Use Exercise 5 to determine two other solutions of the equation.

7. Exercise 1 treats the case when the divisor is of degree 1, in which case we find that the remainder is a constant. Can we talk about division by polynomials of degree exceeding 1? Consider the possibility of dividing the polynomial

$$f(t) = 4t^5 - 3t^4 - 7t^2 + 6$$

by the polynomial

$$g(t) = t^3 + 7t^2 + 3t - 2.$$

If we think of the "size" of a polynomial as being measured by its degree, then, as with numbers, we can take away a multiple of $g(t)$ which leaves as a remainder, a "smaller" quantity which will in turn be subsequently divided by $g(t)$. In the context of polynomials, "smaller" means "of lower degree". Subtracting $4t^2 g(t)$ will leave such a remainder.

(a) Verify that $f(t) - 4t^2 g(t) = -31t^4 - 12t^3 + t^2 + 6$.

(b) Take away a multiple of $g(t)$ from the right hand side of (a) to leave a remainder of lower degree. Continue on in this fashion until a remainder of degree less than that of $g(t)$ is obtained.

(c) Show that the process of parts (a) and (b) can be written in the form of the long division algorithm:

$$
\begin{array}{r}
4t^2 - \phantom{0}31t + \phantom{0}205 \phantom{6} \\
t^3 + 7t^2 + 3t - 2 \,\overline{)\, 4t^5 - \phantom{0}3t^4 + \phantom{0}0t^3 - \phantom{0}7t^2 + \phantom{0}0t + \phantom{0}6} \\
4t^5 + 28t^4 + 12t^3 - \phantom{0}8t^2 \phantom{+ 00t + 6} \\
\hline
-\,31t^4 - 12t^3 + \phantom{0}t^2 + \phantom{0}0t + \phantom{0}6 \\
-\,31t^4 - 217t^3 - 93t^2 + 62t \phantom{+ 6} \\
\hline
205t^3 + 94t^2 - 62t + \phantom{0}6 \\
205t^3 + 1435t^2 + 615t - 410 \\
\hline
-\,1341t^2 - 677t + 416 \\
\end{array}
$$

(d) From the algorithm in (c), read off the quotient and the remainder for the equation

$$f(t) = g(t)q(t) + r(t).$$

Check your answer by directly computing the right hand side.

(e) The algorithm in (c) can be more clearly presented by suppressing the variable to obtain

$$
\begin{array}{r}
4 \quad -31 \quad\ 205 \phantom{6} \\
1\ 7\ 3\ -2 \,\overline{)\, 4 \quad -3 \quad\ 0 \quad\ -7 \quad\ 0 \quad\ 6} \\
4 \quad\ 28 \quad 12 \quad -8 \phantom{\quad 0 \quad 6} \\
\hline
-31 \quad -12 \quad\ 1 \phantom{\quad 0 \quad 6} \\
-31 \quad -217 \quad -93 \quad 62 \phantom{\quad 6} \\
\hline
205 \quad\ 94 \quad -62 \quad\ 6 \\
205 \quad 1435 \quad 615 \quad -410 \\
\hline
-1341 \quad -677 \quad 416 \\
\end{array}
$$

A further compression of this is *Horner's Method of Synthetic Division* which can be regarded as a generalization of his method for division by a binomial $(t - c)$ and which in this case takes the form:

| 1 | 4 | −3 | 0 | −7 | 0 | 6 |
|---|---|----|---|----|---|---|
| −7 |   | −28 | −12 | 8 |   |   |
| −3 |   |   | 217 | 93 | −62 |   |
| 2 |   |   |   | −1435 | −615 | 410 |
|   | 4 | −31 | 205 | −1341 | −677 | 416 |

Explain how to perform the algorithm, how to read off the quotient, and remainder and why it works.

8. Make up a number of long division problems involving polynomials. Solve them by the long division algorithm and by Horner's Method of Synthetic Division, and check the results.

9. Establish:

   *Division Theorem.* Let $f$ and $g$ be two polynomials over a field $\mathbf{F}$ and suppose that $\deg g \geq 1$. Then there are polynomials $q$ (quotient) and $r$ (remainder) such that

   $$f = gq + r \quad \text{and} \quad \deg r < \deg g.$$

10. The Division Theorem was formulated for polynomials over a field. The case of dividing the polynomial $3t^2 + t + 1$ by $2t - 1$ in $Z[t]$ shows that it does not always hold for polynomials over an integral domain. Formulate and prove a modified version of the theorem in this case.

11. In the case of dividing a polynomial $f(t)$ by the binomial $(t - c)$, the remainder can be given by a formula $f(c)$ involving the polynomial $f$ and the coefficients of the divisor. Derive a formula for the remainder when $f(t)$ is divided by $(t - a)(t - b)$.

12. Let $\mathbf{F}$ be a field and let $\mathbf{F}[x, y]$ denote the ring of polynomials in the variable $x$ and $y$ with coefficients in $\mathbf{F}$. Suppose $f(x, y)$ belongs to $\mathbf{F}[x, y]$. Apply the Factor Theorem to the ring $\mathbf{F}[x]$ to show that $f(x, x) = 0$ if and only if $(x - y)$ is a factor of $f(x, y)$. More generally, show that $y - g(x)$ divides $f(x, y)$ if and only if $f(x, g(x)) = 0$, for $g(x)$ in $\mathbf{F}[x]$.

13. Consider the symmetric homogeneous polynomial

    $$f(x, y, z) = x^2 y^3 + x^3 y^2 + x^2 z^3 + x^3 z^2 + y^2 z^3 + y^3 z^2.$$

    To find a representation of this polynomial in terms of the elementary symmetric polynomials

    $$s_1(x, y, z) = x + y + z$$

    $$s_2(x, y, z) = xy + yz + zx$$

    $$s_3(x, y, z) = xyz,$$

    proceed as follows:

(a) Set $z = 0$ and obtain that $f(x, y, 0) = (x + y)x^2 y^2$.

(b) Consider the polynomial

$$g(x, y, z) = f(x, y, z) - s_1(x, y, z)[s_2(x, y, z)]^2.$$

Show that $g(x, y, 0) = g(x, 0, y) = g(0, y, z) = 0$ and deduce that $xyz$ is a factor of $g(x, y, z)$.

(c) Determine a polynomial $h(x, y, z)$ for which

$$g(x, y, z) = xyz\, h(x, y, z).$$

Is $h(x, y, z)$ symmetric and homogeneous? What is the degree of $h(x, y, z)$?

(d) Write $h(x, y, z)$ as a polynomial in $s_1$, $s_2$ and $s_3$.

(e) Write $f(x, y, z)$ as a polynomial in $s_1$, $s_2$ and $s_3$.

14. Carry out the procedure of Exercise 13 on the other two polynomials given in Exercise 1.5.8.

15. *Gauss' Theorem on Symmetric Functions.* In Exercises 1.5.5 and 1.5.8, the representation of a symmetric polynomial in terms of the elementary symmetric polynomials was carried out for specific examples of low degree. In this exercise, we will outline the proof of this result in general.

Let $t_1, t_2, \ldots, t_n$ be $n$ variables and let $s_1, s_2, \ldots, s_n$ be the elementary symmetric functions of these variables; namely, $s_i$ is the sum of the $\binom{n}{i}$ possible products of $i$ of the variables $t_i$ $(1 \leq i \leq n)$. Then

*any symmetric polynomial in the variables $t_i$ can be expressed as a polynomial in the variable $s_i$ $(1 \leq i \leq n)$.*

(a) It is enough to prove the result for homogeneous polynomials. We use induction. Verify that, trivially, the result holds for all polynomials of degree 0 and for all polynomials of a single variable.

(b) Suppose as an induction hypothesis, that the result holds for

   (i) all polynomials of degree $< k$ and any number of variables;

   (ii) all polynomials of degree $k$ and $n - 1$ or fewer variables;
   where $k \geq 1$ and $n \geq 2$.

Let $p(t_1, t_2, t_3, \ldots, t_n)$ be a homogeneous symmetric polynomial of degree $k$. Show that $p(t_1, t_2, \ldots, t_{n-1}, 0)$ is a homogeneous symmetric polynomial of $n - 1$ variables which, by the induction hypothesis, can be written in the form $q(u_1, u_2, \ldots, u_{n-1})$, where $q$ is a polynomial in the elementary symmetric functions $u_j$ of the $n - 1$ variables $t_i$ $(1 \leq i \leq n - 1)$.

(c) In (b), argue that

$$p(t_1, \ldots, t_n) - q(s_1, s_2, \ldots, s_{n-1})$$

is a symmetric polynomial which vanishes for $t_n = 0$, and hence vanishes when any of the variables $t_i$ is individually set equal to 0.

(d) Show that the polynomial $p - q$ in (c) either vanishes or can be written as the product of $t_1 t_2 \ldots t_n$ and a homogeneous symmetric polynomial of degree $k - n$, and use this fact along with the induction hypothesis to complete the proof of Gauss' Theorem.

## Explorations

**E.19. Chromatic Polynomials.** One of the most notorious problems of all time is the Four Color Problem. Suppose that you are given any map drawn upon a sphere; every region is such that you can pass from any point in the region to any other without having to leave the region (technically, any two points in a region can be connected by an arc in the region). One wishes to color this map in such a way that any two regions which have a common boundary line are colored differently. With some maps, such as those like a checkerboard, two colors will be enough. Other maps, such as the map of Canada, will require at least three. However, no one has ever been able to find a map for which more than four colors were needed. The Four Color Problem is to show that no such map exists.

The problem came to light first in 1852, when Francis Guthrie, a student at the University of London, posed it to his brother Frederick, who in turn passed it on to Augustus de Morgan. Evidently, it did the rounds among mathematicians for a number of years, for in 1878, Arthur Cayley mentioned at a meeting of the London Mathematical Society that he was unable to solve it. In the following two years, independent "proofs" were published by P. G. Tait and A. B. Kempe. However, P. J. Heawood discovered errors in these in 1890, although he did prove a fairly general result about coloring maps. Interest grew in the problem and it spawned many new techniques, but a solution eluded an ever-increasing circle of mathematicians. Eventually, in 1976, a complicated proof of the conjecture requiring extensive computer resources was found by Haken and Appel. A survey of map colorings appears in G. Ringel, *Map Color Theorems* (Springer, 1974).

In tackling a problem like the Four Color conjecture, it is customary to reformulate the situation. Represent each region by a *node* or *vertex* (you can think of this as the capital of the region); join two vertices by edges if and only if their corresponding regions have a boundary line in common.

Such an array of vertices and edges is called a *graph*. Some examples of graphs are the following:

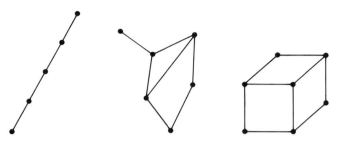

The graphs involved in map coloring problems can be drawn on a plane or a sphere, but there are some graphs which cannot be drawn on a plane surface without the edges intersecting in points other than vertices. Such a graph is

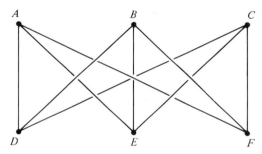

in which each of the vertices $A$, $B$, $C$ is connected to each of the vertices $D$, $E$, $F$.

The question of coloring regions on a map can be reinterpreted as coloring vertices on a graph in such a way that any two vertices joined by an edge are colored differently. Recall that a graph, in this context, is a collection of points or vertices some pairs of which are connected by edges. To study coloring problems in a systematic way, we define for a graph $G$ the *chromatic function* $C_G(t)$. This is the number of ways in which the graph $G$ can be colored with no more than $t$ colors so that two vertices joined by an edge are colored differently. The Four Color Conjecture says that, if $G$ is a graph which can be presented on the surface of a sphere with no crossing of edges except at vertices, then $C_G(4)$ is always at least 1.

(a) Suppose that $n_r$ is the number of ways a graph $G$ with a finite number $n$ of vertices can be colored with exactly $r$ distinct colors. Show that

$$C_G(t) = \sum_{r=0}^{n} \binom{t}{r} n_r.$$

Thus the chromatic function of a graph with finitely many vertices is a polynomial.

(b) Show that, if $G$ is a graph with exactly $n$ vertices and no connecting edges, then

$$C_G(t) = t^n.$$

(c) Show that, if $G$ is a graph with exactly $n$ vertices, any two of which are connected by an edge, then

$$C_G(t) = t(t-1)(t-2)\cdots(t-n+1).$$

(d) The chromatic function of a graph can be found by adding together the chromatic functions of related graphs. Let $G$ be a graph containing two vertices $a$ and $b$ not connected by an edge. From $G$, form a graph $E$ by connecting $a$ and $b$ by an edge and a graph $F$ by replacing $a$ and $b$ by a vertex $c$ (not already in $G$) which is to be connected to any vertex in $G$ which joins either $a$ or $b$. Show that

$$C_G(t) = C_E(t) + C_F(t).$$

Deduce from this that one can express any chromatic function as the sum of chromatic functions of the form $t(t-1)\cdots(t-k+1)$ corresponding to graphs every pair of whose points is joined by an edge.

(e) Each of the following polynomials is the chromatic function of a graph with four vertices. Determine the graphs:

$$t^4 - 6t^3 + 11t^2 - 6t, \quad t^4 - 5t^3 + 8t^2 - 4t, \quad t^4 - 4t^3 + 5t^2 - 2t,$$

$$t^4 - 4t^3 + 6t^2 - 3t, \quad t^4 - 3t^3 + 3t^2 - t, \quad t^4 - 2t^3 + t^2, \quad t^4 - t^3, \quad t^4.$$

(f) Consider the graph with vertices $A$, $B$, $C$, $D$, $E$, $F$ diagrammed above. Show that this graph cannot be colored with 0 or 1 colors, but can be colored with 2 colors in exactly two ways. Show that its chromatic polynomial is

$$t(t-1)(t^4 - 8t^3 + 28t^2 - 47t + 31).$$

(g) What is the chromatic polynomial of the graph made up of the vertices and edges of a cube? of each of the other four platonic solids (tetrahedron, octahedron, dodecahedron, icosahedron)?

**E.20. The Greatest Common Divisor of Two Polynomials.** Formulate a definition of common divisor of two polynomials. What is meant by *greatest common divisor* in this context? There is some ambiguity in what the greatest common divisor should be; we can remove it by insisting that it be monic. Devise a Euclidean algorithm analogous to that of Exercise 1.6.1 for determining the greatest common divisor. Will this algorithm lead to a representation of the form $uf + vg$ ($u$ and $v$ polynomials) for the greatest common divisor? Is it true that every common divisor divides the greatest common divisor? Try out your ideas on the pairs:

(a) $2t^3 + 9t^2 + 8t - 5$ and $t^2 + 5t + 6$

(b) $2t^3 + 9t^2 + 8t - 4$ and $t^2 + 5t + 6$.

**E.21.** Investigate formulae for the remainder when a polynomial $f(t)$ is divided by

(a) $(t - c)^k$

(b) $(t - a_1)(t - a_2)(t - a_3) \cdots (t - a_m)$

(c) $(t - a)^r (t - b)^s$.

## 2.3   The Derivative

The algorithm given in Section 1 for expanding a given polynomial $p(t)$ in terms of powers of $(t - c)$ for a constant $c$ is a mechanical method which does not give much insight into the structural significance of the coefficients. We would like to be able to describe them in terms of the polynomial $p(t)$ and the constant $c$. Surprisingly, this is done through the introduction of a concept which most students encounter in quite a different domain— the calculus. Let us begin with two observations before continuing to the exercises.

(1) $p(t)$ is the sum of monomials $a_k t^k$. If we have an expansion of each monomial in terms of powers of $(t - c)$, then the expansion of $p(t)$ is the sum of the expansions of the monomials.

(2) We can write $p(t) = q(t)(t - c) + p(c)$. If we can obtain an expansion for $q(t)$, then we can insert it into this equation to get one for $p(t)$. The constant $q(c)$ occurs as the coefficient of $(t - c)$ in the expansion for $p(t)$; can we express this in terms of $p$? Can the coefficients of higher powers of $(t - c)$ be similarly identified?

## Exercises

1.  Construct Horner's Table for the expansion of the following polynomials in terms of ascending powers of $(t - c)$:

    (a) $t^2$

    (b) $t^3$

    (c) $t^4$.

    Check your result by expanding $t^k = [c + (t - c)]^k$.

2.  Verify, by Horner's method or otherwise, that the first two terms in the expansion of $t^m$ in terms of powers of $(t - c)$ are

    $$c^m + mc^{m-1}(t - c).$$

3. Show that the first two terms in the expansion of a polynomial

$$p(t) = a_n t^n + a_{n-1} t^{n-1} + \cdots + a_1 t + a_0$$

in terms of ascending powers of $(t - c)$ are

$$p(c) + p'(c)(t - c)$$

where $p'(t)$ is the polynomial $p'(t) = n a_n t^{n-1} + (n-1) a_{n-1} t^{n-2} + (n-2) a_{n-2} t^{n-3} + \cdots + 2 a_2 t + a_1$.

4. *The derivative.* For any polynomial $p(t)$, the polynomial $p'(t)$ defined in Exercise 3 is called the *derivative* of $p(t)$. The process of obtaining the derivative of a polynomial is called *differentiation.*

   (a) Verify that the derivative of the polynomial $3t^5 - 7t^4 + 6t^2 - 5t - 3$ is $15t^4 - 28t^3 + 12t - 5$.

   (b) Find the derivative of the polynomial $4t^{13} - 3t^8 - 5t^7 + 4t^3 + 76t$.

5. *Properties of the derivative.* Establish the following properties of the derivative, where $p$ and $q$ are polynomials and $k$ is a constant:

   (a) $(p + q)'(t) = p'(t) + q'(t)$. Extend this to an arbitrary sum of polynomials.

   (b) $(kp)'(t) = kp'(t)$.

   (c) $(pq)'(t) = p'(t)q(t) + p(t)q'(t)$ in the special case $p(t) = t^r$ and $q(t) = t^s$.

   (d) $(pq)'(t) = p'(t)q(t) + p(t)q'(t)$ for arbitrary polynomials $p$ and $q$. Extend this to a product of more than two polynomials.

   (e) the derivative of $p(t)^r$ is $rp(t)^{r-1}p'(t)$ for an arbitrary positive integer $r$ (use (d)).

   (f) $(p \circ q)'(t) = p'(q(t))q'(t)$.

6. Verify the properties of the derivative given in Exercise 5 for the special case

$$p(t) = 3t^2 - 4t + 2 \quad \text{and} \quad q(t) = 4t^3 - 2t^2 + 6t + 1.$$

7. Since differentiation of a polynomial leads to another polynomial, we can apply the operation of taking the derivative repeatedly. Thus, if $p'(t)$ is the derivative of $p(t)$, the *second derivative* of $p(t)$ is $p''(t) = (p')'(t)$. In general, we can define

$$p^{(0)}(t) = p(t)$$
$$p^{(1)}(t) = p'(t)$$
$$p^{(2)}(t) = p''(t)$$
$$p^{(k)}(t) = (p^{(k-1)})'(t) \quad \text{for } k \geq 3.$$

(a) Verify that if $u(t) = t^5 - 4t^3 + 2t^2 - 7$, then $u'(t) = 5t^4 - 12t^2 + 4t$, $u''(t) = 20t^3 - 24t + 4$, $u^{(3)}(t) = 60t^2 - 24$, $u^{(4)} = 120t$ and $u^{(5)}(t) = 120$.

(b) Show that, for any polynomial $p(t)$ and any positive integer $r$ not exceeding $\deg p$, $\deg p^{(r)} = (\deg p) - r$.

(c) Show that, for any polynomial $p(t)$ of positive degree $n$, $p^{(n)}$ is a constant polynomial and $p^{(r)} = 0$ for $r > n$.

(d) Derive a formula for $(pq)''(t)$ and generalize it to a formula for $(pq)^{(r)}(t)$.

8. Show that for any positive integers $m$ and $k$, with $m \leq k$, the $m$th derivative of $(t - c)^k$ is equal to

$$k(k - 1)(k - 2) \cdots (k - m + 1)(t - c)^{k-m}.$$

9. We are now in a position to show that every polynomial has an expansion in terms of powers of $(t - c)$ and to identify the coefficients. The result is:

*Taylor's Theorem.* Let $p(t)$ be any polynomial of degree $n$ and $c$ be a constant. Then

$$p(t) = p(c) + p'(c)(t - c) + \frac{p''(c)}{2!}(t - c)^2 + \frac{p^{(3)}(c)}{3!}(t - c)^3$$

$$+ \cdots + \frac{p^{(k)}(c)}{k!}(t - c)^k + \cdots$$

where the sum on the right has at most $n + 1$ nonzero terms.

The right-hand side is called the *Taylor expansion of $p(t)$ about $c$.*

Verify the theorem for the polynomial $t^4 + t^2 - 3t + 7$ expanded in terms of powers of $(t - 3)$, checking your answer against the result of Exercise 1.8.

Establish Taylor's Theorem in the following steps:

(a) Use the Division Theorem to establish that $p(t)$ can be written in the form

$$p(t) = c_n(t - c)^n + p_1(t) \quad \text{where} \quad \deg p_1 \leq n - 1.$$

(b) By repeated use of the Division Theorem on the remainders resulting from the procedure in (a), show that

$$p(t) = c_0 + c_1(t - c) + c_2(t - c)^2 + c_3(t - c)^3 + \cdots.$$

(c) Differentiate both sides of the equation in (b) repeatedly. Set $t = c$ to obtain

$$p(c) = c_0$$
$$p'(c) = c_1$$
$$p''(c) = 2!c_2$$
$$p^{(3)}(c) = 3!c_3$$

and in general

$$p^{(k)}(c) = k!c_k \quad \text{for} \quad 0 \leq k \leq n.$$

Substitute in (b) to obtain the result.

10. What does the Taylor Expansion of $p(t)$ about $c$ amount to when $c = 0$?

11. Form the Taylor Expansions for $p(t)$ and $q(t)$ about $c$. Multiply them together and verify that the result is

$$(pq)(c) + (pq)'(c)(t - c) + \frac{1}{2!}(pq)''(c)(t - c)^2 + \cdots.$$

12. (a) Use Taylor's Theorem to establish the Binomial Expansion

$$(1 + t)^n = 1 + nt + \binom{n}{2} t^2 + \cdots + \binom{n}{k} t^k + \cdots + t^n.$$

(b) Prove that

$$(a + b)^n = a^n + na^{n-1}b + \cdots + \binom{n}{k} a^{n-k}b^k + \cdots + b^n.$$

13. Another approach to use in expanding $p(t)$ in terms of powers of $(t - c)$ is to make the substitution $t = c + s$, and compute $p(c + s)$ in ascending powers of $s$. Do this for the polynomial $t^3 - 4t^2 + 7t + 2$ and check that the coefficient of $s^r$ is $p^{(r)}(c)/r!$ for each value of $r$.

14. For several polynomials and values of $c$ of your choice, find the Taylor expansion and check your answer by (i) Horner's Method, (ii) a substitution of the type $t = c + s$, (iii) multiplying out and adding the terms of the expansion.

15. *Multiplicity of zeros.* In Exercise 2.2, we found that $c$ is a zero of a polynomial $p(t)$ if and only if $(t - c)$ is a factor of $p(t)$. Using this result as a basis, we can sharpen the idea of a zero.

(a) Show that, for any number $c$ and any nonzero polynomial $p(t)$, we can find a nonnegative integer $r$, not exceeding the degree of $p(t)$, such that
$$p(t) = (t - c)^r q(t)$$
where $q(t)$ is a polynomial with $q(c) \neq 0$. Thus, $(t - c)^r$ divides $p(t)$ while $(t - c)^{r+1}$ does not. The number $r$ is called the *multiplicity* of $c$ as a zero of $p(t)$. If $c$ is not a zero then $c$ has multiplicity 0. A zero of multiplicity one is called a *simple* zero.

(b) Let $p(c) = 0$. Show that the multiplicity of $c$ as a zero of $p(t)$ exceeds 1 if and only if $c$ is a zero of $p'(t)$. In this situation, show that the multiplicity of $c$ as a zero of $p(t)$ exceeds the multiplicity of $c$ as a zero of $p'(t)$ by 1.

(c) Show that $(t - c)$ is a zero of positive multiplicity $r$ if and only if $p(c) = p'(c) = \ldots = p^{(r-1)}(c) = 0$ and $p^{(r)}(c) \neq 0$.

(d) What is the term of lowest degree in the Taylor expansion of $p(t)$ about $c$ if $c$ is a zero of multiplicity $r$?

16. Let $p$ be a polynomial and $c$ a constant. Prove or disprove: if $p(c) = p''(c) = 0$, then $c$ is a zero of $p$ of multiplicity at least three.

# Explorations

**E.22. Higher Order Derivatives of the Composition of Two Functions.** Let $p$ and $q$ be any two polynomials. Is there a general formula for the $k$th derivative $(p \circ q)^{(k)}(t)$ of their composition? To deal with this question, let us introduce some notation:

$p_k$ to denote $p^{(k)}(q(t))$, the $k$th derivative of $p$ evaluated at $q(t)$;
$q_k$ to denote $q^k(t)$, the $k$th derivative of $q$ at $t$.
Verify that

$$(p \circ q)'(t) = p_1 q_1$$
$$(p \circ q)''(t) = p_2 q_1^2 + p_1 q_2$$
$$(p \circ q)'''(t) = p_3 q_1^3 + 3p_2 q_1 q_2 + p_1 q_3.$$

Compute derivatives of the next few higher orders and look for patterns. For example, try to get the profile of a general term without regard to the exact value of the coefficient; relate the subscript for $p$ to the powers of the derivatives of $q$ in each product (can you explain the relationship). What is the nature and the value of the coefficient of the term with the factor $p_{k-1}$ in the development of the $k$th derivative?

**E.23. Partial Derivatives.** The ring $\mathbf{F}[x, y]$, of polynomials in two variables over a field $\mathbf{F}$ can be thought of in two ways:

(i) as a ring of polynomials in the variable $x$ over $\mathbf{F}[y]$;

(ii) as a ring of polynomials in the variable $y$ over $\mathbf{F}[x]$.

Corresponding to these, we can consider two types of differentiation of a polynomial $f(x, y)$:

(i) *partial differentiation with respect to* $x$ resulting in the *partial derivative* $f_x(x, y)$, with $y$ treated as a constant.

(ii) *partial differentiation with respect to* $y$ resulting in the *partial derivative* $f_y(x, y)$, with $x$ treated as a constant.

For example, if $f(x, y) = 5x^3y^2 + 3x^2y - 2x^3 + 4y - 7$,

$$f_x(x, y) = 15x^2y^2 + 6xy - 6x^2$$

$$f_y(x, y) = 10x^3y + 3x^2 + 4.$$

Just as in the case of polynomials of one variable, we can consider derivatives of higher order. For example, we can define

$$f_{xy}(x, y) = (f_x)_y(x, y), \quad f_{xx}(x, y) = (f_x)_x(x, y),$$

with $f_{yx}$ and $f_{yy}$ defined similarly. What would these second order partial derivatives be for the examples?

(a) Formulate and prove a conjecture concerning the relationship between $f_{xy}$ and $f_{yx}$.

(b) Define partial derivatives of the third order. How many distinct possibilities are there?

(c) Define partial derivatives of the $k$th order, for any positive integer $k$.

(d) Show that, for any polynomial of two variables, the $k$th order partial derivatives all vanish for $k$ sufficiently large. How is the minimum such value of $k$ related to the degree of the polynomial?

(e) We can formulate a version of Taylor's Theorem in which $f(x, y)$ can be expanded about $(a, b)$ in the form:

$$f(x, y) = c_{00} + c_{10}(x - a) + c_{01}(y - b) + c_{11}(x - a)^2 + c_{12}(x - a)(y - b)$$

$$+ c_{22}(y - b)^2 + \cdots.$$

Write down the form of the terms of higher degree and determine the coefficients in terms of the partial derivatives of $f(x, y)$ at $(a, b)$.

(f) Generalize the results of this section for polynomials of more than two variables.

**E.24. Homogeneous Polynomials.** In Section 1.5, we defined a polynomial to be homogeneous of degree $d$ if each of its terms had the same degree $d$. For a polynomial of two variables, this is equivalent to requiring that $f(tx, ty) = t^d f(x, y)$ (see Exercise 1.5.2). Write down a number of homogeneous polynomials of various degrees and compute for each the quantity $x f_x + y f_y$. What does this equal when $\deg f = 1$? $\deg f = 2$? Make a conjecture and prove it. Does the property that you have found characterize homogeneous polynomials (i.e. if a polynomial has the property, must it be homogeneous)? What is the generalization for more than two variables?

**E.25. Cauchy–Riemann Conditions.** Consider a polynomial $f(z)$ over C of the complex variable $z$. If we make the substitution $z = x + yi$, and separate out the real and imaginary parts, we can express $f(z)$ in the form $u(x, y) + iv(x, y)$, where $u$ and $v$ are polynomials in $\mathbf{R}[x, y]$. For example, if $f(z) = 3z^2 + (2 + i)z - (2 - 3i)$, verify that

$$
\begin{aligned}
f(x + yi) &= 3(x^2 + 2xyi - y^2) + (2 + i)(x + yi) - (2 - 3i) \\
&= [3(x^2 - y^2) + 2x - y - 2] + [6xy + x + 2y + 3]i,
\end{aligned}
$$

so that, in this case $u(x, y) = 3(x^2 - y^2) + 2x - y - 2$

$$
v(x, y) = 6xy + x + 2y + 3.
$$

Thus, each complex polynomial $f(z)$ corresponds to two real polynomials $u(x, y)$, $v(x, y)$. What pairs $\{u, v\}$ of polynomials arise in this way? Is it possible to find a corresponding $f$ for any given pair, or must there be some relation connecting $u$ and $v$?

To look at a simple example, show that it is not possible to find a complex polynomial $f(z)$ for which $u(x, y) = x$ and $v(x, y) = 0$. (Such a polynomial would have to satisfy $f(x + iy) = x$ for all real $x$ and $y$.) If $u(x, y) = x$, what are the possibilities for $v(x, y)$?

It turns out that there are two simple equations connecting the partial derivatives $u_x$, $v_x$, $u_y$, $v_y$. By looking at the above example, as well as other polynomials of low degree, make a conjecture. Now prove it, noting that essentially you have to check your conjecture for the polynomial $z^n$ for each positive integer $n$.

Compute the second order derivatives $u_{xx}$, $u_{yy}$, $v_{xx}$, $v_{yy}$ for the above example, as well as for other polynomials. Look for patterns and make conjectures.

Suppose that you are given a polynomial $u(x, y)$ in $\mathbf{R}[x, y]$. Investigate whether it is always possible to find a polynomial $f(z)$ in $\mathbf{C}[z]$ such that $f(x + yi) = u(x, y) + iv(x, y)$, for some real polynomial $v(x, y)$. For example, you might look at $u(x, y) = 0$, $xy$ or $x^2$. If such a polynomial $v(x, y)$ exists, how many possibilities are there?

The relationship connecting the first order partial derivatives of $u$ and $v$ is not just a matter of idle curiosity. The natural generalization of polynomials is a class of functions $f(z)$ defined for a complex variable $z$ which can be expressed by an infinite series of monomials involving powers of $z$. The real and imaginary parts of the functions in this class can be characterized by the Cauchy–Riemann Conditions (in which the notion of derivative is realized through the medium of limits). A second order differential equation satisfied by the real and imaginary parts of these functions is one that has an important part to play in physics. Partly for this reason, functions of a complex variable have a useful role in this science.

**E.26. The Legendre Equation.** In applied mathematics, the differential equation

$$(1 - x^2)y'' - 2xy' + n(n + 1)y = 0 \qquad \ldots (1)$$

(where $y$ is a function of $x$) plays an important role. What sort of solutions does this equation have?

To find out, differentiate the equation $r$ times to obtain

$$(1 - x^2)y^{(r+2)} - 2(r + 1)xy^{(r+1)} + (n - r)(n + 1 + r)y^{(r)} = 0. \qquad \ldots (2)$$

Thus, if $n$ is a positive integer and $r = n$, $z = y^{(r+1)}$ satisfies the equation

$$(1 - x^2)z' - 2(r + 1)xz = 0. \qquad \ldots (3)$$

Show that (3) is not satisfied by any nonzero polynomial $z$ in the variable $x$. Deduce that any polynomial solution $y$ of (1) has degree not exceeding $n$.

Equation (1) in fact does have a polynomial solution $y = P_n(x)$. Check this for $n = 1, 2, 3, 4$. To find it more generally, observe that, from (2),

$$P_n^{(r+2)}(0) = -(n - r)(n + 1 + r)P_n^{(r)}(0). \qquad \ldots (4)$$

It follows from (4) that $P_n^{(r+2)}(0) = 0$ for $r = n, n+2, n+4, \ldots$. Suppose we ensure that $P_n^{(k)}(0) = 0$ for $k \geq n + 1$ by setting $P_n^{(n-1)}(0) = P_n^{(n-3)}(0) = \cdots = P_n^{(n-2i-1)}(0) = \cdots = 0$. Then, by Taylor's Theorem, we have that

$$P_n(x) = \frac{1}{n!}P_n^{(n)}(0)x^n + \frac{1}{(n - 2)!}P_n^{(n-2)}(0)x^{n-2} + \cdots .$$

Suppose that $P_n^{(n)}(0) = n!$. Use (4) to obtain the remaining coefficients. Verify that the polynomial so obtained is a solution of (1).

## 2.4   Graphing Polynomials

One picture is worth a thousand words. This adage is especially true in mathematics in dealing with the behaviour of functions. The graphs of real polynomials can provide at a glance valuable information about their zeros and degrees. It can be a useful tool in analyzing results about polynomials in such areas as the theory of approximation of functions by polynomials. This section will require knowledge from a first course in differential calculus. Let us review some of the terminology required:

Let $f(x)$ be a polynomial defined on **R**. We say that $f(x)$ is *increasing* on an interval $[a, b] = \{x : a \le x \le b\}$ if, for each pair of values $u$, $v$ within the interval such that $u \le v$, $f(u) \le f(v)$. It can be shown that $f$ is increasing on the interval if and only if its derivative is nonnegative there.

$f(x)$ is *decreasing* on $[a, b]$, if, for $a \le u < v \le b$, $f(u) \ge f(v)$. This is equivalent to asserting that its derivative is nonpositive on the interval.

$f(x)$ has a *maximum* at the point $c$ if there is some small interval with $c$ in its interior such that $f(x) \le f(c)$ whenever $x$ lies in the interval. Necessarily, at each maximum, $f'(c) = 0$.

$f(x)$ has a *minimum* at the point $c$ if on some small interval with interior point $c$, $f(x) \ge f(c)$ for each $x$ in the interval. Again, this implies that $f'(c) = 0$.

$c$ is a *critical point* for $f$ if $f'(c) = 0$. At a critical point, $f$ could have a maximum, a minimum or neither a maximum nor a minimum.

We will use the fact that each polynomial $f(x)$ is *continuous* in $x$. This means that small changes in the value of $x$ give rise to small changes in $f(x)$. Thus, a graph of a polynomial is a smooth curve without any breaks or corners. One consequence of this is that somewhere on each interval $[a, b]$, $f(x)$ assumes every value which lies between $f(a)$ and $f(b)$.

## Exercises

1. Sketch the graph of a typical constant polynomial.

2. Sketch the graph of a polynomial of the form $ax + b$, where $a$ and $b$ are real with $a$ nonzero. Deduce from the graph that this polynomial assumes each real value exactly once.

3. (a) Sketch the graphs of the polynomials $x^2$ and $(x - k)^2$.

   (b) Using the representation

   $$ax^2 + bx + c = a \left( x + \frac{b}{2a} \right)^2 - \left( \frac{b^2 - 4ac}{4a} \right)$$

   sketch the graph of the polynomial $ax^2 + bx + c$, where $a$, $b$, $c$ are real and $a$ is nonzero. Distinguish the cases in which the leading coefficient and the discriminant are separately positive and negative. Determine on the graph all maxima and minima for the function.

   (c) On the same axes as in (b), sketch the graph of the derivative of the quadratic. Relate the values of the derivative to the behaviour of the quadratic function.

   (d) Deduce that no polynomial over **R** of degree 2 can assume all possible real values.

(e) Show that, if a quadratic polynomial has two real roots, then its derivative must have its root between them.

4. Sketch the graphs of the following cubics. For each, on the same axes, sketch the graphs of its first and second derivatives.

   (a) $x^3$

   (b) $x^3 + 8$

   (c) $x^3 - 8$

   (d) $x^3 + 12x$

   (e) $x^3 - 12x$

   (f) $x^3 + ax$; distinguish the cases that $a$ is positive and negative

   (g) $x^3 + ax + b$.

5. Consider the general cubic over $\mathbf{R}$, $ax^3 + bx^2 + cx + d$. Show that there is a change of variable of the form $s = x + k$, which will render it in the form $as^3 + ms + n$ for some real numbers $m$ and $n$. Use this fact to discuss the graph of the general cubic.

6. Let $f(x)$ be a cubic polynomial. The point $u$ in $\mathbf{R}$ at which $f''$ vanishes is called an *inflection* point. The point $(u, f(u))$ on the graph of the cubic generally separates the convex part of the graph from the concave part. Show that the graph of any cubic is centrally symmetric about its inflection point. (You have to show that, if $(u - v, f(u) - w)$ is on the graph, so also is the point $(u + v, f(u) + w)$.)

7. Use the results of your investigation on cubics to argue that

   (a) every cubic with real coefficients has at least one real zero;

   (b) every cubic with real coefficients assumes every real value at least once.

8. Sketch the graphs of the following quartics. On the same axes for each, sketch its first and second derivatives.

   (a) $x^4$

   (b) $x^4 + 3x^2 + 2$

   (c) $x^4 - 3x^2 + 2$

   (d) $x^4 - 5x - 6$

   (e) $x^4 + 5x - 6$

   (f) $x^4 + 3x^2 - 36x$

   (g) $x^4 + 3x^2 - 36x + 48$

9. Discuss the possible graphs of the general quartic $ax^4 + bx^3 + cx^2 + dx + e$. You may find it helpful to make a translation of coordinates which will eliminate the cubic term.

10. (a) Sketch the graph of the polynomial

$$6x^5 - 15x^4 - 10x^3 + 30x^2 + 10.$$

   How many real zeros does this polynomial have?

   (b) For each nonnegative integer $m$, find the set of values $k$ for which the polynomial

$$6x^5 - 15x^4 - 10x^3 + 30x^2 + k$$

   has exactly $m$ real zeros (i) not counting multiplicity; (ii) counting multiplicity.

11. Prove that every polynomial $p(t)$ of odd degree with real coefficients has at least one real root. Deduce that $p(t) = (t-r)q(t)$ for some real $r$ and polynomial $q(t)$ over **R**.

## Explorations

**E.27.** Consider the graphs of the polynomials you have drawn already. In how many points can such a graph be intersected by a line with equation of the form $y = k$? of the form $y = mx + b$? Make a conjecture concerning this number and the degree of a polynomial. Investigate further using polynomials of higher degree than 4; be sure to sketch the derivative as well and to relate the values of the derivative to the behaviour of the graph of the polynomial.

**E.28. Rolle's Theorem.** The task of finding the zeros of a polynomial becomes more difficult as the degree of the polynomial increases. Accordingly, it is often helpful to be able to relate the zeros of a polynomial to the roots of its derivative, whose degree is lower. The technique we are about to discuss was initiated by the mathematician Michel Rolle in a book called *Traité d'algèbre* published in 1690.

Suppose that $a$ and $b$ are two consecutive zeros of a real polynomial $f(x)$; that is, $f(a) = f(b) = 0$ and $f$ does not vanish between $a$ and $b$. Sketch some possible graphs for $f(x)$ on the interval $[a, b]$, and argue that $f$ must have at least one maximum or minimum in the interior of the interval. Deduce Rolle's Theorem, that between any two zeros of $f(x)$ there is at least one real zero of $f'(x)$.

Suppose that $u$ and $v$ are two consecutive zeros of $f'(x)$. What can be said about the number of real zeros of $f(x)$ between $u$ and $v$? If the derivative

$f'(x)$ has $k$ real zeros, what can be said about the number of real zeros of $f(x)$.

Show that a real polynomial of degree $n$ cannot have more than $n$ real zeros, counting multiplicity.

Rolle's result as stated above can be strengthened. If $f(x)$ has consecutive real zeros at $a$ and $b$, then by the Factor Theorem, we can write $f(x)$ in the form

$$f(x) = (x-a)^r (x-b)^s g(x)$$

where $g(x)$ is a polynomial which does not vanish at any point in the interval $[a, b]$. (Justify this statement.) Show that

$$(x-a)(x-b)f'(x) = f(x)[r(x-b) + s(x-a)$$
$$+ (x-a)(x-b)g'(x)/g(x)].$$

Every zero of $f'(x)$ between $a$ and $b$ is a zero of the function in the square brackets on the right hand side. Now look at the value of this function at $x = a$ and $x = b$, and draw the conclusion that the number of zeros of $f'(x)$ strictly between $a$ and $b$ must be odd, if we adopt the convention of counting each zero as often as its multiplicity indicates.

Assume that the following is the graph of a real polynomial. What can be said about its degree and about the signs of its first three and last three coefficients?

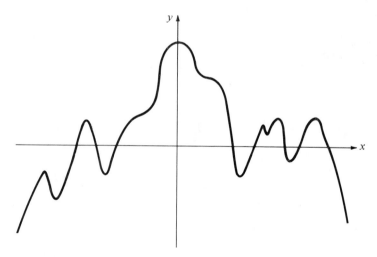

## 2.5   Problems

1. What is the highest multiplicity a root can have for the equation

$$x(x-1)(x-2)\cdots(x-n+1) = k?$$

2. $P_1, P_2, P_3$ are quadratic polynomials with positive leading coefficients and real zeros. Show that, if each pair of them has a common zero, then the trinomial $P_1 + P_2 + P_3$ also has real zeros.

3. Show that, if $n$ is a positive integer greater than 1,

$$\frac{n}{1-x} - \frac{1-x^n}{(1-x)^2}$$

is a polynomial in $x$ of degree $n - 2$, and find its coefficients

   (a) when it is arranged in powers of $x$,

   (b) when it is arranged in powers of $(x - 1)$.

4. If $n$ is a positive integer, prove that

$$(1-x)^{3n} + 3nx(1-x)^{3n-2} + \frac{3n(3n-3)}{1 \cdot 2}x^2(1-x)^{3n-4} + \cdots = (1-x^3)^n.$$

5. Suppose that $ac - b^2 \neq 0$. Consider the equation

$$ax^3 + 3bx^2 + 3cx + d = 0.$$

   (a) Show that the equation has two equal roots if and only if

$$(bc - ad)^2 = 4(ac - b^2)(bd - c^2).$$

   (b) Show that, if the equation has two equal roots, they are each equal to

$$(bc - ad)/[2(ac - b^2)].$$

6. Prove that

$$(n + 1)^{r+1} - (n + 1) = (r + 1)S_r + \binom{r + 1}{2} S_{r-1} + \cdots + (r + 1)S_1$$

   where $S_r = 1^r + 2^r + 3^r + \cdots + n^r$.

7. Find all polynomials $p$ of degree $k$ with real coefficients for which $p(p(t))$ is a positive integer power of $p(t)$.

8. Find all polynomials $p$ and $q$ for which $p(t) = q(p'(t))$.

9. (a) Find all polynomials $p(t)$ of degree not exceeding 3 which commute with their first derivatives, i.e. for which

$$p(p'(t)) = p'(p(t)).$$

   (b) For each integer $n \geq 4$, determine a polynomial of degree $n$ which commutes with its derivative.

10. Let $x^{(n)} = x(x-1)\cdots(x-n+1)$ for $n$ a positive integer and let $x^{(0)} = 1$. Prove that

$$(x+y)^{(n)} = \sum_{k=0}^{n} \binom{n}{k} x^{(k)} y^{(n-k)}.$$

11. Determine a polynomial solution of the differential equation

$$20y''' + 9y'' + 4y' + y = x^3 + 5x^2 - 2x - 2,$$

where $y$ is to be found as a function of $x$.

12. Find a polynomial $f(x)$ of degree 5 such that $f(x) - 1$ is divisible by $(x-1)^3$ and $f(x)$ is itself divisible by $x^3$.

13. If the polynomial $a_3 x^3 + a_2 x^2 + a_1 x + a_0$ $(a_3 \neq 0)$ is the third power of a linear polynomial, prove that

$$9a_0 a_3 = a_1 a_2$$

and

$$a_2^2 = 3a_1 a_3.$$

Prove the converse: if these two conditions are satisfied, then the polynomial is the third power of a linear polynomial.

14. Let $k$ be the smallest positive integer with the property:

There are distinct integers $a$, $b$, $c$, $d$, $e$ such that $p(x) = (x-a)(x-b)(x-c)(x-d)(x-e)$ has exactly $k$ nonzero coefficients.

Find with proof, a set of integers $a$, $b$, $c$, $d$, $e$ for which the minimum is achieved.

15. Define polynomials $f_n(x)$ for $n = 0, 1, 2, \ldots$, by

$$f_0(x) = 1$$

$$f_n(0) = 0 \quad (n \geq 1)$$
$$f'_{n+1}(x) = (n+1)f_n(x+1) \quad (n \geq 0).$$

Find, with proof, the explicit factorization of $f_{100}(1)$ into powers of distinct primes.

16. Find polynomials $f(x)$ such that

$$f(x^2) + f(x)f(x+1) = 0.$$

17. Find all odd monic quintic polynomials over $\mathbf{Z}$ which have at least two integer zeros and take the value $-29670$ when evaluated at $10$. What is the value of the integer zero?

18. Let $f(x)$ be a polynomial of degree at most $n$. Determine the degree of the polynomial

$$f(x) - xf'(x) + (x^2/2!)f''(x) - (x^3/3!)f'''(x) + \cdots$$

$$+ (-1)^n(x^n/n!)f^{(n)}(x).$$

19. Let $r$ be a real number, and let $A$ be the set of polynomials $f$ over $\mathbf{R}$ which satisfy

    (i) $f(0) \geq 0$;
    (ii) if $f(0) = 0$, then $f'(0) = 0$ and $f''(0) \geq 0$;
    (iii) $f(0)f''(0) - f'(0)^2 \geq rf(0)f'(0)$.

    (a) Give an example of a nonconstant polynomial in $A$.
    (b) Prove that, if $c > 0$ and $f$, $g$ are in $A$, then $cf$, $f + g$, $fg$ all belong to $A$.

# Hints

## Chapter 2

1.13. If the zeros of $f(t)$ are known, what are the zeros of $f(t - 3)$?

2.11. The remainder has degree not exceeding 1. Write it in the form $u(t - a) + v(t - b)$.

3.9. (a) Use induction on the degree of $p$. Note that, if $p(t) = a_n t^n + \ldots$, then $\deg(p(t) - a_n(t - c)^n) < \deg p(t)$.

4.6. Translate the graph so that the inflection point is at the origin.

5.1. Let $f(x) = x(x - 1)(x - 2) \cdots (x - n + 1) - k$. Show that $f'(x) = (x - 1)(x - 2)(x - 3) \cdots (x - n + 1) + x(x - 2)(x - 3) \cdots (x - n + 1) + x(x - 1)(x - 3) \cdots (x - n + 1) + \cdots$ and consider the signs of $f'(0)$, $f'(1)$, $f'(2), \ldots$.. What can be deduced about the number of distinct zeros of $f'(x)$ and their multiplicities?

5.2. Let the three polynomials be $a(x - u)(x - v)$, $b(x - u)(x - w)$, $c(x - v)(x - w)$ and examine the sum of these evaluated at $u$, $v$, $w$ in turn.

5.3. Put over a common denominator and check the numerator for a double zero at $x = 1$.

(b) Set $u = x - 1$ and expand binomially.

5.4. $1 - x^3 = (1-x)[(1-x)^2+3x]$. Expand the $n$th power of the expression in square brackets binomially.

5.5. Observe that

$$ax^3 + 3bx^2 + 3cx + d = x(ax^2 + 2bx + c) + (bx^2 + 2cx + d)$$

$$a(ax^3 + 3bx^2 + 3cx + d) = (ax + b)(ax^2 + 2bx + c)$$
$$+ [2(ac - b^2)x - (bc - ad)]$$

$$d(ax^3 + 3bx^2 + 3cx + d) = (cx + d)(bx^2 + 2cx + d)$$
$$+ x^2[(ad - bc)x + 2(bd - c^2)].$$

Note that a root of the equation has multiplicity exceeding 1 if and only if it is a zero of $ax^2 + 2bx + c$.

5.6. The right side is

$$\sum_{j=1}^{r} \sum_{k=1}^{n} \binom{r+1}{j} k^{r+1-j}.$$

Interchange the order of summation, and interpret the $j$-sum as part of a binomial expansion.

5.7. If $p(x)$ is nonconstant, then $p(t) - t^m$ has infinitely many zeros for some value of $m$.

5.9. Differentiate the identity.

5.10. Use induction, noting that $(x + y)^{(m+1)} = (x + y)(x + y - 1)^{(m)}$.

5.11. What must the degree of $y$ be? Differentiate the equation three times and work backwards.

5.13. The zero of the second derivative of the cubic is a zero also of the cubic itself and its first derivative.

5.14. 0 can be at most a simple zero of the quintic.

5.15. Look at the function for small values of $n$ and make a conjecture.

5.16. If $r$ is a zero, so are $r^2$ and $(r - 1)^2$. What are the possible zeros of $f(x)$?

5.17. If $u$ is a zero, then so is $-u$. Thus, the polynomial has the form $x(x - u)(x + u)(x^2 - v)$. Setting $x = 10$ indicates that we should look for two divisors of $-2967$ which sum to 20.

5.18. What does the expansion remind you of? (Change $x$ to $a$.)

# 3

# Factors and Zeros

## 3.1 Irreducible Polynomials

$30 = 6.5$; $t^3 - 6t + 4 = (t-2)(t^2 + 2t - 2)$. Both equations express an element as a product of others. If we disallow the use of $+1$ and $-1$, the factors of 30 are smaller than 30, so that 30 can be written as a nontrivial product of integers in only finitely many ways. Furthermore, factoring further gives $30 = 2 \cdot 3 \cdot 5$ and every factorization of 30 involves products of the primes 2, 3, 5 or their negatives.

For polynomials, degree plays the role of numerical size in restricting the ways in which a polynomial can be written as a product of others. This is a similarity between the domains of integers and polynomials which distinguishes each from the fields of rationals, reals and complex numbers.

Specifically, we ask:

(a) To what extent can a polynomial be decomposed as a product of other polynomials? Is it ever possible to continue factoring the factors we get indefinitely, or must we stop after a finite amount of time?

(b) Is there a notion of "prime" analogous to that for number which can be applied to polynomials?

(c) Can every polynomial be written as a product of these "prime" ones? If so, is such a representation unique up to order of factors?

(d) Can we actually identify the "prime" polynomials?

Let us look in turn at these equations. Over a field, such as $\mathbf{Q}$, $\mathbf{R}$ and $\mathbf{C}$, if $p(t)$ is a polynomial and $c$ is a nonzero constant, then $c^{-1}p(t)$ is also a polynomial over the field. Thus, every polynomial admits *trivial factorizations* of the type

$$p(t) = c \cdot c^{-1}p(t).$$

The constant polynomials play the role of $+1$ and $-1$ for the integers in that they are universal divisors. If we are to give a meaningful analysis of the questions asked, we should ask them only in the context of nontrivial factorizations.

If $p = fg$ is a nontrivial factorization of $p$, then $\deg f$ and $\deg g$ are both strictly less than $\deg p$, and $\deg p = \deg f + \deg g$. We cannot continue to factor indefinitely, and after a finite number of factorizations must arrive at factors which are divisible only by constants and constant multiples of themselves.

Thus, the notion of primeness we require is embodied in this definition: a polynomial $p$ over an integral domain is *irreducible* if and only if

(i) $\deg p \geq 1$ and

(ii) if $p = fg$ for polynomials over the domain, then either $f$ or $g$ is constant.

Whether or not a polynomial can be factored is sensitive to the domain over which it is taken. For example, $t^2 + 1$ can be factored as the product $(t-i)(t+i)$ over the complex field, but it turns out to be irreducible over the reals and the rationals. Thus, we have to consider each domain individually. However, the fact that $\mathbf{Z} \subseteq \mathbf{Q} \subseteq \mathbf{R} \subseteq \mathbf{C}$ means that factorization with respect to one of these domains will have some bearing on factorization with respect to the others.

The study of the solvability of polynomial equations involves looking at fields in which their roots can be found. Knowing the irreducible factors of the polynomials enables us to examine the structure of these fields.

## Exercises

1. (a) Show that the rational $1/2$ can be written as the product of two rationals in infinitely many ways.

   (b) Show that any element of $\mathbf{Q}$ can be written as the product of two others in infinitely many ways. (Thus, a field can have no prime or irreducible elements.)

2. Let $p(t) \in \mathbf{Z}[t]$. Show that the constant polynomial $c$ divides $p(t)$ if and only if $c$ divides every coefficient of $p(t)$.

3. Prove that every linear polynomial $at + b$ over an integral domain is irreducible.

4. Show that every irreducible polynomial over $\mathbf{C}$ is irreducible over $\mathbf{R}$ and that every irreducible polynomial over $\mathbf{R}$ is irreducible over $\mathbf{Q}$.

5. Show that the polynomial $t^2 + 1$ is irreducible over $\mathbf{R}$.

6. Let $p(t)$ be any polynomial over an integral domain and let $k$ belong to the domain. Define $q(t) = p(t - k)$. Show that the polynomial $q(t)$ is irreducible over the domain if and only if $p(t)$ is irreducible over the domain.

7. Let $c$ be an integer.

(a) Show that, if $c$ is positive, $t^2 + c$ is irreducible over $\mathbf{Z}$, $\mathbf{Q}$ and $\mathbf{R}$, but reducible over $\mathbf{C}$.

(b) Suppose that $c = -d^2$ for some integer $d$. Show that $t^2 + c$ is reducible over $\mathbf{Z}$, $\mathbf{Q}$, $\mathbf{R}$ and $\mathbf{C}$.

(c) Suppose that $c$ is negative, but not the negative of an integer square. Show that $t^2 + c$ is irreducible over $\mathbf{Z}$ and $\mathbf{Q}$, but reducible over $\mathbf{R}$ and $\mathbf{C}$.

8. Discuss the irreducibility of the polynomial $at^2 + bt + c$ over $\mathbf{Z}$, $\mathbf{Q}$, $\mathbf{R}$ and $\mathbf{C}$, where $a$, $b$, $c$ are integers.

9. Show that the polynomial $\frac{5}{4}t^2 - \frac{31}{45}t - \frac{8}{5}$ is reducible over $\mathbf{Q}$ and write it as a product of linear factors.

10. Since $\mathbf{Q}$ is the smallest field which contains the ring $\mathbf{Z}$, one might wonder whether there is some connection between reducibility over $\mathbf{Q}$ and over $\mathbf{Z}$. Consider first a polynomial whose coefficients are integers; obviously, if it is reducible over $\mathbf{Z}$, it is reducible over $\mathbf{Q}$. On the face of it, it is not clear that the converse is true. The following exercises will deal with this issue.

(a) Let $f(t)$ be a polynomial with rational coefficients. Show that $f(t)$ can be written in the form $(c/d)g(t)$ where $g(t)$ is a polynomial with integer coefficients whose greatest common divisor is 1.

(b) If $g(t)$ is reducible over $\mathbf{Z}$, show that $f(t)$ is reducible over $\mathbf{Q}$.

(c) Suppose that $f(t)$ is reducible over $\mathbf{Q}$. Show that $g(t)$ can be written in the form

$$g(t) = (a/b)g_1(t)g_2(t)$$

where $a/b$ is a fraction in its lowest terms and $g_1$ and $g_2$ are polynomials over $\mathbf{Z}$ such that the coefficients of each have greatest common divisor 1.

(d) If it can be shown, in (c), that $|b| = 1$, it will follow that $g(t)$ is reducible over $\mathbf{Z}$. To obtain a proof by contradiction of this fact, suppose that $p$ is any prime which divides $b$. Let $u_r t^r$ and $v_s t^s$ be the terms of lowest degree of $g_1$ and $g_2$ respectively whose coefficients are not divisible by $p$. Show that the coefficient of $t^{r+s}$ in $g_1(t)g_2(t)$ is not divisible by $p$, and deduce that the coefficient of this term in $g(t)$ cannot be an integer.

(e) Conclude that, in (a), $f(t)$ is irreducible over $\mathbf{Q}$ if and only if $g(t)$ is irreducible over $\mathbf{Z}$.

(f) Let $h(t)$ be a polynomial over $\mathbf{Z}$. Show that $h(t)$ is irreducible over $\mathbf{Q}$ if and only if it is irreducible over $\mathbf{Z}$.

11. Let $p(t)$ be a monic polynomial over $\mathbf{Z}$. Show that, if $r$ is a rational root of $p(t)$, then $r$ must be an integer.

12. *The Eisenstein Criterion.* Given a polynomial over $\mathbf{Z}$, it is usually difficult to decide immediately whether it is irreducible. Nevertheless, it is possible to determine conditions which will decide the issue in a large number of cases and which will permit the easy construction of irreducible polynomials. One important result is the following *Eisenstein Criterion:*

> Suppose that the polynomial $h(t) = c_n t^n + c_{n-1} t^{n-1} + \cdots + c_1 t + c_0$ has integer coefficients $c_i$ and that there is a prime integer $p$ for which

   (a) $p$ is not a divisor of the leading coefficient $c_n$;

   (b) $p$ is a divisor of every other coefficient $c_0, c_1, \ldots, c_{n-1}$;

   (c) $p^2$ does not divide the constant coefficient $c_0$.

> Then the polynomial $q(t)$ is irreducible over $\mathbf{Z}$.

To get a handle on the proof of this, consider the special case that $h(t)$ is a cubic satisfying the conditions. Suppose that $q(t)$ is reducible. Then it can be written in the form

$$c_3 t^3 + c_2 t^2 + c_1 t + c_0 = (b_1 t + b_0)(a_2 t^2 + a_1 t + a_0).$$

Write out the $c_i$ in terms of the $a_i$ and $b_i$, and argue that $p$ must divide exactly one of $a_0$ and $b_0$, say $a_0$. Deduce that $p$ must accordingly divide $a_1$ and $a_2$, yielding a contradiction. Now give a proof of the criterion for polynomials of arbitrary degree.

13. Use the Eisenstein Criterion with a suitable prime to show that $2t^4 + 21t^3 - 6t^2 + 9t - 3$ is irreducible over $\mathbf{Z}$.

14. Find a linear polynomial over $\mathbf{Z}$ which does not satisfy the Eisenstein Criterion.

15. Show that the polynomial $t^2 + t + 1$ does not satisfy the Eisenstein Criterion for any prime, yet is irreducible over $\mathbf{Z}$.

16. Let $h(t) = c_{2m+1} t^{2m+1} + c_{2m} t^{2m} + \cdots + c_1 t + c_0$ be a polynomial of odd degree $2m + 1 \geq 3$ over $\mathbf{Z}$. Suppose that, for some prime $p$,

   (a) $p \nmid c_{2m+1}$;

   (b) $p \mid c_i \quad (m + 1 \leq i \leq 2m)$;

   (c) $p^2 \mid c_j \quad (0 \leq j \leq m)$;

   (d) $p^3 \nmid c_0$.

Show that $f(t)$ is irreducible. (Eugen Netto, *Mathematische Annalen* **48** (1897)).

17. Let $r$ be a positive integer. Show that $t^r - 2$ is irreducible over $\mathbf{Q}$. Is the polynomial irreducible over $\mathbf{R}$?

18. For each pair $r$, $n$ of positive integers with $2 \leq r \leq n + 1$, show that there is a polynomial of degree $n$ irreducible over $\mathbf{Q}$ with exactly $r$ terms.

19. Let $p$ be prime integer. Show that the polynomial $t^{p-1} + t^{p-2} + \cdots + t + 1$ does not satisfy the Eisenstein Criterion for any prime, but that it is transformed by the substitution $t = 1 + s$ to a polynomial in $s$ which does satisfy the Criterion for the prime $p$. Deduce that the polynomial is irreducible. Check the details when $p$ is given the values 3, 5, and 7.

20. Find the smallest three integers $n$ for which the polynomial $t^{n-1} + t^{n-2} + \cdots + t + 1$ is reducible.

21. Suppose $f(t)$ and $g(t)$ are nonzero polynomials over an integral domain $D$ contained in $\mathbf{C}$, and that $g(t)$ is irreducible over $D$. Suppose also that there is a complex number $w$ (not necessarily in $D$) for which $f(w) = g(w) = 0$. Prove that, for some polynomial $h(t)$ over $D$, $f(t) = g(t)h(t)$.

22. Let $g(t)$ be an irreducible polynomial over an integral domain $D$ contained in $\mathbf{C}$. Prove that every complex zero of $g(t)$ is simple.

## 3.2 Strategies for Factoring Polynomials over Z

It is useful to know how to factor polynomials. Not only is factoring helpful in solving equations, but it is often possible to read off information from a factorization which would otherwise be hidden.

Even when it is known that a polynomial over $\mathbf{Z}$ is reducible, it can be quite a challenge to actually obtain its factors. While there are systematic ways of doing so, they are generally complicated and long. Consequently, it is desirable to have a variety of techniques to make the task manageable. The exercises in this section will introduce some of these. With experience, these techniques can be used with discretion and flexibility for efficient factorization.

### Exercises

1. (a) Consider the quadratic polynomial $6t^2 + 2t - 20$. Determine two integers $u$ and $v$ for which $u + v = 2$ and $uv = -120$. Verify that

$6t + u$ and $vt - 20$ are both integer multiples of the same linear polynomial. Write the given polynomial in the form

$$6t^2 + ut + vt - 20$$

and thence factor it.

(b) For the quadratic $at^2 + bt + c$, show how the determination of integers $u$ and $v$ with $u + v = b$ and $uv = ac$ can lead to a factorization of the quadratic into linear factors over **Z**.

(c) Factor each of the following quadratics and determine the values of $t$ for which it is negative:

$$28t^2 + 57t + 14$$

$$20t^2 + 39t - 44.$$

2. (a) Consider the polynomial $a^k - b^k$. Use the Factor Theorem to deduce that $a - b$ is a factor, and thence write the given polynomial as the product of two factors.

(b) Let $k$ be an odd integer. Use the Factor Theorem (Exercise 2.2.12) to deduce that $a + b$ is a factor of $a^k + b^k$ and write this polynomial as a product of two factors.

3. Sometimes a polynomial can be manipulated into a standard form for factorization. Write the following polynomials as difference of squares and thence factor them over **Z**. Determine their zeros.

(a) $4t^2 - 20t - 11$

(b) $5t^2 - 6t + 1$

(c) $t^4 - 47t^2 + 1.$

4. (a) Show that a monic cubic reducible over **Z** must have an integer zero.

(b) Given that the cubic $t^3 - 8t^2 + 33t - 42$ is reducible over **Z**, factor it.

5. In factoring a polynomial, it is often useful to recall that the degree of the polynomial is the sum of the degrees of its factors. Prove that, if a polynomial of degree $n$ is reducible, then it must have at least one factor whose degree does not exceed $n/2$.

6. One way to factor a polynomial is by the method of *undetermined coefficients*. A factorization of a certain form is assumed, and equations satisfied by the coefficients of the factors is set up. Consider for example the problem of factoring the polynomial

$$t^8 + 98t^4 + 1.$$

Because of the symmetry of the coefficients, one might try to find two symmetrically related factors, each of degree 4.

(a) Show that, if there is a factorization of the form

$$t^8 + 98t^4 + 1 = (t^4 + at^3 + bt^2 + ct + d)(t^4 + kt^3 + mt^2 + nt + r)$$

then we must have $r = d = 1$ or $r = d = -1$

$$0 = a + k$$

$$0 = m + ak + b$$

$$0 = n + ma + bk + c$$

$$98 = 2d + an + bm + ck$$

$$0 = ad + bn + cm + kd$$

$$0 = bd + cn + dm$$

$$0 = d(c + n).$$

(b) Show that $k = -a$, $n = -c$, $a(b - m) = 0$, $c^2 = (b + m)d = a^2 d$.

(c) Show that $a \neq 0$, so that $d = 1$, $c^2 = a^2$, $b = m$.

(d) Determine a factorization of $t^8 + 98t^4 + 1$ as a product of two polynomials of degree 4.

7. Test the following polynomials for irreducibility over $\mathbf{Q}$. Factor all the reducible polynomials as far as you can.

(a) $7t - 8$

(b) $2t^2 + 2t - 1$

(c) $4t^2 + 4t - 1$

(d) $28t^2 + 11t - 24$

(e) $28t^2 - 11t + 24$

(f) $2t^3 + 3t^2 - 21t - 6$

(g) $(t^2 + 2t)^2 - 5(t^2 + 2t) + 6$

(h) $t^4 + 2t^3 + t^2 + t + 1$

(i) $3t^4 - 2t^3 - t^2 - 3t - 1$

(j) $4t^5 - 15t^3 + 5t^2 + 15t - 9$

(k) $t^6 - t^4 - t^2 + 1$

(l) $t^3 - t^2 - 24t - 36$

(m) $t^3 - 7t^2 + 13t - 15$

(n) $t^5 - t^3 - 3t^2 - 2t - 1$

    (o) $t^5 - 3t^4 - t^2 - 4t + 14$

    (p) $t^8 - 98t^4 + 1$.

8. Let $f$ be a nonzero homogeneous polynomial of several variables. Prove that, if $f$ is reducible, then each of its factors must be homogeneous.

9. Let $f$ be a symmetric polynomial of several variables. If $f$ is reducible, must each of its factors be symmetric? Justify your answer.

10. Factor each of the following polynomials.

    (a) $5(a - b)^2 - 4a^2 + 4ab$

    (b) $x^2y - y^3 - x^2z + y^2z$

    (c) $x^2y - xy^2 + y^2z - yz^2 + z^2x - zx^2$

    (d) $x^3 + y^3 + z^3 - 3xyz$

    (e) $bc(b - c) + ca(c - a) + ab(a - b)$

    (f) $x(y^2 - z^2) + y(z^2 - x^2) + z(x^2 - y^2)$

    (g) $yz(y + z) + xz(x + z) + xy(x + y)$

11. Let $n$ be a nonnegative integer and define the polynomial

$$p_n(x, y, z) = x^n(z - y) + y^n(x - z) + z^n(y - x).$$

    (a) Verify that $p_n(x, y, z) = 0$ if any two of $x$, $y$, $z$ are equal.

    (b) Deduce from (a) that $p_0(x, y, z)$ and $p_1(x, y, z)$ are the zero polynomials.

    (c) If $n \geq 2$, show that

$$p_n(x, y, z) = (x - y)(y - z)(z - x)q_n(x, y, z)$$

    where $q_n(x, y, z)$ is a homogeneous symmetric polynomial of degree $n - 2$. Show that the coefficients of $x^{n-2}$, $y^{n-2}$, $z^{n-2}$ in $q_n(x, y, z)$ are each 1.

    (d) Identify $q_n(x, y, z)$ for $n = 2, 3, 4$.

    (e) It is possible to get a shadow of a possible factorization by setting one of the variables equal to 0 and factoring the resultant polynomial of fewer variables. Factor $p_n(x, y, 0)$.

    (f) Factor $p_5(x, y, z)$ and $p_6(x, y, z)$.

12. Factor $(x + y + z)^k - x^k - y^k - z^k$ for $k = 3, 5, 7$.

13. Is it true in general that if a polynomial with integer coefficients is irreducible over $\mathbf{Z}_m$ for some $m$, it must be irreducible over **Z**?

14. Show that each of the following polynomials is irreducible over $\mathbf{Z}_m$ for some $m$. Can you deduce from this that it is irreducible over $\mathbf{Z}$?

    (a) $t^2 + t + 1$

    (b) $49t^2 + 35t + 11$

    (c) $124t^3 - 119t^2 + 35t + 64$.

15. (a) Factor $63t^4 - 2t^3 - 79t^2 + 52t - 10$ over $\mathbf{Z}_4$, $\mathbf{Z}_5$, $\mathbf{Z}_7$, $\mathbf{Z}_9$.

    (b) Use the factorizations of the polynomial in (a) over the finite domains to determine a factorization over $\mathbf{Z}$.

16. To demonstrate how cumbersome a "sure-fire" method can be, consider the problem of factoring the quintic polynomial

$$f(t) = 10t^5 + 3t^4 - 38t^3 - 5t^2 - 6t + 3.$$

One strategy is to note that any factorization over $\mathbf{Z}$ leads to a numerical factorization of the possible values that $f(t)$ can assume. Thus, knowing, say, $f(1)$ determines a finite set of values which any factors might assume at 1.

    (a) Verify that $f(-1) = 35$, $f(0) = 3$, $f(1) = -33$.

    (b) Supposing that $f(t)$ is reducible over $\mathbf{Z}$, we can assume that there is a factor $g(t)$ of degree at most 2. Verify that, if $g(-1) = u$, $g(0) = v$, $g(1) = w$, then

$$2g(t) = (w + u - 2v)t^2 + (w - u)t + 2v.$$

    (c) Show that $u \mid 35$, $v \mid 3$, $w \mid 33$, so that there are $8 \times 4 \times 8 = 256$ possible choices of $(u, v, w)$ to examine in determining $g(t)$.

    (d) Show that $w + u - 2v$ is an even divisor of 20.

    (e) With no loss of generality, we can assume $v = 1$ or $v = 3$ (why?). Show that, if $v = 1$, the possible values of $w + u$ are $-18$, $-8$, $-2$, $0$, $4$, $6$, $12$, $22$, and that if $v = 3$, the possible values of $w + u$ are $-14$, $-4$, $2$, $4$, $8$, $10$, $16$, $26$.

    (f) Use (c), (d), (e) to find candidates for a factor $g(t)$. Does one of them work?

## Exploration

**E.29.** Let $a$, $b$, $n$ be positive integers. Investigate under what conditions the polynomial

$$t^2 - t + a$$

is a factor of $t^n + t + b$ over $\mathbf{Z}$.

**E.30.** A sequence of polynomials $u_n(t)$ is defined by the recurrence:

$$u_0(t) = 0 \qquad u_1(t) = 1$$

$$u_n(t) = tu_{n-1}(t) - u_{n-2} \quad (n \geq 2).$$

Write out and factor the first few terms of the sequence. Look for any interesting relationships among the polynomials in the sequence. Evaluate the polynomials in the sequence at $t = 1, 2, 3, 4$. Here are a number of facts for you to verify:

(1)
$$u_n(t) = t^{n-1} - \binom{n-2}{1} t^{n-3} + \binom{n-3}{2} t^{n-5}$$
$$- \binom{n-4}{3} t^{n-7} + \cdots$$

(2) $u_n(t) = (t^2 - 4)^{-1/2}(y^n - y^{-n})$ where $y = \frac{1}{2}[t + (t^2 - 4)^{1/2}]$

(3)
$$u_n(t) = 2^{-(n-1)}\left[ \binom{n}{1} t^{n-1} + \binom{n}{3} (t^2 - 4)t^{n-3} \right.$$
$$\left. + \binom{n}{5} (t^2 - 4)^2 t^{n-5} + \cdots \right]$$

(4) $u_n(t) = \dfrac{\sin n\theta}{\sin \theta}$ where $t = 2\cos\theta$. What are the zeros of $u_n(t)$?

(5) $u_n^2 - u_k^2 = u_{n-k}u_{n+k}$ $(0 \leq k \leq n)$

(6) $u_{n+1}^2 - u_n^2 = u_{2n+1}$ $(0 \leq n)$

(7) $u_n^2 = u_{n-1}u_{n+1} + 1$ $(1 \leq n)$

(8) $u_1 + u_3 + \cdots + u_{2n-1} = u_n^2$ $(1 \leq n)$

(9) $u_n + u_{n+2k} = (u_{k+1} - u_{k-1})u_{n+k}$ $(0 \leq n; 1 \leq k)$

(10) $u_n + u_{n+2k+1} = (u_{k+1} - u_k)(u_{n+k} + u_{n+k+1})$ $(0 \leq k, n)$

(11) $u_{2k} = (u_{k+1} - u_{k-1})u_k$ $(1 \leq k)$

(12) $u_{2k+1} = (u_{k+1} - u_k)(u_{k+1} + u_k)$ $(0 \leq k)$

(13) $u_n = \frac{1}{2}[tu_{n-1} + ((t^2 - 4)u_{n-1}^2 + 4)^{1/2}]$ $(1 \leq n)$

(14) If $m \mid n$, then $u_m \mid u_n$ (as polynomials with integer coefficients)

(15) Show that

$$\frac{u_1^3 + u_2^3 + \cdots + u_n^3}{u_1 + u_2 + \cdots + u_n} = \frac{(y^{n+1} + 1 + y^{-(n+1)})(y^n + 1 + y^{-n}) - 3(t+1)}{(t^2 - 4)(t+1)}$$

Deduce from this that $u_1^3 + u_2^3 + \cdots + u_n^3$ is divisible by $u_1 + u_2 + \cdots u_n$ $(1 \le n)$.

Define polynomials $v_n$ and $w_n$ as follows:

$$v_0 = 0 \qquad v_1 = 1$$

$$v_{2m} = v_{2m-1} + v_{2m-2} \quad (m \ge 1)$$
$$v_{2m+1} = (t-2)v_{2m} + v_{2m-1} \quad (m \ge 1)$$

$$w_0 = 2 \qquad w_1 = 1$$
$$w_{2m} = (t-2)w_{2m-1} + w_{2m-2} \quad (m \ge 1)$$
$$w_{2m+1} = w_{2m} + w_{2m-1} \quad (m \ge 1)$$

(16) $v_{n+2} = tv_n - v_{n-2} \quad w_{n+2} = tw_n - w_{n-2} \ (n \ge 2)$

(17) $v_{2m} = u_m \ (m \ge 0)$

(18) $v_{2m+1}(t) = (-1)^m w_{2m+1}(-t) \ (m \ge 0)$

(19) $u_n = v_n w_n = \frac{1}{2}(v_{n+1}w_{n-1} + v_{n-1}w_{n+1})$

(20) When $n = 2^r$, $u_n = p_1 p_2 \cdots p_r$ where $p_1(t) = t$ and $p_k = p_{k-1}^2 - 2$ $(k \ge 1)$.

## 3.3 Finding Integer and Rational Roots: Newton's Method of Divisors

A first step in factoring polynomials over $\mathbf{Q}$ is to use the Factor Theorem to locate linear factors by finding rational zeros. It is straightforward to see that all but finitely many rationals can be rejected as possible zeros. Special techniques will narrow down the possibilities even further.

### Exercises

In these exercises, $q(t)$ will represent the polynomial

$$c_n t^n + c_{n-1} t^{n-1} + c_{n-2} t^{n-2} + \cdots + c_1 t + c_0$$

with integer coefficients $c_i$.

1. (a) Show that the rational number $a/b$ is a zero of $q(t)$ if and only if

$$c_n a^n + c_{n-1} a^n b + c_{n-2} a^{n-2} b^2 + \cdots + c_1 ab^{n-1} + c_0 b^n = 0.$$

   (b) If the zero $a/b$ of $q(t)$ is written in lowest terms, show that $c_n$ is a multiple of $b$ and that $c_0$ is a multiple of $a$.

2. Prove that any rational zero of a *monic* polynomial with integer coefficients must in fact be an integer.

3. For each polynomial, determine all of its rational zeros. Recall that, having found one rational zero, you can divide the polynomial by a suitable linear factor and deal with a polynomial of lower degree to determine the other rational zeros.

   (a) $5t^3 - 4t^2 + 3t + 2$
   (b) $6t^3 + 13t^2 - 22t - 8$
   (c) $3t^4 + 5t^3 + 2t^2 - 6t - 4$
   (d) $t^4 - t^3 - 32t^2 - 62t - 56$.

4. If $a/b$ is a rational zero of $q(t)$ written in lowest terms. Then, as we have seen, $a \mid c_0$. Another way of expressing this is to say that $-a$ is a divisor of $q(0)$. There is a useful generalization to this:

   *if $q(t)$ is a polynomial over $\mathbf{Z}$ with rational zero $a/b$, then $bt - a$ is a divisor of $q(t)$ over $\mathbf{Z}$, and, for each integer $m$, $bm - a$ must be a divisor of $q(m)$.*

   In this exercise, this result will be first applied and then demonstrated.

   (a) Consider the polynomial $6t^3 + 13t^2 - 22t - 8$. Show that any positive rational zero must be one of

   $$1, \ 1/2, \ 1/3, \ 1/6, \ 2, \ 2/3, \ 4, \ 4/3, \ 8, \ 8/3.$$

   Evaluate the polynomial at $t = -2, -1, 1, 2$, and apply the result with these values of $m$ to eliminate all but two of the rationals in the list as a possible zero of the polynomial.

   (b) Show that, if $a/b$ is a zero of the polynomial, then

   $$q(t) = (bt - a)(r_{n-1} t^{n-1} + r_{n-2} t^{n-2} + \cdots + r_1 t + r_0)$$

   where the coefficients $r_n, r_{n-1}, \ldots, r_1, r_0$ are rational numbers satisfying

   $$c_n = b r_{n-1}$$

$$c_{n-1} = br_{n-2} - ar_{n-1}$$

$$c_{n-2} = br_{n-3} - ar_{n-2}$$

$$\cdots$$

$$c_2 = br_1 - ar_2$$

$$c_1 = br_0 - ar_1$$

$$c_0 = -ar_0.$$

(c) In (b), suppose that $a/b$ is written in lowest terms. It is known that $a$, $b$ and the $c_i$ are integers, and also that $a \mid c_0$ and $b \mid c_n$. By looking at the equations relating the $r_i$ and the $c_i$ in turn from the top, prove that

$$r_{n-1} \quad \text{is an integer}$$

$$br_{n-2} \quad \text{is an integer}$$

$$b^2 r_{n-3} \quad \text{is an integer}$$

$$\cdots$$

$$b^{n-1} r_0 \quad \text{is an integer.}$$

and conclude that each $r_i$ is a rational whose denominator divides $b^{n-1}$. Similarly, by looking at the equations in turn from the bottom, show that each $r_i$ is a rational whose denominator divides $a^{n-1}$. Deduce from this each $r_i$ is an integer.

(d) Complete the proof of the result.

5. *Newton's Method of Divisors.* The set of equations involving $c_i$, $r_i$, $a$ and $b$ in Exercise 4 is the basis of an algorithm known as *Newton's Method of Divisors.* We wish to check whether $a/b$ is a root of $q(t)$. Write in a row the coefficients of the polynomial, beginning with the constant coefficient:

$$c_0 \quad c_1 \quad c_2 \quad c_3 \quad \cdots \quad c_n.$$

Divide $c_0$ by $a$. If the result is not an integer, then we do not have a zero. If the result is an integer, let $c_0/a = s_0$ (this is $-r_0$ in Exercise 4); write this integer under $c_1$. Draw a line across. Beneath the line, put the sum of $c_1 1$ and $bs_0$:

| $c_0$ | $c_1$ | $c_2$ | $\cdots$ |
|-------|-------|-------|----------|
|       | $s_0$ |       |          |

$$c_0 \quad c_1 + bs_0$$

Divide $c_1 + bs_0$ by $a$. If the result is not an integer, $a/b$ is not a zero. Otherwise, let the quotient be $s_1$, and write $s_1$ under $c_2$. Beneath the line, put $c_2 + bs_1$. Continue on to get a table:

| $c_0$ | $c_1$ | $c_2$ | $c_3$ | $\cdots$ |
|---|---|---|---|---|
| | $s_0$ | $s_1$ | $s_2$ | |
| $c_0$ | $c_1 + bs_0$ | $c_2 + bs_1$ | $c_3 + bs_2$ | $\cdots$ |

Each $s_i$ is equal to $(c_i + bs_{i-1})/a$. Stop if some $s_i$ fails to be an integer, for then $a/b$ is not a zero. If all the $s_i$ turn out to be integers, then for $a/b$ to be a zero, $c_n$ must equal $-bs_{n-1}$, and the last term below the line must be 0.

(a) Justify this algorithm.

(b) The method is mainly used to check for integer zeros ($b = 1$), which explains why the coefficients are listed in ascending rather than descending order in the top row of the table. Devise an alternative algorithm which takes the coefficients in the opposite order.

(c) Two candidates for rational zeros of $2t^3 - 11t^2 + 2t + 15$ are 3 and 3/2. Verify that the respective tables for these are

| 15 | 2 | −11 | 2 | | 15 | 2 | −11 | 2 |
|---|---|---|---|---|---|---|---|---|
| | 5 | | | | | 5 | 4 | −1 |
| 15 | 7 | | | | 15 | 12 | −3 | 0 |

Construct the tables for the candidates 5 and 5/2. Identify zeros of $2t^3 - 11t^2 + 2t + 15$ and factor this polynomial over **Z**.

6. Find all rational zeros of the following polynomials:

(a) $4t^3 - 22t^2 + 7t + 15$

(b) $40t^3 + 25t^2 + 9t - 9$

(c) $5t^2 - 12t + 4$

(d) $t^3 - 9t^2 + 11t + 21$

(e) $8t^3 + 20t^2 - 18t - 45$

(f) $6t^4 + t^3 - 66t^2 + 30t + 56$.

7. Write as a product of irreducible factors over **Z**:

$$18t^5 - 48t^4 + 23t^3 + 174t^2 - 171t - 60.$$

8. Find all the zeros of the polynomial

$$24t^5 + 143t^4 - 136t^3 + 281t^2 + 36t - 140.$$

9. We can check whether $a$ is a zero of a polynomial $q(t)$ over **Z** by evaluating $q(a)$ by Horner's Method. In the case $q(t) = 4t^3 - 22t^2 + 7t + 15$, $a = 5$, we get the table for Horner's and Newton's Methods respectively:

| 4 | −22 | 7 | 15 |   | 15 | 7 | −22 | 4 |
|---|-----|---|----|---|----|---|-----|---|
|   | 20 | −10 | −15 |  |    | 3 | 2 | −4 |
| 4 | −2 | −3 | 0 |   | 15 | 10 | −20 | 0 |

Is there any connection between these tables? Explain.

10. In using Newton's Table to check a negative rational zero, what difference does it make to consider the numerator to be negative and the denominator to be positive rather than the other way around?

11. Consider the linear polynomial $ct + d$. If $a/b$ is a zero of this, then $a \mid d$ and $b \mid c$. Consider rationals which satisfy these conditions, and explain what happens in the bottom line of the table for Newton's Method of Divisors for the unsuccessful candidates.

12. Let $\theta = 2\pi/5$.

    (a) Verify that $\cos\theta/2 + \cos 2\theta = 0$.

    (b) Show that $x = \cos\theta$ satisfies the equation
    $$x = 2(4x^4 - 4x^2 + 1) - 1.$$

    (c) Factor $8x^4 - 8x^2 - x + 1$ over **Z**, and deduce that $\cos\theta$ is a zero of a quadratic polynomial over **Z**.

    (d) Determine $\cos\theta$.

## Exploration

**E.31.** Find all integers $n$ for which the zeros of the quadratic polynomial
$$nt^2 + (n + 1)t - (n + 2)$$
are rational.

## 3.4  Locating Integer Roots: Modular Arithmetic

What is a quick way to see that the polynomial $t^2 - 131t + 267$ cannot possibly have an integer zero? Such a zero must be even or odd. If $t$ is given an even value, then $t^2$ and $-133t$ are even, and the polynomial must take an odd value. Similarly, if $t$ is given an odd value, then the value of the polynomial is again odd. Thus, no substitution for $t$ will give an even value for the polynomial and the polynomial will always fail to vanish for integer $t$. How can we extend this type of argument?

In the example, we made use of the fact that 0 is even. In fact, 0 is the only integer which is divisible by every other integer, and in particular by every prime. Accordingly, if a polynomial $q(t)$ over $\mathbf{Z}$ has an integer zero $r$, then $q(r) \equiv 0 \pmod{m}$ for each $m$. On the one hand, we can find candidates for integer zeros among the solutions of this congruence for suitable values of $m$. On the other hand, if we can find an integer $m$ for which the congruence has no solution, then there can be no integer zero of the polynomial.

In this section, we will be concerned with solving congruences of the form $q(t) \equiv 0 \pmod{m}$. It turns out that we can do this by solving the congruence when $m$ is a prime power and piecing together the solutions for prime powers to discover the solution for an arbitrary modulus.

# Exercises

1. Consider the following congruence:

$$t^2 - 9t - 36 \equiv 0 \pmod{m}$$

(a) Show that the solutions of the congruence for $m = 8$ are the same as the solutions of the congruence

$$t^2 - t - 4 \equiv 0 \pmod{8}.$$

Verify that the only solutions of the congruence modulo 8 are $t \equiv 4$ and $t \equiv 5$. Write out all the integer solutions of the congruence between 0 and 39 inclusive.

(b) Solve the congruence $t^2 - 9t - 36 \equiv 0 \pmod{5}$. Write out all the integer solutions of the congruence between 0 and 39 inclusive.

(c) Show that every solution of the congruence $t^2 - 9t - 36 \equiv 0 \pmod{40}$ is a solution of the congruences in (a) and (b). Use this fact to write out all the solutions of the congruence modulo 40 between 0 and 39 inclusive.

(d) Use the result of (c) to guess the zeros of the polynomial $t^2 - 9t - 36$. Check your guesses.

2. Let $q(t)$ be a polynomial over $\mathbf{Z}$. Show that, if $a \equiv b \pmod{m}$, then $q(a) \equiv q(b) \pmod{m}$, for any positive integer $m$.

3. Let $q(t)$ be a polynomial over $\mathbf{Z}$, and let $m$ be a positive integer which is a product of prime powers $p^k$. Argue that every solution of the congruence

$$q(t) \equiv 0 \pmod{m}$$

is also a solution of the congruence

$$q(t) \equiv 0 \ (\mathrm{mod}\, p^k).$$

4. Find all the positive integers less than 100 which leave a remainder 1 when divided by 3, remainder 2 when divided by 4 and remainder 2 when divided by 5. Perform this task in the following way: (1) write out in increasing order those numbers congruent to 2 modulo 5; (2) cross out those which are not congruent to 2 modulo 4; (3) cross out those which are not congruent to 1 modulo 3.

   How many answers are there? Suppose we remove the restriction that the number be positive and less than 100; what would the answers be?

5. Find a positive integer not exceeding 1000 which leaves a remainder 3 upon division by 7, 4 upon division by 11 and 2 upon division by 13.

6. Suppose that the positive integer $m$ is the product of two integers $u$ and $v$ with greatest common divisor 1. Let $a$ and $b$ be any two integers. Prove that there exists exactly one integer $c$ such that

$$0 \le c \le m - 1$$

$$c \equiv a \ (\mathrm{mod}\, u) \qquad c \equiv b \ (\mathrm{mod}\, v).$$

(You will need the result of Exercise 1.6.6(d).)

7. *Chinese Remainder Theorem.* Let $m = m_1 m_2 \cdots m_r$ be the product of $r$ integers $m_i$, each pair of which has greatest common divisor 1. Suppose $a_1, a_2, \ldots, a_r$ are any $r$ integers. Show that there is exactly one integer $c$ for which

$$0 \le c \le m - 1$$

$$c \equiv a_i \ (\mathrm{mod}\, m_i).$$

One way to prove this is to use induction on $r$, building on Exercise 6. However, another proof can be devised along the following lines:

   (a) Any number which is divisible by all the numbers $m_j$ except $m_i$ has the form $t_i m_1 m_2 \cdots \hat{m}_i \cdots m_r$ (where the hat indicates a deleted entry).

   (b) $t_i$ can be chosen in such a way that the number in (a) is congruent to 1 modulo $m_i$ (see Exercise 1.6.6). Thus, we can find an integer $c_i$ for which

$$c_i \equiv 0 \ (\mathrm{mod}\, m_j) \quad \text{when } j \ne i$$

$$c_i \equiv 1 \ (\mathrm{mod}\, m_i).$$

(c) Let $c = c_1 a_1 + c_2 a_2 + c_3 a_3 + \cdots + c_r a_r$.

8. For which positive values of the integer $n$ is $n^8 - n^2$ not divisible by 504?

9. (a) Solve the following congruence modulo 4 and modulo 9:

$$8t^2 - 7t + 9 \equiv 0.$$

(b) Argue that, by the Chinese Remainder Theorem, there should be two incongruent solutions of the congruence in (a) modulo 36.

(c) Find the solutions of the congruence modulo 36.

10. How many positive integers less than 48 satisfy each of the systems of congruences:

(a) $x \equiv 5 \pmod 6$, $x \equiv 4 \pmod 8$
(b) $y \equiv 5 \pmod 6$, $y \equiv 3 \pmod 8$
(c) $z \equiv 3 \pmod 6$, $z \equiv 5 \pmod 8$?

Use this exercise to argue that in general the Chinese Remainder Theorem is false if the condition that the factors of $m$ are pairwise coprime is dropped.

11. Solve the congruence $t^2 + 2 \equiv 0 \pmod{243}$ by following these steps:

(a) Noting that $243 = 3^5$, argue that any solution of the congruence satisfies $t^2 + 2 \equiv 0 \pmod 3$.

(b) Verify that $t^2 + 2 \equiv 0 \pmod 3$ is satisfied by $t \equiv 1 \pmod 3$.

(c) We now turn to the congruence $t^2 + 2 \equiv 0 \pmod 9$. Show that if $t = 1 + 3u$ is a solution to this congruence, then

$$3 + 6u \equiv 0 \pmod 9$$

which implies that $1 + 2u \equiv 0 \pmod 3$. Thus, $u = 1 + 3v$, and $t = 4 + 9v$.

(d) Show that, if $t = 4 + 9v$ and $t^2 + 2 \equiv 0 \pmod{27}$, then

$$18 + 72v + 81v^2 \equiv 0 \pmod{27}.$$

Reduce this to the equivalent congruence $1 + v \equiv 0 \pmod 3$. Thus, $t = 22 + 27w$.

(e) Continue the process to obtain, in turn, a solution of

$$t^2 + 2 = 0$$

modulo 81 and modulo 243. Check your answer.

(f) A second solution of the congruence $t^2 + 2 \equiv 0 \pmod 3$ is $t \equiv 2$ (mod 3). Find a solution of $t^2 + 2 \equiv 0 \pmod{243}$ of the form $t = 3k + 2$.

12. Determine the number of incongruent solutions for each of the following congruences:

    (a) $2t^3 + t + 3 \equiv 0 \pmod{1000}$

    (b) $2t^3 + t + 3 \equiv 0 \pmod{83349}$.

13. Solve the congruence: $3x^3 - 4x^2 + 10x - 3 \equiv 0 \pmod{675}$. For one of the solutions, evaluate the polynomial and verify that the value is divisible by 675.

14. Consider the polynomial equation

$$t^5 - 4t^4 - 411t^3 - 452t^2 - 3322t - 828 = 0.$$

    (a) Argue that the absolute value of any root is less than 5000.

    (b) Find those values of $t$ between $-5000$ and $5000$ for which the value of the polynomial is a multiple of 5000. Argue that any integer root of the equation must be one of these values and thus find all its integer roots.

15. Using congruences, find all integer roots of the equation

$$2t^4 + 20t^3 + 19t^2 - t - 90 = 0.$$

## Explorations

**E.32. Little Fermat Theorem.** Verify that $n^3 \equiv n \pmod 3$ and $n^5 \equiv n \pmod 5$ for each integer $n$. These are particular instances of a general result:

$$n^p \equiv n \pmod p$$

whenever $p$ is prime and $n$ is an integer.

One way to see this is to follow an argument given by L. Euler (1707–1783). Since the result is clear for $p = 2$, we suppose that $p$ is an odd prime and prove the result by induction on $n$. Observe that $\binom{p}{k}$ is a multiple of $p$ for $1 \leq k \leq p - 1$.

The case $n = 1$ is obvious. Euler gave two arguments for the $n = 2$ case. First, expand the right side of $2^p = (1 + 1)^p$ binomially and write as a congruence modulo $p$. Alternatively, one can get a slightly stronger result by using

$$2^{p-1} - 1 = (1 + 1)^{p-1} - 1 = \sum_{k=1}^{p-1} \binom{p-1}{k}$$

and

$$\binom{p-1}{k} + \binom{p-1}{k+1} = \binom{p}{k+1}$$

for $k = 1, 3, 5, \ldots, p-2$.

The induction step follows easily from

$$(1+n)^p - (1+n) = \sum_{k=1}^{p-1} \binom{p}{k} n^k + (n^p - n).$$

**E.33. Hensel's Lemma.** In Exercise 4.11, we showed how to solve a congruence to a prime power by starting with a solution modulo the prime and then increasing the exponent by steps of 1. Taylor's Theorem is the basis of a more efficient method which allows us to solve the congruence in successive stages in which the exponent is doubled. Suppose $p^k$ is a prime power and the polynomial congruence to be solved is

$$f(t) \equiv 0 \pmod{p^k}. \qquad (*)$$

Suppose $h$ and $k$ are positive integers such that $1 \leq h < k \leq 2h$, and that a solution $u$ is known to the congruence

$$f(t) \equiv 0 \pmod{p^h}.$$

Then any number of the form $u + mp^h$ also satisfies this congruence. We try to find a number $v$ of this form which will satisfy $(*)$. Thus, we can think of $u$ as a "first approximation" to a solution to $(*)$.

Taylor's Theorem can be employed to express $f(v)$ in terms of $f(u)$:

$$f(v) = f(u) + (v-u)f'(u) + (v-u)^2[(1/2)f''(u)$$
$$+ (1/6)f'''(u)(v-u) + \cdots].$$

If $v$ is to satisfy $(*)$ and $v \equiv u \pmod{p^h}$, argue that we must have

$$0 \equiv f(u) + (v-u)f'(u) \pmod{p^k}.$$

Conversely, show that if we can determine $v \equiv u \pmod{p^h}$ such that

$$0 \equiv f(u) + (v-u)f'(u) \pmod{p^{2h}},$$

then $v$ is a solution to $(*)$.

Let us see how this can be used to solve the congruence

$$t^2 + 2 \equiv 0 \pmod{243}.$$

The congruence $t^2 + 2 \equiv 0 \pmod 3$ is satisfied by $t = 1$. Making the substitution $p = 3$, $h = 1$, $f(t) = t^2 + 2$, $u = 1$, we find that $f(u) = 2$, $f'(u) = 2$ and the congruence for $v$ becomes

$$0 \equiv 3 + 2(v-1) \pmod 9.$$

Verify that the unique solution of this is $v \equiv 4 \pmod 9$.

For the next step, let $h = 2$ and $u = 4$, so that $f(u) = 18$ and $f'(u) = 8$ and the congruence for $v$ becomes

$$0 \equiv 18 + 8(v - 4) \pmod{81}$$

or

$$8v \equiv 14 \pmod{81}.$$

This is a congruence of the form studied in Exercise 1.6.6. Since 8 and 81 are coprime, there is a unique solution modulo 81. By making use of the Euclidean algorithm (Exercise 1.6.2) or otherwise, obtain the solution $v \equiv 22 \pmod{81}$.

For the final step, let $h = 4$ and $u = 22$. Show that this leads to the congruence

$$0 \equiv 486 + 44(v - 22) \pmod{81^2}$$

or

$$44v \equiv 482 \pmod{81^2}.$$

Obtain the solution $v = 2695$. Conclude that $t^2 + 2 \equiv 0 \pmod{3^8}$ is satisfied by $t = 2695$, and obtain the solution to the same congruence $\pmod{3^5}$ as required.

Let us return to the general situation. Establish the following result:

Suppose $1 \leq h < k \leq 2h$, $p$ is prime and $f(t)$ is a polynomial over $\mathbf{Z}$. Suppose further that $u$ is an integer for which

(i) $f(u) \equiv 0 \pmod{p^h}$

(ii) $f'(u) \not\equiv 0 \pmod p$.

Then there is an integer $w$ for which $wf'(u) \equiv 1 \pmod{p^{2h}}$ and the number $v = u - wf(u)$ satisfies the congruence

$$f(v) \equiv 0 \pmod{p^{2h}}.$$

Why is the condition $f'(u) \not\equiv 0 \pmod p$ imposed? To see that some condition like (ii) is needed, verify that

$$t^2 + t + 1 \equiv 0 \pmod 3 \text{ is solvable}$$

$$t^2 + t + 1 \equiv 0 \pmod 9 \text{ is not solvable.}$$

This result is the heart of the proof of Hensel's Lemma:

Let $p$ be a prime and let $f(t)$ be a polynomial over $\mathbf{Z}$. Suppose $u_1$ is a solution of the congruence $f(t) \equiv 0 \pmod p$ such that $f'(u_1) \not\equiv 0 \pmod p$. Then for each positive integer $k$, there is a solution $u_k$ of the congruence $f(t) = 0 \pmod{p^k}$.

Solve the congruence $t^4 + t^2 + t + 1 \equiv 0 \pmod{256}$. [Answer: 149].

## 3.5  Roots of Unity

The factorization of $t^n - 1$ was studied by C.F. Gauss (1777–1855), who wanted to find out the condition on $n$ in order that a regular $n$-gon could be constructed using ruler and compasses.

From de Moivre's Theorem, it is possible to locate all the zeros of this polynomial as equally spaced points on the circumference of the unit circle in the complex plane (i.e. as vertices of a regular $n$-gon). From this it is a simple step to derive the decomposition of $t^n - 1$ into linear factors over $\mathbf{C}$. These factors can then be combined to yield factorizations over $\mathbf{R}$ and $\mathbf{Z}$.

In this section, we will denote $t^n - 1$ by $P_n(t)$.

### Exercises

1. *Roots of unity.* Let $n$ be a positive integer, and let $r(\cos\theta + i\sin\theta)$ be a root of the polynomial $P_n(t)$. Use de Moivre's Theorem (Exercise 1.3.8) to show that

$$r^n(\cos n\theta + i\sin n\theta) = 1.$$

Take absolute values of both sides and deduce that $r = 1$. Show that $n\theta$ must be an integer multiple of $2\pi$.

Show that a complete set of zeros of $P_n(t)$ consists of the complex numbers,

$$1,\ \cos 2\pi/n + i\sin 2\pi/n,\ \cos 4\pi/n + i\sin 4\pi/n, \ldots,$$

$$\cos\frac{2(n-1)\pi}{n} + i\sin\frac{2(n-1)\pi}{n}.$$

These are called the $n$th roots of unity.

Draw in the complex plane the unit circle (center 0 and radius 1), and indicate on this circle the location of the $n$th roots of unity.

2. Show that the factorization of $P_n(t)$ over $\mathbf{C}$ into a product of irreducible polynomials is given by

$$\prod_{k=0}^{n-1}\left(t - \left(\cos\frac{2k}{n} + i\sin\frac{2k}{n}\right)\right).$$

3. Factor over $\mathbf{C}$ the following polynomials: $P_2$, $P_3$, $P_4$, $P_6$, $P_8$. Express all coefficients in the form $a + bi$ where $a$, $b$ are real.

4. By a suitable pairing of linear factors over $\mathbf{C}$ of the polynomial $P_n(t)$, show that we can obtain a factorization of $P_n(t)$ as a product of irreducible polynomials over $\mathbf{R}$ as follows:

$$\text{for } n \text{ odd,} \quad P_n(t) = (t-1) \prod_{k=1}^{(n-1)/2} (t^2 - 2\cos(2k\pi/n)t + 1)$$

$$\text{for } n \text{ even,} \quad P_n(t) = (t-1)(t+1) \prod_{k=1}^{(n-2)/2} (t^2 - 2\cos(2k\pi/n)t + 1)$$

In particular, obtain a factorization of $P_3(t)$ and $P_5(t)$ over $\mathbf{R}$.

5. Show that, among the 8 distinct 8th roots of 1, there are

   (i) one square root of 1 other than 1 itself;

   (ii) two fourth roots of 1 which are not square roots of 1;

   (iii) four eighth roots of 1 which are not fourth roots of 1.

6. Which 12th roots of 1 are roots of lower degree? Make a table showing each 12th root of 1 and the minimum exponent to which it must be raised to yield 1.

7. A complex number $\zeta$ is a *primitive $n$th root of unity* if and only if $\zeta^n = 1$ but $\zeta^k \neq 1$ for each integer $k$ with $1 \leq k \leq n-1$. That is, $n$ is the smallest exponent to which $\zeta$ can be raised to yield 1.

   (a) Verify that there are 4 primitive 8th roots and 4 primitive 12th roots of unity.

   (b) Let $w$ be an $n$th root of unity. Show that there exists a positive number $m$ such that (i) $m \mid n$ and (ii) $w$ is a primitive $m$th root of unity.

8. For positive integer $n$, let $\zeta_n = \cos(2\pi/n) + i\sin(2\pi/n)$.

   (a) Show that the zeros of $P_n(t)$ are precisely the powers of $\zeta_n$.

   (b) Show that $1 + \zeta_n + \zeta_n^2 + \cdots + \zeta_n^{n-1} = 0$.

   (c) Prove that the primitive $n$th roots of 1 are the numbers $\zeta_n^a$ where $1 \leq a \leq n-1$ and $\gcd(a, n) = 1$. (Test this for specific $n$, such as $n = 12$.)

   (d) For $p$ a prime, show that every $p$th root of unity is primitive except 1 itself.

   (e) Show that, for $n \geq 3$, the number of primitive $n$th roots of unity is even.

9. The *n*th *cyclotomic polynomials* $Q_n(t)$ is the product of all the linear polynomials $(t - \zeta)$ where $\zeta$ ranges over the primitive *n*th roots of unity. Show that

$$P_n(t) = \prod_{d|n} Q_d(t)$$

where the product is taken over all the divisors *d* of *n*.

(It turns out that the coefficients of $Q_n(t)$ are integers and that these polynomials are irreducible over **Q**.)

10. Verify that

   (a) $Q_2(t) = t + 1$
   (b) $Q_4(t) = t^2 + 1$
   (c) $Q_6(t) = t^2 - t + 1$
   (d) $Q_8(t) = t_4 + 1$
   (e) $Q_p(t) = t^{p-1} + t^{p-2} + \cdots + t + 1$ when *p* is prime.

11. Compute the cyclotomic polynomials $Q_9$, $Q_{10}$, $Q_{12}$, $Q_{14}$, $Q_{15}$ and $Q_{16}$.

12.   (a) If *k* is even, show that $Q_{2k}(t) = Q_k(t^2)$.
   (b) If *k* is odd and exceeds 2, show that $Q_{2k}(t) = Q_k(-t)$.

13. Find the *n*th roots of $-1$ and factor the polynomial $t^n + 1$ over **C**, **R** and **Q**.

14. Let $c = r(\cos\theta + i\sin\theta)$ be a complex number. Find the *n*th roots of *c* and factor the polynomial $t^n - c$ over **C**. In particular, find the *n*th roots of 2 and factor $t^n - 2$ over **C**, **R** and **Q**.

15. Show that, if $k \mid n$, then $P_k(t) \mid P_n(t)$ over **C**, **R** and **Q**. Test the conjecture: if *k* and *m* are divisors of *n*, then $P_k(t)P_m(t) \mid P_n(t)$.

16. Let $\zeta$ be a primitive 5th root of unity. Show that $u = \zeta^2 + \zeta^3$ and $v = \zeta^1 + \zeta^4$ are zeros of the polynomial $t^2 + t - 1$.

17. Let $\zeta$ be a primitive 7th root of unity. Show that $u = \zeta^3 + \zeta^5 + \zeta^6$ and $v = \zeta + \zeta^2 + \zeta^4$ are zeros of the quadratic $t^2 + t + 2$.

18. Let *p* be any prime. We say that *a* is a *quadratic residue* modulo *p* if there is some number *x* such that $x^2 \equiv a \pmod{p}$. Otherwise, we call *a* a quadratic nonresidue.

   (a) Verify that the quadratic residues modulo 5 are 1 and 4 and those modulo 7 are 1, 2, 4.
   (b) Find the quadratic residues modulo 11, 13, 17.

(c) Let $p$ be a prime and let $\zeta$ be a primitive $p$th root of unity. Define

$$u = \Sigma\{\zeta^a \;:\; a \text{ is a quadratic nonresidue modulo } p, 1 \le a < p\}$$

$$v = \Sigma\{\zeta^a \;:\; a \text{ is a quadratic residue modulo } p, 1 \le a < p\}.$$

For the cases $p = 11$, 13, 17, show that $u$ and $v$ are the zeros of a quadratic polynomial of the form $t^2 + t + k$, where $k$ is an integer.

(This has significance for ruler and compasses constructions in the plane. Since such a construction locates points as intersections of straight lines and circles, the coordinates of constructible points are found by solving linear and quadratic equations whose coefficients belong to the smallest field containing the coordinates of points already given or constructed. The result of this exercise implies that the points $u$ and $v$ are constructible in the complex plane once the unit circle is given. In the case $p = 17$, it can be shown that one can solve a succession of quadratic equations, each with coefficients expressible in terms of the roots of its predecessors, until one finally obtains the root $\zeta_{17}$ itself. As a consequence there exists a ruler-and-compasses construction for a regular 17-gon in the plane. This result, due to Gauss, holds when 17 is replaced by any prime, such as 257, which is 1 plus a power of 2. It is this result which accounts for the original interest in cyclotomic polynomials.)

19. (a) Show that the coefficient of $x^n$ in the expansion of $f(x) = (x^6 + x^5 + x^4 + x^3 + x^2 + x)^2$ is the number of ways of rolling a total of $n$ with two distinguishable ordinary cubical dice.

The remainder of the exercise is devoted to assigning numbers to the faces of the two dice in such a way that the number of ways of rolling a total of $n$ is the same as for two ordinary dice.

(b) Verify the factorization

$$f(x) = x^2(x+1)^2(x^2+x+1)^2(x^2-x+1)^2.$$

(c) We wish to write $f(x) = g(x)h(x)$ where $g(x) \ne h(x)$ and

$$g(x) = x^a(x+1)^b(x^2+x+1)^c(x^2-x+1)^d$$

with $0 \le a, b, c, d \le 2$. If we can arrange that $g(x)$ and $h(x)$ are each equal to the sum of six not necessarily distinct terms of the form $x^k$, then the numbers $k$ can be used to label the faces of the dice. Argue that, for the desired labeling, we must have that $g(1) = h(1) = 6$ and $a = b = c = 1$. Determine $g(x)$ and $h(x)$.

(d) Explain how to construct *Sicherman* dice, two nonstandard cubical dice whose faces are assigned numbers in such a way that the number of ways of rolling $n$ is the same as for two ordinary dice.

## Explorations

**E.34. Degree of the Cyclotomic Polynomials.** Let $\phi(n)$ be the degree of the polynomial $Q_n(t)$. This is equal to the number of positive integers $a$ which are coprime with $n$, i.e. for which $1 \le a \le n$ and $\gcd(a, n) = 1$.

Let us compute $\phi(24)$. To find the numbers $a$ between 1 and 24 which are coprime with 24, we sieve out the multiples of those primes dividing 24. We remove 12 multiples of 2 and 8 multiples of 3. However, 4 multiples of 6 have been removed twice. Consequently, only $12 + 8 - 4 = 16$ numbers have been sieved out and there are 8 remaining: 1, 5, 7, 11, 13, 17, 19, 23. Thus $\phi(24) = 8$.

If there are three primes dividing $n$, the situation is more complex. Take the case $n = 60$ for instance. Its prime divisors are 2, 3, 5, and in crossing out multiples of these, we encounter each multiple of 6, 10 and 15 twice and each multiple of 30 three times. To fix our ideas, assign a weight 1 to each number to begin with. Every time we cross a number out, its weight is reduced by 1; every time it is restored, its weight is increased by 1. Our task is to cross out and restore in such a way that in the end, all multiples of 2, 3 and 5 have weight 0 and all other numbers have weight 1. Show that this can be achieved by the following procedures:

cross out all multiples of 2, 3 and 5 in turn
restore all multiples of two of the primes, i.e. of 6, 10, 15, in turn
cross out all multiples of three primes.

Thus we start out with 60 numbers, cross out $30 + 20 + 12$, restore $10 + 6 + 4$ and cross out 2, leaving 16 numbers with weight 1. What are these numbers?

To generalize this, we need the *Möbius function* $\mu(n)$:

$\mu(1) = 1$; $\mu(n) = 0$ if $n$ is divisible by any square exceeding 1;
$\mu(n) = (-1)^k$ if $n$ is the product of exactly $k$ distinct primes.
Prove the following:

(a) If $p$ is prime, then $\phi(p) = p - 1$; $\phi(p^k) = p^{k-1}(p-1)$

(b) $\phi(n) = \sum_{d|n}(n/d)\mu(d) = \sum_{d|n} d\mu(n/d)$ where the sum is taken over all the divisors of $n$. Verify this formula for $n = 24$ and $n = 60$.

(c) If $\gcd(m, n) = 1$, then

$$\phi(mn) = \phi(m)\phi(n)$$

$$\mu(mn) = \mu(m)\mu(n).$$

(d) Suppose $n$ is the product of powers of distinct primes $p$. Show that

$$\phi(n) = n \prod_{p|n} \left(1 - \frac{1}{p}\right).$$

(e) $n = \sum_{d|n} \phi(d)$. Verify this for $n = 12, 24, 60$.

The Möbius function has a role to play in determining the polynomials $Q_n$ in terms of $P_k$. Justify the equations

$$Q_6(t) = \frac{P_1(t)P_6(t)}{P_2(t)P_3(t)}$$

$$Q_9(t) = \frac{P_9(t)}{P_3(t)}.$$

Generalize to general $n$. Use the formula obtained to check the degree of $Q_n(t)$, and to obtain $Q_{pq}(t)$ when $p$ and $q$ are distinct primes.

**E.35. Irreducibility of the Cyclotomic Polynomials.** It has already been observed that the cyclotomic polynomials corresponding to primes are irreducible. Investigate irreducibility of $Q_n(t)$ when $n$ is composite.

**E.36. Coefficients of the Cyclotomic Polynomials.** If $n$ is equal to a prime power or twice a prime power, it is easy to check that the coefficients of $Q_n(t)$ are $+1$ or $-1$. Does this remain true when $n$ is the product of two primes? Is it true in general?

**E.37. Little Fermat Theorem Generalized.** The positive integers less than 24 which are coprime to 24 are 1, 5, 7, 11, 13, 17, 19, 23. Suppose we take any number coprime with 24, say 7, and multiply each number in this list by it, reducing the result modulo 24. The list of products in order is 7, 11, 1, 5, 19, 23, 13, 17. Thus, multiplication by 7 simply permutes the number in the list. To appreciate the significance of this, let us turn to the general situation.

Let $m$ be a positive integer, $k = \phi(m)$ (the function defined in Exploration **E.34**) and $a_1, a_2, \ldots, a_k$ be those positive integers less than $n$ which are coprime with $n$. Let $n$ be any integer coprime with $m$. Prove the following:

(i) for each $i$, there is an index $j$ for which $a_j \equiv na_i \pmod{m}$;

(ii) if $a_r \neq a_s$, then $na_r \neq na_s \pmod{m}$;

(iii) for each $i$, there is an index $j$ for which $a_i \equiv na_j \pmod{m}$;

(iv) if the numbers in the set $\{na_1, na_2, \ldots, na_k\}$ are each replaced by their remainders upon division by $m$, we get precisely the numbers in the set $\{a_1, a_2, \ldots, a_k\}$;

(v) $a_1 a_2 a_3 \cdots a_k \equiv n^k (a_1 a_2 \cdots a_k) \pmod{m}$.

Deduce that $n^{\phi(m)} \equiv 1 \pmod{m}$, and obtain the Little Fermat Theorem as formulated in Exploration **E.32** as a corollary.

## 3.6  Rational Functions

For any field **F**, the set of polynomials **F**[$t$] with the usual addition and multiplication is a commutative ring like **Z**. Just as **Z** can be embedded in the field **Q**, **F**[$t$] is contained in the field of *rational functions*. These are expressions of the form $p(t)/q(t)$ where $p$ and $q$ are polynomials over **F** and $q(t)$ is not the zero polynomial. Addition, subtraction, multiplication and division of rational functions are carried out as for rational numbers. An important difference between polynomials and rational functions is that it is not always possible to evaluate a rational function at every point of its underlying field **F**. A rational function $p(t)/q(t)$ assumes the value $p(c)/q(c)$ when $t = c$, except when $c$ is a zero of $q$, in which case the value is left undefined. For example, $(3t+2)/(t^2-4)$ is a rational function over **Q** whose value is undefined at 2 and $-2$.

## Exercises

1. Show that every rational function $f(t)$ can be written in the form $p(t) + g(t)$, where $p(t)$ is a polynomial (possibly 0) and $g(t)$ is a rational function the degree of whose numerator is smaller than the degree of the denominator.

2. Show that the rational function

$$\frac{at + b}{(t - m)(t - n)}$$

can be written in the form

$$\frac{A}{t - m} + \frac{B}{t - n}$$

for some constants $A$ and $B$ determined by

$$A(t - n) + B(t - m) = at + b.$$

Justify the assertion that the appropriate values of $A$ and $B$ can be found by making the substitutions $t = m$ and $t = n$. Find $A$ and $B$ by this method and check that the values obtained are correct.

3. Write the rational function

$$\frac{t + 14}{(t - 1)(t + 4)}$$

---

in the form

$$\frac{A}{t-1} + \frac{B}{t+4}$$

for suitable constants $A$ and $B$.

4.  (a) Let $p(t)/q(t)$ be a rational function for which the degree of the numerator is less than that of the denominator and the denominator $q(t)$ is the product of $k$ distinct linear factors $t - a_i$ ($1 \le i \le k$). Show that $p(t)/q(t)$ can be written in the form

$$\sum_{1 \le i \le k} \frac{c_i}{(t - a_i)}$$

for suitable constants $c_i$.

(b) Prove that, in (a), the $c_i$ are uniquely determined by the formula

$$c_i = p(a_i)/q'(a_i).$$

(c) Deduce from (a) a representation of that polynomial $f(t)$ of degree less than $k$ for which $f(a_i) = b_i$, where, for $1 \le i \le k$, $b_i$ are assigned values.

5.  In each case, use a partial fraction representation of $a_k$ to determine $\Sigma\{a_k : k = 2, 3, \ldots, n\}$, where $a_k$ is equal to

(a) $1/[k(k-1)]$

(b) $1/[k^3 - k]$

(c) $\dfrac{1}{(kx+1)((k-1)x+1)}$.

6.  The partial fraction decomposition can be extended to the situation in which $q(t)$ is a product of irreducible factors some of whose degrees are greater than 1. In this case, for each such factor $v(t)$, there is a summand of the form $u(t)/v(t)$ where $\deg u(t) < \deg v(t)$.

(a) Verify that $t^4 - 108t + 243 = (t-3)^2(t^2 + 6t + 27)$.

(b) The rational function $\dfrac{t^3 + t^2 + 15t - 27}{t^4 - 108t + 243}$ can be written in the form

$$\frac{A}{t-3} + \frac{B}{(t-3)^2} + \frac{Ct + D}{t^2 + 6t + 27}.$$

Determine the constants $A$, $B$, $C$, $D$ and check your answer.

7.  Express the rational function

$$f(t) = \frac{7t^2 - 2t + 3}{t^4 - 3t^3 + t^2 - 3}$$

as a sum of partial fractions, one associated with each real irreducible factor of the denominator.

8. For $2 \leq n \leq 5$, write out the partial fraction decomposition of $1/(t^n - 1)$ over $Q$ and $C$. What is the decomposition over $C$ for an arbitrary positive integer $n$?

# Explorations

**E.38. Principal Parts and Residues.** What happens in the partial fraction decomposition if the denominator $q(t)$ of a rational function $p(t)/q(t)$ has repeated irreducible factors? Suppose $q(t) = (t-c)^k r(t)$ where $r(c) \neq 0$. Then $p(t)/q(t)$ can be written in the form

$$\frac{u(t)}{(t-c)^k} + \frac{s(t)}{r(t)}$$

where $\deg u(t) < k$. To see this, note that by the Euclidean algorithm, there are polynomials $g(t)$ and $h(t)$ such that

$$1 = g(t)(t-c)^k + h(t)r(t).$$

(Consult Exercise 1.6.2 for the numerical case and Exploration E.20.) Dividing by $q(t)$ and doing some manipulating yields the required representation.

By using the Taylor expansion of $u(t)$ in terms of $t - c$, show that, for suitable coefficients $a_i$,

$$\frac{u(t)}{(t-c)^k} = \frac{a_k}{(t-c)^k} + \frac{a_{k-1}}{(t-c)^{k-1}} + \cdots + \frac{a_1}{t-c}.$$

This is called the *principal part* of the rational function at $c$. For $t$ close to $c$, the numerical behaviour of the rational function is approximated by that of its principal part. The coefficient $a_1$ of $(t-c)^{-1}$ is called the *residue* at $c$. Both the principal part and the residue play an important role in the theory of functions of a complex variable; the residue can be used to compute definite integrals which often cannot be evaluated by elementary means.

In the special case that $k = 1$,

$$q(t) = (t-c)(q'(t) + (t-c)u(t))$$

for some polynomial $u(t)$, where $q'(c) \neq 0$. Show that the residue of $p(t)/q(t)$ at $c$ is equal to $p(c)/q'(c)$.

More generally, if $q(t) = v(t)^k r(t)$, where $v(t)$ is irreducible and gcd $(v(t), r(t)) = 1$, we can write

$$\frac{p(t)}{q(t)} = \frac{w(t)}{(v(t))^k} + \frac{s(t)}{r(t)} \quad \text{where } \deg w < k \deg v.$$

Show, by dividing $w(t)$ by $v(t)$, that there are polynomials $a_i(t)$ $(1 \leq i \leq k)$, for which $\deg a_i < \deg v$ and

$$w(t) = a_k(t) + a_{k-1}(t)v(t) + a_{k-2}(t)v(t)^2 + \cdots + a_1(t)v(t)^{k-1},$$

so that

$$\frac{w(t)}{v(t)^k} = \frac{a_k(t)}{v(t)^k} + \frac{a_{k-1}(t)}{v(t)^{k-1}} + \cdots + \frac{a_1(t)}{v(t)}.$$

## 3.7 Problems on Factorization

1. Factor over $\mathbf{Z}$:

   (a) $4x^4 + 1$

   (b) $x^4 - 20x^2 + 4$

   (c) $x^2y^2 - (x + y)xy + x + y - 1$

   (d) $x^2y^2z^2 - (x + y + z)xyz + xy + yz + zx - 1$

   (e) $(b - c)^3 + (c - a)^3 + (a - b)^3$

   (f) $a^2(b + c) + b^2(c + a) + abc - c^3$

   (g) $bc(b + c) + ca(c + a) + ab(a + b) + 2abc$

   (h) $bc(b + c) + ca(c - a) - ab(a + b)$

   (i) $x^{10} + x^5 + 1$

   (j) $2x^3 + 6xy^2 + z^3 - 3x^2z + 3y^2z$

   (k) $x^2+y^2+z^2+x^2y^2z^2-2x^2yz-2xy^2z-2xyz^2+2xy+2xz+2yz-4$

   (l) $x^2(y^3 - z^3) + y^2(z^3 - x^3) + z^2(x^3 - y^3)$

   (m) $(a + b + c)^4 - (b + c)^4 - (c + a)^4 - (a + b)^4 + a^4 + b^4 + c^4$

   (n) $(bc + ca + ab)^3 - b^3c^3 - c^3a^3 - a^3b^3$

   (o) $2(bc + ca + ab)^2 - a^2(b + c)^2 - b^2(c + a)^2 - c^2(a + b)^2$

   (p) $6(x^5 + y^5 + z^5) - 5(x^2 + y^2 + z^2)(x^3 + y^3 + z^3)$

   (q) $2(x^4 + y^4 + z^4 + w^4) - (x^2 + y^2 + z^2 + w^2)^2 + 8xyzw$

   (r) $x^3y^3 + y^3z^3 + z^3x^3 - x^4yz - xy^4z - xyz^4$

   (s) $(x^2 + y^2 + z^2)(x + y + z)(x + y - z)(y + z - x)(z + x - y) - 8x^2y^2z^2$

   (t) $x(y + z)^2 + y(z + x)^2 + z(x + y)^2 - 4xyz$

   (u) $a^4 + b^4 - c^4 - 2a^2b^2 + 4abc^2$

2. Show that the two equations

$$x^4 - x^3 + x^2 + 2x - 6 = 0$$

$$x^4 + x^3 + 3x^2 + 4x + 6 = 0$$

   have a pair of complex roots in common.

3. Find a polynomial $p(x)$ such that $[p(x)]^5 - x$ is divisible by $(x - 1)$ $(x - 2)(x - 3)$.

4. Find values of $a$ and $b$ which will make

$$(ax + b)(x^5 + 1) - (5x + 1)$$

divisible by $x^2 + 1$. Check your answer.

5. (a) If $ax^3 + bx + c$, where $a \neq 0$, $c \neq 0$, has a factor of the form

$$x^2 + px + 1,$$

show that

$$a^2 - c^2 = ab.$$

(b) In this case, prove that $ax^3 + bx + c$ and $cx^3 + bx^2 + a$ have a common quadratic factor.

6. If the quadratic function $3x^2 + 2pxy + 2y^2 + 2ax - 4y + 1$ can be resolved into factors linear in $x$ and $y$, prove that $p$ must be a root of the equation

$$p^2 + 4ap + 2a^2 + 6 = 0.$$

7. A monic cubic polynomial over $\mathbf{Z}$ has the property that one of its zeros is the product of the other two. Show that it must be reducible over $\mathbf{Z}$.

8. For what integer $a$ does $x^2 - x + a$ divide $x^{13} + x + 90$?

9. Show that

$$b^2(a - b)(c - b)\{(a - b)^2 + (c - b)^2\} - ab^2c(a^2 + c^2) + b^5(a - b + c)$$

is the cube of a polynomial.

10. Determine all values of the parameters $a$ and $b$ for which the polynomial

$$x^4 + (2a + 1)x^3 + (a - 1)^2x^2 + bx + 4$$

can be factored into a product of two monic quadratic polynomials $p(x)$ and $q(x)$ such that the equation $q(x) = 0$ has two different roots $r$ and $s$ with $p(r) = s$ and $p(s) = r$.

11. Prove that $2^{12}$ divides

$$3(81^{n+1}) + (16n - 54)9^{n+1} - 320n^2 - 144n + 243.$$

12. Let $f(x) = x^n + x^a + 1$ be a polynomial over $\mathbf{Z}_2$ such that $0 < a < n$. Show that, if $f(x)$ has any repeated factors, then $f(x)$ is a perfect square.

13. (a) Determine a necessary and sufficient condition on the integers $b$ and $c$ such that both the polynomials

$$t^2 + bt + c \quad \text{and} \quad t^2 + bt + c + 1$$

are reducible over $\mathbf{Z}$.

(b) Determine integers $b$ and $c$ such that both the polynomials

$$3t^2 + bt + c \quad \text{and} \quad 3t^2 + bt + c + 1$$

are reducible over $\mathbf{Z}$.

14. Let $n$ be a positive integer, $p$ be a prime, $q = p^n$ and $m = (q^p - 1)/(q - 1)$. Prove that $x^q - x - 1$ divides $x^m - 1$ over $\mathbf{Z}_p$ if and only if $n = 1$.

15. Factor over $\mathbf{Z}$: $(x^4 - 1)^4 - x - 1$.

16. Prove that the polynomial $(x + y)^n - x^n - y^n$ is divisible by

(a) $x^2 + xy + y^2$ where $n \equiv 5 \pmod 6$
(b) $(x^2 + xy + y^2)^2$ when $n \equiv 1 \pmod 6$
(c) $x^3 + 2x^2y + 2xy^2 + y^3$ when $n$ is a prime exceeding 3.

17. For each positive integer $k$, show that $t^5 + 1$ is a factor of

$$(t^4 - 1)(t^3 - t^2 + t - 1)^k + (t + 1)t^{4k-1}.$$

18. If a positive integer $m$ has a prime factor greater than 3, show that $4^m - 2^m + 1$ is composite.

19. Determine all values of the positive integer $n$ for which $4^n + n^4$ is prime.

20. Suppose that $m$ is a positive odd integer exceeding 3. Prove that

$$\frac{2^{2m} + 1}{5}$$

is a composite integer.

21. Let $f(x, y)$ be a symmetric polynomial. Show that, if $(x - y)$ is a factor of $f(x, y)$, then so is $(x - y)^2$.

22. Prove that, for any positive integer $n$ exceeding 1, the equation $1 + 2x + 3x^2 + \cdots + nx^{n-1} = n^2$ has a rational root between 1 and 2.

## 3.8   Other Problems

1. Suppose that $x^2 = yx - 1$ and $y^2 = 1 - y$. Show that $x^5 = 1$ but that $x \neq 1$.

2. Is
$$\frac{a^3 + b^3}{a^3 + (a-b)^3} = \frac{a+b}{a + (a-b)}$$
correct?

3. For which integers $x$ is $x^4 + x^3 + x^2 + x + 1$ a perfect square?

4. Let $a$ and $b$ be rational. Can $t^5 - t - 1$ and $t^2 + at + b$ ever have a common complex zero?

5. Find $a$ and $b$ so that the equations $x^3 + ax^2 + 11x + 6 = 0$ and $x^3 + bx^2 + 14x + 8 = 0$ may have two roots in common.

6. Suppose that $a_1, a_2, \ldots, a_{2n}$ are distinct integers such that
$$(x - a_1)(x - a_2)(x - a_3) \cdots (x - a_{2n}) + (-1)^{n-1}(n!)^2 = 0$$
has an integer root $r$. Show that
$$2nr = a_1 + a_2 + a_3 + \cdots + a_{2n}.$$

7. Find all integer values of $m$ for which the polynomial
$$t^3 - mt^2 - mt - (m^2 + 1)$$
has an integer zero.

8. Prove that, if $f(t)$ is a polynomial with integer coefficients and there exists a positive integer $k$ such that none of the integers $f(1), f(2), \ldots, f(k)$ is divisible by $k$, then $f(t)$ has no integer zeros.

9. Let $f$ be a monic polynomial over $\mathbf{Z}$ and suppose that there are four distinct integers $a$, $b$, $c$, $d$ for which
$$f(a) = f(b) = f(c) = f(d) = 12.$$
Show that there is no integer $k$ for which $f(k) = 25$.

10. Let the zeros $m$, $n$, $k$ of $t^3 + at + b$ be rational. Prove that the zeros of $mt^2 + nt + k$ are rational.

11. Find constants $a$, $b$, $c$, $d$, $p$, $q$ for which
$$a(x - p)^2 + b(x - q)^2 = 5x^2 + 8x + 14$$
$$c(x - p)^2 + d(x - q)^2 = x^2 + 10x + 7.$$

12. Determine the range of $w(w+x)(w+y)(w+z)$ where $x$, $y$, $z$, $w$ are real numbers which satisfy

$$x + y + z + w = x^7 + y^7 + z^7 + w^7 = 0.$$

13. If $x + y + z = 0$, prove that

$$\frac{x^5 + y^5 + z^5}{5} = \frac{x^3 + y^3 + z^3}{3} \cdot \frac{x^2 + y^2 + z^2}{2}.$$

14. If

$$by/z + cz/y = a$$
$$cz/x + ax/z = b$$
$$ax/y + by/x = c,$$

show that

   (a) $x^{-3} + y^{-3} + z^{-3} + x^{-1}y^{-1}z^{-1} = 0$
   (b) $a^3x^3 + b^3y^3 + c^3z^3 + abcxyz = 0$
   (c) $a^3 + b^3 + c^3 = 5abc$.

15. If $x + y + z = 0$, show that

$$\left\{ \frac{y-z}{x} + \frac{z-x}{y} + \frac{x-y}{z} \right\} \left\{ \frac{x}{y-z} + \frac{y}{z-x} + \frac{z}{x-y} \right\} = 9.$$

16. If $x(1 - mzy/x^3) = y(1 - mxz/y^3) = z(1 - myx/z^3)$ with $x, y, z$ unequal, prove that each quantity is equal to $x + y + z - m$.

17. Suppose $a$, $b$, $c$, $d$ are integers, that $r$ is a zero of $P(x) = x^3 + ax^2 + bx - 1$, $r+1$ is a zero of $y^3 + cy^2 + dy + 1$, and that $P(x)$ is irreducible over $\mathbf{Q}$. Express another zero $s$ of $P(x)$ as a function of $r$ which does not explicitly involve $a$, $b$, $c$ or $d$.

18. Let $n = 2m$, where $m$ is an odd integer greater than 1. Let $\theta = \cos(2\pi/n) + i\sin(2\pi/n)$. Express $(1 - \theta)^{-1}$ explicitly as a polynomial in $\theta$, i.e. $a_k\theta^k + a_{k-1}\theta^{k-1} + \cdots + a_1\theta + a_0$, with integer coefficients $a_i$.

19. Let $F$ be a finite field having an odd number $m$ of elements. Let $p(x)$ and $q(x)$ be irreducible polynomials over $F$ of the form $x^2 + bx + c$.

   (a) Prove that $q(x) = p(x + h) - k$ for some $h$ and $k$ in $F$.
   (b) For how many elements $k$ in $F$ is $p(x) + k$ irreducible?

20. Show that the product of four consecutive terms of an arithmetic progression of integers plus the fourth power of the common difference is a perfect square. Give a nontrivial example in which this quantity is actually a perfect fourth power.

21. Let $f$, $g$, $h$ be polynomials over $\mathbf{R}$ or $\mathbf{C}$. Suppose that $f = g^m h$, where $g$ does not divide $h$.

    (a) Show that $g^{m-1}$ divides $f'$, the derivative of $f$.

    (b) The example $g = (t-1)^3$, $f = (t-1)^3(t^3-1)^2$ shows that it is possible for $g^m$ to divide $f'$. Is it possible for $g^{m+1}$ to divide $f'$?

22. Show that, for every positive integer $n$, there is a polynomial $f(x)$ of degree $n$ and a related polynomial $g(x)$ for which

$$[f(x)]^2 - 1 = (x^2 - 1)[g(x)]^2.$$

23. Put $\{m\}! = (x^m - 1)(x^{m-1} - 1) \cdots (x - 1)$ for $m \geq 1$ and $\{0\}! = 1$.
Show that

$$\frac{\{2m\}!\{2n\}!}{\{m+n\}!\{m\}!\{n\}!}$$

is a polynomial with rational coefficients.

24. If $x + y + z = xyz$, show that

$$\frac{2x}{1-x^2} + \frac{2y}{1-y^2} + \frac{2z}{1-z^2} = \frac{2x}{1-x^2} \cdot \frac{2y}{1-y^2} \cdot \frac{2z}{1-z^2}.$$

25. Suppose that

$$x_2 x_3 + y_2 y_3 = x_3 x_1 + y_3 y_1 = x_1 x_2 + y_1 y_2 = 1$$

and

$$d_1 = x_2 y_3 - x_3 y_2$$
$$d_2 = x_3 y_1 - x_1 y_3$$
$$d_3 = x_1 y_2 - x_2 y_1.$$

Show that $d_1 + d_2 + d_3 = d_1 d_2 d_3$.

26. (a) For which integers $a$, $b$ does the quadratic $t^2 - at + b$ have a zero which is a root of unity?

    (b) Show that, if $t^2 - (a^2 - 2b)t + b^2$ has a zero which is a root of unity, then so does $t^2 - at + b$ (where $a, b \in \mathbf{Z}$).

27. Suppose that $a$, $b$, $c$ are nonzero integers and $u$, $v$ are roots of unity for which $u^2 \neq 1$, $v^2 \neq 1$, and $au + bv + c = 0$. Show that $|a| = |b| = |c|$.

28. Show that $x = \sin(\pi/14)$ is a root of the equation

$$8x^3 - 4x^2 - 4x + 1 = 0.$$

29. A regular polygon of seven sides is inscribed in a circle of unit radius. Prove that the length of a side of the polygon is a root of the equation

$$x^6 - 7x^4 + 14x^2 - 7 = 0$$

and state the geometrical significance of the other roots.

30. Let $f(n) = n^3 + 396n^2 - 111n + 38$. Prove that $f(n) \equiv 0 \pmod{3^m}$ has precisely nine solutions modulo $3^m$ for all integers $m \geq 5$.

31. Show that it is not possible to find polynomials $f(t)$, $g(t)$ over **C** which are coprime such that

$$\frac{f(t+1)}{g(t+1)} - \frac{f(t)}{g(t)} = \frac{1}{t}.$$

# Hints

## Chapter 3

1.5. Suppose $t^2 + 1 = (at + b)(ct + d)$. Derive a contradiction if $a$, $b$, $c$, $d$ are real.

1.8. Reducibility over **Z** requires care. If $at^2 + bt + c$ has rational zeros, it can be written in the form $(u/v)(pt + q)(rt + s)$, where $\gcd(u, v) = \gcd(p, q) = \gcd(r, s) = 1$. Must $v$ be equal to 1?

1.16. Suppose $h(t)$ is reducible and can be written as the product of $\Sigma a_i t^i$ and $\Sigma b_i t^i$. Consider the two cases: (1) $p$ divides only one of $a_0$ and $b_0$; (2) $p$ divides each of $a_0$ and $b_0$. In case (1), compare with the proof of the Eisenstein Criterion. In case (2), show by induction that $p$ must divide $a_k$ and $b_k$ for $0 \leq k \leq m$; look at the coefficients of $t^{2k}$ and $t^k$.

1.21. Consider the greatest common divisor of $f(t)$ and $g(t)$. Consult Exercise 1.6.2 and Exploration E.20.

2.7. (d) Write 11 as the sum of integers of opposite parity whose product is $-24 \cdot 28 = -2^5 \cdot 3 \cdot 7$.

(f) Apply the Eisenstein Criterion.

(i) Factor $3t^4 - 2t^3 - t^2$.

(j) (k) Look for an integer zero.

(n) Eliminate the possibility of a linear factor. Try the method of undetermined coefficients.

(o) Assume a factorization $(t^2 + at + b)(t^3 + ct^2 + dt + e)$. Show that $b$ and $e$ have opposite parity and that, in fact, $b$ is even and $e$ is odd. This reduces the pair $(b, e)$ to two possibilities, up to sign.

(p) Write as a difference of squares.

2.8. Let $g = u + v$ and $h$ be two polynomials with $0 \leq \deg v < \deg u$. Show that $gh$ cannot be homogeneous. Recall Exercise 1.7.4(d).

2.10. (a) (b) Write as the sum of two polynomials which have a factor in common.

(c) (e) (f) Use the Factor Theorem.

(d) If reducible, there must be a homogeneous linear factor. Try a symmetric one.

(g) If reducible, there must be a homogeneous linear factor. Then for some $a$, $b$, the polynomial must vanish under the substitution $z = ax + by$.

2.11. Express $p_n(x, y, z)$ as a polynomial in $z$. Use long division by $z^2 - (x + y)z + xy$; alternatively, experiment with some numerical values of $x$ and $y$ to get a handle on a possible factorization.

2.12. Use the Factor Theorem to find linear factors. What is the factorization when $z = 0$?

2.13. What is the significance of the divisibility of the leading coefficient by $m$?

2.14. Test small moduli.

3.4. (b) If $a/b$ is a root, then the long division algorithm in which $q(t)$ is divided by $bt - a$ should yield integer coefficients at every stage in the quotient.

4.12. Note that $2t^3 + t + 3 = (t + 1)(2t^2 - 2t + 3)$.

5.12. (a) Show that $\zeta$ is a primitive $2k$th root of unity iff $\zeta^2$ is a primitive $k$th root.

(b) Show that $\zeta$ is a primitive $k$th root iff $-\zeta$ is a primitive $2k$th root.

6.4. Use induction on $k$. Choose $c_1$ to express

$$\frac{p(t)}{q(t)} - \frac{c_1}{t - a_1}$$

in the form $p_1(t)/q_1(t)$ where $q_1(t)$ has the zeros $a_2, \ldots, a_k$.

6.7. We seek a representation of the form

$$\frac{A}{t} + \frac{B}{t - 3} + \frac{Ct + D}{t^2 + 1}.$$

6.8. For example

$$\frac{1}{t^5 - 1} = \frac{1}{t - 1} + \frac{Bt^3 + Ct^2 + Dt + E}{t^4 + t^3 + t^2 + t + 1},$$

for suitable $A, B, C, D, E$.

7.1. (c) Rearrange terms; collect terms with factor $x + y$.

(d) Set $z = 1$ and compare with (c). Alternatively, set $xy = 1$.

(e) Set $b = c$.

(g) Set $a + b = 0$.

(i) Set $x$ equal to an imaginary cube root of unity.

(j) Try a substitution $z = ax$.

(k) Express as a difference of squares.

(m) Set $c = 0$, $a + b + c = 0$. Four linear factors are easily found.

(p) Factor the polynomial determined by setting $z = 0$. Alternatively, express in terms of the elementary symmetric functions $x + y + z$, $xy + yz + zx$, $xyz$ (see Exercise 2.2.13).

(q) The polynomial vanishes when $(x, y, z, w) = (1, 1, 1, 1)$. What possible linear factors does this suggest?

(s) Write as a polynomial in $z^2$. Is there a substitution for $z^2$ as a function of $x$ and $y$ which will make the polynomial vanish?

(u) Factor as a difference of squares.

7.3. What should $p(1)$, $p(2)$, $p(3)$ be?

7.4. The polynomial should vanish when $x = i$.

7.5. (a) The zeros of the quadratic are reciprocals; the sum of the zeros of the cubic is 0. Use Exercise 1.5.6.

(b) The zeros of $cx^3 + bx^2 + a$ are the reciprocals of the zeros of $ax^3 + bx + c$.

7.6. Use the method of undetermined coefficients. It can be arranged that the constant term in each linear factor is 1.

7.8. $x^2 - x + a$ must divide $x^{13} + x + 90$ when $x = 0, 1$. The case $a = -2$ can be eliminated almost at once; factor the quadratic.

7.9. To get an idea of what the cube root might be, look at the situation in which any one of the variables is set equal to 0. Make a conjecture and check it out using the Factor Theorem. Can you manipulate the polynomial directly to reveal that it is a cube?

7.10. Let $u = r + s$ and $v = rs$. Express the quadratic factors in terms of $u$ and $v$.

7.11. Let $x = 9^{n+1}$ and try to factor the quadratic in $x$. Can $320n^2 + 144n - 243$ be factored over $\mathbf{Z}$?

7.12. Remember, $1 + 1 = 0$. Since a repeated factor of a polynomial divides its derivative, examine the greatest common divisor of $x^n + x^a + 1$ and its derivative.

7.14. The following facts will be useful to keep in mind:

(1) $pr = 0$ for each $r$ in $\mathbf{Z}_p$;

(2) $x^q - x - 1 \mid x^m - 1$ iff $x^m - 1 \equiv 0 \pmod{x^q - x - 1}$. Thus setting $x^q = x + 1$ should lead to $x^m - 1 \equiv 0$;

(3) $(y + z)^p = y^p + z^p$;

(4) $x^m = x \cdot x^q \cdot \cdots \cdot x^{q^{p-1}}$ and $x^q = x = 1$, $x^q = (x+1)^q = x^q + 1 = x + 2, \ldots$;

(5) By Fermat's Theorem, $x^p - x$ has $p$ zeros $0, 1, 2, \ldots, p - 1$. What does this say about its factorization over $\mathbf{Z}_p$?

7.16. Let $x = ty$. Check when the zeros of $t^2 + t + 1$ are zeros of $(t+1)^n - t^n - 1$ and its derivative.

7.21. If $f(x, y) = (x - y)q(x, y)$, what is the relationship between $q(x, y)$ and $q(y, x)$?

7.22. Look at the cases $n = 2, 3, 4$ and make a conjecture.

8.3. Observe that for any integer $x$, the left side lies between $(x^2 + x/2)^2$ and $(x^2 + x/2 + 1)^2$.

8.6. The integers $r - a_i$ are nonzero and distinct. Arrange them in ascending order of absolute value and examine their product.

8.7. If $t$ is an integer root, then $m^2 + (t^2 + t)m - (t^3 - 1) = 0$ is solvable for integer $m$. The discriminant $(t^2 + 3t - 4)^2 + (24t - 20)$ must be a perfect square not strictly between $(t^2 + 3t - 3)^2$ and $(t^2 + 3t - 5)^2$.

8.8. If $m$ is an integer zero, then $f(m - ck) \equiv 0 \pmod{k}$ for each integer $c$.

8.9. Identify four linear factors of $f(t) - 12$. What can be said about the factorization of the integer $f(k) - 12$?

8.10. What can be said about $m + n + k$?

8.11. Solve the system for $(x - p)^2$ and $(x - q)^2$; use a discriminant condition. Compare coefficients. Guess.

8.12. Use Exercise 2.12 to factor $0 = -w^7 - x^7 - y^7 - z^7$.

8.13. Use Exercise 2.12. Note that $2(x^2+y^2+z^2+xy+yz+zx) = x^2+y^2+z^2$. Alternatively, set $z = -(x+y)$ and simplify the three terms.

8.14. (a) Use the third equation to eliminate $c$ from the other two.

8.16. If $u$ is the desired quantity, then $x^3 - y^3 + mz(x-y) = u(x^2 - y^2)$. Note that $x \neq y$. Find a similar equation for $y$ and $z$.

8.17. Any polynomial for which $r$ is a zero must be divisible by $P(x)$. Use this fact to obtain an equation relating $a$ and $b$.

8.19. (a) The significance of the condition that $m$ is odd is that $u + u \neq 0$ for any element $u$ in $\mathbf{F}$. Consequently the equation $2x = k$ is always solvable for $x$. Choose $h$ so that $p(x+h)$ and $q(x)$ have the same linear coefficient, and then choose $k$.

(b) The reducible quadratics are easy to count—one for each pair of zeros in $\mathbf{F}$. Use (a) to show that the number of irreducibles is the same for each linear coefficient of the quadratic.

8.21. Write $f$ and $g$ as a product of irreducibles and examine the exponents.

8.22. The first few cases are $(f(x), g(x)) = (x, 1)$, $(2x^2 - 1, 2x)$, $(4x^3 - 3x, 4x^2 - 1)$. Make a conjecture about $f(x)$. Consult Exercise 1.3.8. Recall that $\cos^2 n\theta - 1 = -\sin^2 n\theta$.

8.23. Write $\{m\}!$ as a product of cyclotomic polynomials. What is the exponent of $Q_d(x)$?

8.24. $2x/(1-x^2)$ suggests a substitution $x = \tan u$, etc. Look at $\tan(u + \cdots)$. Interpret the condition and the conclusion. Alternatively, put the left side over a common denominator and express the numerator as a polynomial in the elementary symmetric functions.

8.25. Multiply each $d_i$ by two suitable expressions equal to 1 and add.

8.26. (a) If $u$ is an imaginary zero of the quadratic, so is $\bar{u} = u^{-1}$. What is the sum and product of the zeros?

8.27. A second equation relating $u$ and $v$ is obtained by taking complex conjugates. Note that $\bar{u} = u^{-1}$, $\bar{v} = v^{-1}$.

8.28. Use de Moivre's Theorem and the fact that $\sin \pi/2 = 1$ to derive the required equation.

# 4

# Equations

## 4.1 Simultaneous Equations in Two or Three Unknowns

A man standing on a railway bridge 32 meters long observes a train coming towards the bridge at 120 km per hour. It turns out that whichever way he runs at his top speed of 15 km per hour, he will reach the end of the bridge at the same time as the train. How far from the end of the bridge closest to the train is he?

This typical textbook problem can be solved by the introduction of variables and setting up equations which relate them. For example, if we let $x$ be the distance to the end of the bridge nearest the train and $y$ the distance to the other end in meters, we can derive the *simultaneous system* of equations

$$x + y = 32 \qquad y - x = 4$$

both of which must be satisfied by $x$ and $y$.

This system can be solved in a straightforward way. Using one equation, we can solve for $y$ in terms of $x$ and use this to obtain from the other equation a single equation in $x$. However, some problems involve equations of higher degree and it is not so easy to untangle the variables. In this section, a few techniques for handling a simultaneous system will be reviewed. Unless otherwise specified, $x$, $y$, $z$ will denote variables and the other letters constants.

## Exercises

1. Solve the system of equations

$$x + y = 16$$
$$x + z = 20$$
$$y + z = 22.$$

2. Consider the system

$$a_1 x + b_1 y + c_1 z = 0$$

$$a_2 x + b_2 y + c_2 z = 0.$$

Suppose that neither equation is a constant multiple of the other.

(a) Manipulate the first equation multiplied by $c_2$ and the second multiplied by $c_1$ to obtain a single equation relating the variables $x$ and $y$ alone.

(b) Show that, for any solution $x$, $y$, $z$,

$$x : y : z = (b_1 c_2 - b_2 c_1) : (c_1 a_2 - a_1 c_2) : (a_1 b_2 - a_2 b_1).$$

(c) Prove that the set of solutions of the system is given by

$$x = (b_1 c_2 - b_2 c_1)k$$

$$y = (c_1 a_2 - a_1 c_2)k$$

$$z = (a_1 b_2 - a_2 b_1)k$$

where $k$ is any number.

For convenience we can use the display

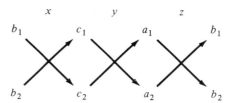

as a mnemonic. In vectorial terms, the solution set of the two equations consists of the set of vectors $(x, y, z)$ which are orthogonal to the vectors $(a_1, b_1, c_1)$ and $(a_2, b_2, c_2)$. Such vectors must be multiples of the cross product of these two vectors.

3. Solve the system

$$
\begin{aligned}
x + 2y + 3z &= 0 \\
x - y + z &= 0 \\
x^2 + y^2 + z^2 &= 152 \ .
\end{aligned}
$$

4. Consider the pair of equations, where $ap \neq 0$:

$$at^2 + bt + c = 0$$

$$pt^2 + qt + r = 0.$$

(a) Show that, if the equations have a root $u$ in common, then

$$u^2 : u : 1 = (br - cq) : (cp - ar) : (aq - bp).$$

(b) Prove that the equations have a root in common if and only if

$$(cp - ar)^2 = (br - cq)(aq - bp).$$

5. (a) Find the conditions on $a$ and $b$ that

$$t^2 - 2t + a = 0$$

$$t^2 - 8t + b = 0$$

have a root in common.

(b) Determine a particular numerical pair $(a, b)$ for which the equations in (a) have a root in common, and check your result by showing that the two quadratics have a common factor.

6. Suppose that $xyz \neq 0$ and that

$$p = x - yz/x, \quad q = y - zx/y, \quad r = z - xy/z, \quad a/x + b/y + c/z = 0.$$

(a) Prove that

$$0 = pxy + qyz + rzx = q/x + r/y + p/z$$

$$0 = pxz + qxy + ryz = r/x + p/y + q/z.$$

(b) Eliminate $x, y, z$ from the given system of equations to obtain an equation in the remaining variables.

7. Show that the solutions of the system

$$x + y + z = a$$
$$xy + yz + zx = b$$
$$xyz = c$$

are given by the roots of the cubic equation

$$t^3 - at^2 + bt - c = 0$$

taken in some order.

8. Solve the system
$$x + y + z = 12$$
$$xy + yz + zx = 41$$
$$xyz = 42 \ .$$

9. By expressing $x^2 + y^2 + z^2$ and $x^3 + y^3 + z^3$ in terms of the elementary symmetric functions (Exercise 1.5.8), find a system of the type in Exercise 7 equivalent to each of the systems and thence solve it:

(a)
$$\begin{aligned} x + y + z &= 15 \\ x^2 + y^2 + z^2 &= 83 \\ x^3 + y^3 + z^3 &= 495 \end{aligned}$$

(b)
$$\begin{aligned} x + y + z &= 2 \\ x^2 + y^2 + z^2 &= 14 \\ xyz &= -6 \ . \end{aligned}$$

10. By factoring the left side of the first equation, or otherwise, solve the system
$$\begin{aligned} x^2 + 3xy - 4y^2 &= 0 \\ x^2 + xy + y^2 &= 7y^3 - 4. \end{aligned}$$

11. (a) Suppose that $r$, $a_i$, $b_i$ $(i = 1, 2, \ldots, n)$ are numbers for which $a_i/b_i = r$ for each $i$. Let $a = \Sigma a_i$, $b = \Sigma b_i$. Show that $a/b = r$.

   (b) If
$$\frac{x}{11 - y} = \frac{y}{6 - z} = \frac{z}{7 - x} = 2$$
   use (a) to determine $x + y + z$, and solve this equation for $x$, $y$, $z$.

## 4.2  Surd Equations

In solving an equation in a single variable $x$, we begin with the assumption that $x$ satisfies the equation and deduce that $x$ must be one of a number of possibilities. If a polynomial equation is involved, we are content to accept all of these as valid solutions. However, strictly speaking, the solution of the equation is not properly complete until the possible solutions have been checked by substitution into the equation.

While, for polynomial equations, all of the putative solutions turn out to be valid, for surd equations, more care is needed. The manipulations for solving surd equations often lead to more general equations, not equivalent to their predecessors, so that in effect information about the solution is lost. The result is that only some of the values turned up by the analysis may satisfy the original equation. The remaining values which do not satisfy the equation are said to be *extraneous*. This phenomenon will be illustrated in the exercises.

For surds involving real numbers, $\sqrt[k]{x}$ denotes the unique real number $y$ for which $y^k = x$ when $k$ is an odd integer. However, when $k$ is an even integer, $\sqrt[k]{x}$ is defined only for $x \geq 0$ and denotes the unique *nonnegative* real number $y$ for which $y^k = x$.

# Exercises

1. Consider the equation $\sqrt{x+2} + \sqrt{x-3} = 1$.

   (a) As part of a strategy to obtain an equation without surds, derive an equation in which the two surds are on opposite sides of the equal sign.

   (b) Square the two sides of the equation in (a) to obtain a new equation.

   (c) Rewrite the equation in (b) so that only the surd term appears on one side.

   (d) Derive a final equation by squaring the two sides of the equation in (c).

   (e) Show that, if the given equation has a solution, then $x = 7$.

   (f) Verify that $x = 7$ does not satisfy the given equation.

   (g) Check which of the equations (d), (c), (b), (a) is satisfied by $x = 7$.

   (h) Consider the given equation along with the equations in (a), (b), (c), (d). For each adjacent pair of equations, state which of the pair implies the other. Explain how the extraneous solution $x = 7$ arises.

   (i) Make up an equation similar to the given equation for which $x = 7$ *is* a solution.

2. Show how to solve a surd equation of the form

$$k\sqrt{ax+b} + cx + d = 0$$

   where $k$, $a$, $b$, $c$, $d$ are real and $ak \neq 0$, by making the substitution $y = \sqrt{ax+b}$. Determine conditions under which it has 0, 1, 2 real roots.

3. Solve the equations

   (a) $\sqrt{x+4} = x - 2$

   (b) $4\sqrt{x+1} = x + 4$

   (c) $4\sqrt{x+1} + x + 4 = 0$.

4. Solve the equations

   (a) $\sqrt{x^2 - 3} = 3x - 5$

   (b) $\sqrt{x^2 - 3} = 5 - 3x$.

5. (a) Verify that, if $w = \sqrt[3]{u} + \sqrt[3]{v}$, then

$$w^3 = u + v + 3w\sqrt[3]{uv}.$$

   (b) Solve the equation

$$\sqrt[3]{14 + x} + \sqrt[3]{14 - x} = 4.$$

6. Solve the equation

$$x^2 + 18x + 30 = 2\sqrt{x^2 + 18x + 45}.$$

7. (a) Without trying to solve it, explain why the equation

$$1 - x + \sqrt{x^2 + 9} = 0$$

   has no real solution.

   (b) For which values of the parameter $b$ does the equation

$$1 - x + \sqrt{x^2 + b} = 0$$

   have a real solution? Verify that the solution satisfies the equation.

## 4.3 Solving Special Polynomial Equations

Since antiquity, it has been known how to solve problems which we now recognize as quadratic equations. Lacking a convenient algebraic notation and having no notion of imaginary number, early mathematicians gave their solutions in the form of algorithms or geometric constructions which were applicable only in special cases. Although some equations of higher degree were handled by Middle Eastern mathematicians around 1000 AD, interest in these rose markedly in the sixteenth century when Tartaglia, Cardan and Ferrari discovered the means of solving cubic and quartic equations in general. During the next 250 years, attempts to solve general equations of higher degree failed, although the theory of equations was consolidated with the help of modern notation and a number system which included surds and imaginaries. In particular, the evidence pointed strongly towards the proposition that every complex polynomial equation had a root and that the number of roots, counting multiplicity, was equal to the degree.

Finally, at the outset of the nineteenth century, Ruffini and Abel established that the roots of a general equation of degree greater than 4 could not be expressed in terms of the coefficients as could those of equations of lower degree. Thus it would not be possible to prove the *Fundamental*

*Theorem of Algebra* that every complex polynomial had a zero by actually expressing a zero in terms of the coefficients. However, already in 1797, Gauss had established this result by other means.

We will review some of the methods for solving polynomial equations, and then turn to the Fundamental Theorem. While a proper formulation of the proof of this theorem requires quite advanced mathematics, it is possible to discuss the strategy behind the proof in such a way that the reader, even without the necessary background in topology, is nevertheless convinced of its plausibility.

## Exercises

1. In Exercise 2.2.4, it was shown that the number of zeros of a nonconstant polynomial over an integral domain (and, in particular, over a field) cannot exceed its degree. This is true, even if each zero is counted as often as its multiplicity. In general, a polynomial need not have as many zeros as its degree might indicate. Provide examples of quadratic polynomials over the following fields which have no zeros in the field: $\mathbf{Q}$, $\mathbf{R}$, $\mathbf{Z}_2$, $\mathbf{Z}_3$, $\mathbf{Z}_5$, $\mathbf{Z}_7$.

2. Consider the case of polynomials over the field of complex numbers. Verify that the number of zeros of a polynomial counting multiplicity is equal to the degree of the polynomial in the following cases:

    (a) the degree does not exceed 4;

    (b) the polynomial is a reciprocal polynomial of degree not greater than 9;

    (c) the polynomial is $t^n - c$, for some positive integer $n$ and complex number $c$.

3. Occasionally, trigonometry can be used to find the roots of a high degree polynomial equation. For example, one problem in the American Mathematical Monthly asked for the roots of the equation

$$f_n(y) = 2a$$

    where

$$f_n(y) = y^n - ny^{n-2} + [n(n-3)/2!]y^{n-4} + \cdots$$
$$+ (-1)^r [n(n-r-1)(n-r-2)\cdots(n-2r+1)/r!]y^{n-2r} + \cdots,$$

    in particular when $a = 1$.

    (a) Solve the equation in the cases $a = 1$ and $n = 2, 3, 4, 6, 8$.

    (b) Verify that $\cos\theta + i\sin\theta$ and $\cos\theta - i\sin\theta$ are the roots of the quadratic equation $t^2 - yt + 1 = 0$, where $y = 2\cos\theta$.

(c)  Use $2\cos n\theta = (\cos\theta + i\sin\theta)^n + (\cos\theta - i\sin\theta)^n$ to verify that

$$2\cos 2\theta = (2\cos\theta)y - 2 = y^2 - 2$$

$$2\cos 3\theta = (2\cos 2\theta)y - 2\cos\theta = y^3 - 3y$$

and, in general, that

$$2\cos n\theta = (2\cos(n-1)\theta)y - 2\cos(n-2)\theta$$

is a polynomial in $y$ for $n \geq 4$.

(d)  Verify that

$$\frac{n}{r}\binom{n-r-1}{r-1} - \frac{n-1}{r}\binom{n-r-2}{r-1} = \frac{n-2}{r-1}\binom{n-r-2}{r-2}$$

for $3 \leq r \leq n/2$.

(e)  Verify that $f_n(y) = yf_{n-1}(y) - f_{n-2}(y)$ for $n \geq 3$.

(f)  Show that $2\cos n\theta = f_n(2\cos\theta)$, and deduce that, if $|a| \leq 1$, the equation is satisfied by $y = 2\cos\theta$, where $\cos n\theta = a$. [See Exploration **E.30** for a similar sequence of polynomials.]

(g)  To handle the case in which $|a| \geq 1$, we go back to (b) and begin with the observation that $u$ and $1/u$ are the roots of the equation $t^2 - yt + 1 = 0$, where $y = u + 1/u$. In a similar way, show that $2(u^n + u^{-n}) = f_n(u + u^{-1})$, and deduce that the equation is satisfied by $y = u + u^{-1}$ where $u^{2n} - au^n + 1 = 0$.

(h)  How many roots does the equation have?

4.  One strategy in solving polynomial equations is to make a transformation of the equations into a special form which will be easier to handle. For example, the method of completing the square (Exercise 1.2.1) permits the reduction of a general quadratic equation to one of the form $t^2 - c = 0$.

Suppose that $p(t)$ is a polynomial of degree $n$. Show that there is a constant $k$ such that $p(t - k)$ is a constant multiple of a polynomial of the form

$$t^n + a_{n-2}t^{n-2} + a_{n-3}t^{n-3} + \cdots$$

in which the coefficient of $t^{n-1}$ vanishes.

Show that each solution of the equation $p(t) = 0$ is of the form $s - k$ where $t = s$ is a solution of the equation $p(t - k) = 0$.

5.  Because of the transformation of Exercise 4, in order to handle the general quintic equation, it suffices to deal with quintic polynomials of the form

$$t^5 + at^3 + bt^2 + ct + d.$$

We might try to factor it as the product of a quadratic and a cubic:

$$(t^2 + ut + v)(t^3 - ut^2 + wt + z).$$

By comparing coefficients, obtain four equations for $u$, $v$, $w$, $z$ in terms of $a$, $b$, $c$, $d$. Eliminate the variables $v$, $w$, $z$ and obtain a polynomial equation for $u$. What is the degree of this equation? Can the value of $u$ be found by solving an equation for a power of $u$ whose degree is less than 5?

6. If a polynomial over a field does not have a zero in the field, then it is possible to find a larger field containing the coefficients which also contains a zero. Let us look at the situation when the polynomial is quadratic and the coefficient field is **Q**.

    Suppose $d \neq 1$ is an integer which is not divisible by any perfect square except 1. Then $\sqrt{d}$ is nonrational and so the polynomial $t^2 - d$ has no zero in **Q**. However, we can extend **Q** to a larger field by "adjoining" $\sqrt{d}$. Let

    $$\mathbf{Q}(\sqrt{d}) = \{a + b\sqrt{d} : a, b \in \mathbf{Q}\}.$$

    (a) Verify that $\sqrt{3}$, $1 + 2\sqrt{3}$ and $(1/3)(7 - 4\sqrt{3})$ are members of $\mathbf{Q}(\sqrt{3})$, but that $\sqrt{2}$, $3^{1/3}$ and $i$ are not.

    (b) Is $i$ a member of $\mathbf{Q}(\sqrt{-3})$?

    (c) Show that $\mathbf{Q}(i) \neq \mathbf{C}$.

    (d) Verify that $\mathbf{Q}(\sqrt{d})$ is closed under the operations of addition, subtraction and multiplication.

    (e) The *surd conjugate* of $a + b\sqrt{d}$ is defined to be $a - b\sqrt{d}$. Verify that the product of any number in $\mathbf{Q}(\sqrt{d})$ with its surd conjugate is rational, and deduce that the reciprocal of any element in $\mathbf{Q}(\sqrt{d})$ is also in $\mathbf{Q}(\sqrt{d})$.

    (f) Show that $\mathbf{Q}(\sqrt{d})$ is the smallest field which contains all of **Q** along with the number $\sqrt{d}$.

    (g) Write $t^2 - d$ as a product of linear factors over $\mathbf{Q}(\sqrt{d})$.

    (h) Determine integers $b$ and $c$ such that $bc \neq 0$ and $t^2 + bt + c$ is irreducible over **Q** and reducible over $\mathbf{Q}(\sqrt{d})$.

7. The role played by **Q** in Exercise 6 can be played by any field **F**. Thus, $\mathbf{F}(\sqrt{d}) = \{a + b\sqrt{d} : a, b \in \mathbf{F}\}$.

    (a) Verify that $\mathbf{C} = \mathbf{R}(\sqrt{-1})$.

    (b) Let $\mathbf{F} = \mathbf{Q}(\sqrt{2})$. Show that $\mathbf{F}(\sqrt{3})$ consists of numbers of the form $a + b\sqrt{2} + c\sqrt{3} + d\sqrt{6}$, where $a, b, c, d \in \mathbf{Q}$. Determine $(a + b\sqrt{2} + c\sqrt{3} + d\sqrt{6})^{-1}$. (Observe that $\mathbf{Q}(\sqrt{2})(\sqrt{3}) = \mathbf{Q}(\sqrt{3})(\sqrt{2})$, so it makes sense to denote this extension by $\mathbf{Q}(\sqrt{2}, \sqrt{3})$ and to regard it as the field obtained from $\mathbf{Q}$ by adjoining both $\sqrt{2}$ and $\sqrt{3}$.)

8. It is possible to adjoin to a field $\mathbf{F}$ zeros of polynomials of degree higher than 2. For example, $\mathbf{F}(2^{1/4})$ is the set of elements of the form

$$a + 2^{1/4}b + 2^{1/2}c + 2^{3/4}d$$

where $a, b, c, d \in \mathbf{F}$.

   (a) Show that each element of $\mathbf{F}(2^{1/4})$ can be written in exactly one way in the form specified.

   (b) Verify that $\mathbf{F}(2^{1/4})$ is closed under addition, subtraction and multiplication.

   (c) Show that
$$(a + 2^{1/4}b + 2^{1/2}c + 2^{3/4}d)^{-1}$$
   also belongs to $\mathbf{F}(2^{1/4})$ and deduce that $\mathbf{F}(2^{1/4})$ is a field.

   (d) In particular, determine $(1+2^{3/4})^{-1}$ and $(1+2^{1/4}+2^{1/2}+2^{3/4})^{-1}$. Check your answer.

[**Remark:** Extending an arbitrary field $\mathbf{F}$ begs the question of the existence of the radical in some field larger than $\mathbf{F}$. For example, in forming $\mathbf{Q}(\sqrt{3})$, we know that $\sqrt{3}$ exists as a real number and so $\mathbf{Q}(\sqrt{3})$ is a subset of $\mathbf{R}$. However, $\mathbf{F}$ need not be a number field in order for the extension to be definable. In general, one adjoins a number making use only of the fact that it is to satisfy a polynomial equation over $\mathbf{F}$. It is in this spirit that, for example, we need not ask what $\sqrt{-1}$ *is*, but just know that it is *something* which satisfies the equation $x^2 + 1 = 0$, i.e. something which yields $-1$ when squared.]

9. (a) Let $\mathbf{F}$ be any field contained in $\mathbf{R}$, and let $f(t)$ be a cubic polynomial over $\mathbf{F}$. Suppose $d$ but not $\sqrt{d}$ is a member of $\mathbf{F}$ and that $\mathbf{F}(\sqrt{d})$ contains a zero of $f(t)$. Prove that $f(t)$ is reducible over $\mathbf{F}$ and accordingly $\mathbf{F}$ contains a zero of $f(t)$.

   (b) Suppose $g(t)$ is a cubic polynomial over $\mathbf{Q}$ and that $\mathbf{F}_0, \mathbf{F}_1, \ldots, \mathbf{F}_n$ is a sequence of fields contained in $\mathbf{R}$ such that $\mathbf{F}_0 = \mathbf{Q}$, and for $i = 1, 2, \ldots, n$, $\mathbf{F}_i = \mathbf{F}_{i-1}(\sqrt{d_i})$ where $d_i \in \mathbf{F}_{i-1}$. Prove that, if $g(t)$ has a zero in $\mathbf{F}_n$, then $g(t)$ has a rational zero.

[**Remark:** This is used to establish the celebrated result that there is no general ruler-and-compasses construction for trisecting an angle.]

10. Let $f(t) = t^3 - 3t + 1$.

(a) Solve the equation $f(t) = 0$ by Cardan's Method (Exercise 1.4.4). To find $u^3$ and $v^3$, we need access to the field $\mathbf{Q}(\sqrt{-3})$ obtained by adjoining $\sqrt{-3}$ to $\mathbf{Q}$, and finally $t = u + v$ is found in the field obtained by further adjoining a cube root of $(1/2)(-1+\sqrt{-3})$. Show that the roots of the equation as well as $\sqrt{-3}$ are all contained in the smallest field which contains $\mathbf{Q}$ along with $\zeta$, a primitive ninth root of unity. Denote this field by $\mathbf{Q}(\zeta)$.

(b) Verify that $\zeta^6 + \zeta^3 + 1 = 0$ and prove that $\mathbf{Q}(\zeta)$ is an integral domain. In fact, $\mathbf{Q}(\zeta)$ is a field, the hard part to show being that it is closed under the taking of reciprocals. The general argument for doing this runs as follows: Let $g(t)$ be any polynomial of degree less than 6 over $\mathbf{Q}$ (so that $g(\zeta)$ is a typical element of $\mathbf{Q}(\zeta)$). Since $t^6+t^3+1$ is irreducible over $\mathbf{Q}$, the greatest common divisor of $g(t)$ and $t^6 + t^3 + 1$ is 1. By the Euclidean algorithm, we can find polynomials $u(t)$ and $v(t)$ over $\mathbf{Q}$ such that

$$u(t)(t^6 + t^3 + 1) + v(t)g(t) = 1.$$

Set $t = \zeta$ to obtain $v(\zeta)g(\zeta) = 1$. Use this technique to determine $\zeta^{-1}$ and $(\zeta^3 + \zeta)^{-1}$ as polynomials of $\zeta$.

(c) With $\zeta$ denoting a primitive ninth root of unity, verify that the zeros of $f(t)$ are $\zeta+\zeta^8$, $\zeta^2+\zeta^7$, $\zeta^4+\zeta^5$ by (i) direct substitution, (ii) showing that the coefficients of $f(t)$ are suitable symmetric functions of the zeros.

(d) The field $\mathbf{Q}(\zeta)$ contains nonreal numbers. However, all of the zeros of $f(t)$ are real. Argue that the smallest field which contains $\mathbf{Q}$ along with the zeros of $f(t)$ is contained in $\mathbf{R}$ and is thus not $\mathbf{Q}(\zeta)$.

(e) Show that, if $u$ is any zero of $f(t)$, then the other two zeros are $u^2 - 2$ and $2 - u - u^2$. Deduce that

$$\mathbf{Q}(u) = \{a + bu + cu^2 : a, b, c \in \mathbf{Q}\}$$

is the smallest field containing $\mathbf{Q}$ and the zeros of $f(t)$.

# Explorations

**E.39. Solving by Radicals.** Perhaps it is surprising to be told that a certain mathematical procedure is impossible, that it can never be found regardless of the time and energy expended in the search. Yet, it can be shown beyond any doubt, that, because of the underlying structure of the number system, there is no general method like those for quadratics, cubics and quartics of finding the roots of polynomial equations of arbitrary degree. To formulate the type of solution we mean, let us look at polynomials of low degree.

To find the zero of $at + b$, we simply divide one coefficient by another; thus, the roots can be found within the field containing the coefficients by means of a field operation (division) performed on the coefficients.

The quadratic is more complicated. Finding the zeros often leads us outside of the smallest field containing the coefficients, but in a special way—through taking square roots. Thus, with the field operations of addition, subtraction, multiplication and division, as well as the taking of square roots, the root of any quadratic is accessible from its coefficients, and the standard quadratic formula displays this explicitly.

In solving a general cubic equation, we had to first solve a quadratic and then take cube roots. Thus, we can obtain the roots of a cubic equation from its coefficients, provided we allow the field operations as well as the extraction of square and cube roots. Finally, any method for solving a quartic equation involved the field operations and the extractions of square and cube roots.

Keeping these cases in mind, we say that a polynomial equation $p(t) = 0$ is *solvable by radicals* if and only if the roots of $p(t) = 0$ are determinable from the coefficients by means of the field operations and the extraction of $k$th roots for certain integers $k$ performed in some order. A *radical* is any number of the form $c^{1/k}$. There exist polynomials of the fifth and higher degree which are not solvable by radicals. The analysis of this theorem requires theory of groups, fields and vector spaces beyond the scope of this book, but the range of ideas can be indicated by means of an example.

The quartic equation

$$t^4 - 4t^3 + 6t^2 - 4t - 1 = 0$$

can be rewritten as $(t - 1)^4 = 2$. It has the four roots $t_1 = 1 + 2^{1/4}$, $t_2 = 1 - 2^{1/4}$, $t_3 = 1 + 2^{1/4}i$, $t_4 = 1 - 2^{1/4}i$, where $2^{1/4}$ denotes the positive fourth root of 2. None of these roots lies in **Q**, the smallest field which contains the coefficients of the polynomial. However, there is a way of telling how much we have to add to **Q** in order to get the roots.

We begin with the observation that two rational numbers, $a$ and $b$, are distinguishable in the sense that there are polynomial equations over **Q** which are satisfied by one but not by the other. Such an equation would be $t - a = 0$, which is satisfied by $a$ but not by $b$. However, $\sqrt{3}$ and $-\sqrt{3}$ are indistinguishable in the sense that any polynomial equation over **Q** which has one of these numbers as a root must also have the other (try to disprove this statement). In general, we look at various subsets of the roots of a polynomial and examine how these can be distinguished from others by polynomial equations of several variables. Let us see how this works out in the example before us.

The polynomial $t^4 - 4t^3 + 6t^2 - 4t - 1 = (t-1)^4 - 2$ is irreducible over $\mathbf{Q}$, and therefore cannot be factored as a product of polynomials of strictly lower degree with rational coefficients. Consequently, there are no linear, quadratic or cubic polynomials over $\mathbf{Q}$ with some of the $t_i$ as zeros. Furthermore, any polynomial over $\mathbf{Q}$ of degree at least 4 which has any $t_i$ as a zero must be divisible by $t^4 - 4t^3 + 6t^2 - 4t - 1$ and therefore have all the $t_i$ among its zeros. Accordingly, there is no polynomial in one variable over $\mathbf{Q}$ which is satisfied by some but not all of the $t_i$.

Let us take the $t_i$ a pair at a time. Verify that $t_1 + t_2$ and $t_3 + t_4$ are both equal to 2 while $t_1 + t_3$, $t_1 + t_4$, $t_2 + t_3$, $t_2 + t_4$ all differ from 2. Thus the equation

$$u + v - 2 = 0$$

is satisfied by $(u, v) = (t_1, t_2), (t_3, t_4)$, but not by other pairs of roots. This shows that there are polynomial equations over $\mathbf{Q}$ which are satisfied by some but not all of the pairs of the roots. Intuitively, this suggests that the four roots are not as far from being in $\mathbf{Q}$ as they might be.

We now introduce the *group* of the quartic equation. To do so, we describe first the permutations associated with the elements $t_1$, $t_2$, $t_3$, $t_4$: there are the $4! = 24$ ways of reordering the four elements by replacing each by another.

The identity permutation, denoted by $\epsilon$, leaves each $t_i$ fixed.

There are six permutations which interchange two of the roots and leave the remaining two alone. These will be denoted by (12), (13), (14), (23), (24), (34); the numbers between the parentheses indicate the indices of the roots to be interchanged.

There are eight permutations which cyclically interchange three of the roots and leave the remaining one alone. These are (123), (124), (132), (134), (142), (143), (234), (243). For example, (142) corresponds to replacing $t_1$ by $t_4$, replacing $t_4$ by $t_2$, and replacing $t_2$ by $t_1$.

There are three permutations which interchange pairs of the four roots. These are (12)(34), (13)(24), (14)(23). For example, (13)(24) interchanges $t_1$ and $t_3$, and interchanges $t_2$ and $t_4$.

There are six permutations which cyclically interchange all four of the roots. These are (1234), (1243), (1324), (1342), (1423), (1432). Thus, (1342) replaces $t_1$ by $t_3$, $t_3$ by $t_4$, $t_4$ by $t_2$ and $t_2$ by $t_1$.

A product of two permutations is defined by applying the permutations in succession. For example the product of (123) and (1243) is effected as follows:

replace $t_1$ by $t_2$ (by (123)), then $t_2$ by $t_4$ (by (1243)), for a net replacement of $t_1$ by $t_4$;

replace $t_2$ by $t_3$, then $t_3$ by $t_1$, for a net replacement of $t_2$ by $t_1$;

replace $t_3$ by $t_1$, then $t_1$ by $t_2$, for a net replacement of $t_3$ by $t_2$;

replace $t_4$ by $t_3$.

Thus we can write $(123) \cdot (1243) = (1432)$. Unlike a product of numbers, this product is not commutative; sometimes, taking the factors in a different order gives a different result. For example, $(1243) \cdot (123) = (1324) \neq (123) \cdot (1243)$.

Verify the following: $(12)(34) \cdot (132) = (234)$; $(13) \cdot (234) = (1423)$; $(12) \cdot (1432) = (243)$.

The set of 24 permutations along with this operation of "multiplication" constitutes the group of all permutations of four numbers. Of these 24, we select a certain subset in the following manner to make up the *group* of the polynomial. Consider all possible polynomial equations with *rational* coefficients of the form $P(t_1, t_2, t_3, t_4) = 0$ which are satisfied by the $t_i$. For example, verify that the following hold:

(a) $t_1 + t_2 + t_3 + t_4 = 4$

(b) $t_1 t_2 t_3 t_4 = -1$

(c) $t_1^2 + t_2^2 + t_3^2 + t_4^2 = 4$

(d) $t_1 + t_2 = 2$

(e) $t_3 + t_4 = 2$

(f) $(t_1 - t_2)^4 = 32$

(g) $(t_2 - t_3)^4 = -8$

(h) $(t_3 - t_4)^4 = 32$

(i) $(t_1 - t_4)^4 = -8$

(j) $(t_1 - t_2)^4 + 4(t_2 - t_3)^4 = 0.$

Equations (a), (b), and (c) are symmetrical in the $t_i$, and remain valid no matter how we permute the variables. However, for the others, there are some permutations of the roots which will render them false. For example, (d) remains valid under the permutation $(13)(24)$ (which converts it to (e)), but not under the permutation $(123)$, since $t_2 + t_3$ is not equal to 2.

The group of the polynomial $t^4 - 4t^3 + 6t^2 - 4t - 1$ over the field $\mathbf{Q}$ (the smallest field which contains the coefficients) is the set of all permutations which preserve the validity of any equation over $\mathbf{Q}$ of the form $P(t_1, t_2, t_3, t_4) = 0$. Denote this group by $G$. Show that $G$ always contains $\epsilon$, does not contain $(123)$, and contains along with any two permutations, their product (in either order).

It can be shown that in fact there are eight permutations in $G$. With this information, it is not too hard to see what they are. Consider the equation (d). Argue that any permutation which replaces $t_1$ by $t_2$ must also replace $t_2$ by $t_1$, so that $(123)$, $(124)$, $(1234)$ and $(1243)$ do not belong to $G$. In a similar way, eliminate (132), (142), (1342), (1432), (13), (134), (234), (14), (23), (24), (143), (243). Thus, $G$ consists of the permutations

$$\epsilon, (12), (34), (12)(34), (13)(24), (14)(23), (1324), (1423).$$

We enlarge the base field (the smallest containing the coefficients) by adjoining a radical to $\mathbf{Q}$: the hope is that by adjoining enough radicals, we will eventually obtain a large enough field to contain not only the coefficients, but also the roots as well. In the present example, we can achieve this by adjoining first the radical $i = \sqrt{-1}$, and then the radical $2^{1/4}$.

The smallest field $\mathbf{Q}(i)$ containing $\mathbf{Q}$ and the radical $i$ is

$$\{a + bi : a, b \in \mathbf{Q}\}.$$

Just as we did for $\mathbf{Q}$, we can now define the group of our polynomial over $\mathbf{Q}(i)$ as the set of all permutations which preserve the validity of polynomial equations of the roots whose coefficients are allowed to lie in $\mathbf{Q}(i)$. The family of admissible equations will now be larger than before; for example, it will include

(k) $(t_1 - t_2)(t_3 - t_4)^3 = -32i$

(l) $(t_1 - t_2)^3(t_3 - t_4) = 32i$.

Accordingly, the group over $\mathbf{Q}(i)$ will be smaller; the equations (k) and (l) can be used to argue that it will not contain (12), (34), (13)(24) and (14)(23). However, it turns out that each of the permutations $\epsilon$, (1423), (12)(34), (1324) converts the two equations above as well as every other valid equation over $\mathbf{Q}(i)$ into a valid equation.

The field $\mathbf{Q}(i)$ is a *splitting field* for the polynomial $t^2 + 1$, being the smallest field that contains both the coefficients and the roots of $t^2 + 1$, or alternatively, being the smallest field in which the polynomial can be "split" into linear factors. Show that the group of $t^2 + 1$ over $\mathbf{Q}$ consists of the two permutations of its zeros, namely the identity permutation and the permutation which interchanges them. Call this group $G_1$. This group can be tied in with the groups of the original polynomial over $\mathbf{Q}$ and $\mathbf{Q}(i)$ in the following way. Partition the group $G$ into two subsets:

$$H_1 = \{\epsilon, (1423), (12)(34), (1324)\}$$

$$H_2 = \{(12), (34), (13)(24), (14)(23)\}.$$

Observe that $H_1$ is the group over $\mathbf{Q}(i)$. Verify that, if each of the permutations in $H_1$ is applied to $(t_1 - t_2)^3(t_3 - t_4)/32$, we obtain a polynomial in the $t_i$ equal to $i$. However, verify also that each of the permutations in $H_2$ gives a polynomial equal to $-i$. This actually holds for other polynomials in the roots which are equal to $i$ as well. Thus, we can think of the sets $H_1$ and $H_2$ as corresponding to the two elements of $G_1$, and we can express the relationship between the groups $G$, $G_1$ and $H_1$ by the notation: $G_1 \cong G/H_1$.

Finally, we subject the base field to a second enlargement by adjoining $2^{1/4}$ to $\mathbf{Q}(i)$ to get the field $\mathbf{Q}(i, 2^{1/4})$, which is the splitting field of the quartic polynomial. This field consists of numbers of the form

$$a + 2^{1/4}b + 2^{1/2}c + 2^{3/4}d,$$

where $a, b, c, d \in \mathbf{Q}(i)$.

The group $K_1$ of the quartic polynomial over $\mathbf{Q}(i, 2^{1/4})$ consists solely of the identity permutation. For example, any permutation must keep $t_1$ fixed since $t_1 = 1 + 2^{1/4}$ is a polynomial equation over $\mathbf{Q}(i, 2^{1/4})$ satisfied by the roots.

There are two groups of interest: $G/H_1$ and $H_1$, which correspond to the adjunction of the two radicals $i$ and $2^{1/4}$ to the base field. The main idea to grasp is that in some sense we can describe the degree of symmetry displayed by the roots of a polynomial equation by means of a certain group of permutations, and that this group can be broken down into component parts through successive adjoinings to the base field for the polynomial. In general, suppose we have a polynomial $p(t)$ over a base field $\mathbf{F}_0$ for which the equation is solvable by radicals. This means that the splitting field for the polynomial is the culmination of a sequence of intermediate fields, each of which is obtained from its predecessor by adjoining all $k$th roots of some number $a$:

$$\mathbf{F}_0 \le \mathbf{F}_1 \le \mathbf{F}_2 \le \cdots \le \mathbf{F}_{s-1} \le \mathbf{F}_s$$

where, for each $j \ge 1$, $\mathbf{F}_j = \mathbf{F}_{j-1}(a_j^{1/k_j})$ with $a_j \in \mathbf{F}_{j-1}$. Compute the groups of $p(t)$ over these fields, $G_j$ being the group over $\mathbf{F}_j$. Then

$$G_0 \ge G_1 \ge G_2 \ge \cdots \ge G_{s-1} \ge G_s$$

where the last group consists of the identity permutation alone. Each field $\mathbf{F}_j$ is the splitting field of a polynomial of the form $t^k - a$ over $\mathbf{F}_{j-1}$; the group of this polynomial over $\mathbf{F}_{j-1}$ is $H_j$. We can write $H_j \cong G_{j-1}/G_j$ to indicate that there is a close relationship among the three groups.

It can be shown that the groups $H_j$ arising from adjunctions of radicals are characterized by a special property, and this in turn imposes a restriction on the structure of $G$. However, it is possible to find polynomials with degree as low as 5 whose groups do not satisfy the restriction, and therefore whose equations cannot be solved by radicals.

What is the group associated with the polynomial $t^3 - 3t + 1$ (Exercise 3.10)?

## E.40. Constructions Using Ruler and Compasses.

An early topic in many Euclidean geometry courses is ruler-and-compasses constructions. This reflects the ancient Greek interest in the so-called Three Famous Problems of Antiquity, namely

(1) given an arbitrary angle, to construct an angle of one third the magnitude using only straightedge and compasses (Trisecting an Angle);

(2) given the side of a cube, to construct the side of a cube of twice the volume (Duplication of the Cube);

(3) given the radius of a circle, to construct the side of a square whose area is equal to that of the circle (Squaring the Circle).

After mastering the method of bisecting a general angle, many students spend long hours trying to find a method of angle trisection and often have trouble believing that the job is impossible–not because they are not clever enough, but because of the intrinsic mathematical structure. Let us see why this is so.

If there were a general trisection method, then it would work in particular for an angle of 60 degrees. This would mean that we could construct an angle of 20 degrees, or equivalently, construct a right-angled triangle with hypotenuse of length 1 and one side of length $\cos 20°$. The problem is the following:

> Given a segment of length 1, is it possible to construct a segment of length $x = \cos 20°$, with the following operations permitted
>
> (1) choice of an arbitrary point
> (2) construction of a straight line through two specified points
> (3) construction of a circle with a specified center and radius
> (4) determination of points of intersection of two straight lines, two circles or a line and a circle?

If we introduce cartesian coordinates in the plane, the coordinates of the points of intersection in (4) can be found by solving linear or quadratic equations whose coefficients lie in the smallest field determined by the coefficients of the equations of the lines or circles involved.

Suppose we begin with the points $(0, 0)$ and $(1, 0)$ (determining a segment of length 1) and construct other points successively using lines and circles determined by points already constructed. Then the coordinates of each point so constructed would lie in some field $\mathbf{F}_n$ which is the last in a chain of quadratic extensions (as described in Exercise 3.9).

Can we construct the point $(x, 0)$, where $x = \cos 20°$? Using the formula relating $\cos 3\theta$ and $\cos \theta$, verify that $x$ satisfies the equation $8x^3 - 6x - 1 = 0$, and that the polynomial on the left side is irreducible over $\mathbf{Q}$. Now apply Exercise 3.9.

Give a similar argument to show that duplication of the cube is impossible.

## 4.4   The Fundamental Theorem of Algebra: Intersecting Curves

The *Fundamental Theorem of Algebra* states that *every polynomial of positive degree over* **C** *has at least one complex zero*. Since there is no algorithm which will allow us to construct such a root in general, the proof of this theorem has to be tackled indirectly in a way which exploits the structure of the complex plane. We look at the complex numbers geometrically. Each complex number $z = x + yi$ can be represented by a point $(x, y)$ in the Cartesian plane. A polynomial equation in the variable $z$ is equivalent to a pair of real equations in the variables $x$ and $y$, whose loci are curves in the plane. The intersections of these curves correspond to solutions of the polynomial equation. Thus, the proof of the fundamental theorem depends on ensuring that certain curves intersect. The exercises in this section will examine the situation when the polynomial has low degree and suggest how one proceeds with the task in general.

## Exercises

1. Let $a, b$ be real numbers with $a \neq 0$. Show that, if $z = x + yi$, with $x$ and $y$ real, the complex equation

$$az + b = 0$$

   is equivalent to the simultaneous real system

$$ax + b = 0 \qquad ay = 0$$

   in the sense that any solution of one corresponds to a solution to the other. Solve the real system graphically and argue that there is always a unique solution.

2. Illustrate graphically the solutions in the complex plane of the following equations:

   (a) $3z + 4 = 0$

   (b) $(2 + i)z + (-3 + 4i) = 0$.

3. Let $a = p + qi$ and $b = r + si$ be two complex numbers with $a \neq 0$. Find a real system of two simultaneous equations equivalent to the complex equation $az + b = 0$. Solve the system graphically and argue that it always has a unique solution.

4. (a) Let $a$, $b$, $c$ be real numbers, with $a \neq 0$. Show that the substitution $z = x + yi$ permits a reformulation of the complex equation

$$az^2 + bz + c = 0$$

as a simultaneous real system

$$ax^2 + bx - ay^2 + c = 0 \qquad \text{(H)}$$

$$y(2ax + b) = 0. \qquad \text{(L)}$$

(b) Show that the locus of (L) is a perpendicular pair of straight lines intersecting in the point $(-b/2a, 0)$.

(c) Verify that (H) can be rewritten in the form

$$(x + b/2a)^2 - y^2 = (1/2a)^2(b^2 - 4ac).$$

(d) If $b^2 - 4ac = 0$, show that the locus of (H) is a pair of perpendicular straight lines intersecting in $(-b/2a, 0)$. Sketch the graphical solution of the two equations.

(e) Suppose $b^2 - 4ac \neq 0$. Show that the locus of (H) is a hyperbola whose center is $(-b/2a, 0)$ and whose asymptotes are the lines $x - y + b/2a = 0$ and $x + y + b/2a = 0$.

(f) If $b^2 - 4ac > 0$, show that the loci of (H) and (L) intersect in two points on the $x$-axis. Sketch the graphical solution of the two equations.

(g) If $b^2 - 4ac < 0$, show that the loci of (H) and (L) intersect in two distinct points on the line $x = -b/2a$ which are symmetrical about the real axis. Sketch the graphical solution of the two equations.

(h) Show that (d), (f) and (g) confirm that:

  (i) if $b^2 - 4ac = 0$, the quadratic equation $az^2 + bz + c = 0$ has a single real root;

  (ii) if $b^2 - 4ac > 0$, the quadratic equation has two distinct real roots;

  (iii) if $b^2 - 4ac < 0$, the quadratic equation has two distinct nonreal roots, each the complex conjugate of the other.

(i) On the graphs sketched in (d), (f) and (g) draw a circle whose center is at the origin and whose radius is sufficiently great that its interior contains all the points of intersection of the loci of (H) and (L). Label the points where the locus of (H) intersects the circle by the letter R (as in *real*) and the points where the locus of (L) intersects the circle by the letter I (as in *imaginary*). Verify that there are four points with each of the labels R and I, and that the R-points alternate with the I-points.

5. Illustrate graphically the solutions in the complex plane of the following equations:

(a) $z^2 + 2z + 3 = 0$

(b) $z^2 + 2z - 3 = 0$

(c) $z^2 - 6z + 9 = 0$

(d) $z^2 + 2iz - 1 = 0$

(e) $z^2 + (1 + 2i)z + (-1 + i) = 0$.

6. (a) Let $p$ and $q$ be nonzero reals. Show that the cubic equation

$$z^3 + pz + q = 0$$

is equivalent through the transformation $z = x + yi$ to the system

$$x^3 - 3xy^2 + px + q = 0 \qquad\qquad \text{(A)}$$

$$y(3x^2 - y^2 + p) = 0. \qquad\qquad \text{(B)}$$

(b) Show that the locus of (B) consists of the $x$-axis along with a hyperbola whose center is $(0, 0)$ and whose asymptotes are the lines $y = \sqrt{3}x$ and $y = -\sqrt{3}x$. How does the sign of $p$ determine which of the $x$-axis and the $y$-axis is the transverse axis of the hyperbola?

(c) Verify that the equation (A) can be written in the form

$$y^2 = (1/3)(x^2 + p + q/x).$$

(d) Verify that

$$\sqrt{x^2 + p + (q/x)} - x = [p + (q/x)]/\left[\sqrt{x^2 + p + (q/x)} + x\right],$$

and deduce that the locus of (A) is asymptotic to the pair of straight lines whose equation is $3y^2 = x^2$.

(e) Show that (A) is asymptotic to the $y$-axis.

(f) Sketch the graphs of (A) and (B) on the same axes, indicating where the curves are likely to intersect. Then draw a circle with center at the origin whose radius is large enough that all the points of intersection of (A) and (B) are in its interior. Label all the intersection points of (A) with this circle by the letter R, and all the intersection points of (B) with this circle by the letter I. Verify that the letters R and I alternate.

7. Carry out the procedure of Exercise 6 on the equation $z^3 + z + 1 = 0$. Show that the equivalent real system is

$$x^3 - 3xy^2 + x + 1 = 0 \qquad y(3x^2 - y^2 + 1) = 0.$$

Sketch the graphs of these two curves, and verify that the cubic equation has one real solution between $-1$ and $0$ and two nonreal complex conjugate solutions.

8. Solve graphically the equations:

    (a) $z^3 - 7z + 6 = 0$

    (b) $z^3 - z - 1 = 0$

    (c) $z^3 + i = 0$.

9. (a) Show that the transformation $z = x + yi$ applied to the equation

$$z^4 - 2z^2 + 3z - 2 = 0$$

yields the simultaneous real equations

$$x^4 - 6x^2y^2 + y^4 - 2x^2 + 2y^2 + 3x - 2 = 0 \qquad \text{(C)}$$

$$4x^3y - 4xy^3 - 4xy + 3y = 0. \qquad \text{(D)}$$

  (b) Show that (C) can be rewritten in the form

$$y^2 = (3x^2 - 1) \pm \sqrt{8x^4 - 4x^2 - 3x + 3}.$$

When $-2 < x < 1$, show that there is one positive value for $y^2$ giving two real values of $y$. When $x < -2$ or $x > 1$, show that there are two distinct positive values of $y^2$ giving four real values of $y$. When $x = -2$, and $x = 1$, show that there are two values of $y^2$, one of which is zero, and that there are three values of $y$.

  (c) Find $\tan\theta$, $\tan^2\theta$, $\tan 3\theta$, $\tan^2 3\theta$ for $\theta = \pi/8$, and determine the asymptotes of the curves (C) and (D).

  (d) Sketch the loci of the equations (C) and (D), and on the same axes indicate a circle with center at the origin and radius sufficiently large that the interior of the circle contains all the intersection points of the curves (C) and (D). Label all the intersection points of the circle and the locus of (C) with R and of the circle and the locus of (D) with I. Verify that the points R and I alternate.

10. Let $n$ be a positive integer. Consider the polynomial equation

$$z^n + a_{n-1}z^{n-1} + a_{n-2}z^{n-2} + \cdots + a_1 z + a_0 = 0.$$

Suppose that $z = x + yi$ transforms this equation to $u(x, y) + iv(x, y) = 0$, where $u$ and $v$ are polynomials over $\mathbf{R}$.

We wish to examine where a circle with centre at the origin and a very large radius $r$ intersects each of the loci $u(x, y) = 0$ and $v(x, y) = 0$. Let $(r\cos\theta, r\sin\theta)$ be a typical point on this circle. Use de Moivre's Theorem (Exercise 1.3.8) to show that

$$u(r\cos\theta, r\sin\theta) = r^n \cos n\theta + \text{terms of lower degree in } r$$

$$v(r\cos\theta, r\sin\theta) = r^n \sin n\theta + \text{terms of lower degree in } r.$$

Since $r$ is very large, the terms involving the highest powers of $r$ in the expansions of $u$ and $v$ will be dominant, and the terms of lower degree will be negligible by comparison except where $\cos n\theta$ and $\sin n\theta$ are near zero. Consequently, as $\theta$ increases from 0 to $2\pi$, it is to be expected that $u(r\cos\theta, r\sin\theta)$ changes sign roughly as $r^n \cos n\theta$ and $v(r\cos\theta, r\sin\theta)$ changes sign roughly as $r^n \sin n\theta$.

Sketch the graphs of $\cos n\theta$ and $\sin n\theta$ as functions of $\theta$ over the domain $0 \le \theta < 2\pi$, and verify that each has $2n$ zeros, those of one function interlacing those of the other.

If we mark with an R the intersection of the largest circle and the locus of $u(x,y) = 0$ and with an I the intersection of the large circle and the locus of $v(x,y) = 0$, argue that the points R and I will be approximately evenly spaced around the circumference of the circle and the R-points will alternate with the I-points.

The final step is to argue that the part of the locus of $u(x,y) = 0$ inside the circle consists of a number of curves connecting pairs of R-points and the part of the locus of $v(x,y) = 0$ inside the circle consists of a number of curves connecting pairs of I-points, and then to deduce that the two loci must inevitably intersect. Check this in the case of the quintic: indicate on a circle ten points R and ten points I alternating with the R-points; join pairs of points with the same letter and check the plausibility that some line joining 2 R-points must intersect some line joining 2 I-points.

# 4.5   The Fundamental Theorem: Functions of a Complex Variable

A second approach to the Fundamental Theorem involves the idea of curves winding around the origin. To handle this, we need some way of visualizing the action of functions of complex variables. In the case of real functions, this is done by sketching graphs in the plane. However, since the space of complex numbers has two real dimensions, we would need a four-dimensional space in which to construct the graph of a complex function of a complex variable. We avoid this by looking at two complex planes, one for the domain of the function and the other for the range. Suppose $h(z)$ is a function of the complex variable $z$ and $w = h(z)$. For each $z$ in the complex plane of the domain (the $z$-plane), we plot the corresponding point $w = h(z)$ in the plane of the range (the $w$-plane). We write $z = x + yi$ and $w = u + vi$.

This in itself is not very useful. To get a sense of how the function $h$ behaves, it is better to envisage $z$ as a moving point in the $z$-plane and to consider how the image point $h(z)$ moves correspondingly through the $w$-plane. Thus, as $z$ traces out a certain curve, one would expect $h(z)$ to trace out a curve in the $w$-plane. By looking at the images of special curves, we can analyze the behaviour of $h$. In particular, if a curve in the $z$-plane passes through a zero of $h(z)$, its image in the $w$-plane must pass through the origin. Thus, the problem of showing that $h$ has a zero reduces to the problem of showing that some image curve must pass through the origin.

## Exercises

1.  Let $h(z) = z^2 - (3+i)z + (1-2i)$.

    (a) Draw the real and imaginary axes of two complex planes, to be labelled the $z$-plane and the $w$-plane. In the $z$-plane plot each of the points $0$, $1$, $i$, $1+i$, $5+i$, $-2$, and in the $w$-plane plot the images of these points under $h$.

    (b) Let $x$ be real, and show that $h(x) = (x^2 - 3x + 1) - (x+2)i$. Show that the image of the real axis under the mapping $h$ is the curve given parametrically in the $w$-plane by

    $$u = x^2 - 3x + 1$$

    $$v = -(x+2).$$

    Verify that this image curve is a parabola with equation $u = v^2 + 7v + 11$. Sketch this parabola, and indicate by an arrow the direction in which $h(z)$ moves along the parabola as $z$ moves in the positive direction along the real axis.

2.  Consider the polynomial $p(z) = z^2 + z + 1$. This polynomial has two zeros, both on the circle in the $z$-plane of radius $1$. We study the impact of this fact on the images of circles of various radii under the mapping $z \longrightarrow p(z)$.

    (a) Verify that

    $$p(r\cos\theta + ir\sin\theta) = r(\cos\theta + i\sin\theta)(2r\cos\theta + 1) + (1 - r^2).$$

    Deduce that, for any point $z$, its image $p(z)$ in the $w$-plane can be found by following a dilatation with center $0$ and magnification factor $(2r\cos\theta + 1)$ by a translation $1 - r^2$ in the direction of the positive real axis.

    (b) Verify that $p(0) = 1$ and that

    $$|p(r\cos\theta + ir\sin\theta) - 1| \le r(2r+1) + r^2 = r(3r+1).$$

(c) Deduce from (b), that if $|z| < 1/3$, then $p(z)$ is contained within the circle of center 1 and radius 2/3. More generally, deduce from (b) that the closer a circle is to the origin in the $z$-plane, the closer its image in the $w$-plane will be to the point 1.

(d) Suppose $r > 1/2$. Verify that, as $\theta$ increases from 0, the point $p(r\cos\theta + ir\sin\theta)$ starts from a point $1 + r + r^2$ on the real axis and moves in a counterclockwise direction about the origin, staying above the real axis as long as $2r\cos\theta + 1$ is positive. Show that the image of the circle of radius $r$ under the mapping $z \longrightarrow p(z)$ crosses the real axis at the points $1 + r + r^2$, $1 - r^2$, $1 - r + r^2$ and again at $1 - r^2$ when the argument of $z$ is 0, some angle between $(1/2)\pi$ and $\pi$, $\pi$, and some angle between $\pi$ and $(3/2)\pi$, respectively.

(e) Suppose $r < 1/2$. Show that the real axis is intersected only twice by the image of the circle of radius $r$, and argue that this image is a small loop which does not intersect itself.

(f) Let $C_r$ be the circle of radius $r$ in the $z$-plane and $D_r$ its image in the $w$-plane under the mapping $z \longrightarrow p(z)$. Sketch $D_r$ for

   (i) $0 < r < 1/2$
   (ii) $r = 1/2$
   (iii) $1/2 < r < 1$
   (iv) $r = 1$
   (v) $1 < r$ (say $r = 4$).

   When $1/2 < r$, verify that $D_r$ has two loops.

(g) Verify that $D_4$ lies in the annulus $\{w : 8 < |w| < 24\}$.

(h) Let $r$ be a fixed radius. Imagine a vector drawn in the $w$-plane from 0 to a point on $D_r$. As $z$ traces around $C_r$ in a counterclockwise direction, the vector joining 0 to $p(z)$ will rotate. Verify that, if $0 < r < 1$, this vector will move back and forth without completing even a single rotation around 0, while, if $1 < r$, the vector will make two complete circuits of the origin.

3. Let $p(z) = z^2 + 2z + 1$.

   (a) Verify that $p(r\cos\theta + ir\sin\theta) = 2r(\cos\theta + i\sin\theta)(r\cos\theta + 1) + (1 - r^2)$.

   (b) Carry out an analysis of the image curves of $C_r$ as in Exercise 2, and verify that, as $r$ decreases, the value for which the inner loop of the image $D_r$ disappears is the same as the value for which $D_r$ passes through the origin. Explain the significance of this.

4. Let $p(z) = z^2 + 3z + 2$.

(a) Verify that $p(r \cos \theta + ir \sin \theta) = r(\cos \theta + i \sin \theta)(2r \cos \theta + 3) + (2 - r^2)$.

(b) Analyze the image $D_r$ of the curves $C_r$ for the values $r > 2$, $r = 2$, $r = 3/2$, $r = 1$ and $r < 1$. Verify that, as $r$ decreases, the inner loop contracts to a point while $D_r$ still makes at least one circuit of the origin, and disappears for $r = 3/2$.

5. Let $p(z) = z^3 - 7z + 6$.

(a) Verify that $p(r \cos \theta + ir \sin \theta) = r(\cos \theta + i \sin \theta)(4r^2 \cos^2 \theta - r^2 - 7) + (-2r^3 \cos \theta + 6)$.

(b) Show that $p(r \cos \theta + ir \sin \theta)$ is real exactly when $\theta = 0$, $\theta = \pi$ or when $\cos^2 \theta = (1/4) + (7/[4r^2])$. Deduce that, when $r$ is very large, the image of the circle of center 0 and radius $r$ crosses the real axis when $\theta$ is equal or close to one of the angles $0$, $\pi/3$, $2\pi/3$, $\pi$, $4\pi/3$, $5\pi/3$.

(c) Suppose $r^2 < 7/3$. Show that

(i) $\operatorname{Im} p(r \cos \theta + ir \sin \theta)$ and $\operatorname{Im}(r \cos \theta + ir \sin \theta)$ have opposite signs;

(ii) the image of the circle of radius $r$ meets the real axis only twice;

(iii) the image does not make a circuit of the origin when $r < 1$ and makes one circuit of the origin when $1 < r$.

(d) Suppose $r^2 > 7/3$. Show that the image of the circle of radius $r$ crosses the real axis 6 times when $\theta = 0$, $\theta_r$, $\pi - \theta_r$, $\pi$, $\pi + \theta_r$, $2\pi - \theta_r$, where $\theta_r = \arccos((1/4)(1 + 7r^{-2}))^{1/2}$.

(e) Verify that the image of the circle $C_r$ makes

(i) no circuit of the origin when $0 < r$;

(ii) one circuit of the origin when $1 < r < 2$;

(iii) two circuits of the origin when $2 < r < 3$;

(iv) three circuits of the origin when $3 < r$.

6. Let $h(z) = z^n$. Show that, if $z$ makes one counterclockwise circuit of the origin along the circle $|z| = r$, then $h(z)$ makes $n$ counterclockwise circuits of the origin.

7. Let $p(z) = z^n + a_{n-1}z^{n-1} + a_{n-2}z^{n-2} + \cdots + a_1 z + a_0$. Suppose that $r$ is any positive real number exceeding 1 for which

$$r \geq 2\{|a_{n-1}| + |a_{n-2}| + |a_{n-3}| + \cdots + |a_1| + |a_0|\}.$$

(a) Show that, if $|z| = r$, then

$$|z^n - p(z)| \leq \{|a_{n-1}| + |a_{n-2}| + \cdots + |a_1| + |a_0|\}r^{n-1} \leq (1/2)r^n.$$

(b) Use (a) to argue that, when $|z| = r$, then $p(z)$ is in the annulus with center 0, inner radius $(1/2)r^n$ and outer radius $(3/2)r^n$.

(c) Use (a) to argue further that, as $z$ makes one counterclockwise circuit of the circle with center 0 and radius $r$, $p(z)$ must make $n$ circuits of the origin.

8. Sketch a proof of the fundamental theorem along the following lines.

(a) It suffices to prove the theorem for monic polynomials $p(z)$.

(b) If $p(0) = 0$, the theorem holds. It suffices therefore to establish the result when $p(0) \neq 0$.

(c) If $r$ is sufficiently small, then as $z$ traces around $C_r$, its image $p(z)$ will not make a circuit of the origin.

(d) If $r$ is sufficiently large, then as $z$ traces around $C_r$, its image $p(z)$ will make at least one circuit of the origin.

(e) Let $r$ grow slowly from small to large values. How will the image $D_r$ of $C_r$ vary? Why is it plausible to infer that for some value of $r$, $D_r$ must contain the origin?

## 4.6 Consequences of the Fundamental Theorem

The fundamental theorem has a number of consequences about the factorization of a polynomial and the extent to which a polynomial can be determined by certain of its values.

## Exercises

1. Let $p(t)$ be a polynomial over $\mathbf{C}$ with positive degree $n$. By the Fundamental Theorem it has at least one complex zero, but, it is known that there are no more than $n$ distinct zeros (Exercise 2.2.4). Let these be $t_1, t_2, \ldots, t_k$ where $1 \leq k \leq n$. Show that there are positive integers $m_i$ and a constant $c$ for which

   (i) $m_1 + m_2 + \cdots + m_k = n$;
   
   (ii) $p(t) = c(t - t_1)^{m_1}(t - t_2)^{m_2} \cdots (t - t_k)^{m_k}$.

2. Prove that a polynomial over $\mathbf{C}$ is irreducible if and only if its degree is 1. [A field with the property that only linear polynomials are irreducible is said to be *algebraically closed;* thus, this result states that $\mathbf{C}$ is an algebraically closed field. Over such a field, every nonconstant polynomial has a zero.]

3. Let $p(t)$ be a polynomial over $\mathbf{R}$. $p(t)$ can have zeros which are either real or nonreal; however, the complex conjugate of any zero is also a zero (Exercise 1.3.14).

(a) If $s_i$ is a nonreal zero of $p(t)$, prove that the polynomial

$$t^2 - (2\text{Re } s_i)t + |s_i|^2$$

is a polynomial irreducible over $\mathbf{R}$ which divides $p(t)$.

(b) Prove that $p(t)$ can be written as a product of real irreducible linear and quadratic polynomials.

(c) Deduce that a polynomial over $\mathbf{R}$ is irreducible if and only if it is linear or quadratic of the form $at^2 + bt + c$ with $b^2 - 4ac < 0$.

4. Let $p(t)$ be a polynomial over $\mathbf{C}$ of positive degree $n$. Prove that $p(t)$ assumes every complex value at least once and assumes all but finitely many complex values $n$ times.

5. Show that every real polynomial of odd degree has an odd number of real zeros, counting multiplicity, and deduce that such a polynomial has at least one real zero.

6. Let $r_1, r_2, \ldots, r_n$ be all the roots (each repeated as often as its multiplicity indicates) of the complex polynomial $a_n t^n + a_{n-1} t^{n-1} + \cdots + a_1 t + a_0$. Show that

$$r_1 + r_2 + \cdots + r_n = -a_{n-1}/a_n$$

$$r_1 r_2 r_3 \cdots r_n = (-1)^n a_0/a_n,$$

and that the sum of all possible products of $k$ of the roots is equal to $(-1)^{n-k} a_{n-k}/a_n$.

7. Suppose that a complex polynomial $p(t)$ can be factored in two ways:

$$p(t) = \prod_{i=1}^{m}(t - a_i) = \prod_{j=1}^{n}(t - b_j).$$

Show that $m = n$ and that the $a_i$'s are the same as the $b_j$'s in some order.

8. Let $a_0, b_0, a_1, b_1, \ldots, a_n, b_n$ be $2(n + 1)$ complex numbers, with the $a_i$ distinct. Show that there is at most one polynomial $p(t)$ of degree not exceeding $n$ for which $p(a_i) = b_i$ $(0 \le i \le n)$.

9. Suppose that a polynomial $p(t)$ over $\mathbf{C}$ has the following properties:

(i) the multiplicity of 1 as a zero of $p(t)$ is even (possibly 0);

(ii) If $r$ is a zero of $p(t)$, then $1/r$ is a zero with the same multiplicity as $r$.

Prove that $p(t)$ is a reciprocal polynomial (see Exercises 1.4.13 and 1.4.16).

10. Let $f(t)$ be a polynomial of degree $n$ over **Z**. Show that, if $f(t)$ assumes prime integer values for $2n + 1$ distinct values of $t$, then $f(t)$ is irreducible over **Z**.

## Explorations

**E.41. Zeros of the Derivative.** Suppose a polynomial $f(t)$ of degree $n$ over **R** has $n$ real zeros. Since every zero of $f(t)$ of multiplicity $m$ is a zero of $f'(t)$ of multiplicity $m - 1$, and since, by Rolle's Theorem, there is a zero of $f'(t)$ between any distinct pair of consecutive zeros of $f(t)$, it follows that $f'(t)$ has at least $n - 1$ real zeros counting multiplicity. Since $\deg f'(t) = n - 1$, it follows that all the zeros of $f'(t)$ are real.

The fact that the reality of all zeros of a polynomial implies the reality of all zeros of its derivative can be formulated in a way which leads to an interesting generalization. If $u$ is the smallest and $v$ the largest zero of $f(t)$, then not only the zeros of $f(t)$ but also those of $f'(t)$ lie in the closed interval $[u, v]$.

Now suppose that $f(t)$ is an arbitrary polynomial over **C** of degree $n$ with zeros $z_1, z_2, \ldots, z_n$ (not necessarily all distinct). These are represented by points in the complex plane. Let $P$ be the smallest convex polygonal region with boundary which contains them; some of the zeros will be vertices, others may lie on edges while the remainder will be in the interior. The diagram illustrates a possible situation:

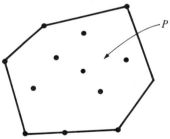

Observe that this polygonal region can be represented as the intersection of half-planes, namely those portions of the complex plane which lie on one side of the lines containing the edges of the polygon.

The result we want is that all the zeros of the polynomial $f'(t)$ also lie in the polygonal region $P$.

Verify that this result is true for any polynomial of degree 2. What are the possible shapes of $P$ in this case?

To prove the result in general, we require some preliminary facts which will help reduce the situation to the special case that two zeros are real and the remainder lie in the upper half plane.

(1) If $S_k$ is the set of zeros of $f^{(k)}(t)$ $(k = 0, 1, \ldots, n-1)$, then the set of zeros of the polynomial $g^{(k)}(t)$, where $g(t) = f(t+w)$, is $S_k - w = \{z - w : z \in S_k\}$.

(2) With $S_k$ as in (1), the set of zeros of $h^{(k)}(t)$, where $h(t) = f(wt)$, is $w^{-1}S_k = \{w^{-1}z : z \in S_k\}$.

(3) If $r_1, r_2, \ldots, r_k$ are the zeros of $f(t)$ with respective multiplicities $m_1, m_2, \ldots, m_k$, then

$$\frac{f'(t)}{f(t)} = \frac{m_1}{t - r_1} + \cdots + \frac{m_k}{t - r_k}.$$

From (1) and (2), we see that the result holds for $f(t)$ iff it holds for either $f(t+w)$ or $f(wt)$ for fixed $w \neq 0$. By suitable choice of $w_1$ and $w_2$, we can arrange that any two zeros representing vertices of $P$ correspond to two real zeros $u$ and $v$ of the polynomial $q(t) = f(w_1(t+w_2))$ and that the remaining zeros of $q(t)$ lie in the upper half plane, i.e. if $q(w) = 0$, then Im $w \geq 0$.

From (3), any zero of $f'(t)$ which is not an $r_i$ must make the rational function on the right side vanish. We show that the right side will not vanish for $t = z$ when Im $z < 0$. Look at each term individually, and observe that $\text{Im}(z - r_i) < 0$, so that $\text{Im}(m_i(z - r_i)^{-1}) > 0$. Hence $\text{Im}(f'(z)/f(z))$ is positive.

Now go back to the function $f(t)$ and argue that all of the zeros of $f'(t)$ lie in the same halfplane determined by each edge of $P$ as contain the zeros of $f(t)$. Hence, all the zeros of $f'(t)$ must lie in $P$.

# 4.7 Problems on Equations in One Variable

1. Solve $x^4 + a^4 = 4ax(x^2 + a^2)$.

2. Solve the equation

$$(x^2 - x - 2)^4 + (2x + 1)^4 = (x^2 + x - 1)^4$$

3. Solve $(x^2 - 4)(x^2 - 2x) = 2$.

4. Solve the equation $(x + a)(x + 2a)(x + 3a)(x + 4a) = b^4$.

5. For which values of $a$, $b$, $c$ does the equation

$$\sqrt{x + a\sqrt{x} + b} + \sqrt{x} = c$$

have infinitely many real zeros.

6. Let $k$ be a positive real number. Solve for real $x$.

$$\sqrt{x + \sqrt{2x - 1}} + \sqrt{x - \sqrt{2x - 1}} = k.$$

7. Let $a, b$ be positive. Solve the equation

$$\sqrt{2ab + 2ax + 2bx - a^2 - b^2 - x^2} = \sqrt{ax - a^2} + \sqrt{bx - b^2}.$$

8. Solve $(x^2 + 3x + 2)(x^2 + 7x + 12) + (x^2 + 5x - 6) = 0.$

9. (a) Let $p(t) = at^2 + bt + c$. Suppose that $u \neq v$ and that $p(u) = u$, $p(v) = v$. Let $q(t) = p(p(t)) - t$. Show that $u$ and $v$ are two zeros of the quartic polynomial $q(t)$ and determine a quadratic whose zeros are the other two zeros of $q(t)$.

   (b) Apply (a) to solve

   $$(t^2 - 3t + 2)^2 - 3(t^2 - 3t + 2) + 2 - t = 0.$$

10. Solve for real $x$:

$$\frac{x - \sqrt{x^2 - 1}}{\sqrt{x + \sqrt{x^2 - 1}}} = \sqrt[4]{x^2 - 1}\left[\sqrt{x^2 + x} - \sqrt{x^2 - x}\right].$$

11. Show that $4\cos^2(\pi/14)$ is the greatest root of the equation

$$x^3 - 7x^2 + 14x - 7 = 0.$$

12. Consider the equation $x^4 = (1 - x)(1 - x^2)^2$. Show that if either $1 - x^2 = -x^3$ or $1 - x^2 = x$ holds, then the equation is satisfied. Deduce that

$$x^4 - (1 - x)(1 - x^2)^2 = (x^3 - x^2 + 1)(x^2 + x - 1).$$

13. Find a real solution to the equation

$$(x^2 - 9x - 1)^{10} + 99x^{10} = 10x^9(x^2 - 1).$$

# 4.8 Problems on Systems of Equations

Solve the following systems of equations:

1. $x^4 + y^2 - xy^3 - 9x/8 = 0$
   $y^4 + x^2 - yx^3 - 9y/8 = 0$  (for real $x$, $y$).

2. $x/a + b/y + c/z = a/x + y/b + c/z = a/x + b/y + z/c = 1$.

3. $x + y + z = 0$
   $x^2 + y^2 + z^2 = 6ab$
   $x^3 + y^3 + z^3 = 3(a^3 + b^3)$.

4. $x + y - z = 2$
   $x^2 + y^2 - z^2 = 8 - 2xy$
   $x^3 + y^3 - z^3 = 86 - 3xyz$.

5. $xy + yz + zx = a^2 - x^2 = b^2 - y^2 = c^2 - z^2$.

6. $31x^2y^2 - 7y^4 - 112xy + 64 = 0$
   $x^2 - 7xy + 4y^2 + 8 = 0$.

7. $x^2 - (y - z)^2 = a^2$
   $y^2 - (z - x)^2 = b^2$
   $z^2 - (x - y)^2 = c^2$
   in which $a$, $b$, $c$ are constants different from 0.

8. Let $a$, $b$, $c$ be different real numbers. Show that the only real solution of the system of equations

$$x + y + z = 0$$

$$ax + by + cz = 0$$

$$x^3 + y^3 + z^3 = 3(b - c)(c - a)(a - b)$$

is $x = b - c$, $y = c - a$, $z = a - b$.

9. $x(y + z) = a$
   $y(z + x) = b$
   $z(x + y) = c$
   where it is understood that the greatest of $a$, $b$, $c$ is less than the sum of the other two.

10. (a) $u + v = (1/u) + w = (1/v) + (1/w)$
    (b) $u + v + (1/uv) = (1/u) + w + (u/w) = (1/v) + (1/w) + vw = uv + (w/u) + (1/vw)$.

11. $x^2 + y^2 = 13$      $x^3 + y^3 = 35$.

12. $x + y + z = 3a$
$x^2 + y^2 + z^2 = 14 + 2a + 5a^2$
$xyz = 6 - 10a - 4a^2$.

13. $a(y - z) + b(z - x) + c(x - y) = 0$
$(x - y)(y - z)(z - x) = d^3$
$x + y + z = e$.

14. $x(y + z)^2 = 1 + a^3$
$x + y = 3/2 + z$
$yz = 3/16$.

15. $x^2 + 2xy - y^2 = ax + by$
$x^2 - 2xy - y^2 = bx - ay$   ($a, b$ are real constants).

16. $x^2 + w^2 + v^2 = a^2$   $vw + u(y + z) = bc$
$w^2 + y^2 + u^2 = b^2$   $wu + v(z + x) = ca$
$v^2 + u^2 + z^2 = c^2$   $uv + w(x + y) = ab$
($a, b, c$ are positive reals; $x, y, z, u, v, w$ to be real).

17. $u + v = a$
$ux + vy = b$
$ux^2 + vy^2 = c$
$ux^3 + vy^3 = d$   ($a, b, c, d$ are nonzero constants).

18. $x + y + z = 5$
$x^2 + y^2 + z^2 = 9$
$xy + u + vx + vy = 0$
$yz + u + vy + vz = 0$
$zx + u + vz + vx = 0$.

19. $by + cz = (y - z)^2$
$cz + ax = (z - x)^2$
$ax + by = (x - y)^2$.

20.
$$\frac{(ab + 1)(x^2 + 1)}{(x + 1)} = \frac{(a^2 + 1)(xy + 1)}{(y + 1)}$$
$$\frac{(ab + 1)(y^2 + 1)}{(y + 1)} = \frac{(b^2 + 1)(xy + 1)}{(x + 1)}.$$

21. $x^3 + y^3 + z^3 = 8$
$x^2 + y^2 + z^2 = 22$
$\frac{1}{x} + \frac{1}{y} + \frac{1}{z} = \frac{-z}{xy}$.

22. Prove that $(x, y) = (1, 2)$ is the unique real solution of

$$x(x + y)^2 = 9$$
$$x(y^3 - x^3) = 7.$$

23. Given that none of $|a|$, $|b|$, $|c|$ is equal to 1, and that

$$x^2 = y^2 + z^2 - 2ayz$$
$$y^2 = z^2 + x^2 - 2bzx$$
$$z^2 = x^2 + y^2 - 2cxy,$$

show that

$$\frac{x^2}{1 - a^2} = \frac{y^2}{1 - b^2} = \frac{z^2}{1 - c^2}.$$

What happens if $|a| = 1$?

24. $x^2 + 2y^2 + 3z^2 = 36$
$3x^2 + 2y^2 + z^2 = 84$
$xy + xz + yz = -7.$

25. Find all real $a$ for which there exist nonnegative reals $x_i$ for which

$$\sum_{k=1}^{5} kx_k = a$$

$$\sum_{k=1}^{5} k^3 x_k = a^2$$

$$\sum_{k=1}^{5} k^5 x_k = a^3.$$

26. Determine all real $p$ for which the system $x+y+z = 2$, $yz+zx+xy = 1$, $xyz = p$ has a real solution.

27. $xy = 2$
$(3 + 6y/(x - y))^2 + (3 - 6y/(x + y))^2 = 82.$

28. Show that the real solutions $(x_i, y_i, z_i)$ of

$$(x - 5)^2 + (y - 2)^2 + (z - 6)^2 = 49$$
$$(x - 11)^2 + (y - 7)^2 + (z - 2)^2 = 49$$
$$38x - 56y - 13z = 0$$

satisfy $x_1 + x_2 = 16$, $y_1 + y_2 = 9$, $z_1 + z_2 = 8$.

29. Given that $a \geq 1$ and $b$ are real numbers, prove that the system

$$y = x^3 + ax + b$$

$$z = y^3 + ay + b$$

$$x = z^3 + az + b$$

has exactly one real solution.

30. Solve

$$x(x - y - z) = a$$

$$y(y - z - x) = b$$

$$z(z - x - y) = c.$$

31. If $x + x^{-1} = a$, $y + y^{-1} = b$ and $a^2 + b^2 + z^2 = 4 + abz$, determine $z$ in terms of $x$ and $y$.

32. If

$$\frac{x^2}{a^2} + \frac{y^2}{b^2} + \frac{z^2}{c^2} = 1$$

and

$$\frac{ux}{a^2} = \frac{vy}{b^2} = \frac{wz}{c^2},$$

show that $(x/u + y/v + z/w)^2 = a^2/u^2 + b^2/v^2 + c^2/w^2$.

33. Let $a$, $b$, $c$ be positive numbers such that $\sqrt{a} + \sqrt{b} + \sqrt{c} = \sqrt{3}/2$. Prove that the system

$$\sqrt{y - a} + \sqrt{z - a} = 1$$

$$\sqrt{z - b} + \sqrt{x - b} = 1$$

$$\sqrt{x - c} + \sqrt{y - c} = 1$$

has exactly one real solution.

34. Solve for $x$, $y$, $z$:

$$yz = a(y + z)$$

$$xz = b(x + z)$$

$$xy = c(x + y).$$

# 4.9  Other Problems

1. For which complex numbers $a$ is the mapping

$$z \longrightarrow z + az^2$$

one-one on the closed unit disc ($|z| \le 1$)?

2. The mapping $z \longrightarrow z^2$ maps a straight line in the complex plane onto a parabola. Identify the vertex of the parabola.

3. Find necessary and sufficient conditions on $p$, $q$, $r$ that the roots of

$$x^3 + px^2 + qx + r = 0$$

are the vertices of an equilateral triangle in the complex plane.

4. Find, in terms of $a$, $b$, $c$, a formula for the area of a triangle in the complex plane whose vertices are the roots of

$$x^3 - ax^2 + bx - c = 0.$$

5. Show that a necessary and sufficient condition that a real cubic equation $ax^3 + bx^2 + cx + d = 0$ have one real and two pure imaginary roots is that $bc = ad$ and $ac > 0$.

6. Let $a$, $b$, $c$, $d$ be complex numbers, all with absolute value equal to unity. Prove that, in the unit circle with center 0, the polynomial $az^3 + bz^2 + cz + d$ has a maximum absolute value not less than $\sqrt{6}$.

7. Let $p(x)$ be a polynomial of positive degree $n$ and let $0 \le m \le n$. Suppose that $c_0, c_1, c_2, \ldots, c_n$ are constants for which

$$c_0 + c_1 x + c_2 x^2 + \cdots + c_{n-m-1} x^{n-m-1} + c_{n-m} p(x)$$

$$+ c_{n-m+1} p(x+1) + \cdots + c_n p(x+m) = 0$$

identically. Show that $c_0 = c_1 = \cdots = c_n = 0$.

8. $f(t)$ is a polynomial of degree $n$ over $\mathbf{C}$ such that a power of $f(t)$ is divisible by a power of $f'(t)$, i.e. $[f(t)]^p$ is divisible by $[f'(t)]^q$ for some positive integers $p$ and $q$.

   Prove that $f(t)$ is divisible by $f'(t)$ and that $f(t)$ has a single zero of multiplicity $n$.

9. The nonconstant polynomials $p(t)$ and $q(t)$ over $\mathbf{C}$ have the same set of numbers for their zeros, but with possibly different multiplicities. The same is true of the polynomials $p(t) + 1$ and $q(t) + 1$. Prove that $p(t) = q(t)$.

10. Determine a monic cubic polynomial over $\mathbf{Q}$ one of whose zeros is $1 - 2^{1/3} + 2^{2/3}$.

## Hints

### Chapter 4

1.1. Add the equations and determine $x + y + z$ first.

1.3. Use Exercise 2 and the first two equations to obtain $x : y : z$.

1.4. Set $t = u$ and consider the two equations as a linear system in the "variables" $u^2$, $u$, 1.

1.6. (a) Multiply the first equation by $xy$, $xz$, respectively. Treat the other equations similarly.

1.11. (a) Add the equations $a_i = rb_i$.

3.3. (c) $t^{k+2} = yt^{k+1} - t^k$.

3.4. Use the quadratic and cubic cases to predict what $k$ should be.

3.8. We wish to find a number $p + 2^{1/4}q + 2^{1/2}r + 2^{3/4}s$ such that $(a + \cdots)(p + \cdots)$ is an element of $\mathbf{F}$. Begin by multiplying $(a + 2^{1/2}c) + 2^{1/4}(b + 2^{1/2}d)$ by a number to yield a product with no terms in $2^{1/4}$. Now multiply by a second factor which disposes of the terms in $2^{1/2}$.

3.9. If $a + b\sqrt{d}$ is a zero of $f(t)$, then so also is $a - b\sqrt{d}$. Use the Factor Theorem to determine a quadratic factor of $f(t)$ over $\mathbf{F}$. Or, the third zero is the sum of the zeros less $2a$.

6.10. If $p = qr$ is prime, then either $q$ or $r$ is 1. How often can a polynomial assume the values $\pm 1$?

7.1. Write equation as difference of squares equals 0.

7.2. Let $u = x^2 + x - 1$, $v = x^2 - x - 2$. Write as equation in $u$ and $v$. Factor.

7.3. Let $x = 1 - t$.

7.4. Write the left side as $(u - a^2)(u + a^2)$, where $u = x^2 + 5ax + 5a^2$.

7.7. Square both sides and write in terms of $y = x[x - (a + b)]$.

7.9. Use the Factor Theorem to find the form of $p(t) - t$ and also of $p(p(t)) - p(t)$. Note that $p(p(t)) - t$ is the sum of these two quantities.

8.1. Begin by eliminating the terms with coefficient $9/8$ and determine simple possible relations between $x$ and $y$.

8.2. The equations lead to relations of the form $u + v = u^{-1} + v^{-1}$; study this first.

8.5. Let $u = a^2 - x^2 = \cdots$, and write $xy + yz + zx$ in terms of $u$.

8.6. One strategy is to find a homogeneous equation of the fourth degree in $x$ and $y$ which might be factored. Multiply the second equation by $14xy$. Note also that the second equation can be used to give an expression for 64.

8.7. Factor the left sides as differences of squares. Let $u = -x + y + z$, etc.

8.9. Solve first for $xy$, $yz$ and $zx$.

8.10. (a) $1 = (u + v - u^{-1})(u + v - v^{-1})$.

(b) Equate the four members in pairs. Eliminate denominators and manipulate terms to find factors which might be cancelled out.

8.11. Solve for $u = x + y$, $v = xy$.

8.13. If $a$, $b$, $c$ can not all be the same, the first equation and $(y - z) + (z - x) + (x - y) = 0$ lead to a determination of $(y - z) : (z - x) : (x - y)$.

8.14. The last two equations can be used to determine $(y + z)^2 = (y - z)^2 + 4yz$ in terms of $x$.

8.15. Note that $(x + yi)^2 = (x^2 - y^2) + 2xyi$. What is $(x + yi)[(a + b) + (a - b)i]$? Be careful; $x$ and $y$ need not be real.

8.16. Express the difference of two expressions for $b^2 c^2$ as a sum of squares.

8.20. Dispose of the case $a = b$ first. Make a change of variable $x = (1 + u)/(1 - u)$, $y = (1 + v)/(1 - v)$; $u$ and $v$ each satisfy a quartic equation whose roots are obvious.

8.25. For any quadratic polynomial $p(t)$, $ap(a) = \Sigma kp(k^2)x_k$.

8.28. If $(x_1, y_1, z_1)$ and $(x_2, y_2, z_2)$ satisfy the system, then $(x_1 + x_2, \ldots)$ should satisfy the linear equation. Observe that $(x, y, z) = (16, 9, 8)$ satisfies the linear equation.

8.33. Let $T$ be an equilateral triangle of side 1. There is a one-one correspondence between points $P$ inside the triangle and positive reals $a$, $b$, $c$ for which $\sqrt{a} + \sqrt{b} + \sqrt{c} = \frac{1}{2}\sqrt{3}$, and the distances from $P$ to the respective sides are $\sqrt{a}$, $\sqrt{b}$, $\sqrt{c}$. Also, $\sqrt{y - a}$ is the side length of a right triangle with hypotenuse $\sqrt{y}$ and other side $\sqrt{a}$.

9.5. Note that the conditions imply that $ct + d = a^{-1}c(at + b)$.

9.6. Noting that the relation $1 + \omega + \omega^2 = 0$ is a good way of disposing of excess terms, calculate $|p(z)|^2 + |p(\omega z)|^2 + |p(\omega^2 z)|^2$, where $\omega$ is an imaginary cube root of unity.

9.8. Write the derivative $f'(t)$ as a product of irreducible powers which divide $f(t)$ and a polynomial $g(t)$. Does $g(t)$ have any zeros in common with $f(t)$? What is the degree of $g(t)$?

9.9. By way of reconnoitring, look first at the possibility that $p(t)$ has only simple zeros. What can be said of the zeros of $(p - q)(t)$? of $p'(t)$?

# 5

# Approximation and Location of Zeros

## 5.1  Approximation of Roots

The population of a biological species in a favorable environment will increase steadily from year to year. Often, we can assume that the annual increase will be given by a factor $a$, so that if $x$ is this year's population and $f(x)$ next year's, $f(x) = ax$, where $a > 1$. This equation may model the situation quite well for small populations, but as the population grows, factors arise which tend to limit it: nonavailability of food, greater visibility to predators, conflict arising from crowded conditions. Mathematically, this can be handled by introducing a "second-order" term in the population function:

$$f(x) = ax - bx^2 \quad (b > 0).$$

Thus for large values of $x$, $f(x)$ will be less than $x$ and the population will decrease. There will be an intermediate level of population, $x_0$, for which the factors promoting increase are balanced by those promoting decrease. This level can be found by solving the quadratic equation $x = f(x)$.

Those familiar with the habits of June bugs, spruce budworm and lemmings are aware of another phenonenon. For these, there is no "steady state" population which persists from year to year. Rather, their numbers cycle between large and small values. Can we model this using some function $f(x)$ given above?

If we imagine a two-year cycle of a lean year with population size $x_1$ followed by a glut year with larger population size $x_2$, then these two numbers will satisfy

$$x_2 = f(x_1) \quad x_1 = f(x_2)$$
$$x_i \neq f(x_i) \quad \text{for } i = 1, 2.$$

Thus, $x_1$ and $x_2$ will be roots of the quartic equation

$$x = f(f(x)) = a^2 x - (ab + ba^2)x^2 + 2ab^2 x^3 - b^2 x^4.$$

Whether one model successfully represents the cycling population turns on whether $a$ and $b$ can be chosen so that the quartic has two positive real roots apart from $x = (a-1)/b$ which are less than $x = a/b$ (the population level which leads to extinction).

This quartic equation is easy to solve explicitly. However, if we look at the possibility of a three-year cycle in which there are distinct population sizes $x_1$, $x_2$ and $x_3$ for which

$$x_2 = f(x_1) \quad x_3 = f(x_2) \quad x_1 = f(x_3),$$

then the $x_i$ will be roots of the polynomial equation

$$x = f(f(f(x)))$$

which is of the eighth degree. Two obvious roots are $x = 0$ and the positive value of $x$ for which $x = f(x)$. But when the factors corresponding to these roots are removed, this still leaves us with an equation of the sixth degree to solve. Even if it were possible to solve this equation, it might be at the expense of a fair bit of grind, and the answer might be in a form from which it is hard to find out much about the roots, even whether they are relevant to the problem at hand. What is really needed is a collection of techniques which will yield useful information in an efficient way.

Many mathematical situations give rise to polynomial equations. If the degree is higher than four, it is usually hopeless to think of solving them exactly. However, this may not be necessary, and techniques which allow us to locate the roots in a rough sense may be enough to support the analysis we wish to make.

Information sought about roots falls into two main categories:

(i) numerical approximation

(ii) location.

Under the first heading are treated methods which yield roots to desired numerical accuracy. Under the second are discussed methods which will enable us to say whether there are real or nonreal roots, positive or negative roots, or roots within a given region of the complex plane.

In this section, we will examine methods of numerical approximation.

## Exercises

1. *The method of bisection.* Let $p(t)$ be any polynomial over $R$. Suppose further that $p(a)$ is negative while $p(b)$ is positive. The graph of $p$ is a continuous curve which joins the point $(a, p(a))$ below the $x$-axis to the point $(b, p(b))$ above the $x$-axis. Accordingly, it will cross the axis somewhere between $a$ and $b$.

   (a) Argue that $p(t)$ has at least one real zero between $a$ and $b$.

   (b) Show that, either $(a + b)/2$ is a zero of $p(t)$, or else $p((a+b)/2)$ differs in sign from either $p(a)$ or $p(b)$.

(c) Show that, if $(a + b)/2$ is not a zero of $p(t)$, then there is a zero either between $a$ and $(a + b)/2$ or else between $(a + b)/2$ and $b$.

(d) Let $n$ be a given number. Determine an algorithm which will produce an interval of length no more than $(b - a)2^{-n}$ which contains a zero of $p(t)$.

2. For each of the following polynomials, find consecutive integer values of the variable at which the values of the polynomial differ in sign. Between these values, determine by the method of Exercise 1, a zero to an accuracy of two decimal places

(a) $t^4 - t^3 - t^2 - t - 1$

(b) $2x^3 - 9x^2 + 12x + 7$

(c) $2t^6 - 7t^5 + t^4 + t^3 - 12t^2 - 5t + 1$

3. *Linear interpolation.* Refer to the diagram

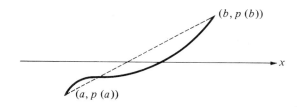

(a) Find the equation of the straight line joining the points $(a, p(a))$ and $(b, p(b))$, and determine its intersection with the $x$-axis.

(b) Explain why it is reasonable to take

$$a - \frac{p(a)(b - a)}{p(b) - p(a)}$$

as a first approximation to a zero of $p(t)$.

(c) Devise a modification of the Method of Bisection using (b) and test it out on the polynomials in Exercise 2.

4. *Horner's Method.* The Horner's algorithm provides a systematic way of approximating zeros. Let $p(t) = t^4 - 3t^3 + 7t^2 - 15t + 1$ and consider the following chain of Horner's tables:

| 1 | −3 | 7 | −15 | 1 |
|---|---|---|---|---|
|  | 2 | −2 | 10 | −10 |

| 1 | −1 | 5 | −5 | −9 |
|---|---|---|---|---|
|  | 2 | 2 | 14 |  |

| 1 | 1 | 7 | 9 |  |
|---|---|---|---|---|
|  | 2 | 6 |  |  |

| 1 | 3 | 13 |  |  |
|---|---|---|---|---|
|  | 2 | ↓ | ↓ | ↓ |

| 1 | 5 | 13 | 9 | −9 |
|---|---|---|---|---|
|  | 0.5 | 2.75 | 7.875 | 8.4375 |

| 1 | 5.5 | 15.75 | 16.875 | −0.5625 |
|---|---|---|---|---|
|  | 0.5 | 3.00 | 9.375 |  |

| 1 | 6.0 | 18.75 | 26.25 |  |
|---|---|---|---|---|
|  | 0.5 | 3.25 |  |  |

| 1 | 6.5 | 22 |  |  |
|---|---|---|---|---|
|  | 0.5 | ↓ | ↓ | ↓ |

| 1 | 7 | 22 | 26.25 | −0.5625 |
|---|---|---|---|---|
|  | 0.02 | 0.1404 | 0.4428 | 0.5339 |

| 1 | 7.02 | 22.1404 | 26.6928 | −0.0286 |
|---|---|---|---|---|

. . .

(a) Argue that $p(t)$ has a zero between 2 and 3.

(b) Use Horner's Table to verify that, if $u = t-2$, then $p(t) = q(u) = u^4 + 5u^3 + 13u^2 + 9u - 9$.

(c) Show that $q(u)$ has a zero between 0.5 and 0.6, and therefore that $p(t)$ has a zero between 2.5 and 2.6.

(d) Justify the foregoing Horner's tables as an attempt to estimate $2.52 = 2 + 0.5 + 0.02$ as a zero of $p(t)$. Carry the table further to obtain an accuracy of three decimal places.

5. *Newton's Method.* A method which can be regarded as an infinitesimal version of that of Exercise 3 and a formalization of that of Exercise 4 can be formulated with the help of Taylor's Theorem. We suppose that we are given a number $u$ which is believed to be close to an

unknown zero $r$ of $p(t)$. This method will enable us to (often) find a better approximation to $r$.

(a) Use Taylor's Theorem to argue that $0 = p(r)$ is approximately equal to $p(u)+p'(u)(r-u)$, and therefore $r$ will be approximately equal to $u - p(u)/p'(u)$.

Thus, we can start with an approximation $u_1 = u$, and for $i = 2, 3, 4, \ldots$, move to successive approximations $u_i$, where

$$u_i = u_{i-1} - p(u_{i-1})/p'(u_{i-1}).$$

(b) There is a zero of the polynomial $2t^6 - 7t^5 + t^4 + t^3 - 12t^2 - 5t + 1$ between $t = 3$ and $t = 4$. Explain and verify the following table created in approximating this zero; give the zero to two decimal places.

| $t$ | $p(t)$ | $p'(t)$ | $p(t)/p'(t)$ |
|---|---|---|---|
| 3 | $-257$ | 139 | $-1.849$ |
| 4.849 | 7594.21 | 13224.74 | 0.574 |
| 4.275 | 2385.60 | 5703.92 | 0.418 |
| 3.857 | 691.35 | 2673.82 | 0.259 |
| 3.598 | 160.02 | 1504.00 | 0.106 |
| 3.492 | 20.18 | 1144.69 | 0.018 |
| 3.474 | 0.073 | 1089.66 | 0.000 |

(c) There is also a zero of the same polynomial between 0.125 and 0.25. Using each of these values as first approximations, determine this zero to three decimal places.

(d) Use Newton's Method to approximate the zeros of the other polynomials in Exercise 2(a) and 2(b).

(e) We can get a geometric picture of Newton's Method. Show that the tangent to the graph of $y = p(x)$ through the point $(u, p(u))$ intercepts the $x$-axis at the point $(u - p(u)/p'(u), 0)$. For each of the following graphs, the initial approximation is indicated. Indicate where the next few approximations will lie.

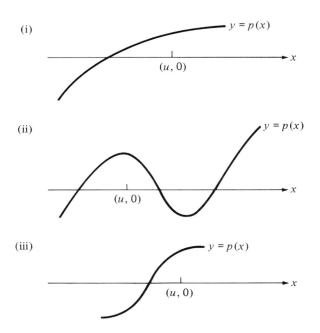

6. Let $c \neq 1$. Show that Newton's Method applied to the polynomial $tc - 1$ and any first approximation yields the zero $c^{-1}$ in one step. With a diagram, explain why this occurs.

7. (a) Let $c > 0$. Show that Newton's Method applied to the polynomial $t^2 - c$ yields from any positive approximation $a_1$, the sequence $\{a_n\}$ of successive approximations to $\sqrt{c}$, where

$$a_{n+1} = \frac{1}{2}(a_n + c/a_n) \quad (n \geq 1).$$

(b) Find an expression which relates the difference $a_{n+1}^2 - c$ to the difference $a_n^2 - c$. Argue that, as $n$ becomes larger, $a_n$ gets closer and closer to $\sqrt{c}$. Show that $\{a_n\}$ is decreasing for $n \geq 2$.

8. Determine to three decimal places all of the real zeros of the polynomial $3t^4 - 2t^3 - t^2 - 3t + 1$. Use all of the methods discussed in these exercises.

9. Newton's Method for successive approximations to the root of an equation is a particular instance of a more general approach:

Given an equation $p(t) = 0$, derive from it an equation of the form $t = f(t)$ with which it shares a solution. Pick a first approximation $u_1$ to the solution, and let $u_n = f(u_{n-1})$ for $n \geq 2$. With suitable choice of $f$ and $u_1$, the sequence $\{u_n\}$ should converge to a solution $r$ of $t = f(t)$ and hence of $p(t) = 0$.

For Newton's Method, one makes the choice $f(t) = t - p(t)/p'(t)$.

Other choices are possible. Consider the polynomial equation

$$t^3 - 4t - 18 = 0.$$

(a) Argue that there is a unique positive root $r$ and that it lies between 3 and 4.

(b) The equation can be rewritten in the form $t = f(t)$, where $4f(t) = t^3 - 18$. Begin with a value $u_1$ (say 3) and define successively $u_n = f(u_{n-1})$ for $n \geq 2$. Does $u_n$ approach a solution of the given equation?

(c) Sketch the graphs of the polynomial $t$ and $f(t)$. The solution of the equation is determined by the point at which the graphs cross. For the values of $u_1$ you used in (b), plot the points $(u_1, u_2), (u_2, u_2), (u_2, u_3), (u_3, u_3), (u_3, u_4), \ldots$. How do you account for the behaviour observed in (b)?

(d) Let $g(t) = (4t + 18)^{1/3}$. Show that the given equation is equivalent to the equation $t = g(t)$. Let $v_1$ be equal to 3, 4 or some other value of your choice, and define $v_n = g(v_{n-1})$ for $n \geq 2$. Does the sequence $\{v_n\}$ appear to approach a limit?

(e) Sketch the graphs of the functions $t$ and $g(t)$, and plot the points $(v_1, v_2), (v_2, v_2), (v_2, v_3), (v_3, v_3), \ldots$. Indicate on the $x$-axis the points $v_1, v_2, v_3, v_4, \ldots$.

(f) Let $v > r$ where $3 < r < 4$ and $r^3 = 4r + 18$. Verify that

$$v - g(v) = \frac{v^3 - (4v + 18)}{v^2 + v(4v + 18)^{1/3} + (4v + 18)^{2/3}}$$

and

$$g(v) - r = \frac{4(v - r)}{(4v + 18)^{2/3} + (4v + 18)^{1/3}(4r + 18)^{1/3} + (4r + 18)^{2/3}}.$$

Deduce that $r < g(v) < v$ and that $g(v) - r < (1/2)(v - r)$. Argue that if $v_1 = v$, then the sequence $\{v_n\}$ should converge towards the root $r$. Hence determine to five decimal places a solution of the given polynomial equation.

# Explorations

**E.42. Convergence of Newton Approximations.** It is not always the case that Newton's Method approximates the intended zero of a polynomial. The difference between successive approximations to a zero of $p(t)$ is given by an expression of the form $p(u)/p'(u)$, which may be large if $p'(u)$ is small (the graph is quite flat at the approximation) or $p(u)$ is large (the approximation is too far from the zero).

Consider, for example the polynomial $t^3 - 11t^2 + 24t = t(t-3)(t-8)$. Sketch the graph of this polynomial. Examine the sequence of approximations which begin respectively with 4, 5, 5.5, 5.7, 6.

One may well wonder what happens if the polynomial has no real zeros. Experiment with the polynomial $t^2 + 1$. Show that if we begin with the approximation $\cot\theta$, then the next approximation is $\cot 2\theta$. You might also observe that if the first approximation is $\operatorname{Re} w/\operatorname{Im} w$ for some $w \in \mathbf{C}$, then the next one is $\operatorname{Re} w^2/\operatorname{Im} w^2$.

**E.43. Newton's Method According to Newton.** In his monograph, *Analysis by equations of an infinite number of terms*, Isaac Newton provided a method of showing how a function can be written as a power series. As an illustration, he approximated a solution of a numerical equation. Study the following passage and compare it to the procedure described in Exercise 5:

> Let the Equation $y^3 - 2y - 5 = 0$ be proposed to be resolved: and let 2 be a number which differs from the Root sought, by less than a tenth Part of itself. Then I put $2 + p = y$, and I substitute this Value in Place of it in the Equation, and thence a new Equation arises, viz. $p^3 + 6p^2 + 10p - 1 = 0$, whose Root $p$ is to be sought for, that it may be added to the Quotient: viz. thus (neglecting $p^3 + 6p^2$ upon the Account of their smallness) $10p - 1 = 0$, or $p = 0,1$ is near the Truth; therefore I write 0, 1 in the Quotient, and then suppose $0, 1 + q = p$, and this it's [sic!] value I substitute, as formerly, whence results $q^3 + 6, 3q^2 + 11, 23q + 0, 061 = 0$.
>
> And since $11, 23q + 0, 061 = 0$ comes near to the Truth, or since $q$ is almost equal to $-0,0054$ (viz. by dividing until as many Figures arise as there are places betwixt the first Figures of this and the principal Quotient) I write $-0,0054$ in the lower part of the Quotient, since it is negative.
>
> And then supposing $-0,0054 + r = q$, I substitute this as formerly, and thus the Operation is continued as far as you please. But if I desire to continue the Work only to twice as many Figures as there are in the Quotient except one, instead of $q$ I substitute $-0,0054 + r$ into this $6, 3q^3 + 11, 23q + 0, 061$, viz. neglecting its first Term ($q^3$) upon the Account of it's Smallness,

and there arises $6, 3r^2 + 11, 16196r + 0, 000541708 = 0$ almost, or (rejecting $6, 3r^2$) $r = -0, 00004853$ almost, which I write in the negative part of the Quotient. Finally, subducting the negative Part of the Quotient from the affirmative, I have $2, 09455147$ the Quotient sought.

$$y^3 - 2y - 5 = 0$$

$$+2, 10000000$$
$$-0, 00544853$$
$$+2, 09455147 = y$$

| $2 + p = y$ | $+y^3$ | $+8 + 12p + 6p^2 + p^3$ |
|---|---|---|
| | $+2y$ | $-4 - 2p$ |
| | $-5$ | $-5$ |
| | Sum | $-1 + 10p + 6p^2 + p^3$ |

| $0, 1 + q = p$ | $+p^3$ | $+0, 001 + 0, 03q + 0, 3q^2 + q^3$ |
|---|---|---|
| | $+6p^2$ | $+0, 06 + 1, 2 + 6, 0$ |
| | $+10p$ | $+1, +10$ |
| | $-1$ | $-1$ |
| | Sum | $+0, 061 + 11, 23q + 6, 3q^2 + q^3$ |

| $-0.0054 + r = q$ | $+6, 3q^2$ | $+0, 000183708 - 0, 06804r + 6, 3r^2$ |
|---|---|---|
| | $+11\ 23q$ | $-0, 060642 + 11, 23$ |
| | $+0, 061$ | $+0, 061$ |
| | Sum | $+0, 000541708 + 11, 16196r + 6, 3r^2$ |

$$-0, 00004854 + s = r$$

Verify that the method given by Newton is essentially the method given in the Exercises, and work Newton's example in the modern way. Interpret Newton's table.

**E.44. Newton's Method and Hensel's Lemma.** Refer to Explorations E.33 and E.43. In Newton's Method, we start with an approximation $u$ and move to a new approximation $v = u - f(u)/f'(u)$. If we are lucky, we will end up with an approximation which is closer to the desired root, in the sense that $|f(v)|$ will be smaller than $|f(u)|$. Thus, the absolute value is used to give a measure of closeness. To make this more precise, we introduce into $\mathbf{R}$ a distance function $d$ defined by

$$d(a, b) = |a - b|.$$

The smaller the value of $d$, the closer the points $a$ and $b$. This distance function has three fundamental properties:

(i) $d(a, b) = 0$ if and only if $a = b$;

(ii) $d(a,b) = d(b,a)$;

(iii) $d(a,c) \leq d(a,b) + d(b,c)$.

The third property is the *triangle inequality;* roughly speaking, it says that two points which are close to a given point cannot be too far from each other.

There is an analogue of this idea of distance and closeness that we can formulate for integers. Let $p$ be a fixed prime. Beginning with the idea that $0$ is the only integer which is divisible by every prime power $p^k$, we regard an integer as being "close" to zero if it is divisible by a high power of $p$. Accordingly, we define, for the integer $n$,

$$|n|_p = p^{-s}$$

if $p^s$ is the highest power of $p$ which divides $n$.

Verify that, in the case $p = 3$, $|17|_3 = 1$, $|18|_3 = 1/9$, $|972|_3 = 1/243$. What is $|1250|_2$? $|1250|_5$? $|1250|_7$?

We now define the *p-adic distance* between the two integers $m$ and $n$ by

$$d_p(m, n) = |m - n|_p.$$

Verify that $d_p$ satisfies the properties (i), (ii), (iii) listed above.

Suppose we seek an integer root of the polynomial equation $f(x) = 0$ over **Z**. Then we want a succession of approximations $u$ which will make $|f(u)|_p$ smaller and smaller, so that the $p$-adic distance from $u$ to the desired solution will tend to $0$. Following Newton's lead, we use the approximation $u$ to obtain the approximation $v = u - f(u)/f'(u)$.

However, there is a problem with this. All the quantities we deal with are supposed to be integers, and $v$ will not in general be an integer. The problem is that the reciprocal of an integer is not an integer, so we need some way of replacing the quantity $1/f'(u)$ by something which makes sense in this context. Review the Exploration on Hensel's Lemma, and explain how this difficulty is surmounted. Explain why the choice of $v$ ensures that $|f(v)|_p$ will be smaller than $|f(u)|_p$.

**E.45. Continued Fractions: Lagrange's Method of Approximation.** Suppose we know that a polynomial $p(t)$ vanishes for some value $r$ of $t$ between the integers $a$ and $a + 1$. Then $r = a + (1/s)$ where $s > 1$. If in the equation $p(r) = 0$ we make the substitution $r = a + (1/s)$, we get a polynomial equation in $s$: $q(s) = 0$. Let $b$ be an integer for which the root $s$ lies between $b$ and $b + 1$. Write $s = b + (1/u)$, and continue the process, solving a polynomial equation for $u$, placing $u$ between consecutive integers, and so on. We finally end up with

$$r = a + \cfrac{1}{b + \cfrac{1}{c + \cfrac{1}{d + \cfrac{1}{e + \cdots}}}}$$

where $a$, $b$, $c$, $d$, $e$, ... are integers. Such an expression is called a continued fraction. More conveniently, we can write this as $a+1/b+1/c+1/d+1/e+\ldots$, where each fraction bar has everything that follows in the denominator.

Approximations to $r$ can be obtained by taking the part after any of the plus signs to be zero. Verify that this yields the rational approximations:

$$a, (ab + 1)/b, (abc + a + c)/(bc + 1), \ldots.$$

For example, if the equation to be solved is $t^2 - 2 = 0$, we know there is a root between 1 and 2. Set $t = 1 + 1/s$ and get $s^2 - 2s - 1 = 0$, which has a root between 2 and 3. Set $s = 2 + 1/u$ and get $u^2 - 2u - 1 = 0$. Deduce that

$$\sqrt{2} = 1 + 1/2 + 1/2 + 1/2 + 1/2 + \ldots$$

and verify that the successive approximations to $\sqrt{2}$ are 1, 3/2, 7/5, 17/12, 41/29, .... Work these out numerically to see how they approach $\sqrt{2}$. Compare this sequence of approximations with those obtained from Newton's Method (Exercise 5.1.7) and the starting values 1, 3/2, 7/5, 17/12, respectively. (Compare Exploration E.30. When $t = 6$, the sequence $u_n$ is 0, 1, 6, 35, 204, .... Factor the terms of this sequence.)

Find the continued fraction representations of the roots of $t^2 - t - 1 = 0$ and give the first five rational approximations to these roots.

Show that the substitution $t = 1 + 1/s$ in $t^4 - t^3 - t^2 - t - 1 = 0$ yields $3s^4 + 2s^3 - 2s^2 - 3s - 1 = 0$, which has a root between 1 and 2. Substitute $s = 1 + 1/u$ to obtain $u^4 - 11u^3 - 22u^2 - 14u - 3 = 0$. Show that this equation in $u$ has a solution between 12 and 13, and hence obtain the approximation 25/13 for a root of the original equation.

**E.46. Continued Fractions: Another Approach for Quadratics.** The quadratic equation $t^2 + bt + c = 0$ can be rewritten as $t = -b - (c/t)$. Substitute this expression for $t$ into the right side, and repeat the process to obtain a continued fraction expansion for $t$. Try this on the polynomials $t^2 - 2t - 1$, $t^2 - 2t + 1$, $t^2 - t - 1$, $t^2 - 2$, $t^2 - 3t + 2$, $t^2 - t + 1$, $t^2 - t - 6$, $t^2 + t - 6$. Does the related sequence of approximations always converge? If so, to which zero does it converge?

## 5.2 Tests for Real Zeros

How many real zeros does the polynomial $t^5 - 3t^4 - t^2 - 4t + 14$ have? Where are they located? Quickly, we can say there is at least one but no more than five. Can we find out more without having to go to the trouble of solving the equation, even approximately? The answer is yes, there are methods which will allow us to determine whether real roots exist and how many lie inside a given interval without having to pin them down numerically. Three common ones are the Descartes' Rule of Signs, the Fourier–Budan Method and Sturm's Method. Each of these provides more precise information than its predecessor, but at the expense of an increased amount of work.

In the following exercises, $(a, b)$ refers to the open interval $\{x : a < x < b\}$ and $[a, b]$ to the closed interval $\{x : a \leq x \leq b\}$.

## Exercises

1. Show that the polynomial $6t^8 + 5t^6 + 12t^4 + 2t^2 + 1$ has no real zeros.

2. Show that a nontrivial polynomial whose nonzero coefficients are all positive reals can have no real nonnegative zeros.

3. Suppose that $p(t)$ is a polynomial over $\mathbf{R}$ with positive leading coefficient such that $p(k) < 0$ for some real $k$. Show that $p(t)$ has a real zero which exceeds $k$.

4. Show that the real linear polynomial $at + b$ $(ab \neq 0)$ has a positive zero if and only if the coefficients $a$ and $b$ have opposite signs.

5. Let $p(t)$ be a polynomial over $\mathbf{R}$. Show that $r$ is a zero of the polynomial $p(t)$ if and only if $-r$ is a zero of $p(-t)$.

6. Suppose a real polynomial $a_n t^n + a_{n-1} t^{n-1} + \cdots + a_1 t + a_0$ be given. Write down its nonzero coefficients in order, and replace each positive one by $+$ and each negative one by $-$. We say that the coefficients have $k$ sign changes if, as we read along the sequence of $+$ and $-$ signs, there are $k$ places where there is a sign change.

   (a) Verify that the sequence of signs corresponding to $8t^3 + 3t^2 - t + 1$ is $+ + - +$, and that the coefficients have two sign changes.

   (b) Verify that the sequence of signs corresponding to the polynomial $(t - 4)(8t^3 + 3t^2 - t + 1)$ is $+ - - + -$, and that the coefficients have three sign changes.

   (c) Suppose that the coefficients of the polynomial $p(t)$ have $k$ sign changes, and that $r$ is a positive real. Show that the polynomial $(t - r)p(t)$ has at least $k + 1$ sign changes.

7. *Descartes' rule of signs.* Let $p(t)$ be a polynomial over **R**. Prove that $p(t)$ cannot have more positive zeros counting multiplicity than there are sign changes in its coefficients, and cannot have more negative zeros counting multiplicity than there are sign changes in the coefficients of $p(-t)$.

8. Use Descartes' Rule of Signs to verify that $t^{11} + t^8 - 3t^5 + t^4 + t^3 - 2t^2 + t - 2$ has at most 5 positive and 2 negative zeros. Deduce that it has at least 4 nonreal zeros.

9. Verify Descartes' Rule of Signs for each of the following quadratics by actually determining its zeros:

   (a) $8t^2 - 8t + 1$
   (b) $4t^2 - 4t + 1$
   (c) $t^2 - t + 1$.

10. Suppose in the real polynomial $a_n t^n + \cdots + a_1 t + a_0$, $a_n a_0 \neq 0$ and $a_k = a_{k+1} = 0$ for some $k$ with $1 \leq k \leq n - 2$. Prove that the polynomial cannot have all its zeros real.

11. Prove that the real polynomial $a_n t^n + a_{n-1} t^{n-1} + \cdots + a_3 t^3 + t^2 + t + 1$ cannot have all its zeros real.

12. There are various ways of obtaining estimates of an upper bound for the real zeros of a real polynomial. For example, suppose that

$$f(t) = a_n t^n + a_{n-1} t^{n-1} + \cdots + a_p t^p + \cdots + a_1 t + a_0$$

is a polynomial over **R** such that

   (1) $a_n > 0$
   (2) there is at least one negative coefficient
   (3) $p$ is the largest value of $i$ for which $a_i < 0$.

Let $M = \max\{|a_i| : a_i < 0\}$.

   (a) Verify that, if $r > 1$, then

$$f(r) \geq a_n r^n - M(r^{p+1} - 1)(r - 1)^{-1}.$$

   (b) Prove that, if $r > 1 + (M/a_n)^{1/(n-p)}$, then

$$a_n(r - 1)^{n-p} > M$$

   and deduce that $a_n r^n (r - 1) > M r^{p+1}$.

   (c) Deduce from (a) and (b) that if $r > 1 + (M/a_n)^{1/(n-p)}$, then $f(r) > 0$.

(d) Deduce that, if $r$ is any positive zero of $f(t)$, then

$$r < 1 + \sqrt[n-?]{\frac{M}{a_n}}.$$

13. Suppose $f(t) = \Sigma a_k t^k$ is a polynomial of degree $n$ over $\mathbf{R}$ which satisfies the hypotheses of Exercise 12. Let

$$s_i = \Sigma\{a_j : j > i, a_j > 0\},$$

the sum of all positive coefficients of powers of $t$ greater than the $i$th;

$$R = \max\{|a_i|/s_i : a_i < 0\}.$$

(a) Verify that $t^m = (t-1)(t^{m-1} + t^{m-2} + \cdots + t + 1) + 1$.

(b) Consider the polynomial

$$g(t) = 3t^4 + 2t^3 - 8t^2 + t - 7.$$

By (a) applied to powers of $t$ with positive coefficients, verify that

$$g(t) = 3(t-1)t^3 + (-8 + 5(t-1))t^2 + 5(t-1)t + (-7 + 6(t-1)) + 6.$$

Deduce that $g(r) > 0$ if $r - 1 > \max\{8/5, 7/6\} = 8/5$.

(c) Show that, in general, if $r > 1$, then

$$f(r) \geq \Sigma\{[a_i + (r-1)s_i]r^i : a_i < 0\}$$

and deduce that, if $f(r) = 0$, then $r < 1 + R$.

14. Apply the results of Exercises 12 and 13 to determine upper bounds for the real zeros of

(a) $t^{11} + t^8 - 3t^5 + t^4 + t^3 - 2t^2 + t - 2$

(b) $t^7 - t^6 + t^5 + 2t^4 - 3t^3 + 4t^2 + t - 2$.

15. *Rolle's Theorem.* At the end of the seventeenth century, Michel Rolle gave a method of locating intervals which contain zeros of a real polynomial $p(t)$. If $p(t)$ is hard to deal with, determine a real polynomial $q(t)$ whose derivative is equal to $p(t)$, i.e. $q'(t) = p(t)$. It may happen that it is easier to determine where the zeros of $q(t)$ might be, perhaps because $q(t)$ is factorable.

Rolle's result is that, if $a$ and $b$ are distinct real zeros of $q(t)$, then the open interval $(a, b)$ will contain at least one real zero of $p(t)$. The proof of this result is sketched in Exploration **E.28** at the end of Section 2.4.

(a) If $a \neq b$ and $q(a) = q(b)$, show from Rolle's Theorem that $p(t) = q'(t)$ has a zero in the interval $(a, b)$.

(b) Let $q(t)$ have $m$ real zeros counting multiplicity. Show that $p(t)$ has at least $m - 1$ real zeros counting multiplicity.

(c) Let $p(t) = 2t^3 - 3t^2 - t + 1$. Determine a polynomial $q(t)$ for which $q'(t) = p(t)$. If possible, factor $q(t)$ and use the result to determine intervals which contain the zeros of $p(t)$.

(d) Let $a_0, a_1, \ldots, a_n$ be real numbers satisfying $a_0 + a_1/2 + a_2/3 + \cdots + a_n/(n+1) = 0$. Show that the equation $a_0 + a_1 t + \cdots + a_n t^n = 0$ has at least one real root.

16. Let $f(t)$ be any polynomial over $\mathbf{R}$ of degree exceeding 1. Prove that there is a real value of $k$ for which not all the zeros of $f(t) + k$ are real.

17. *Theorem of Fourier and Budan.* A more sophisticated test than Descartes' Rule of Signs is given by the following. Suppose $p(t)$ is a polynomial over $\mathbf{R}$, and that $u$ and $v$ are reals with $u < v$ and $p(u)p(v) \neq 0$. The number of zeros between $u$ and $v$ cannot be greater than $A - B$, where $A$ is the number of changes of sign in the sequence $(p(u), p'(u), p''(u), \ldots)$ and $B$ is the number of changes of sign in the sequence $(p(v), p'(v), p''(v), \ldots)$. If this number differs from $A - B$, it must do so by an even amount.

(a) Let $p(t) = t^5 - t^4 - t^3 + 4t^2 - t - 1$. Form the sequences $(p(u), p'(u), p''(u), \ldots)$ for $u = -2, -1, 0, 1$. Verify that the signs of the terms of the sequences are given in the following table:

$$
\begin{array}{rlll}
u = -2 & : & (- + - + - +) & 5 \text{ changes} \\
u = -1 & : & (+ - - + - +) & 4 \text{ changes} \\
u = 0 & : & (- - + - - +) & 3 \text{ changes} \\
u = 1 & : & (+ + + + + +) & 0 \text{ changes}
\end{array}
$$

Deduce from the Fourier–Budan Theorem that there is exactly one real zero in each of the intervals $(-2, -1)$ and $(-1, 0)$, and either one or three real zeros in the interval $(0, 1)$.

(b) Use the Fourier–Budan Theorem to verify that $8t^2 - 8t + 1$ has one zero in each of the intervals $(0, 1/2)$ and $(1/2, 1)$.

(c) Obtain Descartes' Rule of Signs for positive zeros as a consequence of the Fourier–Budan Theorem.

(d) Show that for the polynomial $t^4 + t^2 + 4t - 3$, Descartes' Rule of Signs gives a sharper estimate for the number of negative zeros than the Fourier–Budan Theorem.

18. Verify the Fourier–Budan Theorem for the following quadratics by computing their roots:

(a) $t^2 - t \quad u = -1 \quad v = 2$

(b) $t^2 - t + 1 \quad u = -1 \quad v = 2.$

19. Let $p(t) = 2t^6 - 7t^5 + t^4 + t^3 - 12t^2 - 5t + 1$.

(a) Verify that

$$
\begin{aligned}
p(t) &= (2t - 7)t^5 + (t^2 - 12)t^2 + (t^2 - 5)t + 1 \\
&= 2t^6 + t(t + 1)(-7t^3 + 8t^2 - 7t - 5) + 1,
\end{aligned}
$$

and deduce that all the real zeros of $p(t)$ lie between $-1$ and $7/2$.

(b) Use the Fourier–Budan Theorem to deduce that the polynomial has

   (i) exactly one zero in the interval $(3, 7/2)$

   (ii) one or three zeros in the interval $(1/8, 1/4)$

   (iii) nil or two zeros in the interval $(-1, -1/2)$.

(c) Calculate $p(-1)$, $p(-1/2)$ and $p(-3/4)$ and refine the conclusion in (b)(iii).

20. Prove the theorem of Fourier–Budan for linear and quadratic polynomials. (In the quadratic case, let the polynomial be $t^2 + bt + c$ and look at the possible arrangements of signs in the sequence $(t^2 + bt + c, 2t + b, b)$ at various points $u$ and $v$. Sketch graphs to illustrate each possibility.)

21. Locate intervals which contain real zeros of the following polynomials. Get as much information as you can about these zeros.

(a) $t^5 - 3t^4 - t^2 - 4t + 14$

(b) $24t^5 + 143t^4 - 136t^3 + 281t^2 + 36t - 140$

(c) $2t^4 + 5t^3 + t^2 + 5t + 2$

(d) $16t^6 + 3t^4 - 3t^3 - 142t^2 - 9t - 21$

(e) $16t^7 - 5t^5 - 97t^4 - 95t^3 - 79t^2 + 36.$

# Explorations

**E.47. Proving the Fourier–Budan Theorem.** The key to proving the Fourier–Budan Theorem is the observation that deleting the first entry of the sequences $(p(u), p'(u), \ldots)$ and $(p(v), p'(v), \ldots)$ gives the corresponding sequences for the derivative. This suggests that we should explore the relationship between the zeros of $p(t)$ and those of $p'(t)$, and the tool for doing this is Rolle's Theorem (see Exploration **E.28**). The proof can be executed by induction on the degree of the polynomials. Suppose it holds for all degrees less than $n = \deg p(t)$. To avoid complications, let us suppose that the derivative does not vanish at either $u$ or $v$.

There are a number of cases to consider:

(1) $p$ and $p'$ have the same sign at $u$ and the same sign at $v$;

(2) $p$ and $p'$ have the same sign at $u$ and opposite signs at $v$;

(3) $p$ and $p'$ have opposite signs at $u$ and the same sign at $v$;

(4) $p$ and $p'$ have opposite signs at both $u$ and $v$.

A second subdivision of cases is whether $p$ has at least one or no zeros in the interval $[u, v]$.

Suppose, to take a special case, that $p$ does have at least one zero in the interval, the smallest being $y$ and the largest $z$, so that $u < y \leq z < v$, and that $p$ and $p'$ have the same sign at $u$ but opposite signs at $v$. Assigning $A$ and $B$ the meanings of Exercise 2.17 and $C$ and $D$ the corresponding meaning for $p'$, establish the following (generous use of graph-sketching is recommended):

(i) $A = C$, $B = D + 1$;

(ii) If $p(t)$ has $k$ zeros counting multiplicity in $[u, v]$, then $p'$ has $k - 1 + 2i$ zeros in $[y, z]$ for some nonnegative integer $i$;

(iii) $p'$ has an odd number of zeros between $u$ and $y$;

(iv) $p'$ has an odd number of zeros between $z$ and $v$.

Now use the induction hypothesis that the number of zeros of $p'$ is less than $C - D$ by an even nonnegative integer.

**E.48. Sturm's Theorem.** The difficulty with the Method of Fourier–Budan is that, while the polynomials $p(t)$ and $p(t) + c$ might have different numbers of real roots in an interval, the taking of the derivative suppresses the constant $c$ and causes information to be lost. A more refined method, due to Sturm, permits us to actually count the number of real roots in any interval, although Sturm's sequences are more troublesome to obtain than Fourier–Budan sequences.

Let $p(t)$ be a given real polynomial. To define the polynomial sequence $(p_0(t), p_1(t), p_2(t), \ldots)$ we need, we make a continued use of the division algorithm in much the same way as is done for the Euclidean algorithm. The $p_i(t)$ are chosen to satisfy:

$$p_0(t) = p(t)$$

$$p_1(t) = p'(t), \quad \text{the derivative of } p(t)$$

$$p_0(t) = p_1(t)q_1(t) - p_2(t), \quad \deg p_2 < \deg p_1$$

$$p_1(t) = p_2(t)q_2(t) - p_3(t), \quad \deg p_3 < \deg p_2$$

$$p_2(t) = p_3(t)q_3(t) - p_4(t), \quad \deg p_4 < \deg p_3$$

and so on until a zero remainder is reached. Thus, at each stage for $k \geq 2$, $p_k(t)$ is the *negative* of the remainder when $p_{k-2}(t)$ is divided by $p_{k-1}(t)$. If there are multiple roots, the last nonzero remainder will not be constant. Call this the Sturm sequence for $p(t)$.

Verify that two such sequences, taken to the last nonzero term, are

$$(t^3 - t + 1, 3t^2 - 1, (2/3)t - 1, -23/4)$$

$$(t^3 - 3t + 2, 3t^2 - 3, 2t - 2).$$

Let $u < v$. Suppose that $U$ is the number of sign changes in the sequence $(p_0(u), p_1(u), p_2(u), p_3(u), \ldots)$ and $V$ is the number of sign changes in the sequence $(p_0(v), p_1(v), p_2(v), p_3(v), \ldots)$. *Sturm's Theorem* says that the number of real roots of $p(t)$ between $u$ and $v$ (with each multiple root counted exactly once) is *exactly* $U - V$.

To see how it works, consider the following example: find the number of roots between $-2$ and $+2$ for each of the following polynomials (i) $t^2 - t$, (ii) $t^2 - t + 1/4$, (iii) $t^2 - t + 1$. First, verify that the Fourier–Budan Theorem does not allow us to distinguish the behaviour of these three polynomials. Now verify that the Sturm sequences of polynomials for the three are respectively:

(i) $(t^2 - t, 2t - 1, 1/4)$

(ii) $(t^2 - t + 1/4, 2t - 1)$

(iii) $(t^2 - t + 1, 2t - 1, -3/4)$.

Use Sturm's Theorem to determine the number of roots between $-2$ and $+2$ for each and check the result directly.

Use Sturm's Theorem to locate the real roots of the polynomial $t^5 - t^4 - t^3 + 4t^2 - t - 1$ and $2t^6 - 7t^5 + t^4 + t^3 - 12t^2 - 5t + 1$ to within intervals bounded by consecutive integers. Approximate each root to two places of decimals.

Obtain from Sturm's Theorem the following corollary: all roots of a monic polynomial are real if and only if all the nonzero polynomials in its Sturm sequence have positive leading coefficients.

Having worked out a couple of Sturm series, you will probably feel the need for some simplification. Note first that, since we are interested only in the sign of the values of certain polynomials, we can replace any polynomial in Sturm's sequence by a positive constant multiple of it. What this means is that we can set up our division at each stage in such a way as to clear all the fractions and deal only with integers.

Secondly, observe that if we divide the terms of Sturm's sequence by the same polynomial, the number of sign changes will not be affected. If $q(t)$ is the greatest common divisor of $p(t)$ and $p'(t)$, verify that the Sturm sequence for $p(t)/q(t)$ is the same as the Sturm sequence of $p(t)$ divided by $q(t)$. Verify also that the zeros of $p(t)/q(t)$ are all simple and are exactly the same as the zeros of $p(t)$. Thus, one can deal with the polynomial $p(t)/q(t)$ which may have degree less than that of $p(t)$.

How does one prove Sturm's Theorem? Argue that it suffices to prove the result when all the zeros of $p(t)$ are simple. In this case, show that Sturm's sequence ends in a constant and that it is not possible for two consecutive polynomials in Sturm's sequence to have a common zero. For convenience, we will assume that no polynomial in Sturm's sequence vanishes at the endpoints $u$ and $v$ of the interval under consideration. We look at several cases:

(i) Suppose that no polynomial $p_i(t)$ has a zero between $u$ and $v$. Argue that each polynomial in Sturm's sequence maintains the same sign on the interval $(u, v)$, and so the number of sign changes in Sturm's sequence evaluated at $u$ and at $v$ are the same.

(ii) Suppose that $r$ lies between $u$ and $v$ and that $r$ is a zero of some $p_i(t)$ $(1 \leq i \leq n-1)$. Note that $r$ must be a simple zero. Show that $p_{i-1}(r)$ and $p_{i+1}(r)$ are nonzero and have opposite signs. Choose the number $s$ sufficiently small that $p_{i-1}$ and $p_i$ do not change sign and $p_i$ has only the root $r$ in the closed interval $[r-s, r+s]$. Show that the number of sign changes in $\{p_{i-1}(t), p_i(t), p_{i+1}(t)\}$ is the same at $r-s$ and $r+s$.

(iii) Suppose that $r$ lies between $u$ and $v$ and that $p(r) = 0$. Show that a small positive real $s$ can be chosen so that

(a) $u < r - s < r < r + s < v$

(b) $p(t)$ has only the simple zero $r$ in the closed interval $[r-s, r+s]$

(c) $p'(t)$ has no zero in the closed interval $[r-s, r+s]$

(d) $p(r-s)$ has sign opposite to each of $p(r+s)$, $p'(r-s)$ and $p'(r+s)$.

Deduce that Sturm's sequence has one more sign change when evaluated at $r - s$ than when evaluated at $r + s$.

Complete the proof of Sturm's Theorem by subdividing the closed interval $[u, v]$ into subintervals at the endpoints of which none of the $p_i$ vanish and within which there is at most one number $r$ which is a zero of any of the $p_i$.

**E.49. Oscillating Populations.** In the beginning of this chapter, the function $f(x) = ax - bx^2$ was introduced. We can think of $x$ as representing a population density for some species in a certain area, so that its value will lie between 0 and 1. Consider the special case $a = b$. Thus, let

$$f(x) = ax(1 - x) \quad (0 \leq x \leq 1).$$

Sketch the graph of the equation $y = f(x)$ when $a = 1/2, 1, 2, 3, 4, 8$. On the same axes, sketch the graph of $y = x$.

If $a$ is sufficiently small, then $f(x) < x$ for all $x > 0$, and the population modelled by this function is headed for extinction. Show that, if $a > 1$, then the function $f(x)$ has a unique *fixed point* $u$ for which $u = f(u)$ and $0 < u < 1$.

Is this fixed point stable or unstable? It will be stable if, when we start the population size at a value close to $u$, it will, as the years pass, approach ever more closely to $u$. Otherwise it will be unstable. There is a convenient way of picturing the situation.

Consider the example $a = 3/2$. In this case, the fixed point of $f(x)$ is $1/3$, and we can sketch the graph below.

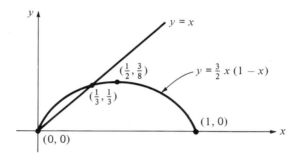

Suppose that we begin with a population density $x_0$, which we will mark on the $x$-axis. To find the point on the axis corresponding to the density $x_1 = f(x_0)$ one year later,

(i) draw the vertical line $x = x_0$; it will meet the graph of the equation $y = f(x)$ at the point $(x_0, x_1)$;

(ii) draw the horizontal line $y = x_1$ through $(x_0, x_1)$; it will meet the graph of the equation $y = x$ at the point $(x_1, x_1)$;

(iii) draw a perpendicular from $(x_1, x_1)$ to meet the $x$-axis at $(x_1, 0)$.

By repeating this procedure, we can mark on the axes the population densities $x_2, x_3, \ldots$ for successive years. Perform this for initial population densities $1/4$ and $1/2$, and verify in both cases that the population moves monotonically (either steadily increasing or steadily decreasing) towards $1/3$.

When there is stability, it is not always the case that the population varies monotonically. Sometimes it oscillates around the stable position while approaching it. Sketch on the same axes the graphs of $y = (2.5)x(1-x)$ and $y = x$. With an initial population density of $0.5$, compute the density for the next five years. Using your graphs, make a diagram to illustrate how the population oscillates around the fixed point $0.6$.

There are values of $a$ for which the fixed point is an unstable population density. For example, in the case $a = 4$, the fixed point is 0.75. Calculate the population density for ten successive years when the initial density is 0.74. Make a diagram to illustrate the situation. How do you account for the instability? What do you think will happen to the population density in the long run?

For which values of the parameter $a$ will the nonzero fixed point $u$ for the function $f(x)$ represent a stable population density?

Consider the possibility of a population density which oscillates between two values which are the roots other than 0 and $u$ of the equation $x = f(f(x))$. Verify that this equation can be written

$$x[ax - (a - 1)][a^2x^2 - a(a + 1)x + (a + 1)] = 0.$$

Verify that, when $a = 3$, it has a triple root $x = 2/3$, that when $a > 3$, it has three positive real roots with one on either side of $(a - 1)/a$, and when $a < 3$, it has two nonreal roots. Discuss the stability of the fixed point when $a$ is less than, equal to and greater than 3. Compute the derivative of $f$ at the fixed point, and discuss the significance of the value of that derivative.

If you have access to a computer, examine the possibility of a population density which has a $k$-year cycle, for which, if $x_{i+1} = f(x_i)$, we have $x_k = x_0$, but $x_0, x_1, x_2, \ldots, x_{k-1}$ all differ.

## 5.3   Location of Complex Roots

Certain classes of difference (recursion) and differential equations have associated with them a polynomial whose roots give information about the properties of the solutions. In Exporation **E.50** we will see for instance, that if the zeros of the related polynomial lie in the interior of the unit disc in the complex plane, then the $n$th term of a recursion will tend to 0 as $n$ grows large. For differential equations, it is often useful to know whether or not there are any zeros in the right half plane (i.e. for which the real part is positive). One approach is to adopt the idea used in our second proof of the Fundamental Theorem of Algebra (Section 4.5), that if a polynomial has a zero in the region surrounded by a closed curve, then the image of that curve under the action of the polynomial makes at least one circuit of the origin.

In the exercises, we will first look at estimates of the size of the zeros, and then sample techniques for locating zeros within certain regions of the complex plane.

## Exercises

1. Let $w$ be any zero of the polynomial $a_n t^n + a_{n-1} t^{n-1} + \cdots$. Show that

   (a) $w = a_n^{-1}[-a_{n-1} - a_{n-2}w^{-1} - \cdots - a_1 w^{-n+2} - a_0 w^{-n+1}]$

   (b) $|w| \le |a_n|^{-1} \sum_{k=0}^{n} |a_k|$.

2. Let $z_0, z_1, z_2, \ldots, z_n$ be complex numbers. Show that

   (a) $|z_1 + z_2 + \cdots + z_n| \le |z_1| + |z_2| + \cdots + |z_n|$

   (b) $|z_0 + z_1 + \cdots + z_n| \ge |z_0| - |z_1| - |z_2| - \cdots - |z_n|$.

   When does equality occur?

3. *Cauchy's estimate.* Let $p(t) = a_n t^n + \cdots + a_0$ and let $K$ be the maximum of the absolute values of $a_0/a_n,\ a_1/a_n, \ldots, a_{n-1}/a_n$. Show that, if $w$ is a zero of $p(t)$, then

   (a) $0 = a_n w^n [1 + (a_{n-1}/a_n)w^{-1} + \cdots + (a_1/a_n)w^{-n+1} + (a_0/a_n)w^{-n}]$

   (b) $|w| < K + 1$.

4. (a) Show that, if $w$ is a nonvanishing zero of the polynomial $p(t)$, then $w^{-1}$ is a root of the polynomial $t^n p(t^{-1})$.

   (b) Use (a) and the result of Exercise 3 to determine a lower bound for the absolute value of a nonzero root of a polynomial equation.

5. (a) Let $n \ge 1$, and suppose $a_i\ (0 \le i \le n)$ are complex numbers with $a_0 \ne 0$. Show that the polynomial

   $$t^n - |a_{n-1}|t^{n-1} - |a_{n-2}|t^{n-2} - \cdots - |a_1|t - |a_0|$$

   has a unique positive zero $r$.

   (b) Prove that every zero of the polynomial

   $$t^n + a_{n-1}t^{n-1} + \cdots + a_1 t + a_0$$

   satisfies $|w| \le r$, where $r$ is defined as in (a).

6. The polynomial $at + b$ has the zero $-b/a$. In (c), this will be generalized to an estimate for the zeros of polynomials of higher degree. In what follows, $n \ge 2$,

   $$g(t) = b_n t^n + \cdots + b_1 t + b_0$$

   $$f(t) = a_n t^n + \cdots + a_1 t + a_0.$$

   (a) Let $0 < b_n \le b_{n-1} \le b_{n-2} \le \cdots \le b_1 \le b_0$. Show that every zero $w$ of the polynomial $g(t)$ must satisfy $|w| > 1$.

(b) Let $b_n \geq b_{n-1} \geq \cdots \geq b_0 > 0$. Show that every zero $w$ of the polynomial $g(t)$ must satisfy $|w| < 1$.

(c) Let $a_i$ be all positive and let $u$ be the minimum and $v$ the maximum of the quantities $a_{n-1}/a_n$, $a_{n-2}/a_{n-1}, \ldots, a_1/a_2$, $a_0/a_1$. Show that every zero $w$ of the polynomial $f(t)$ must satisfy $u < |w| < v$.

(d) Verify that all the zeros $w$ of $7t^4 + 8t^3 + 2t^2 + 3t + 1$ satisfy $1/4 < |w| < 3/2$.

7. Use the exercises to obtain estimates on the absolute values of the zeros of the following polynomials. Compare the sharpness of the different techniques.

(a) $t^2 - t + 1$

(b) $t^5 - t^4 - t^3 + 4t^2 - t - 1$

(c) $t^{11} + t^8 - 3t^5 + t^4 + t^3 - 2t^2 + t - 2$

(d) $3t^2 - t + 4$.

8. *Schur–Cohn Criterion.*

(a) Show that both zeros $w$ of the real polynomial $t^2 + bt + c$ satisfy $|w| < 1$ if and only if $|b| < 1 + c < 2$.

(b) Show that all the zeros $w$ of the real polynomial $t^3 + bt^2 + ct + d$ satisfy $|w| < 1$ if and only if

$$|bd - c| < 1 - d^2 \quad |b + d| < |1 + c|.$$

9. A polynomial over the complex field is said to be *stable* if and only if every one of its zeros has a negative real part. (The terminology arises from applications; stability of a polynomial corresponds to physical or biological stability of some system giving rise to the polynomial.)

(a) Show that a real stable polynomial must be the product of a real constant and factors of the form $t + r$ and $t^2 + bt + c$, where $r$, $b$ and $c$ are positive.

(b) Show that the signs of all the coefficients of a real stable polynomial must be the same.

(c) Show that a linear or quadratic real polynomial is stable if and only if all the coefficients are of the same sign.

(d) Give an example of a cubic polynomial whose coefficients are all positive, but which is not stable.

10. We obtain a criterion for stability of the cubic polynomial

$$f(t) = at^3 + bt^2 + ct + d \quad (a > 0).$$

(a) Show that, if all the zeros of $f(t)$ have negative real part, then $a$, $b$, $c$, $d$ are all positive. Henceforth, we will assume that this condition holds.

(b) Show that, if all the zeros of $f(t)$ are real, then all must be negative and $bc - ad > 0$.

(c) Show that, if $bc \leq ad$, then $f(t)$ must have a nonreal zero.

(d) Suppose that $f(t)$ has two nonreal zeros $u \pm vi$ ($v \neq 0$). Verify that, for $t = u \pm vi$, $f(t) = 0$ is equivalent to

$$au^3 + bu^2 + cu + d = v^2(3au + b)$$

and

$$3au^2 + 2bu + c = av^2;$$

eliminate $v^2$ to obtain

$$8a^2u^3 + 8abu^2 + 2(b^2 + ac)u + (bc - ad) = 0.$$

Show that

(i) if $bc > ad$, then $u < 0$;

(ii) if $bc = ad$, then $u = 0$;

(iii) if $bc < ad$, then $u > 0$.

(e) Show that a cubic polynomial $at^3 + bt^2 + ct + d$ is stable if and only if all its coefficients are nonzero with the same sign and $bc > ad$.

[This condition has a nice generalization, known as the *Routh–Hurwitz Criteria*, to polynomials of degree $n$. These can be conveniently expressed using determinants.]

11. *Nyquist diagram.* To test the stability of a given polynomial, first determine a positive real $M$ which is greater than the absolute values of its zeros. Let $C$ be the curve consisting of that portion of the imaginary axis consisting of points $yi$ for which $|y| \leq M$ and the semicircle of center 0 and radius $M$ lying to the right of the imaginary axis. The image $p(C)$ of this curve is called the *Nyquist diagram* for the polynomial.

(a) Show that the polynomial is stable if and only if its Nyquist diagram does not make a circuit of the origin.

(b) Verify that all of the zeros of $z^2 + 2z - 3$ lie inside the circle of center 0 and radius 6. Show that this polynomial maps the $y$-axis (imaginary axis) onto the parabola $y^2 + 4x + 12 = 0$ with vertex $(-3, 0)$ (the point $(x, y)$ is identified with the complex number $x + yi$). Verify that the image of the point $6(\cos\theta + i\sin\theta)$ is

$(6\cos\theta + 1)12(\cos\theta + i\sin\theta) - 39$. Sketch the Nyquist diagram for this polynomial (taking $M = 6$) and verify that it winds once around the origin. Deduce that the quadratic is not stable.

12. Discuss stability of the following polynomials:

  (a) $z^3 + 2$

  (b) $z^3 + 2z^2 + 3z + 1$

  (c) $z^4 + 3z^3 + z^2 + z + 8$.

# Exploration

**E.50. Recursion Relations.** The Fibonacci sequence was introduced in Exploration **E.14**; it is given by the recursion relations

$$F_1 = F_2 = 1 \quad F_{n+1} = F_n + F_{n-1} \quad (n = 2, 3, 4, \ldots).$$

While any term of the sequence can be found using these equations, the process is both tedious and sensitive to a miscalculation, since an error at any point would contaminate the computation from then on. One would like to have a formula from which the $n$th term can be found directly.

For some sequences, such a formula is easy to find. For example, let $S_n$ be the sum of the positive integers up to $n$. This sequence can be defined recursively by $S_1 = 1$, $S_n = S_{n-1} + n$. However, one can directly compute each $S_n$ by the familiar formula $S_n = (1/2)n(n + 1)$. For the Fibonacci sequence, there, too, is a formula, but one would have to be clever indeed to guess it. However, with the proper approach, it can be found in a straightforward way.

Often, we can solve a problem by looking to a simpler one for guidance. Either we might find a stepping stone to the more complex situation or at least get a better idea of the ingredients of a solution. What would happen if we had a sequence for which each term depended only on its immediate predecessor, to wit

$$u_{n+1} = ru_n \quad (n = 2, 3, 4, \ldots)$$

for some number $r$ independent of $n$? This sequence is geometric and the $n$th term is $u_1 r^{n-1}$. Can we reduce the Fibonacci sequence to this one?

With $r$ a number to be specified later, we can write the recursion relation for $F_n$ as:

$$F_{n+1} + (r - 1)F_n = rF_n + F_{n-1} = r(F_n + r^{-1}F_{n-1}).$$

If we pick $r$ so that $r - 1 = r^{-1}$, then $u_n = F_{n+1} + (r - 1)F_n$ will be a geometric progression. Verify that

$$F_{n+1} + (r - 1)F_n = (F_2 + (r - 1)F_1)r^{n-1} = r^n \quad \text{for } n \geq 2.$$

Verify that there are two possible values of $r$: $r_1 = (1/2)(1 + \sqrt{5})$, $r_2 = (1/2)(1 - \sqrt{5})$. Eliminate $F_{n+1}$ from the two equations

$$F_{n+1} + (r_1 - 1)F_n = r_1^n$$

$$F_{n+1} + (r_2 - 1)F_n = r_2^n$$

to obtain that

$$F_n = (1/\sqrt{5})[r_1^n - r_2^n].$$

Check that this formula actually works for $1 \le n \le 5$.

Use a pocket calculator to work out the values of $r_1$ and $r_2$. Argue that $F_{n+1}/F_n$ gets closer and closer to $r_1$ as $n$ becomes larger (so that $F_n$ becomes close to being a geometric progression).

There are a number of striking relations satisfied by the terms of the Fibonacci sequence which can be established with the aid of the formula for the general term. For example, show that:

(i) $F_n^2 + F_{n+1}^2 = F_{2n+1}$  $(n \ge 1)$

(ii) $F_{n+1}F_{n-1} - F_n^2 = (-1)^n$  $(n \ge 2)$

(iii) $F_{n+1}^3 + F_n^3 - F_{n-1}^3 = F_{3n}$  $(n \ge 2)$

(iv)

$$\sum_{k=1}^{n} \binom{2n}{2k} F_{2k} = \frac{1}{2}(F_{4n} - F_{2n})  \quad (n \ge 1)$$

(v) if $m|n$, then $F_m|F_n$.

Other recursion relations can be handled in a similar way. Work out the general term for the recursion $x_n = 3x_{n-1} - 2x_{n-2}$ $(n \ge 3)$, given that the initial terms are (a) $x_1 = 1$, $x_2 = 3$; (b) $x_1 = 1$, $x_2 = 2$.

Sometimes the recursion gives rise to a polynomial equation with a double root. Show that, if $x_{n+1} = 2ax_n - a^2 x_{n-1}$ $(n \ge 2)$, the form $x_{n+1} - (2a - r)x_n = r(x_n - (2a - r)x_{n-1})$ is possible only for $r = a$. How would you modify the technique to deal with this situation? Obtain $x_n = a^{n-1}x_1 + (n-1)a^{n-2}(x_2 - ax_1)$.

Now consider a sequence for which the general term depends on its three immediate predecessors, such as for example one whose terms satisfies

$$v_{n+1} = 6v_n - 11v_{n-1} + 6v_{n-2}  \quad (n \ge 3).$$

Show that, if we rewrite this in the form

$$(v_{n+1} + av_n + bv_{n-1}) = r(v_n + av_{n-1} + bv_{n-2}),$$

then $r^3 = 6r^2 - 11r + 6$ and $(r, a, b) = (1, -5, 6), (2, -4, 3), (3, -3, 2)$. Use this information to set up a system of three linear equations to solve for $v_{n+1}$, $v_n$, $v_{n-1}$ in terms of $v_1$, $v_2$, $v_3$.

In general, for fixed $k$, a recursion of the type

$$v_{n+1} = a_0 v_n + a_1 v_{n-1} + \cdots + a_k v_{n-k}$$

can be written in the form

$$(v_{n+1} + b_0 v_n + \cdots) = r(v_n + b_0 v_{n-1} + \cdots)$$

provided $r$ is a root of the equation $r^{k+1} = a_0 r^k + \cdots + a_k$. Knowing the roots of this equation will provide us information on the general behavior of the sequence. In the special case that all its roots $r_1, r_2, \ldots, r_k$ are distinct, it can be shown that

$$v_n = c_1 r_1^n + c_2 r^2 + \cdots + c_k r^k$$

where the coefficients $c_i$ depend on $v_1, v_2, \ldots, v_k$. Thus, if all $r_i$ are real and less than 1 in absolute value, $v_n$ will become closer and closer to zero. In any case, as $n$ grows larger, $|v_{n+1}/v_n|$ tends to the largest $|r_i|$.

For each of the following recursions, write out the first 10 or so terms to get some idea of how it behaves, whether increasing or decreasing without bound, tending to zero, oscillating boundedly or unboundedly. In each case, there is a certain polynomial whose zeros can be used to give a formula for the general term of the recursion. Find the zeros of the polynomials and consider how the behaviour of the sequence is related to the following properties of the zeros:

(a) the zero of largest absolute value is positive and exceeds 1

(b) the zero of largest absolute value is negative and is less than 1

(c) the zero of largest absolute value lies outside the unit circle

(d) the zero of largest absolute value lies on the unit circle

(e) all zeros have absolute value less than 1

(f) all zeros are $n$th roots of unity for some integer $n$.

(i) $u_0 = 1 \quad u_1 = 2 \quad u_{n+1} = u_n - u_{n-1} \quad (n \geq 1)$

(ii) $u_0 = 2 \quad u_1 = 3 \quad u_{n+1} = 3u_n - 2u_{n-1} \quad (n \geq 1)$

(iii) $u_0 = 3 \quad u_1 = 2 \quad u_{n+1} = 3u_n - 2u_{n-1} \quad (n \geq 1)$

(iv) $u_0 = u_1 = 1 \quad 6u_{n+1} = 7u_n - 2u_{n-1} \quad (n \geq 1)$

(v) $u_0 = u_1 = u_2 = 1 \quad u_{n+1} = 2u_n - u_{n-1} + 2u_{n-2} \quad (n \geq 2)$

(vi) $u_0 = u_1 = u_2 = 1 \quad 2u_{n+1} = u_n - 2u_{n-1} + u_{n-2} \quad (n \geq 2)$

(vii) $u_0 = u_1 = 1 \quad u_2 = 2 \quad u_{n+1} = u_n - 2u_{n-1} + 2u_{n-2} \quad (n \geq 2)$.

## 5.4   Problems

1. Prove that, for complex $k$ the polynomial $t^3 - 3t + k$ never has more than one real zero in the closed interval $[0, 1]$.

2. Prove that the positive root of
$$x(x + 1)(x + 2) \cdots (x + n) = 1$$
is less than $1/n!$.

3. Prove that all roots but one of the equation
$$nx^n = 1 + x + x^2 + \cdots + x^{n-1}$$
have absolute value less than 1.

4. Show that $x^4 - 5x^3 - 4x^2 - 7x + 4 = 0$ has no negative roots.

5. Let $f_0(t) = t$ and $f_n(t) = f_{n-1}(t)^2 - 2$ for $n \geq 1$. Show that the equation $f_n(t) = 0$ has $2^n$ real roots.

6. Let $a$ and $b$ be unequal real numbers. Prove that
$$(a - b)t^n + (a^2 - b^2)t^{n-1} + \cdots + (a^{n+1} - b^{n+1}) = 0$$
has at most one real root.

7. Let $a_1 > a_2 > a_3 > a_4 > a_5 > a_6$, and let $p = a_1 + \cdots + a_6$, $q = a_1 a_3 + a_3 a_5 + a_5 a_1 + a_2 a_4 + a_4 a_6 + a_6 a_2$, $r = a_1 a_3 a_5 + a_2 a_4 a_6$. Show that all the zeros of $2t^3 - pt^2 + qt - r$ are real.

8. Let $a, b, c > 0$. Show that the equation
$$x^3 - (a^2 + b^2 + c^2)x - 2abc = 0$$
has a unique positive root $u$ which satisfies
$$(2/3)(a + b + c) \leq u < a + b + c.$$

9. The equation $(x - a_1)(x - a_2) \cdots (x - a_n) = 1$ where $a_i \in R$, has $n$ real roots $r_i$. Find the minimum number of real roots of the equation
$$(x - r_1)(x - r_2) \cdots (x - r_n) = -1.$$

10. Let $p(x)$ be a polynomial of degree $n$ with real roots $a_1, a_2, a_3, \ldots, a_n$. Suppose the real number $b$ satisfies
$$|b - a_1| < |b - a_i| \quad (2 \leq i \leq n).$$
Prove that
$$|p(b)| \geq 2^{-n+1}|p'(a_1)(b - a_1)|.$$

11. Prove that the roots of the cubic equation
$$t^3 - (a + b + c)t^2 + (ab + bc + ca - d^2 - e^2 - f^2)t$$
$$+ (ad^2 + be^2 + cf^2 - abc - 2def) = 0$$
are real when $a$, $b$, $c$, $d$, $e$, $f$ are real.

12. Determine each real root of
$$x^4 - (2.10^{10} + 1)x^2 - x + 10^{20} + 10^{10} - 1 = 0$$
correct to four decimal places ($\cdot$ denotes multiplication).

13. Consider polynomials $ax^2 - bx + c$ with integer coefficients which have two distinct zeros in the open interval $(0, 1)$. Exhibit with a proof the least positive integer value of $a$ for which such a polynomial exists.

14. The sequence $\{q_n(x)\}$ of polynomials is defined by
$$q_1(x) = 1 + x \quad q_2(x) = 1 + 2x$$
and, for $m \geq 1$ by
$$q_{2m+1}(x) = q_{2m}(x) + (m + 1)x q_{2m-1}(x)$$
$$q_{2m+2}(x) = q_{2m+1}(x) + (m + 1)x q_{2m}(x).$$
Let $x_n$ be the largest real solution of $q_n(x) = 0$. Prove that $\{x_n\}$ is an increasing sequence whose limit is 0.

15. Assuming that all the zeros of the cubic $t^3 + at^2 + bt + c$ are real, show that the difference between the greatest and the least of them is not less than $(a^2 - 3b)^{1/2}$ nor greater than $2(a^2 - 3b)^{1/2}$.

16. How many roots of the equation $z^6 + 6z + 10 = 0$ lie in each quadrant of the complex plane?

17. Show that
$$4x^{16}(x + 1)^{64} - 3x^9(x + 1)^{27} + 2x^4(x + 1)^8 - 1 = 0$$
has at most 14 positive roots.

18. For which real values of $a$ do all roots of $z^3 - z^2 + a = 0$ satisfy $|z| \leq 1$?

19. Let the zeros $a$, $b$, $c$ of $f(t) = t^3 + pt^2 + qt + r$ be real, and let $a \leq b \leq c$. Prove that, if the interval $(b, c)$ is divided into *six* equal parts, a zero of $f'(x)$ will be found in the *fourth* part, counting from the end $b$. What will be the form of $f(t)$ if the root in question of $f'(t) = 0$ falls at either end of the fourth part?

20. Let $n_1 < n_2 < \cdots < n_k$ be a set of positive integers. Prove that the polynomial $1 + z^{n_1} + z^{n_2} + \cdots + z^{n_k}$ has no zeros inside the circle $|z| < (1/2)(\sqrt{5} - 1)$.

21. For which complex values of $a$ do all the zeros of $z^3 + 12(1 + i\sqrt{3})z + a$ lie on a straight line?

22. Show that $1 + t + t^2/2! + t^3/3! + \cdots + t^{2n}/(2n)! = 0$ has no real roots.

23. Prove that a polynomial $p(t)$ for which $p(t)$ is real when $t$ is real and nonreal when $t$ is nonreal must be linear.

24. Let $p(t) = a_0 + a_1 t + \cdots + a_n t^n$ be a real polynomial of degree exceeding 1, such that

$$0 < a_0 < - \sum_{k=1}^{[n/2]} [1/(2k+1)]a_{2k}.$$

   Prove that $p(t)$ has a real zero $r$ such that $|r| < 1$.

25. $p(z)$ is a complex polynomial whose zeros can be covered by a closed circular disc of radius $R$. Show that the zeros of $np(z) - kp'(z)$ can be covered by a closed circular disc of radius $R + |k|$, where $n$ is the degree of $p(z)$, $k$ is any complex number and $p'(z)$ is the derivative of $p(z)$.

26. Find all the zeros of $az^{p+q} - bz^p + b - a$ $(0 < a < b)$ which satisfy $|z| = 1$.

27. Suppose that $-1 \le u \le 1$. Prove that each root of the equation

$$x^{n+1} - ux^n + ux - 1 = 0$$

   has modulus 1.

28. Show that, for all integers $k \ge 0$,

$$(n+1)^{-k}x^n + n^{-k}x^{n-1} + \cdots + 2^{-k}x + 1 = 0$$

   has no real root if $n$ is even and exactly one root if $n$ is odd.

29. Let $a_1, a_2, \ldots, a_n$ be nonzero reals with $a_1 < a_2 < \ldots < a_n$. Show that the following equation holds for $n$ real values of $x$:

$$\frac{a_1}{a_1 - x} + \frac{a_2}{a_2 - x} + \cdots + \frac{a_n}{a_n - x} = n,$$

   if all the $a_i$ have the same sign. What happens if $a_1 < 0 < a_n$?

30. Let $k > 0$. Show that, if $|a_i| < k$ $(1 \le i \le n)$, then

$$1 + a_1 z + a_2 z^2 + \cdots + a_n z^n = 0$$

   has no root with $|z| < 1/(k+1)$. Is the converse true?

31. Show that every zero $z$ of the complex polynomial

$$f(z) = z^n + a_{n-1}z^{n-1} + \cdots + a_1 z + a_0$$

satisfies $-b \le \operatorname{Re} z \le a$, where $a$ and $b$ are the unique positive roots of the equations

$$x^n + (\operatorname{Re} a_{n-1})x^{n-1} - |a_{n-2}|x^{n-2} - |a_{n-3}|x^{n-3} - \cdots - |a_1|x - |a_0| = 0$$

$$x^n - (\operatorname{Re} a_{n-1})x^{n-1} - |a_{n-2}|x^{n-2} - |a_{n-3}|x^{n-3} - \cdots - |a_1|x - |a_0| = 0.$$

32. Suppose that $p(t)$ has $n$ distinct real zeros exceeding 1. Show that

$$q(t) = (t^2 + 1)p(t)p'(t) + t[p(t)^2 + (p'(t))^2]$$

has at least $2n - 1$ distinct real zeros.

33. Let $p(x)$ be a polynomial of degree $n$ with only real zeros and real coefficients. Show that

$$(n - 1)[p'(x)]^2 - np(x)p''(x) \ge 0.$$

34. Show that there exist infinitely many monic polynomial equations over $\mathbf{Z}$ of degree $n$ such that $n-1$ of the roots occur within a specified interval, however small.

35. Let $m$ be a positive integer and define the real polynomials $f(x)$ and $g(x)$ by

$$(1 + ix)^m = f(x) + ig(x).$$

Prove that, for arbitrary real numbers $a$ and $b$, not both zero,

$$af(x) + bg(x)$$

has only real zeros.

# Hints

## Chapter 5

2.10. How many sign changes can there be altogether in the polynomial at $t$ and at $-t$?

2.11. Multiply by $t - 1$ and apply Exercise 10.

3.5. (b) Note that the polynomial in (a) is positive when $t > r$.

3.6. (a) A standard way to isolate the difference of the coefficients is to multiply the polynomial by $1 - t$. Show that, if $|w| \leq 1$ and $w \neq 1$, then $|(1 - w)g(w)| \geq b_0 - (b_0 - b_1) - (b_1 - b_2) - \cdots - (b_{n-1} - b_n) - b_n$.

(b) Use the result of Exercise (a) along with Exercise 4(a).

(c) The conditions imply that $0 < a_n u^n \leq a_{n-1} u^{n-1} \leq \cdots \leq a_1 u \leq a_0$.

3.8. (a) Consider two cases, according as the zeros are real or nonreal. Observe that, if $1 + c \leq |b|$, then both zeros are real. Examine $(1+c) \pm b$ as a function of the zeros.

(b) This solution needs some careful arguing. Again, look at two cases according as one or three zeros are real. Write out $1 \pm b + c \pm d$ and $1 - c + bd - d^2$ in terms of the zeros. It may be useful to recognize the values of the polynomials at $t = -1$ and $t = 1$.

4.1. If $u$ and $v$ are distinct real zeros, find a relation between them which indicates that $0 < u, v < 1$ cannot occur. Alternatively, look at what the graph of the polynomial must be.

4.2. Show that the left side exceeds 1 at $x = 1/n!$. Express $(x + k)$ as $k(1 + x/k)$.

4.3. Multiply by $x^{-n}$.

4.4. Separate the even and odd powers of $x$; this is a "quickie".

4.5. Prove more generally by induction that $f_n(t) = k$ has $2^n$ real roots when $|k| < 2$.

4.6. Multiply the equation by $(t-a)(t-b)$. Rolle's Theorem will be useful. (Exploration E.28).

4.7. Rewrite the polynomial as $(t-a_1)(t-a_3)(t-a_5)+(t-a_2)(t-a_4)(t-a_6)$.

4.8. The hard part is to show that $f(2(a + b + c)/3) \leq 0$. This is equivalent to showing that $10\Sigma a^3 - 6\Sigma a^2 b + 6abc$ is nonnegative. Since this quantity vanishes when $a = b = c$, try to write it in the form $A(b - c)^2 + B(c - a)^2 + C(a - b)^2$. $A$, $B$ and $C$ will of course be linear and homogeneous in $a$, $b$, $c$.

4.9. Factor $\Pi(x - a_i) - 1$.

4.10. Factor $p(t)$ and compute $p'(a_1)$. Note that $|a_i - a_1| \leq |a_i - b| + |b - a_1|$.

4.11. Let $a \leq b \leq c$ and $u \leq v$, where $u$ and $v$ are the zeros of $(t - a)(t - c) - e^2$. $f(u)$ can be written as a perfect square.

4.12. The left side can be written as a quadratic in $(x^2 - 10^{10})$.

4.13. What is the sign of the value of the polynomial at 0 and 1? What about the discriminant? What about the product of the zeros? Use $(a-1)(c-1) \geq 0$ to show $b \geq 5$.

4.14. Use induction. Show that $x_{2m+2} > -1/(m+1)$ by plugging $x = -1/(m+1)$ into both equations.

4.16. Let $z = r(\cos\theta + i\sin\theta)$ and write the equation as a real system in $r$ and $\theta$. Let $0 < \theta < \pi$. What is the sign of $\sin 6\theta$?

4.17. Differentiate. Factor out the highest power of $x$ and $x+1$. Differentiate some more. The strategy is to use Rolle's Theorem to relate the positive roots of the equation to the roots of an equation of the form $u(x)x^m(x+1)^m = 0$, where the degree of $u(x)$ can be identified.

4.18. First look at the case in which all zeros are real. For the situation in which there are two nonreal zeros $u \pm iv$ and one real zero $r$, use the relationship between zeros and coefficients to obtain a system for $r$, $u$ and $u^2 + v^2$. Eliminate $r$ and $u$ and get a cubic equation for $u^2 + v^2$ which involves $a$.

4.19. Make a linear change of variable and deal with a polynomial whose zeros are $-u$, $-v$, $v$ where $u \geq v \geq 0$. Identify the endpoints of the interval which should contain a zero of the derivative.

4.20. What quadratic over $\mathbf{Z}$ has $(1/2)(\sqrt{5}-1)$ as zero? Deal with the case $n_1 \geq 2$ first. If $n_1 = 1$, multiply the polynomial by $1 - z$.

4.21. What is the sum of the zeros? If the zeros are on a straight line, what is the relationship between this line and the origin?

4.22. The polynomial has a minimum. Its derivative vanishes there.

4.23. What does the property imply concerning the zeros of $p(t)$? Look at $p(t) + k$ for all real values of $k$.

4.24. Let $q'(t) = p(t)$ and $q(0) = 0$. Show that $q(1) < q(-1)$ and that $q(t)$ is increasing at $t = 0$. Sketch the graph of $q(t)$.

4.25. Look at $n - kp'(z)/p(z)$.

4.26. What roots of unity will satisfy the equation? Suppose $|z| = 1$. Note that $bz^p = az^{p+q} + (b-a)$. Take absolute values.

4.27. If there is a root with modulus unequal to 1, there is a root $z$ with modulus exceeding 1. For such a root, $z^n(z - u) = (1 - uz)$. Let $z = r(\cos\theta + i\sin\theta)$ and look at the square of the absolute values of both sides.

4.28. Use induction.

4.29. Multiply up the denominators and evaluate the two sides at $x = a_i$.

4.31. If $\operatorname{Re} z > a$, then $|z + a_{n-1}| \geq \operatorname{Re}(z + a_{n-1}) > a + \operatorname{Re}(a_{n-1})$.

4.32. $q(t)$ can be factored.

4.33. Use induction. Multiply the left side by $n - 1$.

4.34. It is easy to find a polynomial of degree $n - 1$ over $\mathbf{Z}$ whose zeros are prescribed rationals. If such a polynomial is perturbed by adding a sufficiently small multiple of $t^n$, the zeros will not move much.

4.35. Express $af(x) + bg(x)$ in terms of $(1 + ix)^m$ and $(1 - ix)^m$. What can be said about the absolute value of the ratio of these two quantities when $x$ is a zero?

# 6

# Symmetric Functions of the Zeros

## 6.1 Interpreting the Coefficients of a Polynomial

Does the polynomial $t^6+2t^5+3t^4-4t^3+5t^2+6t+7$ have any nonreal zeros? Some methods for answering this question have already been discussed, but there is another approach which exploits the relationship between the coefficients and the roots. If the zeros $t_1, t_2, \ldots, t_6$ were real, then

$$Q = t_1^2 + t_2^2 + \cdots + t_6^2$$

would have to be positive. Is there a way of determining $Q$ without having to go to all the trouble of finding the zeros?

$Q$ is a symmetric function of the zeros. In Exercise 4.6.6, it was indicated that, as a result of the Fundamental Theorem of Algebra, the elementary symmetric functions of the zeros are expressible in terms of the coefficients. In Exercise 2.2.15, Gauss' Theorem that every symmetric polynomial can be written in terms of the elementary symmetric functions was presented. As a consequence, we can obtain an expression for $Q$ in terms of the coefficients and, thus, check its sign. If $Q$ turns out to be negative, we can be sure that not all the zeros are real. If $Q$ turns out to be nonnegative, then there may or may not be nonreal zeros.

Already, for polynomials of low degree, we have exploited the relationship between zeros and coefficients (see, for example, Exercise 1.2.16 and 4.1.7-9). With the Fundamental Theorem in hand, we can generalize this to polynomials of arbitrary degree.

Recall that, if $t_i$ $(1 \leq i \leq n)$ are the zeros of $a_n t^n + \cdots + a_1 t + a_0$ and $s_r$ is the $r$th elementary symmetric function of the $t_i$, then

$$s_r = (-1)^r a_{n-r}/a_n.$$

## Exercises

1. Find a cubic equation whose roots are the squares of the roots of the equation $x^3 - x^2 + 3x - 10 = 0$.

2. Let $u$, $v$, $w$ be the zeros of the cubic polynomial $4t^3 - 7t^2 - 3t + 2$. Determine a cubic polynomial whose zeros are $u-(1/vw)$, $v-(1/wu)$, $w - (1/uv)$.

3. Let $m$, $n$, $p$, $q$ be the zeros of the quartic polynomial $t^4-3t^3+2t^2+t-1$. Without determining any of $m$, $n$, $p$, $q$ explicitly, determine a sextic polynomial whose zeros are $mn$, $mp$, $mq$, $np$, $nq$, $pq$. Check your answer in an independent way.

4. Find the monic polynomial whose zeros are the reciprocals of those of the polynomial $t^3 - 2t^2 + 6t + 5$. Find the polynomial of degree 3 over $\mathbf{Z}$ with these zeros for which the coefficients have greatest common divisor 1.

5. Let $p(t) = \Sigma a_r t^r$ be a polynomial over $\mathbf{C}$. Using the relationship between zeros and coefficients, verify that a polynomial whose zeros are the reciprocals of those of $p(t)$ is

$$a_0 t^n + a_1 t^{n-1} + a_2 t^{n-2} + \cdots + a_{n-1} t + a_n = t^n p(1/t).$$

6. Let $p(t) = \Sigma a_r t^r$ have zeros $t_i$. Determine a polynomial whose zeros are $kt_i$ $(1 \le i \le n)$.

7. (a) The sum of the zeros (counting multiplicity) of a polynomial is 0. Prove that the sum of the zeros of the derivative is also 0.

   (b) If $z_1, \ldots, z_n$ are the zeros of a polynomial $p(t) = t^n + a_{n-2}t^{n-2} + \cdots$ and $w_1, \ldots, w_{n-1}$ are the zeros of $p'(t)$, prove that $n\Sigma w_i^2 = (n-2)\Sigma z_i^2$.

8. (a) Show that, if all the zeros of a polynomial $p(t) = \Sigma a_r t^r$ are real, then $a_{n-1}^2 \ge 2a_{n-2}a_n$ and $a_1^2 \ge 2a_0 a_2$. Show that the converse is not true.

   (b) Use (a) to verify that not all the zeros of the polynomial $t^6 + 2t^5 + 3t^4 - 4t^3 + 5t^2 + 6t + 7$ are real.

   (c) By making use of Rolle's Theorem, strengthen the result of (a) to: if the zeros are real, then $(n - 1)a_{n-1}^2 \ge 2na_n a_{n-2}$. Give an example to show that the converse is not true for $n \ge 3$.

9. Consider the cubic polynomial $p(t) = t^3 + at^2 + bt + c$, with $a$, $b$, $c$ real. Suppose its zeros are $x$, $y$, $z$.

   (a) Verify that $xy = z^2 + az + b$ and that

   $$(x - y)^2 = -[3z^2 + 2az - (a^2 - 4b)] = (a^2 - 3b) - p'(z).$$

(b) Prove that, if $x$, $y$, $z$ are real, then $a^2 \geq 3b$ and that $z$ (and, by symmetry, $x$ and $y$ as well) must lie in the closed interval $[u, v]$, where

$$u = (1/3)\left(-a - 2\sqrt{a^2 - 3b}\right)$$

$$v = (1/3)\left(-a + 2\sqrt{a^2 - 3b}\right).$$

Note that $p'(u) = p'(v) = a^2 - 3b$.

(c) With the hypotheses of (b), prove that $p(u) \leq 0 \leq p(v)$ and deduce that

$$|2a^3 - 9ab + 27c| \leq 2(a^2 - 3b)^{3/2}$$

and thence that

$$a^2b^2 - 4a^3c + 18abc - 4b^3 - 27c^2 \geq 0.$$

(d) Suppose that $a^2b^2 - 4a^3c + 18abc - 4b^3 - 27c^2 \geq 0$. Show that $a^2 - 3b \geq 0$ and that the cubic $p(t)$ has three real zeros.

[Cf. Problems 1.4.4, 5.4.15, 6.2.5.]

10. Solve the equation $t^4 - t^3 - 7t^2 + 23t - 20 = 0$, given that the product of two of the roots is $-5$.

11. Consider the polynomial equation

$$x^4 + px^3 + qx^2 + rx + s = 0.$$

(a) Prove that the product of two of its roots is equal to the product of the other two iff $r^2 = p^2s$.

(b) Prove that, if the sum of two of its roots is equal to the sum of the other two, then $p^3 + 8r = 4pq$.

(c) Suppose that $p^3 + 8r = 4pq$. Must the sum of two of its roots equal the sum of the other two?

12. In Exercise 1.4.11, the quartic equation

$$t^4 + pt^2 + qt + r = 0 \qquad\qquad \ldots(1)$$

was solved by factoring the left side as $(t^2 + ut + v)(t^2 - ut + w)$, where $u$ satisfies the equation

$$t^6 + 2pt^4 + (p^2 - 4r)t^2 - q^2 = 0. \qquad\qquad \ldots(2)$$

(a) Argue that the roots of (2) can be given as $\pm 2a$, $\pm 2b$, $\pm 2c$, where $a$, $b$, $c$ are selected so that $8abc = -q$.

(b) Let $\alpha = a - b - c$, $\beta = b - c - a$, $\gamma = c - a - b$, $\delta = a + b + c$. Verify that

$$\alpha + \beta + \gamma + \delta = 0$$

$$\alpha\beta + \alpha\gamma + \alpha\delta + \beta\gamma + \beta\delta + \gamma\delta = (\alpha + \beta)(\gamma + \delta) + \alpha\beta + \gamma\delta$$

$$= -2(a^2 + b^2 + c^2)$$

$$\alpha\beta\gamma + \alpha\beta\delta + \alpha\gamma\delta + \beta\gamma\delta = (\alpha + \beta)\gamma\delta + \alpha\beta(\gamma + \delta) = 8abc$$

$$\alpha\beta\gamma\delta = (a^2 + b^2 + c^2)^2 - 4(a^2b^2 + a^2c^2 + b^2c^2).$$

(c) Make use of (2) to show that $a^2$, $b^2$, $c^2$ are the roots of the equation $64t^3 + 32pt^2 + 4(p^2 - 4r)t - q^2 = 0$.

(d) From (b) and (c), show that $\alpha$, $\beta$, $\gamma$, $\delta$ are the roots of (1).

## 6.2   The Discriminant

In the theory of the quadratic polynomial $at^2 + bt + c$, it is possible to detect the presence of a double zero by examining the discriminant $b^2 - 4ac$, a function of the coefficients. For the cubic polynomial $t^3 + pt + q$, the expression $27q^2 + 4p^3$ plays a similar role. When the coefficients are real, the signs of these expressions determine whether all the zeros are real. (See Exercises 1.2.1, 1.4.4.) In Exercise 1.9, it was found that the zeros of $t^3 + at^2 + bt + c$ were real if and only if $a^2b^2 - 4a^3c + 18abc - 4b^3 - 27c^2$ was nonnegative.

For a polynomial of arbitrary degree, a function of the coefficients can be determined which will vanish precisely when the polynomial has a nonsimple zero and, for real polynomials, will be nonnegative if (but not necessarily, only if) all zeros are real. The link between zeros and coefficients will be Gauss' theorem for symmetric functions. (See Exercises 1.5.10, 2.2.15.)

## Exercises

1. Let $p(t) = a_n t^n + \cdots + a_0$ be a polynomial with zeros $t_1, \ldots, t_n$.

   (a) Consider the expression

   $$(t_1 - t_2)(t_1 - t_3) \cdots (t_2 - t_3)(t_2 - t_4) \cdots (t_3 - t_4) \cdots (t_{n-1} - t_n),$$

   which is a product of $\binom{n}{2}$ terms (one for each pair of zeros). Verify that it will vanish exactly when there is a zero of multiplicity exceeding 1, but that it is not symmetric in the zeros.

(b) Let $D(t_1, t_2, \ldots, t_n)$ be the *square* of the expression in (a). This is called the *discriminant* of the polynomial $p(t)$. Verify that it is a homogeneous symmetric polynomial of degree $n(n-1)$ and that it vanishes exactly when there is a zero of multiplicity greater than 1.

(c) Must the discriminant of a polynomial over $\mathbf{R}$ be real?

2. Verify that for the quadratic polynomial $at^2 + bt + c$, the discriminant, as defined in Exercise 1(b) is $(b/a)^2 - 4(c/a)$ (which is equal to the usual discriminant divided by a square).

3. (a) Suppose that $p(t)$ is a polynomial over $\mathbf{R}$ with discriminant $D$ whose zeros are all real. Prove that $D \geq 0$.

(b) Show that the converse of (a) is true for quadratic and cubic polynomials, but not for polynomials of higher degree.

4. Show that if $p(t)$ is a polynomial over $\mathbf{R}$ with all zeros distinct, then

(i) if there are an odd number of pairs of nonreal complex conjugates among the zeros, then $D < 0$;

(ii) if there are an even number of pairs of nonreal complex conjugates among the zeros, then $D > 0$.

5. (a) Find the discriminant of the polynomial $t^3 + pt + q$ and state its relationship to the quantity $27q^2 + 4p^3$.

(b) Find the discriminant of the cubic $t^3 + at^2 + bt + c$ and of the general cubic $a_3 t^3 + a_2 t^2 + a_1 t + a_0$.

6. Find the zeros of each of the following quartics and use them to evaluate their discriminants:

(a) $t^4 - 1$

(b) $t^4 + 5t^2 + 4$.

7. Show that the discriminant of a polynomial over $\mathbf{C}$ is nonzero if and only if the greatest common divisor of the polynomial and its derivative is a nonzero constant.

# Exploration

**E.51.** What is the discriminant of $t^n - 1$?

## 6.3   Sums of the Powers of the Roots

While the determination of an arbitrary symmetric function of the zeros of a polynomial might be quite difficult, it is relatively easy to determine the sums of various powers of the roots by setting up a recursion relation. In the exercises, we suppose that $p(t) = t^n + c_{n-1}t^{n-1} + \cdots + c_1 t + c_0$ is a monic polynomial with zeros $t_1, \ldots, t_n$. Define

$$p_0 = n$$

and

$$p_k = t_1^k + t_2^k + \cdots + t_n^k \quad (k = 1, 2, \ldots).$$

As before, $s_r$ will denote the $r$th elementary symmetric function of the zeros.

## Exercises

1. Let $a$, $b$, $c$ be the roots of the equation

   $$x^3 - 2x^2 + x + 5 = 0.$$

   Find the value of $a^4 + b^4 + c^4$.

2. Verify that $p_1 = s_1$ and that $p_2 = s_1^2 - 2s_2$.

3. It is straightforward to give a recursion relation for $p_k$ when $k \geq n$. Prove that, for $r \geq 0$,

   $$p_{n+r} + c_{n-1}p_{n+r-1} + c_{n-2}p_{n+r-2} + \cdots + c_1 p_{r+1} + c_0 p_r = 0.$$

4. When $k < n$, the recursion relation to express $p_k$ in terms of sums of earlier powers is a little more complicated. Verify that

   (a) $p_1 + c_{n-1} = 0$

   (b) $p_2 + c_{n-1}p_1 + 2c_{n-2} = 0.$

5. The purpose of this exercise is to develop a conjecture concerning $p_3$.

   (a) Observe that $p_3$ is of degree 3 in the $t_i$. Infer that $c_{n-4}, c_{n-5}, \ldots,$ $c_1$, $c_0$ will not likely be involved in an expression of $p_3$.

   (b) On the basis of the information obtained in Exercises 3 and 4, argue that it is reasonable to guess an equation of the form

   $$p_3 + k_1 c_{n-1}p_2 + k_2 c_{n-2}p_1 + k_3 c_{n-3} = 0.$$

(c) Assuming for the moment that (b) is valid, what must the coefficients $k_i$ be? Try the substitutions $(1,0,0,0,\ldots)$, $(1,1,0,0,\ldots)$ and $(1,1,1,0,\ldots)$ for $(t_1, t_2, t_3, t_4, \ldots)$ to obtain three linear equations which must be satisfied by the $k_i$. Solve these equations to obtain $k_1 = k_2 = 1$ and $k_3 = 3$.

6. On the basis of Exercise 3, 4 and 5, it is reasonable to conjecture that, for $k = 1, 2, 3, \ldots, n-1$,

$$p_k + c_{n-1}p_{k-1} + c_{n-2}p_{k-2} + \cdots + c_{n-k+1}p_1 + kc_{n-k} = 0.$$

This result can actually be established by induction on the degree $n$ of $p(t)$ (i.e. the number of the $t_i$).

(a) Verify that the result holds for $n = 2$ and $n = 3$.

(b) Suppose that the result has been established for polynomials of degree up to $n-1$. Let $c_i'$ and $p_i'$ be obtained respectively from $c_i$ to $p_i$ by setting $t_n = 0$. Verify that the monic polynomial with zeros $t_1, t_2, \ldots, t_{n-1}$ is

$$t^{n-1} + c_{n-1}'t^{n-2} + \cdots + c_3't^2 + c_2't + c_1'.$$

Use the induction hypothesis to establish that

$$p_k + c_{n-1}p_{k-1} + \cdots + c_{n-k+1}p_1 + kc_{n-k}$$

(considered as a polynomial in the $t_i$) vanishes for $t_n = 0$. Deduce that it is divisible by $t_n$, and therefore, because of its symmetry is actually divisible by $t_1 t_2 \ldots t_n$. Conclude that, because its degree as a polynomial in the $t_i$ is equal to $k < n$, it must in fact vanish.

7. Let $t^2 + c_1 t + c_0$ have zeros $t_1$ and $t_2$. Write down recursion relations for $p_k = t_1^k + t_2^k$ ($1 \le k \le 5$) and verify them directly. Determine these $p_k$ in terms of $c_1$ and $c_0$ only.

Carry this out for the polynomials $t^2 - 3t + 2$ and $t^2 + t + 1$, and verify your answers directly.

8. Find the sum of the fifth powers of the zeros of $t^3 + 7t^2 - 6t - 1$.

9. Let $z_1, z_2, \ldots, z_n$ be complex numbers for which $z_1^k + z_2^k + \cdots + z_n^k = 0$ for $1 \le k \le n$. Must each $z_i$ vanish?

## Explorations

**E.52.** Use the recursion relations for the $p_k$ to obtain expressions for each of them which involve only the coefficients $c_i$ and none of the other $p_i$.

Another approach to obtaining the recursion relation for the $p_k$ is to make use of infinite series expansions. Observe that the derivative satisfies

$$p'(t) = \sum_{i=1}^{n} \frac{p(t)}{t - t_i} = \frac{p(t)}{t} \sum_{i=1}^{n} \left(1 - \frac{t_i}{t}\right)^{-1}.$$

Make a change of variable $t = 1/s$. Then

$$s^{n-1}p'(1/s) = s^n p(1/s) \sum_{i=1}^{n} (1 + t_i s + t_i^2 s^2 + t_i^3 s^3 + \cdots)$$

or

$$nc_n + (n-1)c_{n-1}s + (n-2)c_{n-2}s^2 + \cdots + 2c_2 s^{n-2} + c_1 s^{n-1}$$

$$= (c_n + c_{n-1}s + c_{n-2}s^2 + \cdots + c_0 s^n) \sum_{i=1}^{n}(1 + t_i s + \cdots).$$

On collecting terms in the various powers of $s$, the right side becomes

$$(c_n + c_{n-1}s + c_{n-2}s^2 + \cdots + c_0 s^n)(n + p_1 s + p_2 s^2 + p_3 s^3 + \cdots)$$

$$= nc_n + (nc_{n-1} + p_1 c_n)s + (nc_{n-2} + p_1 c_{n-1} + p_2 c_n)s^2 + \cdots.$$

Bring all nonzero terms to one side of the equation. On the basis that a power series (like a polynomial) vanishes iff all its coefficients vanish, obtain the recursion relations.

**E.53. A Recursion Relation.** Fix the values of $c_1, c_2, c_3, \ldots$ and consider the sequence $\{p_1, p_2, p_3, \ldots, p_n, \ldots\}$ whose first $n$ terms are given and whose remaining terms satisfy the equation

$$p_{n+r} + c_{n-1}p_{n+r-1} + c_{n-2}p_{n+r-2} + \cdots + c_1 p_{r+1} + c_0 p_r = 0$$

for $r = 1, 2, 3, 4, 5, \ldots$. Assign the values to the first $n$ entries according to the equations

$$p_k = -c_{n-1}p_{k-1} - c_{n-2}p_{k-2} - \cdots - kc_{n-k}.$$

Use the theory of Exploration **E.50** to find a formula for $p_k$ as a sum of $k$th powers. Look at the particular cases $n = 1, 2, 3$.

**E.54. Sum of the First $n$ $k$th Powers.** What is the formula for the sum $1^k + 2^k + \cdots + n^k$, the sum of the first $n$ $k$th powers? Consider the monic polynomial

$$S_n(t) = (t-1)(t-2)(t-3)\cdots(t-n) = t^n + b(n,1)t^{n-1}$$

$$+ b(n,2)t^{n-2} + \cdots + b(n,n)$$

whose roots are the integers $1, 2, \ldots, n$. Verify that the coefficients are given in the following table:

| $n$ | $b(n,0)$ | $b(n,1)$ | $b(n,2)$ | $b(n,3)$ | $b(n,4)$ | $b(n,5)$ | $b(n,6)$ |
|---|---|---|---|---|---|---|---|
| 0 | 1 | | | | | | |
| 1 | 1 | $-1$ | | | | | |
| 2 | 1 | $-3$ | 2 | | | | |
| 3 | 1 | $-6$ | 11 | $-6$ | | | |
| 4 | 1 | $-10$ | 35 | $-50$ | 24 | | |
| 5 | 1 | $-15$ | 85 | $-225$ | 274 | $-120$ | |
| 6 | 1 | $-21$ | 175 | $-735$ | 1624 | $-1764$ | 720 |
| 7 | 1 | $-28$ | 322 | $-1960$ | 6769 | $-13132$ | 13068 |
| 8 | 1 | $-36$ | 546 | $-4536$ | 22449 | $-67284$ | 118124 |

Use this table and the relations developed earlier in this section to find the sum of the first $n$ $k$th powers for $1 \leq n \leq 8$ and $1 \leq k \leq 7$. Look into the possibility of determining a formula for the sum in general. Can you verify the familiar formulae for $k = 1, 2, 3$? Assess for its effectiveness this approach for getting a general formula.

## 6.4   Problems

1.  (a) If the roots of $x^3 + ax^2 + bx + c = 0$ are in arithmetic progression, prove that $2a^3 - 9ab + 27c = 0$.

   (b) If the roots of $x^3 + ax^2 + bx + c = 0$ are in geometric progression, show that $a^3c = b^3$.

2. Given the product $p$ of the sines of the angles of a triangle and the product $q$ of their cosines, show that the tangents of the angles are the roots of the equation

$$qx^3 - px^2 + (1+q)x - p = 0.$$

3. If $a$, $b$, $c$ are the roots of the equation

$$x^3 - x^2 - x - 1 = 0,$$

   (a) show that $a$, $b$, $c$ are all distinct,

   (b) show that
$$\frac{b^n - c^n}{b - c} + \frac{c^n - a^n}{c - a} + \frac{a^n - b^n}{a - b}$$
   is an integer for $n = 1, 2, \ldots$.

4. It is given that the roots of the equation

$$17x^4 + 36x^3 - 14x^2 - 4x + 1 = 0$$

are in harmonic progression. Find these roots.

5. The product of two of the four roots of the quartic equation

$$x^4 - 18x^3 + kx^2 + 200x - 1984 = 0$$

is $-32$. Determine $k$.

6. Three of the roots of $x^4 - px^3 + qx^2 - rx + s = 0$ are $\tan A$, $\tan B$ and $\tan C$, where $A$, $B$, $C$ are angles of a triangle. Determine the fourth root as a function of $p$, $q$, $r$, $s$ alone.

7. If $x$, $y$, $z$ are real and satisfy $x + y + z = 5$ and $yz + zx + xy = 3$, prove that $-1 \le z \le 13/3$.

8. Let $u$, $v$, $w$ be the roots of $x^3 - 6x^2 + ax + a = 0$. Determine all real $a$ for which $(u - 1)^3 + (v - 2)^3 + (w - 3)^3 = 0$. For each $a$ determine the corresponding values of $u$, $v$, $w$.

9. Determine those values of the real number $a$ and positive integer $n$ exceeding 1 for which

$$\sum_{k=1}^{n} \frac{x_k + 2}{x_k - 1} = n - 3$$

where $x_1, x_2, \ldots, x_n$ are the zeros of $x^n + ax^{n-1} + a^{n-1}x + 1$.

10. If $u + v + w = 0$, prove that, for $n = 0, 1, 2, \ldots$,

$$u^{n+3} + v^{n+3} + w^{n+3} = uvw(u^n + v^n + w^n)$$
$$+ (1/2)(u^2 + v^2 + w^2)(u^{n+1} + v^{n+1} + w^{n+1}).$$

11. Let $p$, $q$, $r$, $s$ be the roots of the quartic equation $x^4 - ax^3 + bx^2 - cx + d$. Find the quartic whose roots are $pq + qr + rp$, $pq + qs + sp$, $pr + rs + sp$, $qr + rs + sq$.

12. Find all integer values of $a$ such that all the zeros of $t^4 - 14t^3 + 61t^2 - 84t + a$ are integers.

13. Suppose that $p(z)$ is a polynomial of degree at least 2 whose coefficients are complex numbers and not all real. Prove that the equation $p(z)p(-z) = p(z)$ has roots in both the upper and lower half planes (i.e. for which $\operatorname{Im} z$ take both positive and negative values). Give an example to show that this is false when the degree of $p(z)$ is 1.

14. Determine all polynomials of degree $n$ with each of its $n+1$ coefficients equal to $+1$ or $-1$ which have only real zeros.

15. Suppose that every zero of the polynomial $f(x)$ is simple. If every root of the equation $f'(x) = 0$ be subtracted from every root of the equation $f(x) = 0$, find the sum of the reciprocals of the differences.

16. Show that the roots of the cubic equation

$$64t^3 - 192t^2 - 60t - 1 = 0$$

are $\cos^3(2\pi/7)\sec(6\pi/7)$, $\cos^3(4\pi/7)\sec(2\pi/7)$, $\cos^3(6\pi/7)\sec(4\pi/7)$.

17. Solve for $x$, $y$, $z$, $w$:

$$x + ay + a^2z + a^3w = a^4$$
$$x + by + b^2z + b^3w = b^4$$
$$x + cy + c^2z + c^3w = c^4$$
$$x + dy + d^2z + d^3w = d^4$$

where $a$, $b$, $c$, $d$ are all distinct.

18. Let $x = u(v-w)^2$, $y = v(w-u)^2$, $z = w(u-v)^2$ where $u+v+w = 0$. Eliminate $u$, $v$, $w$ to obtain

$$x^3 + y^3 + z^3 + a(x^2y + x^2z + y^2x + y^2z + z^2x + z^2y) + bxyz = 0$$

for suitable $a$ and $b$.

19. Let $u$, $v$, $w$ be distinct constants. Solve

$$\frac{x}{a+u} + \frac{y}{b+u} + \frac{z}{c+u} = 1$$
$$\frac{x}{a+v} + \frac{y}{b+v} + \frac{z}{c+v} = 1$$
$$\frac{x}{a+w} + \frac{y}{b+w} + \frac{z}{c+w} = 1.$$

# Hints

## Chapter 6

1.4. There are two approaches: either use symmetric functions, or else make a substitution into $t^3 - 2t^2 + 6t + 5 = 0$ to determine the equation that $s = 1/t$ must satisfy.

1.8. (a) Can a quadratic counterexample be found?

1.9. (a) Note that $xyz = -c$. For the latter part, express $(x-y)^2$ in terms of symmetric functions of $x$ and $y$, which are then expressible in terms of $z$.

(b) What does the negativity of $3z^2 + 2az - (a^2 - 4b)$ imply about the relationship between $z$ and the zeros of the quadratic?

(d) What is the sign of $p(t)$ at the zeros of $p'(t)$?

1.11. (a) Express $p$ and $r$ in terms of the roots to get two equations for $(a + b)$ and $(c + d)$, where $a$, $b$, $c$, $d$ are the zeros and $ab = cd$. If $r^2 = p^2 s$, the equation is quasi-reciprocal (Exercise 1.4.17).

(c), (d) Consider the substitution $x = -p/2 - y$.

3.1. $a^3$ and $a^4$ can be expressed in terms of lower powers of $a$.

3.9. Use Exercise 6 to determine the polynomial of degree $n$ with zeros $z_i$.

4.1. (a) Let the zeros be $r - s$, $r$, $r + s$.

4.3. (a) A double root is a zero of the derivative.

(b) Use $b^{k+3} = b^{k+2} + b^{k+1} + b^k$, etc. and induction.

4.4. The reciprocals of terms in harmonic progression are in arithmetic progression.

4.8. Note that $(u - 1) + (v - 2) = -(w - 3)$. Cube this equation.

4.9. Let $y_k = x_k - 1$. What polynomial has zeros $y_k$?

4.11. Let $u = r + s$, $v = rs$, $w = p + q$, $z = pq$ and express the coefficients of the given quartic and the zeros of the required quartic in terms of $u$, $v$, $w$, $z$.

4.12. Look at values assumed by $t^4 - 14t^3 + 16t^2 - 84t$ (which can be easily factored).

4.13. What can be said about the sum of the roots of the equation?

4.14. Where $r_i$ are the roots, look at $\Sigma(r_i^2 + r_i^{-2})$. What is the lower bound for the values of each summand?

4.15. Look at $f(x)/f'(x)$.

4.16. The purported roots have the form $v^3(4v^3 - 3v)^{-1}$, where $v = \cos\theta$ for suitable $\theta$. One strategy is to first determine the cubic equation whose roots are the reciprocals of these. The reciprocals can be simplified to an expression in $v^{-2}$. The cubic equation whose roots are $v$ should be found, so that the elementary functions in the $v^{-2}$ can be determined.

4.17. $a$, $b$, $c$, $d$ are the zeros of the quartic $t^4 - wt^3 - zt^2 - yt - x$.

4.18. $u$, $v$, $w$ are the zeros of a polynomial $t^3 + pt + q$. Express the symmetric functions of $x$, $y$, $z$ in terms of $p$ and $q$.

4.19. Consider the cubic polynomial $(t + a)(t + b)(t + c) - x(t + b)(t + c) - y(t + a)(t + c) - z(t + a)(t + b)$. What are its zeros? What values of $t$ will isolate the coefficients $x$, $y$, $z$?

# 7

# Approximations and Inequalities

## 7.1 Interpolation and Extrapolation

A scientist wishes to know the index of refraction of pure water relative to air for sodium light at a temperature of 54°C. A table provides the following information:

| Temperature (°C.) | Index ($r(T)$) |
|:---:|:---:|
| 20 | 1.33299 |
| 30 | 1.33192 |
| 40 | 1.33051 |
| 50 | 1.32894 |
| 60 | 1.32718 |
| 70 | 1.32511 |
| 80 | 1.32287 |
| 90 | 1.32050 |
| 100 | 1.31783 |

Likely, the number sought lies between 1.32718 and 1.32894. Can we be more precise? Can we find a function which expresses the index in terms of the temperature? If not, can we sensibily approximate a functional relation between the variables by means of a polynomial? We will look at some possible approaches, and assess the effectiveness for this particular situation.

## Exercises

1. One way to find the required index of refraction from the given table is to argue that, as 54 is 4/10 of the way from 50 to 60, the index lies 4/10 of the way from 1.32894 to 1.32718. Verify that this leads to the result

$$1.32894 - (0.4)(0.00176) = 1.32824.$$

Does this seem reasonable?

2. The approach in Exercise 1 amounts to assuming that, between 50° and 60°, the index $r(T)$ is a linear function of $T$:

$$r(T) = aT + b \quad (50 \leq T \leq 60).$$

The coefficients $a$, $b$ are to be determined from the conditions that $r(50) = 1.32894$ and $r(60) = 1.32718$. Verify that this leads to $a = -0.000176$ and $b = 1.33774$. What value of $r(54)$ will this yield?

3. Let $f(x)$ be a real-valued function defined on the closed interval $[a, b] = \{x : a \le x \le b\}$. Show that $f(x)$ is linear (i.e. a polynomial of degree 1) if and only if

$$f(x) = (b - a)^{-1}[(x - a)f(b) + (b - x)f(a)].$$

Verify that, if $x = (1 - t)a + tb$, this condition can be rewritten

$$f(x) = (1 - t)f(a) + tf(b).$$

4. (a) On a graph, plot the points $(T, r(T))$ and join the points by a smooth curve. Observe that the relationship between $T$ and $r(T)$ is probably not linear and argue that the linear approximation has probably underestimated the true value of $r(54)$.

   (b) In an attempt to improve our estimate of $r(54)$, we can try a quadratic formula $r(T) = aT^2 + bT + c$, for values of $T$ near 54. Since there are three coefficients to be found, we use information about $r$ for three values of $T$. We take those values closest to 54, namely 40, 50 and 60. Show that this leads to the formula

   $$r(T) = -0.00000095T^2 - 0.0000715T + 1.33489.$$

   What value of $r(54)$ does this yield?

   (c) The experimental value of $r(54)$ is 1.32827. Is this what you would expect?

5. In the general interpolation situation, we have a function $f$ whose values are known at $n + 1$ points:

   $$f(a_i) = b_i \quad (i = 0, 1, 2, \ldots, n).$$

To estimate its values elsewhere, we could take in its place a polynomial which agrees with the function $f$ at the $a_i$. As in the example of the index of refraction, we try to make the degree of the polynomial as small as possible.

In this connection, we have the result that there is a unique polynomial $p(t)$ such that $\deg p(t) \le n$ and $p(a_i) = b_i$ for $i = 0, 1, 2, \ldots, n$. (cf. Exercise 4.6.8).

There is a straightforward way of constructing this polynomial. First, we need some building blocks. With the help of the Factor Theorem, determine, for each $i = 0, 1, \ldots, n$, that there is a polynomial $p_i(t)$ of degree not exceeding $n$ for which

$$p_i(a_i) = 1 \quad p_i(a_j) = 0 \quad (i \ne j).$$

Then verify that $p(t) = \sum_{i=0}^{n} b_i p_i(t)$ is the desired polynomial. This is called the *Lagrange polynomial* determined by the conditions.

6. Determine the quadratic Lagrange polynomial $q$ which satisfies $q(-2) = 7$, $q(1) = 2$ and $q(3) = 1$. Write it in the form $at^2 + bt + c$.

7. In practice, the Lagrange form is not generally the best way to obtain a polynomial fitting certain data, particularly if many other values of the polynomial are also required.

Consider the sequence: 1, 3, 6, 10, 15, 21, 28, 36, .... Show that these are successive values of a quadratic polynomial evaluated on the positive integers.

We form a *difference table* for this sequence as follows. Write out the terms of the sequence in a column; beside the column, put a second whose entries are the differences between successive terms of the first; form a third column from the second in the same fashion, and continue on. Here is what you get in this case:

$$
\begin{array}{ccc}
1 & & \\
 & 2 & \\
3 & & 1 \\
 & 3 & \quad 0 \\
6 & & 1 \\
 & 4 & \quad 0 \\
10 & & 1 \\
 & 5 & \quad 0 \\
15 & & 1 \\
 & \cdot & \quad \cdot
\end{array}
$$

From the table, decide what terms should follow the entry 36 in the original sequence. Check your guess using the quadratic polynomial.

8. Consider the sequence of values: 2, 11, 35, 85, 175, 322, 546, 870, .... Carry out the procedure of Exercise 7, continuing until a column of zeros is obtained. This will occur in the sixth column of the table.

Let us suppose that the sequence arises from the evaluation of a function $f(n)$ at the successive points 1, 2, 3, .... What do you think the form of this function might be? What do you think the next two terms of the sequence following 870 might be?

What is the $n$th term of the sixth column of the table?

What is the $n$th term of the fifth column of the table?

What is the $n$th term of the fourth column of the table? the third column? the second column? the first column?

Give a function $f(n)$ which reproduces the sequence.

9. Let $f(n)$ be a function defined on the positive integers. As in Exercises 7 and 8, construct a table for the sequence $f(1)$, $f(2)$, $f(3)$, ....

   (a) Verify that the $n$th term of the second column is $f(n+1) - f(n)$. Call this quantity $\Delta f(n)$.

   (b) Verify that the $n$th term of the third column is $[f(n + 2) - f(n + 1)] - [f(n + 1) - f(n)] = f(n + 2) - 2f(n + 1) + f(n)$. Call this quantity $\Delta(\Delta f(n))$ or $\Delta^2 f(n)$.

   (c) Let us denote the $n$th element of the $(k + 1)$th column by

   $$\Delta^k f(n).$$

   Prove that

   $$\Delta^k f(n) = \Delta(\Delta^{k-1} f(n)) = \sum_{i=0}^{k} (-1)^{k-i} \binom{k}{i} f(n + i).$$

10. In general, for any function $f(t)$, we can define three operators

    $$\begin{aligned} If(t) &= f(t) &&: \text{identity} \\ Ef(t) &= f(t + 1) &&: \text{shift} \\ \Delta f(t) &= f(t + 1) - f(t) &&: \text{difference.} \end{aligned}$$

    Any of these operators can be iterated. Thus,

    $$E^k f(t) = E(E^{k-1} f(t))$$

    $$\Delta^k f(t) = \Delta(\Delta^{k-1} f(t)) \quad (k \geq 2).$$

    (a) Verify that, for functions $f$ and $g$ and constant $c$,

    $$\Delta(cf(t)) = c\Delta f(t)$$

    $$\Delta(f + g)(t) = \Delta f(t) + \Delta g(t).$$

    (b) Verify that $E^k f(t) = f(t + k)$.

    (c) When $f(t) = t^2 - 3t + 1$, verify that $If(t) = t^2 - 3t + 1$, $Ef(t) = t^2 - t - 1$, $\Delta f(t) = 2t - 2$ and $\Delta^2 f(t) = 2$.

    (d) The operators $I$, $E$, $\Delta$ can be manipulated like numbers, with iteration playing the role of multiplication and $I$ playing the role of 1. Justify the equations

    $$\Delta = E - I \quad \text{and} \quad E = I + \Delta.$$

    (e) Determine an expression for $\Delta^k f(t)$ in terms of $f(t)$, $f(t+1)$, ... by expanding $\Delta^k = (E - I)^k$ binomially and applying this operator to $f(t)$. Compare with Exercise 9(c).

(f) Manipulating formally, we have $E^k = (I+\Delta)^k$. Expand the right side by the binomial theorem and use it to obtain the result

$$f(t+k) = f(t) + k\Delta f(t) + \binom{k}{2}\Delta^2 f(t) + \binom{k}{3}\Delta^3 f(t) + \cdots .$$

(g) Make the substitution $t = 1$, $k = n - 1$ in (f) to determine the polynomial in $n$ whose values at $n = 1, 2, 3, \ldots$ are respectively

(i) $1, 3, 6, 10, 15, \ldots$

(ii) $2, 11, 35, 85, 175, 322, \ldots$.

11. Consider the function $g(n)$ whose values for $n = 1, 2, 3, 4, 5, 6$ are respectively 6, 50, 225, 735, 1960, 4536. Since we know six values of $g$, we might suppose that $g(n)$ is given by a polynomial of the fifth degree in $n$. Make up a difference table, and determine what this polynomial is.

12. Refer to the table for the index of refraction of water for sodium light given at the beginning of this section. Let $f(t)$ be the index when the temperature is $10t$ degrees Celsius.

(a) Determine: $\Delta f(2)$, $\Delta^3 f(3)$, $\Delta^2 f(5)$.

(b) Neglect fourth order differences, and use

$$\begin{aligned}
f(5.4) &= (I + \Delta)^{1.4} f(4) \\
&= f(4) + (1.4)\Delta f(4) + (1/2)(1.4)(0.4)\Delta^2 f(4) + \cdots
\end{aligned}$$

to approximate the index of refraction for 54°C.

(c) Use $f(5.4) = (I + \Delta)^{-0.6} f(6)$ to approximate the index of refraction for 54°C.

13. Given the following table

| number | natural logarithm |
|---|---|
| 0.5 | −0.69315 |
| 1.0 | 0.00000 |
| 1.5 | 0.40547 |
| 2.0 | 0.69315 |
| 2.5 | 0.91629 |
| 3.0 | 1.09861 |
| 3.5 | 1.25276 |
| 4.0 | 1.38629 |

interpolate to approximate the natural logarithms of 1.25, 0.75, 2.1, 2.71828. Extrapolate to find the natural logarithm of 0, 0.25, 5.0. Use various methods. Check your answers with a pocket calculator or from a table of logarithms.

14. Investigate finding an approximate value of $\sqrt{2}$ from a table of powers of 2 with integer exponents.

15. Find the polynomial of least degree whose values at $-3, -2, -1, 0,$ $1, 2, 3$, are, respectively, 5, 6, 13, 17, 21, 23, 29.

   (a) as the Lagrange polynomial for these data

   (b) using the difference operator.

   Are your answers to (a) and (b) the same. Why?

16. *Factorial powers.* Define the rth *factorial power* of $t$ by

$$t^{(0)} = 1$$

$$t^{(1)} = t$$

$$t^{(r)} = t(t-1)(t-2)\cdots(t-r+1) \quad (r\,\text{factors}) \quad (r \geq 1).$$

   (a) Verify that
$$t^2 = t^{(2)} + t$$
$$t^3 = t^{(3)} + 3t^{(2)} + t.$$

   (b) Express $t^4$ and $t^5$ in terms of factorial powers.

   (c) Show that every (ordinary) power of $t$, and therefore every polynomial in $t$, can be expressed as a linear combination

$$\sum a_r t^{(r)}$$

   of factorial powers. To see how this can be done systematically, consult Exploration **E.18** in Chapter 2.

   (d) Show that, for each $r = 1, 2, \ldots, \Delta t^{(r)} = rt^{(r-1)}$.

17. (a) Show that for any polynomial $f(t)$ of degree $n$, $\Delta^n f(t) \neq 0$ and $\Delta^k f(t) = 0$ for $k \geq n + 1$. Is the converse true, i.e. if $\Delta^{n+1} f(t) = 0$ for some function $f(t)$, must $f(t)$ be a polynomial of degree $n$?

   (b) Deduce from (a), that, for any positive base $b \neq 1$, $b^t$ is not a polynomial in $t$.

   (c) Use a difference table to argue that the $n$th term of the Fibonacci sequence defined in Exploration **E.14** is not a polynomial in $n$.

18. (a) Let $f(t)$ be a polynomial of degree not exceeding $k$ over **C**. Verify that

$$f(t) = \sum_{i=0}^{k} (-1)^{k-i} \binom{t}{i} \binom{t-i-1}{k-i} f(i)$$

where, for any nonnegative integer $m$, and any $u$,

$$\binom{u}{m} = \frac{u(u-1)(u-2)\cdots(u-m+1)}{m!} = \frac{u^{(m)}}{m!}.$$

(b) Let $a_1, a_2, \ldots, a_n$ be arbitrary complex numbers and suppose that $f(t)$ is a polynomial over $\mathbf{C}$ of degree less than $n$. Show that, if $p(t) = (t - a_1)(t - a_2)\cdots(t - a_n)$, then

$$f(t) = \sum_{i=1}^{n} [f(a_i)/p'(a_i)](t - a_1)\cdots(\widehat{t - a_i})\cdots(t - a_n).$$

19. (a) Give an example of a polynomial over $\mathbf{Q}$ whose coefficients are not integers, but which take an integer value for every integer value of the variable.

  (b) Suppose that $f(t)$ is a polynomial of degree $k$ over $\mathbf{C}$ and that $f(0)$, $f(1)$, $f(2), \ldots, f(k)$ are integers. Prove that $f(n)$ is an integer for each integer $n$.

20. Find the polynomial $h$ of least degree for which

$$h(k) = 2^k \quad (k = 0, 1, 2, \ldots, n).$$

What is $h(n+1)$?

# Explorations

**E.55. Building Up a Polynomial.** Consider the following simple-minded way of building up, step by step, a polynomial whose value at $a_i$ is $b_i$ $(0 \le i \le n)$. Let the first guess at the required polynomial be $b_0$. This has the correct value at $a_0$, but not at $a_1$. So we add a correction which will keep the correct value at $a_0$ and make the value at $a_1$ correct: $b_0 + b_{01}(t - a_0)$, for a suitably chosen $b_{01}$. What should the value of $b_{01}$ be? Now we add a further term to make the value of the polynomial correct at $a_2$. So as to not disturb what has been achieved already, this term should be of the form $b_{012}(t - a_0)(t - a_1)$.

At the $k$th stage, we should have a polynomial

$$b_0 + b_{01}(t - a_0) + b_{012}(t - a_0)(t - a_1) + \cdots$$

$$+ b_{012\ldots k}(t - a_0)(t - a_1)\cdots(t - a_{k-1}).$$

What should the expressions for the coefficients $b_{012\ldots k}$ be? Try to formulate them in such a way that it is possible to use some analogue of the difference table.

The polynomial obtained should be the same as the Lagrange polynomial of degree $n$ determined by the data. Investigate whether this is indeed so.

**E.56. Propagation of Error.** In making a certain physical observation, we measure the values of a function as follows:

$$f(1) = f(2) = f(3) = 0, \quad f(4) = 1, \quad f(5) = f(6) = f(7) = 0.$$

We have reason to believe that the function should vanish identically and that $f(4)$ is in error. Keeping $f(4)$ as indicated, construct a difference table. Investigate various ways of interpolating the value of $f(3.5)$ using polynomials which fit some of the data.

**E.57. Summing by Differences.** Let $g(t) = \Delta f(t)$. Show that

$$\sum_{i=a}^{b} g(i) = f(b+1) - f(1).$$

In Exploration **E.18**, this formula was invoked in the case $g(i) = i^{(k)}$. Apply it to find the following sums:

(a) $a + ar + ar^2 + \cdots + ar^{n-1}$

(b) $a + (a + d) + (a + 2d) + \cdots + (a + (n-1)d)$

(c) $\sum_{k=1}^{n} k^m$ where $m = 1, 2, 3, 4, 5$.

For (c), express the power $k^m$ as a sum of terms involving factorial powers. See Explorations **E.18** and **E.54**.

The operators $\Delta$ and $\Sigma$ are analogous to differentiation and integration in calculus. What correspond to the following results?

(i) $\dfrac{d}{dt} \displaystyle\int_a^t f(u)du = f(t)$

(ii) $\displaystyle\int_a^b f'(u)du = f(b) - f(a)$

(iii) $\displaystyle\int_a^b f'(u)g(u)du = f(b)g(b) - f(a)g(a) - \displaystyle\int_a^b f(u)g'(u)du.$

**E.58. The Absolute Value Function.** Let $f(t) = |t|$ for $-1 \leq t \leq 1$, and let $f_n(t)$ be the unique polynomial of degree at most $n$ for which $f_n(t) = f(t)$ when $t$ is one of the $n+1$ equally spaced points $(-1) + \frac{2k}{n}$

$(k = 0, 1, 2, \ldots, n)$. Verify the table

| $n$ | $f_n(t)$ |
|---|---|
| 1 | 1 |
| 2 | $t^2$ |
| 3 | $(3t^2 + 1)/4$ |
| 4 | $(-4t^4 + 7t^2)/3$ |
| 5 | $(-125t^4 + 290t^2 + 27)/192$ |
| 6 | $(81t^6 - 135t^4 + 74t^2)/20$ |

Argue that, in general, $f_n(t)$ is an even function and that $f_n'(0) = 0$.

Investigate various strategies for conveniently obtaining the polynomial $f_n(t)$. For example, one might first determine the polynomial $g_n(t)$ of degree not exceeding $n$ for which $g_n(t) = \max(t, 0)$ at the specified points and note that $f_n(t) = g_n(t) + g_n(-t) = 2g_n(t) - t$. Alternatively, if $n = 2m$ and $h_m(t) = |t|$ for $t = -m, -m + 1, \ldots, -1, 0, 1, \ldots, m$, then $h_{m+1}(t) = h_m(t) + c_m t^2 (t^2 - 1)(t^2 - 4) \cdots (t^2 - m^2)$ for some constant $c_m$ and $f_n(t) = \frac{1}{m} h_m(mt)$.

If $t$ is a point in the closed interval $[-1, 1]$, is it necessarily the case that $\lim_{n \to \infty} f_n(t) = f(t)$?

## 7.2 Approximation on an Interval

Suppose that $f(t)$ is a real-valued function defined for at least some values of $t$ with $a \leq t \leq b$. For example $f(t)$ could be a table of values used by an insurance company to determine premiums or could be a nonpolynomial function given by a formula. How can we accurately compute $f(t)$ for certain values of $t$ between $a$ and $b$? As we have seen in Chapter 2, the values of polynomials are straightforward to compute, so it is worth trying to find a polynomial which closely approximates the function on the given interval.

It is natural to take a polynomial which agrees with $f$ where the value of $f$ can be explicitly determined. But this is fraught with danger. If $f(t) = \sin(2\pi n t)$ for $0 \leq t \leq 1$, the polynomial of least degree which coincides with $f$ when $t = 0, 1/n, 2/n, \ldots, 1$ is 0, but this hardly reflects the behaviour of $f$. Furthermore, a slight change in the value of $f$ at one of the evaluation points may dramatically alter the polynomial which interpolates it. We have no guarantee that making a polynomial close in value to $f$ at one place will ensure that it is close overall.

This difficulty can be circumvented by giving up the requirement of exact agreement at some points in favour of gaining some flexibility for making the approximation close everywhere.

What does it mean for functions to be close on a whole interval? There is not one right answer which applies in all contexts. The "distance" between two functions can be defined in many ways, and we will content ourselves here with a very brief sampler of approximation theory.

## Exercises

1. *Least squares.* Suppose we have a function $f$ whose values are known for certain points $a_i$ within an interval: $f(a_i) = b_i$. Typically, $b_i$ may be the experimentally observed values of one variable when another variable is given the value $a_i$ (eg. indices of refraction corresponding to temperatures). It often happens that the points $(a_i, b_i)$ in the Cartesian plane fall roughly along a straight line, so that it is reasonable to make a linear approximation $p(t) = mt + k$ to $f(t)$. The coefficients $m$ and $k$ are to be chosen to make $p$ "fit" as closely as possible to $f$ at the data points $(a_i, b_i)$.

   The criterion used for least squares approximation is that $m$ and $k$ should be chosen to minimize

   $$\sum_{i=1}^{r} |p(a_i) - f(a_i)|^2 = \sum_{i=1}^{r} (ma_i + k - b_i)^2$$

   (a) Consider the particular case, $f(0) = 1$, $f(1) = 3$, $f(2) = 4$. Show that, according to the criterion, $m$ and $k$ should be chosen to minimize

   $$(k-1)^2 + (m+k-3)^2 + (2m+k-4)^2$$
   $$= 5m^2 + 6mk + 3k^2 - 22m - 16k + 26.$$

   To carry out the minimization, fix $k$ and complete the square for the resulting quadratic in $m$. Deduce a relationship between $m$ and $k$ for the minimum to occur. Now carry out the same procedure reversing the roles of $m$ and $k$ which will yield the minimizing values. Verify that these are $m = 3/2$ and $k = 7/6$. Plot the points $(0, 1)$, $(1, 3)$, $(2, 4)$ and the line $y = mx + k$. Judging with your eye, do you think the line obtained is reasonable?

   (b) Experiment with some other examples.

   (c) What happens when $r = 2$?

   (d) What happens when the points $(a_i, b_i)$ turn out to be collinear?

2. *Alternation.* In Exercise 1, we sought a line of closest fit on the basis of a finite number of values of the function to be approximated. Here, we consider a different setting in which we actually have an expression for $f(t)$ for every $t$ in the interval $[a, b]$. For any approximant $p(t)$ we let the quantity

   $$\max\{|f(t) - p(t)| : a \le t \le b\}$$

measure the degree of closeness between $f(t)$ and $p(t)$. This measures how far apart the values of $f(t)$ and $p(t)$ can get over the whole interval. The approximation problem is to choose $p$ from a set of desired approximants to make this maximum as small as possible. (It is re-emphasized that this is but one of many possible measures of closeness we could have chosen.)

(a) Suppose we ask for our approximant $p(t)$ to be a constant polynomial $c$. Then, the question is: what value of $c$ will make

$$\max\{|f(t) - c| \; : \; a \le t \le b\}$$

as small as possible?

Consider the following graphical representation. Argue that, for the optimum value of $c$, the function $f(t) - c$ should have a maximum of the same absolute value but opposite sign as its minimum.

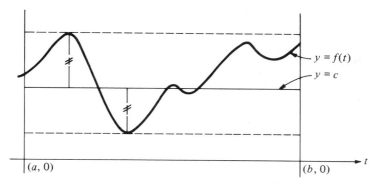

(b) Find the best constant polynomial approximation on the closed interval $[0, 1]$ to each of the following functions:

   (i) $\sin(\pi t)$

   (ii) $t^k$ ($k$ a positive integer)

   (iii) $1/(1 + t)$.

(c) Consider polynomial approximations of degree not exceeding 1. Then we wish to choose $a$ and $b$ in such a way as to minimize

$$\max\{|f(t) - (at + b)| \; : \; 0 \le t \le 1\}.$$

By considering the following sketch, argue that for the optimum choice of $a$ and $b$, $f(t) - (at + b)$ assumes an extreme value at least three times, that all of these extreme values have the same absolute value and that there are at least two changes in sign as we move from one extreme value to the next.

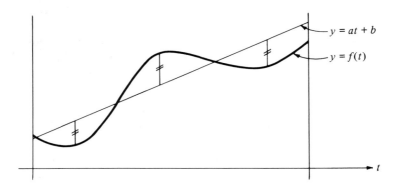

3. Suppose that $f(t)$ is a given function and that $p(t)$ is a polynomial of degree not exceeding $n$. Let

$$K = \max\{|f(t) - p(t)| \ : \ a \leq t \leq b\}.$$

Suppose that there are $n + 2$ distinct points $t_0, t_1, t_2, \ldots, t_{n+1}$ in the interval such that

(i) $|f(t_i) - p(t_i)| = K \quad (0 \leq i \leq n + 1)$

(ii) $f(t_{i+1}) - p(t_{i+1}) = -[f(t_i) - p(t_i)] \quad (0 \leq i \leq n)$

(i.e., the maximum distance between the graphs is achieved with alternate signs at least $n + 2$ times). The diagram illustrates a possible situation when $n = 3$.

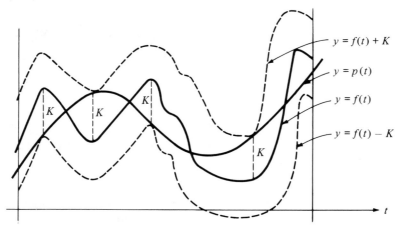

(a) Suppose that $q(t)$ is a polynomial such that

$$\max\{|f(t) - q(t)| \ : \ a \leq t \leq b\} < K.$$

Argue that the graph of $q$ must cross that of $p$ in at least $n+1$ places. (Use a sketch.) Deduce that $\deg q(t) \geq n+1$.

(b) Suppose, instead, for the polynomial in (a) that

$$\max\{|f(t) - q(t)| : a \leq t \leq b\} = K.$$

Argue that either $q(t) = p(t)$ or $\deg q(t) \geq n+1$.

(c) Conjecture a characterization of the polynomial $p(t)$ of degree not exceeding $n$ for which

$$\max\{|f(t) - p(t)| : a \leq t \leq b\}$$

is as small as possible. Do you think that the polynomial $p(t)$ is uniquely determined?

4. (a) Solve the following problem for $k = 1, 2, 3$. Find that polynomial $p_k(t)$ of degree not exceeding $k-1$ which minimizes

$$\max\{|t^k - p_k(t)| : -1 \leq t \leq 1\}$$

over all polynomials of degree not exceeding $k-1$.

(b) Let $C_k = t^k - p_k(t)$. Sketch the graphs of $C_1$, $C_2$, $C_3$. Roughly speaking, what should be graph of $C_k$ look like?

(c) Does this exercise have anything to do with Exercise 1.3.15 and Exploration E.6?

5. *Bernstein polynomials.* Let $f(t)$ $(0 \leq t \leq 1)$ be a function defined on the closed unit interval $[0, 1]$. The *Bernstein polynomial* of order $n$ corresponding to $f(t)$ is defined by

$$B(f, n; t) = \sum_{k=0}^{n} f(k/n) \binom{n}{k} t^k (1 - t)^{n-k}.$$

(a) Verify that

$$B(f, 1; t) = f(0)(1 - t) + f(1)t$$
$$B(f, 2; t) = f(0)(1 - t)^2 + 2f(1/2)t(1 - t) + f(1)t^2$$
$$B(f, 3; t) = f(0)(1 - t)^3 + 3f(1/3)t(1 - t)^2$$
$$+ 3f(2/3)t^2(1 - t) + f(1)t^3.$$

(b) Prove that, if $f(t) = 1$ and $g(t) = t$ for $0 \leq t \leq 1$, then

$$B(f, n; t) = 1 \quad B(g, n; t) = t.$$

(c) What is $B(t^2, n; t)$? These are not all the same. Check the cases $n = 2, 3, 4$ to discover the pattern.

**Comment:** It turns out that

$$\lim_{n \to \infty} B(f, n; t) = f(t)$$

when the function $f(t)$ is continuous in $t$.

Let us get some insight into this result.

Since

$$\Sigma \left( \begin{array}{c} n \\ k \end{array} \right) t^k (1 - t)^{n-k} = 1$$

(why?), we can think of the terms of this sum as nonnegative weights which add up to 1. For each $t$, $B(f, n; t)$ is a weighted average of the values $f(0)$, $f(1/n)$, $f(2/n)$, ..., $f(1)$. When $t = 0$, the only weight which does not vanish is at the $k = 0$ term and $B(f, n; 0) = f(0)$. When $t = 1$, all the weight is at the $k = n$ term and $B(f, n; 1) = f(1)$. For intermediate values of $t$, it turns out that the terms where $k/n$ is close to $t$ are weighted more heavily than the other terms. By the continuity of $f(t)$, when $k/n$ is near $t$, then $f(k/n)$ is close to $f(t)$, so that, in the average, the most weighted terms are those that are about equal to $f(t)$. This weighting becomes more pronounced as $n$ increases, so that $B(f, n; t)$ draws ever closer to $f(t)$ for each value of $t$.

Another way to look at the situation is as follows. Fix $t$ between 0 and 1. Imagine a dartboard of area 1, of which a portion of area $t$ is painted red. Hurl $n$ darts randomly at the board, all piercing the board somewhere. If $k$ of them land in the red area, you receive a prize of $f(k/n)$ dollars. What is your expectation (intuitively, your average winnings) if this game were to be frequently repeated? The probability of getting exactly $k$ darts in the red area is the product of

$\left( \begin{array}{c} n \\ k \end{array} \right)$ the number of ways of selecting the $k$ darts from the $n$ thrown

$t^k$         the probability that all $k$ of these darts land in the red area

$(1 - t)^{n-k}$ the probability that the other $(n - k)$ darts do not land in the red area.

The expectation is the average payoff over all possible occurrences, i.e. the sum of the products of the payoffs and probability of getting them. This is $B(f, n; t)$.

Now increase the number $n$ of darts used in the game. For very large $n$, it is highly probable that the proportion landing in the red area will be close to $t$, so that the expected payoff will be close to $f(t)$.

6.  (a) Show that the mapping $f \longrightarrow B(f, n; t)$ is linear: i.e.

$$B(f + g, n; t) = B(f, n; t) + B(g, n; t)$$

$$B(cf, n; t) = cB(f, n; t)$$

for any functions $f$ and $g$ and any constant $c$.

(b) For $n = 1, 2, 3, 4$, find those values of $k$ and those functions $f$ for which

$$B(f, n; t) = kf(t).$$

[Such functions $f(t)$ are called *eigenfunctions* of the operator $B$ and the corresponding values of $k$ *eigenvalues*.]

## Explorations

**E.59. Taylor Approximation.** (Knowledge of calculus required.) Let a function $f(t)$ be given which is defined for values of $t$ near 0 and which possesses derivatives of all order at $t = 0$. This means that the function has a derivative, its derivative also has a derivative, and so on indefinitely. One can try to approximate $f(t)$ by a polynomial which, up to some order, has exactly the same derivative values as $f(t)$ does when $t = 0$.

Show that a polynomial $p(t)$ of degree not exceeding $n$ is determined uniquely by specifying the $n + 1$ values $p(0), p'(0), \ldots, p^{(n)}(0)$.

Suppose the values $f(0), f'(0), \ldots, f^{(n)}(0)$ are given. What is the polynomial $p(t)$ for which $p^{(k)}(0) = f^{(k)}(0)$ when $0 \leq k \leq n$? We call this the Taylor approximant of order $n$ at the value 0.

Sketch the graphs of $\log(1 + t)$ and $\sin t$ along with their Taylor approximants of orders 1, 2 and 3 at the value 0.

How should the Taylor approximant of order $n$ at the general value $c$ be defined?

**E.60.** In this chapter, we have discussed a number of approximation techniques. We can compare their effectiveness on a particular example. Consider the problem of approximating $x^{1/2}$ for $100 \leq x \leq 200$. On the same axis, with a large scale, sketch the graphs of the following functions:

(a) $x^{1/2}$

(b) the linear function whose values agree with $x^{1/2}$ when $x = 100$ and $x = 196$

(c) the quadratic function whose values agree with $x^{1/2}$ at $x = 100$, $x = 144$ and $x = 196$

(d) the linear polynomial $p(t)$ which minimizes

$$\max\{|t^{1/2} - p(t)| \; : \; 100 \leq t \leq 200\}$$

(e) the Bernstein polynomial of degree 2 (modified to the interval)

(f) the Taylor approximants of degrees 1, 2 and 3 at 100, 121, 144.

In particular, use the above approximants to estimate $(125)^{1/2}$, and compare your results with the true answer.

## 7.3   Inequalities

Often, an important consideration in comparing two functions is to determine when one exceeds the other. One of the most useful inequalities is that between the arithmetic and geometric means of positive quantities. For nonnegative reals $a_1, a_2, \ldots, a_n$, the geometric mean is defined to be

$$(a_1 a_2 \cdots a_n)^{1/n}$$

and the arithmetic mean to be

$$(a_1 + a_2 + \cdots + a_n)/n.$$

The Arithmetic-Geometric Means (AGM) Inequality asserts that

$$(a_1 a_2 \cdots a_n)^{1/n} \le (a_1 + a_2 + \cdots + a_n)/n$$

with equality if and only if all the $a_i$ are equal. This has already been established in the cases $n = 2$ and $n = 3$ (Exercises 1.2.17 and 1.5.9) by showing that $x^2 + y^2 - 2xy \ge 0$ and $x^3 + y^3 + z^3 - 3xyz \ge 0$ when $x, y, z \ge 0$. For inequalities in general, one useful strategy is to write an appropriate function as a sum or product of polynomials known to be positive, such as squares of other polynomials. This was also the basis of one of the arguments (Exercise 1.2.15) used in establishing the Cauchy–Schwarz–Bunjakovsky (CSB) Inequality:

$$\left( \sum_{i=1}^{n} a_i b_i \right)^2 \le \sum_{i=1}^{n} a_i^2 \sum_{i=1}^{n} b_i^2$$

with inequality iff the ratio of the $a_i$ is equal to the ratio of the $b_i$.

In the exercises, we will establish some other useful inequalities.

## Exercises

1.  (a) Write $x^4 + y^4 + z^4 + w^4 - 4xyzw$ in the form $ap^2 + bq^2 + cr^2$ where $a$, $b$, $c$ are positive constants and $p$, $q$, $r$ are polynomials over $\mathbf{R}$ in $x$, $y$, $z$, $w$.

    (b) Establish the AGM inequality for $n = 4$.

2. (a) Verify the identity

$$x^6 + y^6 + z^6 + u^6 + v^6 + w^6 - 6xyzuvw$$
$$= \frac{1}{2}(x^2 + y^2 + z^2)[(y^2 - z^2)^2 + (z^2 - x^2)^2 + (x^2 - y^2)^2]$$
$$+ \frac{1}{2}(u^2 + v^2 + w^2)[(v^2 - w^2)^2 + (w^2 - u^2)^2 + (u^2 - v^2)^2]$$
$$+ 3(xyz - uvw)^2.$$

(b) Establish the AGM inequality for $n = 6$.

3. Suppose it is known that, for a certain positive integer $k$,

$$x_1^k + x_2^k + x_3^k + \cdots + x_k^k \geq k x_1 x_2 \cdots x_k$$

for any nonnegative real $x_i$. Prove that

$$y_1^n + y_2^n + y_3^n + \cdots + y_n^n \geq n y_1 y_2 \cdots y_n$$

for any nonnegative real $y_i$, where $n = 2k$ or $n = 3k$.

4. Deduce from Exercise 3 that the AGM inequality holds for $n = 2^r 3^s$ for $r$, $s$ nonnegative integers.

5. We wish to demonstrate for any positive integer $n$ and real variables $x_i$ that
$$x_1^n + x_2^n + \cdots + x_n^n - n x_1 x_2 \cdots x_n \geq 0.$$
One can follow the procedure suggested by Exercise 4 in which the problem can be reduced to the case in which $n$ is prime. However, for larger values of $n$, it is far from clear whether one can conveniently manipulate the left side into a form which is clearly nonnegative (Exploration **E.61** asks you to investigate the case of $n = 5$). As often happens in mathematics, a more convenient argument can be found if we are prepared to prove a more general result. Accordingly, we introduce weights $w_i$. These are positive real numbers $w_i$ which satisfy

$$w_1 + w_2 + \cdots + w_n = 1.$$

The generalized arithmetic mean with weight **w** is defined by the expression
$$w_1 a_1 + w_2 a_2 + \cdots + w_n a_n$$
and the generalized geometric mean with weight **w** by

$$a_1^{w_1} a_2^{w_2} \cdots a_n^{w_n}.$$

(a) What values of $w_i$ will yield the standard means?

(b) Let $u = w_{n-1}/(w_{n-1} + w_n)$, $v = w_n/(w_{n-1} - w_n)$ and $a = ua_{n-1} + va_n$. Verify that

$$w_1 a_1 + w_2 a_2 + \cdots + w_n a_n = w_1 a_1 + w_2 a_2 + \cdots + w_{n-2} a_{n-2}$$

$$+ (w_{n-1} + w_n)a.$$

(c) Prove, by induction on $n$, that for all nonnegative real $a_i$ and positive weights $w_i$,

$$a_1^{w_1} a_2^{w_2} \cdots a_n^{w_n} \leq w_1 a_2 + \cdots + w_n a_n.$$

6. (Knowledge of calculus required.) With the help of some simple properties of the graph of the logarithm function, a short and general proof of the AGM inequality is possible. We use the notation of Exercise 5. For $x > 0$, let $\log x$ denote the logarithm of $x$ to base $e$. Observe that $D(\log x) = 1/x > 0$ and $D^2(\log x) = -1/x^2 < 0$, so that the graph of the equation $y = \log x$ is increasing and concave.

(a) Sketch the graph of the equation $y = \log x$. Verify that the tangent to the graph through the point $(1, 0)$ has the equation $y = x - 1$ and that the graph lies below its tangent through the whole domain of $x$. Deduce that $\log x \leq x - 1$ for $x > 0$.

(b) Verify that

$$\log(a_1^{w_1} a_2^{w_2} \cdots a_n^{w_n}) = w_1 \log a_1 + \cdots + w_n \log a_n.$$

(c) Let $m = w_1 a_1 + w_2 a_2 + \cdots + w_n a_n$. Show that

$$\log a_i - \log m \leq (a_i/m) - 1 \quad (i = 1, 2, \ldots, n).$$

(d) By multiplying the $i$th equation in (c) by $w_i$ and adding, show that

$$(\Sigma w_i \log a_i) - \log m \leq 0.$$

Deduce from this the generalized AGM inequality.

7. Prove the following

(a) $x + x^{-1} \geq 2$    for $x > 0$

(b) $4x(1 - x) \leq 1$    for $x \in \mathbb{R}$.

8. Let $a_1, \ldots, a_n$ be positive reals. Use the CSB inequality to establish that

$$(a_1 + a_2 + \cdots + a_n)(a_1^{-1} + a_2^{-1} + \cdots + a_n^{-1}) \geq n^2$$

with equality iff all the $a_i$ are equal.

9. (a) Prove by induction on $n$ that, if $x_i \geq -2$ and $x_i$ all have the same sign, then

$$(1 + x_1)(1 + x_2) \cdots (1 + x_n) \geq 1 + x_1 + x_2 + \cdots + x_n.$$

When does equality hold?

(b) Deduce from (a) the *Bernoulli inequality*

$$(1 + x)^n > 1 + nx$$

for nonzero $x \geq -2$ and $n$ a positive integer exceeding 1.

10. *Newton's inequalities.* Let $n$ be a positive integer and suppose that $x_1, x_2, \ldots, x_n$ are positive real numbers. For $r = 1, 2, \ldots, n$, define

$s_r = \Sigma x_1 x_2 \cdots x_r$, the $r$th elementary symmetric function

$u_r = \dfrac{\Sigma x_1 x_2 \cdots x_r}{\dbinom{n}{r}}$, the average of the products of $r$ numbers

$v_r = u_r^{1/r}$.

Observe that $v_1$ is the ordinary arithmetic mean and $v_n$ the ordinary geometric mean of the $x_i$, so that $v_n \leq v_1$. This can be generalized to the chain of inequalities

$$v_n \leq v_{n-1} \leq \cdots \leq v_2 \leq v_1.$$

(a) Let $f(t) = \Pi\{(t - x_i) : i = 1, 2, \ldots, n\}$. Show that

$$f(t) = t^n + \sum_{r=1}^{n}(-1)^r \binom{n}{r} u_r t^{n-r}.$$

(b) Use Rolle's Theorem to argue that, for $k = 1, 2, \ldots, n - 1$, the $k$th derivative $f^{(k)}(t)$ is a polynomial of degree $n - k$ with $n - k$ real positive zeros counting multiplicity.

(c) Verify that $f^{(n-2)}(t) = (n!/2)(t^2 - 2u_1 t + u_2)$ and deduce from the discriminant condition that $u_2 \leq u_1^2$.

(d) Note that

$$\binom{n}{2} \frac{u_{n-2}}{u_n} = \frac{1}{x_1 x_2} + \cdots + \frac{1}{x_{n-1} x_n}$$

$$\binom{n}{1} \frac{u_{n-1}}{u_n} = \frac{1}{x_1} + \cdots + \frac{1}{x_n}$$

and apply (c) to $1/x_i$ to obtain $u_{n-2} u_n \leq u_{n-1}^2$.

(e) The results of (c) and (d) can be generalized to $u_{r-1}u_{r+1} \leq u_r^2$ for $r = 2, 3, \ldots, n-1$. This is established by an induction argument on the number of the $x_i$. The result holds for $n = 3$. Suppose it holds when the number of the $x_i$ does not exceed $n - 1$. From (a), obtain that

$$f'(t) = n\left[t^{n-1} + \sum_{r=1}^{n-1}(-1)^r \binom{n-1}{r} u_r t^{n-1-r}\right].$$

(f) Suppose the zeros of $f'(t)$ are $y_1, \ldots, y_{n-1}$. For $1 \leq r \leq n-1$, let

$$\binom{n-1}{r} z_r = \Sigma y_1 y_2 \cdots y_r,$$

so that the $z_r$ are to the $y_i$ what the $u_r$ are to the $x_i$. Use the argument of (a) to obtain

$$f'(t) = n\left[t^{n-1} + \sum_{r=1}^{n-1}(-1)^r \binom{n-1}{r} z_r t^{n-1-r}\right].$$

Deduce that $u_r = z_r$ for $1 \leq r \leq n-1$. Use the induction hypothesis to obtain $u_{r-1}u_{r+1} \leq u_r^2$ for $2 \leq r \leq n-2$.

(g) Use (c), (d), (e), (f) to obtain for $1 \leq r \leq n-1$,

$$u_2(u_1u_3)^2(u_2u_4)^3 \cdots (u_{r-1}u_{r+1})^r \leq u_1^2 u_2^4 u_3^6 \cdots u_r^{2r},$$

and, hence

$$u_{r+1}^r \leq u_r^{r+1}, \quad \text{i.e. } v_{r+1} \leq v_r.$$

11. Suppose that $a$, $b$, $c$ are nonnegative reals for which $(1 + a)(1 + b)(1 + c) = 8$. We can apply the result of Exercise 10 to show that $abc \leq 1$. Simply show that the left side of the given condition is not less than $(1 + u)^3$ where $u^3 = abc$.

12. Suppose that $a$, $b$, $c$ are positive real numbers. Show that

$$\frac{1}{a} + \frac{1}{b} + \frac{1}{c} \leq \frac{a^8 + b^8 + c^8}{a^3 b^3 c^3}.$$

This can be established by repeated use of the AGM inequality. To get started, write the right side as the sum of three terms of the form $a^5/b^3c^3$ and apply the AGM inequality to the sum of each pair.

## Exploration

**E.61.** Can the polynomial $x^5 + y^5 + z^5 + u^5 + v^5 - 5xyzuv$ be manipulated into a form in which it is clearly seen to be nonnegative for $x$, $y$, $z$, $u$, $v$ nonnegative? Consider the analogous question for other numbers of variables.

# 7.4  Problems on Inequalities

1. Find all real triples $(x, y, z)$ for which $(1-x)^2+(x-y)^2+(y-z)^2+z^2 = 1/4$.

2. Prove that, for real $x$, $y$, $z$,

$$(x + y)z \le (1/2)(x^2 + y^2) + z^2.$$

3. Suppose that $a$ and $b$ are nonzero real numbers and that all the zeros of the real polynomial

$$at^n - at^{n-1} + a_{n-2}t^{n-2} + \cdots + a_2t^2 - n^2bt + b = 0$$

are real and positive. Prove that all the zeros are equal.

4. Show that the real polynomial $t^n + at^{n-1} + bt^{n-2} + \cdots + k$ has at least one nonreal zero if $a^2 - 2b < n(k^{2/n})$.

5. If $x, y, z \ge 0$, prove that

$$x^3 + y^3 + z^3 \ge y^2z + z^2x + x^2y$$

and determine when there is equality.

6. Prove, that, if $x, y, z \ge 0$, then

$$x(y - z)^2 + y(x - z)^2 \ge (x - z)(y - z)(x + y - z).$$

7. Show that, if all the zeros of $at^4 - bt^3 + ct^2 - t + 1$ are positive, then $c \ge 80a + b$.

8. Prove that, for $x \ge 0$ and $n$ a positive integer,

$$x^{n+1} - (n + 1)x + n \ge 0.$$

9. Let $a, b, c, d > 0$ and $c^2 + d^2 = (a^2 + b^2)^3$. Show that

$$a^3/c + b^3/d \ge 1$$

with equality iff $ad = bc$.

10. Suppose $N(x) = 0$ for $x$ negative, $N(0) = 1$ and $N(x) = N(x-6)+x$ for $x$ positive. Show that, for each positive integer $x$, that

$$(x + 1)(x + 5)/12 \le N(x) \le (x^2 + 6x + 12)/12.$$

11. Let $A$, $B$, $C$ be the angles of a triangle. Prove that

$$\tan^2(A/2) + \tan^2(B/2) + \tan^2(C/2) \ge 1.$$

12. Determine $u$ so that the zeros of the polynomial $t^2 - (3u + 1)t + (2u^2 - 3u - 2)$ are real and the sum of their squares is minimal.

13. Let $m$ and $n$ be positive integers and let $x$, $y$ be positive reals. Show that
$$\left(\frac{x+y}{m+n}\right)^{m+n} \geq \left(\frac{x}{m}\right)^m \left(\frac{y}{n}\right)^n.$$

14. Determine the largest value of $y$ such that
$$\frac{1}{1+x^2} \geq \frac{y-x}{y+x}$$

for all $x > 0$.

15. Let $f(t)$ be an irreducible polynomial of degree $n$ exceeding 1 over $\mathbf{Z}$, and suppose that $r$ is a zero of $f(t)$. Show that there is some constant $k$ which depends on $f$ and $r$ such that $|p/q - r| \geq k/q^n$ for each rational $p/q$ written in lowest terms with $q$ a positive integer.

16. Find the maximum and minimum values of
$$\frac{x+1}{xy+x+1} + \frac{y+1}{yz+y+1} + \frac{z+1}{zx+z+1}$$

subject to the conditions that $x, y, z \geq 0$ and $xyz = 1$.

## 7.5  Other Problems

1. Let $n$ be a positive integer greater than 2 and let $f$ be any polynomial of degree not exceeding $n - 2$. If $a_1, a_2, \ldots, a_n$ are any complex numbers and $p(t) = (t - a_1)(t - a_2) \cdots (t - a_n)$, prove that
$$\sum_{i=1}^{n} \frac{f(a_i)}{p'(a_i)} = 0.$$

2. Let $a$, $b$, $c$, $d$ be distinct complex numbers. Show that
(a)
$$\frac{a^4}{(a-b)(a-c)(a-d)} + \frac{b^4}{(b-a)(b-c)(b-d)} + \frac{c^4}{(c-a)(c-b)(c-d)}$$
$$+ \frac{d^4}{(d-a)(d-b)(d-c)}$$
$$= a+b+c+d.$$

(b)

$$\frac{a^5}{(a-b)(a-c)} + \frac{b^5}{(b-a)(b-c)} + \frac{c^5}{(c-a)(c-b)}$$
$$= (a+b+c)^3 - 2(a+b+c)(bc+ca+ab) + abc.$$

3. Let

$$S_1(n) = \sum_{i=1}^{n} i^2$$

$$S_k(n) = \sum_{i=1}^{n} S_{k-1}(i) \quad (k \geq 2).$$

Show that $S_k(n) = n(n+1)\cdots(n+k)(2n+k)/(k+2)!$.

4. Let $m$ and $n$ be integers with $m \geq n \geq 0$, and let $c$ be any constant. Define

$$f(m) = \sum_{k=0}^{m} (-1)^k \binom{m}{k} (m-k+c)^n.$$

Show that $f(m) = 0$ if $m > n$ and that $f(n) = n!$.

5. Let $-1 \leq u \leq 1$. Determine the smallest number $K_u$ which satisfies the condition

$$|g'(u)| \leq K_u$$

whenever $g(t)$ is a polynomial such that $\deg g(t) \leq 2$ and $|g(t)| \leq 1$ for $-1 \leq t \leq 1$.

6. Let $f(x) = (x - x_1)\cdots(x - x_n)$ for $-1 \leq x_i \leq 1$. Prove that there cannot exist numbers $a$, $b$ for which

   (i) $|f(a)| \geq 1$
   (ii) $|f(b)| \geq 1$
   (iii) $-1 < a < 0 < b < 1$.

7. Let $n_i$ $(0 \leq i \leq k)$ be any $k+1$ integers for which $n_0 < n_1 < n_2 < \cdots < n_k$. Show that

$$\prod_{0 \leq i < j \leq k} \frac{(n_j - n_i)}{(j - i)}$$

is an integer.

8. Construct, with proof that the construction works, a polynomial $p(x)$ over $\mathbf{Z}$ such that

$$|p(x) - 0.5| < 1/1981$$

for each $x$ for which $0.19 \leq x \leq 0.81$.

9. Determine all pairs $(p, q)$ of real numbers for which the inequality

$$\left| \sqrt{1 - x^2} - px - q \right| \leq (1/2)(\sqrt{2} - 1)$$

is true for each $x$ for which $0 \leq x \leq 1$.

10. Prove that the polynomial $p$ with degree not exceeding $n$ that assumes the value $y_k$ at the values

$$x_k = \cos u_k \quad \text{where} \quad u_k = [(2k + 1)\pi/(2n + 2)]$$

$(k = 0, 1, 2, \ldots, n)$ is given for $x \neq x_k$ for any $k$ by

$$p(x) = N(x)/D(x)$$

where

$$N(x) = \sum_{k=0}^{n} [(-1)^k y_k \sin u_k/(x - x_k)]$$

and

$$D(x) = \sum_{k=0}^{n} [(-1)^k \sin u_k/(x - x_k)].$$

# Hints

## Chapter 7

1.9. (c) Prove by induction.

1.20. Observe that

$$2^k = 1 + \binom{k}{1} + \binom{k}{2} + \cdots + \binom{k}{k}.$$

2.6. (a) Look at the equation for the eigenfunction when $t = 0, 1$. What does this imply if $k \neq 1$?

3.1. (a) One term is $(x^2 - y^2)^2$.

3.3. For the $n = 2k$ case, pair the $y_i^n$ off and apply the AGM inequality to each pair.

4.1. Apply the CSB inequality to $\{(1-x), (x-y), (y-z), z\}$ and $\{1, 1, 1, 1\}$.

4.2. Express the difference of the two sides as a sum of squares.

4.3. Let $r_i$ be the zeros. Apply the CSB inequality to $\{r_i\}$ and $\{r_i^{-1}\}$.

4.4. Apply the AGM inequality to $\{r_i^2\}$.

4.5, 4.6. Take the difference of the two sides.

4.7. If $a \neq 0$, let $u = p^{-1} + q^{-1}$, $v = r^{-1} + s^{-1}$. Use the AGM inequality to obtain a lower estimate for $c/a - b/a$ in terms of $u$ and $v$.

4.8. Use induction.

4.9. Let $c = ua^3$ and $d = vb^3$. The condition expressed in terms of $u$, $v$ and $w = a^2(a^2 + b^2)^{-1}$ is $u^2w^3 + v^2(1 - w)^3 = 1$. When $ad = bc$, we have that $w = v(u + v)^{-1}$ and the condition becomes $u^2v^3 + v^2u^3 = (u+v)^3$, which is equivalent to $1 = (u^{-1} + v^{-1})^2$. It has to be shown in general that $u^{-1} + v^{-1} \geq 1$. This can be done by looking at $u^2w^3 + v^2(1 - w)^3 - (u^{-1} + v^{-1})^{-2}$.

4.11. Let $A$, $B$ be acute and express the left side in terms of $u = \tan A/2$ and $v = \tan B/2$; apply the AGM inequality.

4.12. Express the sum of squares in terms of $u$ and minimize over those $u$ for which the zeros of the quadratic are real.

4.13. Use the AGM inequality for the set which contains $x/m$ $m$ times and $y/n$ $n$ times.

4.14. The inequality is equivalent to $y \leq \min\{x + 2/x : x \geq 0\}$. The function in the brackets can be minimized using the AGM inequality in a way similar to Problem 13.

4.15. Let $f(t) = (t - r)g(t)$; note that $g(r) \neq 0$ (why?). Determine $M$ so that $|g(t)| \leq M$ for $|t - r| \leq 1$ and observe that $|f(p/q)| \geq 1/q^n$.

5.1. Consider the coefficient of $t^{n-1}$ of the Lagrange polynomial $p(t)$ for which $p(a_i) = f(a_i)$. What is $p(t)$?

5.2. Use Lagrange polynomials.

5.3. Use double induction on $k$ and $n$. The cases $n = 1$ and $k = 1$ are clear. Suppose the result is known for any $n$ and $1 \leq k \leq r - 1$ as well as for $k = r$ and $1 \leq n \leq m - 1$.

5.4. $f(m)$ is the $n$th order difference of a polynomial of degree $n$. Write the sum without the summation sign.

5.5. If $1/2 \leq |u| \leq 1$, write $g(x)$ as a Lagrange polynomial with respect to its values at $x = -1$, $0$, $1$. Then one can get a sharp estimate of $|g'(u)|$ in terms of $|g(-1)|$, $|g(0)|$ and $|g(1)|$. The case $0 \leq |u| \leq 1/2$ is more difficult. Sketch a few graphs; there are essentially two cases to consider according as $g(x)$ is monotone on $[-1, 1]$ or not. Reduce to the situation that $g(x)$ assumes both the values $-1$ and $+1$.

5.6. Begin by a careful study of the quadratic case. What happens if all the $x_i$ have the same sign? Use an induction argument. The case that $x_1 \le a < 0 < b \le x_n$ can be disposed of by looking at the values of the polynomial at $a$ and $b$ after dividing by $(x - x_1)(x - x_n)$. As for the case that $a \le x_1 \le x_n \le b$, note that $p(a)p(b)$ is the product of quadratics $(a - x_i)(b - x_i)$ in the $x_i$. The remaining cases can be handled by similar types of consideration.

5.7. It has to be shown that $G(\mathbf{x}) = \Pi(x_j - x_i)/(j - i)$ takes integer values whenever the components of $\mathbf{x}$ are integers. Argue that it is enough to consider nonnegative integer values and prove by induction on the maximum of the $x_i$. The result is clear if this maximum is $k$. Now fix $k$ of the $k + 1$ variables and consider $G$ as a function of a single variable. Use the fact that a polynomial of degree not exceeding $k$ over $\mathbf{Z}$ which assumes integer values for at least $k + 1$ consecutive integers assumes integer values at each integer point.

5.8. There is an obvious example of a polynomial with integer coefficients which assumes arbitrarily small values on any open subinterval of $[-1, 1]$. Adapt this.

5.9. Draw a diagram.

5.10. Express this polynomial using Lagrange's formula. What are the zeros of the Tchebychev polynomials? Use this fact in your expression for $p(x)$. Look at the special case in which each $y_k$ is 1.

# 8

# Miscellaneous Problems

1. Solve $x(3y - 5) = y^2 + 1$ in integers.

2. Solve
$$a^3 - b^3 - c^3 = 3abc$$
$$a^2 = 2(b + c)$$
simultaneously in positive integers.

3. Find all integer solutions $(x, y, z)$ of the system
$$3 = x + y + z = x^3 + y^3 + z^3.$$

4. Let $f(x) = x^2 + x$. Show that $4f(a) = f(b)$ has no solutions in positive integers.

5. Solve the equation $(x^2 + y)(x + y^2) = (x - y)^3$ for integers $x$, $y$.

6. Consider the diophantine equation $x^3 = y^2 + 4$. Observing that $y^2 + 4 = (y + 2i)(y - 2i)$, solve first the equation $(u + vi)^3 = y + 2i$ for integers $u$, $v$, $y$ and use this to obtain solutions $(x, y)$ in integers to the given equation.

7. Let $a$, $b$, $c$, $d$ be integers with $a \neq 0$. Can $axy + bx + cy + d = 0$ have infinitely many solutions in integers $x$ and $y$?

8. Solve for integers $x$, $y$ the equation
$$x^3 - y^3 = 2xy + 8.$$

9. Determine infinitely many solutions in rational numbers $x$, $y$, $z$, $t$ of the equation:
$$(x + y\sqrt{2})^2 + (z + t\sqrt{2})^2 = 27 + 10\sqrt{2}.$$

10. Determine all integer solutions $(x, y, z)$ of
$$\sqrt[3]{x + \sqrt{y}} + \sqrt[3]{x - \sqrt{y}} = z.$$

11. Find ten rational values of $x$ such that $3x^2 - 5x + 4$ is the square of a rational number.

12. If $x$, $y$, $z$ are rational numbers for which $x^3 + 3y^3 + 9z^3 - 9xyz = 0$, prove that $x = y = z = 0$.

13. Solve each of the following equations for unequal integers $x$, $y$:

    (a) $(x+1)^2 - x^2 - (x-1)^2 = (y+1)^2 - y^2 - (y-1)^2$
    (b) $(x+1)^3 - x^3 - (x-1)^3 = (y+1)^3 - y^3 - (y-1)^3$
    (c) $(x+1)^4 - x^4 - (x-1)^4 = (y+1)^4 - y^4 - (y-1)^4$.

14. Let the polynomial $f(t) = t^n + a_{n-1}t^{n-1} + \cdots + a_1 t + 1$ have nonnegative coefficients and $n$ real zeros. Prove that $f(2) \geq 3^n$.

15. Solve the system of equations:

$$x^2 + 2yz = x$$
$$y^2 + 2xz = z$$
$$z^2 + 2xy = y \ .$$

16. $P$, $Q$, $R$ are three polynomials over $\mathbf{R}$ of degree 3 for which $P(x) \leq Q(x) \leq R(x)$ for all real $x$. For some real $u$, equality holds. Prove that there exists a constant $k$ with $0 \leq k \leq 1$ for which $Q = kP + (1-k)R$. Does this property still hold if $P$, $Q$, $R$ are of degree 4?

17. Find an explicit polynomial $P(a,b)$ such that there is a straight line intersecting the graph of $y = x^4 + ax^3 + bx^2 + cx + d$ in exactly four points if and only if $P(a,b) > 0$.

18. Find the value of the real number $p$ for which the equation

$$x^3 + px^2 + 3x - 10 = 0$$

has three real roots $a$, $b$, $c$ for which $c - b = b - a > 0$.

19. $p(x)$ and $q(x)$ are polynomials which satisfy the identity $p(q(x)) = q(p(x))$ for all real $x$. If the equation $p(x) = q(x)$ has no real solution, show that the equation $p(p(x)) = q(q(x))$ also has no real solution.

20. Let $P(x,y)$ be a polynomial in $x$ and $y$ of degree at most 2. Suppose that $A$, $B$, $C$, $A'$, $B'$, $C'$ are six distinct points in the $xy$-plane such that $A'$ lies on $BC$, $B'$ lies on $AC$ and $C'$ lies on $AB$. Prove that if $P$ vanishes at these six points, then $P$ is identically equal to zero.

21. Show that

$$\frac{c(a-b)^3 + a(b-c)^3 + b(c-a)^3}{c^2(b-a) + a^2(c-b) + b^2(a-c)} = a + b + c.$$

22. Determine a polynomial with integer coefficients one of whose zeros is $(3/5)^{1/7} + (5/3)^{1/7}$.

23. (a) Let $n_1, n_2, \ldots, n_s$ be a set of positive integers exceeding 1, with any pair relatively prime. Let $a_i$ be the number of prime factors of $n_i$ counting repetitions when $n_i$ is written as a product of primes. Define

$$p(x) = \sum_{i=1}^{s} \frac{x^{a_i}}{n_i} - \sum_{i \neq j} \frac{x^{a_i + a_j}}{n_i n_j} + \sum \frac{x^{a_i + a_j + a_k}}{n_i n_j n_k} - \cdots$$

where the right side is the sum of $2^s - 1$ terms. Prove that $p(x)$ is monotonically increasing on the closed interval $[0, 2]$, that $p(x) \leq 1$ there and $p(x)$ can assume the value 1 if and only if one of the $n_i$ is a power of 2.

(b) Let $m$ and $n$ be two positive integers exceeding 1, and let $a$ and $b$, respectively, be the number of prime factors counting repetitions when $m$ and $n$ are written as the product of primes. Suppose the least common multiple of $m$ and $n$ is $k$ and that $c$ is the number of prime factors of $k$. Let

$$p(x) = x^a/m + x^b/n - x^c/k.$$

Show that $p(x)$ is increasing on the closed interval $[0, 2]$, $p(x) \leq 1$ there with equality possible if and only if either $m$ or $n$ is a power of 2.

24. Find polynomials $p(x)$ and $q(x)$ over $\mathbf{Z}$ such that $p(\sqrt{2} + \sqrt{3} + \sqrt{5})/q(\sqrt{2} + \sqrt{3} + \sqrt{5}) = \sqrt{2} + \sqrt{3}$.

25. Consider the equation

$$\sqrt{2p + 1 - x^2} + \sqrt{3x + p + 4} = \sqrt{x^2 + 9x + 3p + 9} \qquad (*)$$

in which $x$ and $p$ are real and the square roots are real and nonnegative. Show that, if $(*)$ holds, then

$$(x^2 + x - p)(x^2 + 8x + 2p + 9) = 0.$$

Hence, find the set of real $p$ such that $(*)$ is satisfied by exactly one real number $x$.

26. What condition must be satisfied by the coefficients $u$, $v$, $w$ of the polynomial

$$x^3 - ux^2 + vx - w$$

in order that the line segments whose lengths are the zeros of the polynomial can form a triangle.

27. Determine necessary and sufficient conditions on $a$, $b$, $c$ such that

$$ax + by + cz = 0$$

    and

$$a\sqrt{1-x^2} + b\sqrt{1-y^2} + c\sqrt{1-z^2} = 0$$

    should admit a real solution $x$, $y$, $z$, with $|x| \le 1$, $|y| \le 1$, $|z| \le 1$.

28. Let the polynomial $x^{10} + \star x^9 + \cdots + \star x + 1$ be given, where the starred coefficients are to be filled in by two players playing alternately until no stars remain. The first player wins if all zeros of the polynomial are nonreal; the second player wins otherwise. Is there a winning strategy for the second player?

29. Let $A$, $B$, $X$, $Y$ be variables subject to $AX - BY = 1$.

    (a) Find explicit polynomials $u$ and $v$ in $A$, $B$, $X$, $Y$ over $\mathbf{Z}$ such that $A^4 u - B^4 v = 1$.

    (b) Show that, for each pair $m$, $n$ of positive integers, there are polynomials $u$, $v$ in $A$, $B$, $X$, $Y$ for which $A^m u - B^n v = 1$.

30. Let $p(x)$ be a polynomial over $\mathbf{R}$ of even degree $n$ for which $p(x) \ge 0$ for all $x$. Prove that $p(x) + p'(x) + \cdots + p^{(n)}(x) \ge 0$ for all $x$.

31. Three positive numbers $x$, $y$, $z$ lie between the least and greatest of three positive numbers $a$, $b$, $c$. If

$$x + y + z = a + b + c \quad \text{and} \quad xyz = abc,$$

    show that, in some order $x$, $y$, $z$ are equal to $a$, $b$, $c$.

32. Prove that every polynomial has a nonzero polynomial multiple whose exponents are all divisible by 1 000 000.

33. Let $p$ and $q$ be polynomials over $\mathbf{C}$ of positive degree. Suppose that $P_k = \{z \in \mathbf{C} : p(z) = k\}$ and $Q_k = \{z \in \mathbf{C} : q(z) = k\}$. Show that, if $P_0 = Q_0$ and $P_1 = Q_1$, then $p = q$.

34. Suppose that $u$, $v$, $w$, $x$, $y$, $z$ are real numbers with $x$, $y$, $z$ all distinct for which the equations

$$u^3 + x^3 = v^3 + y^3 = w^3 + z^3 = a^3$$

    and

$$u(y - z) + v(z - x) + w(x - y) = 0$$

    hold. Show that $uvw + xyz = a^3$.

    What is the situation if $x = y$?

35. Show that the set of real numbers $x$ which satisfy the inequality

$$\sum_{k=1}^{70} \frac{k}{x-k} \geq \frac{5}{4}$$

is a union of disjoint intervals, the sum of whose lengths is 1988.

36. Let $a$, $b$, $c$, $d$, $e$, $f$ be complex numbers for which $\Sigma a = \Sigma a^3 = 0$. Prove that

$$(a+c)(a+d)(a+e)(a+f) = (b+c)(b+d)(b+e)(b+f).$$

37. Eliminate $u$, $v$, $w$ from the equations

$$a = \cos u + \cos v + \cos w$$
$$b = \sin u + \sin v + \sin w$$
$$c = \cos 2u + \cos 2v + \cos 2w$$
$$d = \sin 2u + \sin 2v + \sin 2w.$$

38. Eliminate $\theta$ between

$$x = \cot\theta + \tan\theta$$
$$y = \sec\theta - \cos\theta.$$

39. Show that $x^6 - x^5 + x^4 - x^3 + x^2 - x + 3/4$ has no real zeros.

40. Prove that the local maximum and local minimum values of the real polynomial $x^3 + 3px^2 + 3qx + r$ are given by

$$2p^3 - 3pq + r + 2(p^2 - q)^{3/2}$$

and

$$2p^3 - 3pq + r - 2(p^2 - q)^{3/2}.$$

41. (a) Suppose that

$$\frac{a^2 + b^2 - c^2 - d^2}{a - b + c - d} = \frac{a^2 - b^2 - c^2 + d^2}{a + b + c + d}.$$

Show that

$$\frac{ab - cd}{a - b + c - d} = \frac{bc - ad}{a + b + c + d}.$$

(b) Find a set of integers $a$, $b$, $c$, $d$ which satisfy the equations in (a).

42. Prove that a real polynomial $p(x)$ which assumes rational values for rational $x$ and irrational values for irrational $x$ must be linear.

43. If $a/(bc - a^2) + b/(ca - b^2) + c/(ab - c^2) = 0$, prove that
$a/(bc - a^2)^2 + b/(ca - b^2)^2 + c/(ab - c^2)^2 = 0$.

44. Solve $1 + 1/1 + 1/1 + 1/1 + \cdots + 1/1 + x = x$. (The left side is a continued fraction with $n$ slashes.)

45. Find all polynomials $f(t)$ over $\mathbf{R}$ such that

$$f(t)f(t+1) = f(t^2 + t + 1).$$

46. Is there a set of real numbers $u$, $v$, $w$, $x$, $y$, $z$ satisfying

$$u^2 + v^2 + w^2 + 3(x^2 + y^2 + z^2) = 6$$

$$ux + vy + wz = 2?$$

47. Suppose that $x + y + z = x^{-1} + y^{-1} + z^{-1} = 0$. Show that

$$\frac{x^6 + y^6 + z^6}{x^3 + y^3 + z^3} = xyz.$$

48. Show that, if $p$ is an odd prime and $k$ is a positive integer, then

$$z^p + 1 \mid (z^{p-1} - 1)(z^{p-2} - z^{p-3} + \cdots + z - 1)^k + (z+1)z^{(p-1)k-1}.$$

49. Observe that $8^3 - 7^3 = (2^2 + 3^2)^2$ and $105^3 - 104^3 = (9^2 + 10^2)^2$. Show that, if the difference of two consecutive cubes is a square, then it is the square of the sum of two successive squares.

50. Prove that the only positive solution of

$$x + y^2 + z^3 = 3$$

$$y + z^2 + x^3 = 3$$

$$z + x^2 + y^3 = 3$$

is $(x, y, z) = (1, 1, 1)$.

51. How many real solutions are there to the equation

$$6x^2 - 77[x] + 147 = 0,$$

where $[x]$ denotes the greatest integer not exceeding $x$?

52. Let $k$ be a real parameter. Sketch the possible forms of the graphs of the equation
$$y = x^4 + kx^2 - 2k^2(2k + 1)x,$$
specifying for each form the values of $k$ which give rise to it.

53. A triangle has sides of length 29, 29, 40. Find all other triangles with integer sides with the same perimeter and area.

54. Show that

$$2[(x-y)(x-z)+(y-z)(y-x)+(z-x)(z-y)]$$

can be expressed as the sum of three squares.

55. If $x, y, z > 0$, show that

$$\frac{1}{x(1+y)} + \frac{1}{y(1+z)} + \frac{1}{z(1+x)} \geq \frac{3}{1+xyz}.$$

56. Reduce to lowest terms

$$\frac{(ab-x^2)^2 + (ax+bx-2x^2)(ax+bx-2ab)}{(ab+x^2)^2 - x^2(a+b)^2}.$$

# Explorations

**E.62.** For $n \geq 2$, let $q_n(z)$ be the polynomial $z^{-1}[(1+z)^n - 1 - z^n]$. For which values of $n$ do all its zeros satisfy $|z| = 1$?

**E.63. Two Trigonometric Products.** For small positive integer values of $m$, use a pocket calculator to find the approximate values of the products

$$\cos(\pi/(2m+1))\cos(2\pi/(2m+1))\cos(3\pi/(2m+1))\ldots\cos(m\pi/(2m+1))$$

and

$$\tan(\pi/(2m+1))\tan(2\pi/(2m+1))\tan(3\pi/(2m+1))\ldots\tan(m\pi/(2m+1)).$$

Make a conjecture as to the exact values of these expressions. Can you establish this conjecture, at least for some values of $m$?

One way to approach the situation is to look at the polynomial

$$f(t) = \prod_{k=1}^{m} \left(t - \sec^2 \frac{k\pi}{2m+1}\right).$$

The two products can be found by taking the square roots of the values $|f(0)|$ and $|f(1)|$. With the help of a calculator, make conjectures regarding the coefficient of $f(t)$ for various values of $m$, and try to check your conjecture for as many cases as possible.

**E.64. Polynomials All of Whose Derivatives Have Integer Zeros.** The cubic polynomial $t^3 - 36t^2 + 285t - 250$ and its first and second derivatives all have integer zeros. Find other cubic polynomials with the same property. Do there exist polynomials of higher degree which, along with all their derivatives, have integer zeros?

**E.65. Polynomials with Equally Spaced Zeros.** Let $f(t)$ be a polynomial whose zeros are equally spaced, i.e. form an arithmetic progression. Does this impose any symmetry on its graph? on the zeros of its derivatives? Can anything be said in general about the relative sizes of the local maximum and minimum values of the polynomial?

**E.66. Composition of Polynomials of Several Variables.** If $p(t)$ and $q(t)$ are polynomials in a single variable $t$, then $\deg(p \circ q) = (\deg p)(\deg q)$. One consequence of this is that $(p \circ q)(t) = t$ identically only if the degree of both $p$ and $q$ is 1. Thus, the only polynomials of a single variable which possess a *polynomial* inverse with respect to composition are linear.

What is the situation for more than one variable? Suppose, for example, we consider the mapping

$$\mathbf{f}(x,y) = (f_1(x,y), f_2(x,y))$$

$$\mathbf{g}(x,y) = (g_1(x,y), g_2(x,y))$$

which take the real cartesian plane of points $(x,y)$ into itself, where the component functions $f_i$ and $g_i$ are polynomials in the variables $x$ and $y$. These two mappings can be composed to obtain $\mathbf{f} \circ \mathbf{g}$:

$$\mathbf{f} \circ \mathbf{g}(x,y) = (f_1(g_1(x,y), g_2(x,y)), f_2(g_1(x,y), g_2(x,y))).$$

For example, if $\mathbf{f}(x,y) = (x + y^2, x^3 - 2xy)$ and $\mathbf{g}(x,y) = (xy, x - y)$, then

$$\mathbf{f} \circ \mathbf{g}(x,y) = (x^2 + y^2 - xy, x^3y^3 - 2x^2y + 2xy^2).$$

What can be said about the relationships among the degrees of the polynomials? Is it necessarily the case that the degree of the components of $\mathbf{f} \circ \mathbf{g}$ exceeds the degree of the components of $\mathbf{f}$ and $\mathbf{g}$. Suppose $\mathbf{f}$ and $\mathbf{g}$ are mappings with polynomial components for which $\mathbf{f} \circ \mathbf{g}(x,y) = (x,y)$ identically. Must all the components of $\mathbf{f}$ and $\mathbf{g}$ be of degree 1?

**E.67. The Mandelbrot Set.** Let $c$ be a fixed complex number and define the quadratic polynomial $f(z) = z^2 + c$. Define the following sequence:

$$z_1 = 0 \quad z_n = f(z_{n-1}) \quad \text{for } n \geq 2.$$

on a complex plane, plot the points of this sequence for the cases $c = 0, 1, -1, -2, -1/2, i, 1 + i$.

Depending on the value of $c$, the terms of the sequence can either (i) remain within the confines of some disc of finite radius, or (ii) contain terms of arbitrarily large absolute value. The *Mandelbrot set* $M$ consists of all those values of $c$ for which (i) occurs. Using a computer or pocket calculator, test other values of $c$ to see whether they lie in the set $M$. Plot the points of the Mandelbrot set on the complex plane. There is, of course, the problem of deciding whether or not the terms of a sequence which initially does not wander too far from 0 will eventually remain bounded. Is there any way of deciding how many terms to compute before $c$ can be put in the Mandelbrot set or not?

Alternatively, one can fix a value of $c$ and begin the iteration with different values of $z_1$. Let $P_c$ be the set of $z_1$ for which the sequence remains within some disc of finite radius. What does $P_0$ look like? $P_1$? Try to plot points in $P_c$ for various values of $c$.

**E.68. Sums of Two Squares.** One of the most celebrated results of Leonard Euler (1707-1783) is that every prime of the form $4n + 1$ can be written as the sum of two squares. From a certain identity (which you can derive for yourself by examining the squares of the absolute values of $a + bi$, $c + di$ and the product $(a + bi)(c + di)$), it can be shown that every number which is the product of sums of two squares is itself a sum of two squares. Moreover, every product of factors of the form $2^u$, $p^v$ and $q^{2w}$, where $u$, $v$, $w$ are nonnegative integers and $p$, $q$ are primes with $p \equiv 1 \pmod 4$ and $q \equiv 3 \pmod 4$, can be expressed as the sum of two squares.

There is an algorithm which determines the representation of a prime $4n + 1$ as the sum of two squares. Define a transformation $T$ on number triples as follows:

$$T(x, y, z) = \begin{cases} (x - y - z, y, 2y + z) & \text{if } x > y + z \\ (y + z - x, x, 2x - z) & \text{if } x \le y + z. \end{cases}$$

Verify that the quantity $4xy + z^2$ is left invariant by this transformation.

Start with $(n, 1, 1)$ and apply $T$ repeatedly. For example, when $n = 8, 9$, we obtain the chains

$$\begin{aligned}
(8, 1, 1) &\longrightarrow (6, 1, 3) \longrightarrow (2, 1, 5) \longrightarrow (4, 2, -1) \\
&\longrightarrow (3, 2, 3) \longrightarrow (2, 3, 3) \longrightarrow (4, 2, 1) \\
&\longrightarrow (1, 2, 5) \longrightarrow (6, 1, -3) \longrightarrow (8, 1, -1) \\
&\longrightarrow (8, 1, 1) \longrightarrow \cdots \\
(9, 1, 1) &\longrightarrow (7, 1, 3) \longrightarrow (3, 1, 5) \longrightarrow (3, 3, 1) \\
&\longrightarrow (1, 3, 5) \longrightarrow (7, 1, -3) \longrightarrow (9, 1, -1) \\
&\longrightarrow (9, 1, 1) \longrightarrow \cdots
\end{aligned}$$

If, somewhere in the chain, we find the triple $(r, r, s)$, show that $4n + 1 = (2r)^2 + s^2$. For example, the chain $(9, 1, 1)$ contains $(3, 3, 1)$ and we find that $37 = 6^2 + 1^2$.

If the number $4n+1$ is expressible as the sum of two squares, will the chain beginning with $(n, 1, 1)$ always contain the form $(r, r, s)$? Does repeated application of $T$ to $(n, 1, 1)$ always return to $(n, 1, 1)$? If so, what can be said about the length and symmetry of the cycle? If not, what are the exceptional values of $n$?

**E.69. Quaternions.** A polynomial of degree $n$ over a field has at most $n$ zeros. This result may no longer hold if instead of a field we choose a structure for which not all the field axioms are valid. One such structure was invented by the British mathematician William R. Hamilton in 1843. Its elements are *quaternions*, generalized complex numbers of the form

$$a + bi + cj + dk$$

where $a$, $b$, $c$, $d$ are real and $i$, $j$, $k$ are distinct elements assumed to satisfy the relations

$$i^2 = j^2 = k^2 = -1, \quad ij = k, \quad jk = i, \quad ki = j.$$

We add, subtract and multiply quaternions in much the same way we do complex numbers, and assume that we have access to associativity of addition and multiplication and to the distributive law. For example, verify that we must have $ji = -k$, $ik = -j$, $kj = -i$, $ijk = -1$ and

$$\begin{aligned}(a + bi + cj + dk)(p + qi + rj + sk) &= (ap - bq - cr - sk) \\ &+ (aq + bp + cs - dr)i + (ar + cp + dq - bs)j \\ &+ (as + dp + br - cq)k.\end{aligned}$$

All the field axioms except commutativity of multiplication (Axiom M.2, Section 1.7) hold. By considering the product $(a + bi + cj + dk)(a - bi - cj - dk)$, show that each element for which not all of $a$, $b$, $c$, $d$ are zero has a multiplicative inverse and determine what this inverse is. Do this for several numerical examples and check your work.

An equation of the form $ux = v$ where $u$ and $v$ are quaternions with $u \neq 0$ has a unique solution. The situation for quadratic equations is more interesting. Find all quaternion solutions to the equations $x^2 = 1$ and $x^2 = -1$. Investigate the general quadratic equation $ax^2 + bx + c = 0$.

Are there any quadratic polynomials irreducible over the quaternions? Does the factor theorem hold? Is it true that a polynomial can always be factored as a product of irreducibles? Is such a factorization unique up to the order of the factors?

Investigate the equation $x^n = 1$ for $n \geq 3$.

# Hints

## Chapter 8

1. Divide $y^2 + 1$ by $3y - 5$ to obtain $x$ and clear fractions.

2. The difference of the sides of the first equation can be factored.

3. Factor $(x + y + z)^3 - (x^3 + y^3 + z^3)$.

4. Multiply by 4 and complete the square on both sides.

5. Find a quadratic equation for $y$ in terms of $x$.

7. Multiply by $a$ and find two factors whose product contains the sum of the first three terms.

8. Write $y = x + u$ and apply a discriminant condition to the quadratic in $x$ to obtain an equation in $u$.

9. Obtain two rational equations and use them to find a homogeneous quadratic equation in $x$ and $y$.

10. Let the equation $u + v = z$. Cube this equation and show that $uv$ is an integer $p$. Determine expressions for $x$ and $y$.

11. Set the polynomial equal to $u^2$ and examine the discriminant of the polynomial in $x$.

12. If there is a rational solution, it must be an integer. Let $(u, v, w)$ be such. What can be said about divisibility by 3?

14. Apply the AGM inequality to $|r_1 r_2 \cdots r_k|$ to obtain an inequality for the coefficients $a_i$.

15. Add the three equations to determine $x + y + z$. Take the difference of the second and third equations to determine other simple relations among the variables. Be careful about dividing by a quantity which might be zero.

16. $Q(x) - P(x) = (x - u)F(x)$. What can be said about the degree and sign of $F(x)$?

17. What can be said about the convexity of the graph which can be intersected by a line in four points? What implication does this have for the second derivative of the quartic?

18. The condition $2b = a + c$ can be used to derive an equation for $b$ which does not involve $p$.

19. The hypothesis implies that $p(t) - q(t)$ never changes in sign. Let $t = p(x)$, $t = q(x)$.

20. Let $y = mx + k$ be a side of the triangle. Then $P(x, mx + k)$ has at least three distinct zeros. Use the Factor Theorem.

21. Use the Factor Theorem.

22. Write the number in the form $v = u + u^{-1}$ and determine $u^7 + u^{-7}$ in terms of $v$.

23. (a) $1 - p(x)$ can be written as a product.

24. We want $\sqrt{2} + \sqrt{3} + \sqrt{5}$ to be a zero of $p(x) - (x - \sqrt{5})q(x)$. What is the polynomial of smallest degree in $y = x - \sqrt{5}$ over $\mathbf{Z}$ with the zero $y = \sqrt{2} + \sqrt{3}$?

25. Two squarings with some rearranging leads to a product of two quadratics set equal to zero. Thus the four possible roots can be explicitly identified and analyzed.

26. Look at the sign of the coefficients, the discriminant and the product $(a + b - c)(b + c - a)(c + a - b)$, where $a$, $b$, $c$ are the zeros.

27. Look at cases according as $a$, $b$, $c$ are positive, negative or zero. The hard case is that in which none are zero and the signs differ. Let $x = \cos u$, etc.

28. A real zero can be guaranteed as the sum of the values of the polynomial for two different substitutions of the variable vanishes.

29. Solve the problem for $m = n = 2$ by considering $(AX - BY)^3$.

30. Let $q(x) = p(x) + \cdots$. What is the parity of the degree of $q(x)$? What can be said about the extremum of $q(x)$?

31. Let $f(t)$ and $g(t)$ be cubics with the zeros $x$, $y$, $z$ and $a$, $b$, $c$ respectively. Look at the sign of $(f - g)(t)$ at $a$ and $c$.

32. Let the zeros of the polynomial be $r_i$. Recall the factorization of $x^n - r_i^n$ where $n = 1\ 000\ 000$.

33. Consider the number of distinct zeros of $p(x)$, $q(x)$ and $(p - q)(x)$. How many zeros does $p'(x)$ have counting multiplicity?

34. The system $p + q + r = 0$, $up + vq + wr = 0$, $xp + yq + zr = 0$ has a nontrivial solution for $(p, q, r)$. Show that $bu + cx = bv + cy = bw + cz = 1$ for some $b$ and $c$. Use the conditions $cx = 1 - bu$ and $x^3 = a^3 - u^3$ to obtain a cubic equation satisfied by $u$. The same equation is satisfied by $v$ and $w$.

35. Multiply by the product of the $(x - k)$ to render the difference between the two sides of the inequality in the polynomial form $p(x)$. What is the relation between the endpoints of the interval and the zeros of $p(x)$?

36. Show that the left side multiplied by $(a + b)$ is symmetric in all six variables. You will need to show that $\Sigma abc = 0$.

37. Write $a + bi$, $a - bi$, $c + di$, $c - di$ in terms of the symmetric functions of $x = \cos u + i \sin u$, etc. Use the first three equations to solve for the symmetric functions which can then be plugged into the fourth.

38. Look at $x^2 - y^2$ and $xy$.

40. The local extrema are those $k$ for which $y = k$ is tangent to the graph of the cubic.

41. $2(ab - cd) = [a^2 + b^2 - c^2 - d^2] - [(a - b)^2 - (c - d)^2]$. Factor the difference of squares.

42. All the coefficients must be rational so we can assume that they are integers. Reduce to the monic case with prime constant coefficient, so that a great deal can be said about the rational zeros. Recall the role of the Intermediate Value Theorem in guaranteeing real zeros.

43. Let $u = a/(bc - a^2)$, etc. and look at $vw - u^2$, etc.

44. Explore the situation for small $n$ and make a conjecture.

45. If $r$ is a zero of $f(t)$, what other zeros must there be?

46. Apply the AGM inequality to $u^2 + 3x^2$.

47. Factor $x^3 + y^3 + z^3 - 3xyz$.

48. Every zero of the divisor should make the dividend vanish.

49. The equation is $(x + 1)^3 - x^3 = y^2$. Express as a quadratic in $x$ and complete the square.

50. Take the differences of the equations in pairs.

51. Sketch the graphs of $y = 6x^2$, $y = 77x - 147$ and $y = 77[x] - 147$.

53. Determine the form of a cubic whose zeros are $u = 49 - a$, etc. Heron's formula for the area will be useful.

56. The denominator is a difference of squares. Use the Factor Theorem.

# Answers to Exercises and Solutions to Problems

## Answers to Exercises

### Chapter 1

**1.1.** (a) 5, 1, 0, 7; (b) 5, 3, 2, 8; (c) 3, 0, 0, 4; (d) 2, $-1$, 1, 6.

**1.2.** (a) $t^7$; (b) $2t^3$; (c) 0.

**1.3.** There are seven polynomials. (a) $-\infty$, 0, 0, 0, 0, 0; (b) 4, 0, 0, 3, 0, 3/16; (c) 2, 3, 0, 1, 3, 13/4; (d) 2, 0, $-3$, 8, 0, 7/2; (h) 3, 0, 3, $-4$, 0, $-1$; (i) 4, 1, 0, 8, 1, $-\frac{1}{2}$; (k) 5, 0, 0, 6, 0, 9/16.

**1.7.** 650 078 260 327 823.

**1.8.** 731 765 148 134 177 451 740.

**1.10.** $\deg p \circ q = \deg q \circ p = (\deg p)(\deg q)$.

**1.12.** (a) Let $p(t) = a_n t^n + \cdots + a_1 t + a_0$ be such a polynomial. Then $a_0 = p(0) = 0$. For any real nonzero $c$, $a_n c^{n-1} + \cdots + a_1 = p(c)/c = 0$. (We do not know that the left side vanishes at $c = 0$ without further development; in order to avoid this issue, we need to make a more elaborate argument at this point.) Suppose, if possible, $a_1 \neq 0$. Choose $c$ such that $0 < c < 1$ and

$$2c(|a_2| + \cdots + |a_n|) < |a_1|.$$

Then

$$
\begin{aligned}
|a_n c^{n-1} &+ a_{n-1} c^{n-2} + \cdots + a_1| \\
&\geq |a_1| - [|a_n| c^{n-1} + |a_{n-1}| c^{n-2} + \cdots + |a_2| c] \\
&\geq |a_1| - c[|a_n| + |a_{n-1}| + \cdots + |a_2|] \\
&> |a_1| - (1/2)|a_1| > 0,
\end{aligned}
$$

a contradiction. Hence $a_1 = 0$, and, for all $c \neq 0$, $a_n c^{n-2} + \cdots + a_2 = p(c)/c^2 = 0$. Continue on to show in turn that $a_2, a_3, a_4, \ldots a_n$ all vanish. Thus, $p(t)$ must be the zero polynomial.

With more background, other proofs can be given. For example, the conditions $p(0) = p(1) = \cdots = p(n) = 0$ leads to a system of $n + 1$ linear equations in the unknowns $a_0, a_1, \ldots, a_n$ for which the solution is unique.

Alternatively, the identical vanishing of $p(t)$ implies the same for all of its derivatives, whence Taylor's Theorem identifies all the coefficients as 0.

The reader might wish to reflect on the validity of the following proof, which assumes the Factor Theorem. Let $n$ be the degree of a nonzero polynomial $p$ which vanishes identically. Since $p$ vanishes at $0, 1, 2, 3, \ldots, n$, we can write

$$p(t) = t(t-1)(t-2)\cdots(t-n)q(t),$$

for some polynomial $q$. The degree of the left side is $n$ while that of the right side is at least $n+1$, a contradiction.

**1.13.** (a) If $f(2t) = h(f(t))$, then $\deg f(t) = \deg f(2t) = \deg h(t) \deg f(t)$, so that, either $f$ is a constant $c$ and $h(c) = c$, or else $h(t)$ is linear of the form $ut + v$. If $f(t) = a_n t^n + \cdots + a_1 t + a_0$ with $a_n \neq 0$, we have

$$2^n a_n t^n + 2^{n-1} a_{n-1} t^{n-1} + \cdots + 2a_1 t + a_0$$

$$= u a_n t^n + u a_{n-1} t^{n-1} + \cdots + u a_1 t + (u a_0 + v).$$

Hence $2^k a_k = u a_k$ $(1 \leq k \leq n)$ and $a_0 = u a_0 + v$. Thus, $f(t) = a_n t^n + a_0$ and $h(t) = 2^n t + (1 - 2^n) a_0$ where $n$ is a positive integer, and $a_n$ and $a_0$ are arbitrary constants.

(b) The relation has the form $f(2t) = h(f(t))$, with $f(t) = \sin^2 t$ and $h(t) = 4t(1-t)$. But, since $h$ is quadratic, by (a), $f$ cannot be a polynomial.

**1.14.** *First solution.* If $\log t$ is a polynomial, then, by 13.(a), $\log 2t = \log 2 + \log t$ implies $\log t = at^n + b$. But then $\log 2 = (2^n at^n + b) - (at^n + b) = (2^n - 1)at^n$, which is false.

*Second solution.* Suppose $\log t = f(t)$, a polynomial of degree $n$. Since $f(t^2) = 2f(t)$, comparing degrees yields $2n = n$, whence $n = 0$. But then $\log t$ must be a constant, which is false.

**1.15.** Suppose that $g(t) = a_n t^n + \cdots$ is periodic with positive period $k$ and $a_n \neq 0$. Then $g(t+k) - g(t) = kn a_n t^{n-1} + \cdots$ vanishes identically, so its leading coefficient $kn a_n$ is 0. Hence $n = 0$ and $g$ must be constant.

**1.16.** If $t^{1/3}$ is a polynomial of degree $k$, then $3k = \deg t = 1$, which contradicts $k$ being an integer.

**1.17.** $p(f - g) = 0$. If $f - g$ is not the zero polynomial, then the left side would be a polynomial of nonnegative degree which vanishes identically, a contradiction.

**1.19.** (a) The coefficients of all the odd powers of $t$ are 0.

(b) The coefficients of all the even powers of $t$ are 0.

**1.20.** Suppose $p(t)$ has more than one term and $p(t) = at^n + bt^m + \cdots$, where $n = \deg p$ and $m$ is the next highest nonnegative exponent corresponding to a nonzero coefficient. Equating $p(t^2)$ and $[p(t)]^2$ yields $at^{2n} + bt^{2m} + \cdots = a^2 t^{2n} + 2ab t^{m+n} + \cdots$, whence $2m = m + n$. But this contradicts $m < n$.

Hence, $p(t)$ has a single term, and it can be seen that the polynomials commuting with $t^2$ are precisely the powers of $t$.

**2.1.** (b) 3, 4;

(c) $\dfrac{-b + \sqrt{b^2 - 4ac}}{2a}$ and $\dfrac{-b - \sqrt{b^2 - 4ac}}{2a}$

**2.3.** $m$, $(1 - m^2)/2m$.

**2.5.** $1$, $-1/3$, $0$.

**2.7.** The zeros of $6x^2 - 5x - 4$ are $-1/2$ and $4/3$. Thus $6x^2 - 5x - 4 = (2x + 1)(3x - 4)$ is negative for $-1/2 < x < 4/3$.

**2.8.** The discriminant of $x^2 + (1 - 3k)x + (2 - k)$ is $9k^2 - 2k - 7 = (9k + 7)(k - 1)$, which is nonnegative for $k \geq 1$ and $k \leq -7/9$. These are the only values of $k$ for which the equation is solvable.

**2.9.** The range of the function is the set of $k$ for which

$$(k - 1)x^2 + (3k - 1)x + (2k + 1) = 0$$

is solvable. The discriminant, equal to $(k - 1)^2 + 4$, is always positive so that the equation is solvable for all $k$.

**2.10.** Since $m + n = 5/6$ and $mn = -1/2$, we have that the sum of the roots of the desired quadratic is $(m+n)-(m^2+n^2) = (m+n)-(m+n)^2 + 2mn = -31/36$ and the product of the roots is $mn - (m^3 + n^3) + (mn)^2 = mn - (m + n)^3 + 3mn(m + n) + (mn)^2 = -449/216$. Thus, a quadratic polynomial with the desired roots is $216t^2 + 186t - 449$.

**2.12.** (a) Suppose $r$ is a common nonrational zero of $p(t)$ and $q(t)$. Choose integers $a$ and $b$ for which $(ap + bq)(t)$ is either linear or a constant. If $(ap+bq)(t)$ were linear, it would have rational coefficients but a nonrational zero, which cannot occur. Hence it is a constant. Since $(ap+bq)(r) = 0$, we must have that $(ap + bq)(t)$ is the zero polynomial, from which the result follows.

(b) $t^2 - t$; $t^2 + t$.

**2.13.** Suppose $x = (b - d)/(a - c)$ satisfies $x^2 - ax + b = 0$. Since $(a - c)x = (b - d)$, $-ax + b = -cx + d$, so that $x^2 - cx + d = x^2 - ax + b = 0$ as required.

**2.14.** (a) The quadratic polynomial is always nonnegative and can vanish only if $t = -b_i/a_i$ for each $i$ simultaneously. Hence it has either coincident real zeros or two nonreal zeros.

(b) The desired inequality is a rewriting of the condition that the discriminant of the polynomial in (a) is nonpositive. Equality occurs if and only if the discriminant vanishes, which occurs when there is a single real zero. The condition for equality is that $b_i/a_i$ are equal for all $i$.

**2.16.** The abscissae $x_1$ and $x_2$ of the endpoints of the chord $y = mx + k$ are the roots of the equation $(b^2 + a^2m^2)x^2 + 2a^2kmx + a^2(k^2 - b^2) = 0$. Hence the coordinates $(X, Y)$ of the midpoints of the chord are given by

$$X = -a^2km(b^2 + a^2m^2)^{-1}, \quad Y = mX + k = kb^2(b^2 + a^2m^2)^{-1},$$

from which it follows that the diameter lies along the line $b^2x + ma^2y = 0$. For the case that all the chords are vertical, the locus has equation $y = 0$.

**2.18.** Since the quadratic equation has a nonrational root, $p(u) \neq 0$ for each rational $u$. Since $p(u)$ can be written as a fraction whose denominator is $q^2$, it follows that $1/q^2 \leq |p(u)|$. The right inequality follows from exercise 2(b). For the final assertion, take $M = 1/K$.

**3.1.** Suppose $x$, $y$, $u$, $v$ are real and $x + yi = u + vi$. Rearranging terms and squaring yields $(x - u)^2 = -(v - y)^2$, whence $x - u = v - y = 0$.

**3.2.** (c) $z \longrightarrow wz$ is a central similarity (dilatation) with factor $|w|$ followed by a counterclockwise rotation through angle $\arg w$.

**3.4.** Given $z$, to construct $1/z$: Construct the line $A$ joining $0$ to a point $u$ on the unit circle such that the real axis bisects the angle $z0u$. Let the circle with center $0$ and radius $|z|$ meet the positive real axis at $r$. Let $B$ be the line through $r$ and $u$, and $C$ be the line through $1$ parallel to $B$. The intersection of $A$ and $C$ is the point $1/z$.

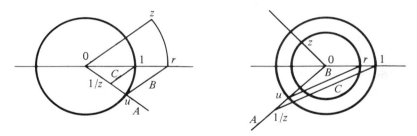

**3.5.** The locus evidently contains $c/w = c\overline{w}/|w|^2$. For $z$ on the locus, $\mathrm{Re}[(z - c/w)w] = 0$, whence $(z - c/w)w = ik$ for some real $k$. The locus is a straight line consisting of points $z = [c + ik]\overline{w}/|w|^2$ $(k \in \mathbf{R})$.

**3.6.** Let $z = x + yi$. For $k = 1$, the locus is a straight line with equation $2x = -1$. For $k \neq 1$, the equation of the locus is

$$x^2 + y^2 = k^2(x^2 + 2x + 1 + y^2).$$

For $k > 1$, the locus is a circle with center $(-k^2/(k^2 - 1), 0)$ and radius $k/(k^2 - 1)$. If $k < 1$, the locus is a circle with center $(k^2/(1 - k^2), 0)$ and radius $k/(1 - k^2)$.

**3.7.** Let the rocks $U$ and $V$ be the points $0$ and $1$ respectively in the complex plane. If $T$ is at $z$, then $P$ is at $iz$ and $Q$ is at $1 + (-i)(z - 1) = (1 + i) - iz$.

The treasure is located at $(1/2)(1+i)$, midway between $P$ and $Q$. Since this does not depend on $z$, the position of the treasure is the same regardless of the position of $T$.

**3.8.** For positive integer $n$, use induction. For $n = -1$, note that $(\cos x - i \sin x)(\cos x + i \sin x) = 1$. For $n = -k < 0$, we have that $(\cos x + i \sin x)^{-k} = (\cos kx + i \sin kx)^{-1} = \cos(-k)x + i \sin(-k)x$.

**3.9.** Each equation implies that $|z| = 1$, so that $z = \cos\theta + i \sin\theta$ for some $\theta$.

(a) $\cos 3\theta + i \sin 3\theta = 1$ implies that $3\theta$ is a multiple of $2\pi$, thus the solutions are $1$, $\cos(2\pi/3) + i \sin(2\pi/3) = \frac{1}{2}(-1 + i\sqrt{3})$, $\cos(2\pi/3) - i \sin(2\pi/3) = \frac{1}{2}(-1 - i\sqrt{3})$.

(b) $1$, $i$, $-1$, $-i$.

(c) $1$, $(1 + i\sqrt{3})/2$, $(-1 + i\sqrt{3})/2$, $-1$, $(-1 - i\sqrt{3})/2$, $(1 - i\sqrt{3})/2$.

(d) $1$, $(1 + i)/\sqrt{2}$, $i$, $(-1 + i)/\sqrt{2}$, $-1$, $(-1 - i)/\sqrt{2}$, $-i$, $(1 - i)/\sqrt{2}$.

In each case, the numbers are equally spaced on the perimeter of the unit circle with center at the origin.

**3.10.** (a) Since $x^2 - y^2 = a$, $2xy = b$, it follows that $x^2$ and $-y^2$ are the roots of

$$t^2 - at - b^2/4 = 0.$$

Let $c^2 = a^2 + b^2$, $c > 0$. Then $x^2 = \frac{1}{2}(c + a)$, $y^2 = \frac{1}{2}(c - a)$. If $b > 0$, $(x, y) = (\pm\sqrt{(c + a)/2}, \pm\sqrt{(c - a)/2})$; if $b < 0$, $(x, y) = (\pm\sqrt{(c + a)/2}, \mp\sqrt{(c - a)/2})$. In the case that $b = 0$, $c = |a|$, and $x = 0$ or $y = 0$ according as $a$ is negative or positive.

(b) $3 - 4i$ and $-3 + 4i$.

**3.11.** (a) We have $2t = -3 \pm \sqrt{-3 + 4i}$. Applying Exercise 10 yields $\sqrt{-3 + 4i} = \pm(1 + 2i)$. The roots are $-2 - i$ and $-1 + i$.

(b) Note that the discriminant is $-7 - 24i = (3 - 4i)^2$. The roots are $2 - 3i$ and $-1 + i$.

**3.12.** We can apply the quadratic formula for the roots of the equation. Exercise 10 shows how the square root of the discriminant can be found.

**3.13.** $|1 + iz| = |1 - iz| \iff (1 + iz)(1 - i\overline{z}) = (1 - iz)(1 + i\overline{z}) \iff i(z - \overline{z}) = -i(z - \overline{z}) \iff z - \overline{z} = 0 \iff z$ is real.

**3.15.** (a) Let $\theta = \arccos x$. The equation is equivalent to

$$\cos(n + 1)\theta + \cos(n - 1)\theta = 2 \cos n\theta \cos \theta.$$

(b) $x$, $2x^2 - 1$, $4x^3 - 3x$, $8x^4 - 8x^2 + 1$.

(c) Since $T_{n+1} = 2xT_n - T_{n-1}$, by induction it can be established that $T_n$ is a polynomial of degree $n$.

(d) By de Moivre's Theorem,

$$\cos n\theta \; + \; i\sin n\theta$$

$$= \sum_{s=0}^{n} \binom{n}{s} \cos^{n-s}\theta \sin^{s}\theta i^{s}$$

$$= \sum_{r=0}^{[n/2]} \binom{n}{2r} \cos^{n-2r}\theta \sin^{2r}\theta (i^2)^r$$

$$+ \sum_{t=0}^{[(n-1)/2]} \binom{n}{2t+1} \cos^{n-2t-1}\sin^{2t+1}\theta i^{2t+1}$$

$$= \sum_{r=0}^{[n/2]} \binom{n}{2r} (-1)^r \cos^{n-2r}\theta (\sin^2\theta)^r + i\{\cdots\}.$$

Equating real parts, setting $\theta = \arccos x$, $\cos\theta = x$, $\sin^2\theta = 1 - x^2$ yields the results.

**4.1.** (a) Let $p(t) = at^3 + bt^2 + ct + d$. Then for any number $r$, $p(t) - p(r) = (t-r)[a(t^2 + rt + r^2) + b(t+r) + c] = (t-r)[at^2 + (ar+b)t + (ar^2 + br + c)]$. The result follows immediately.

(b) One root is 1. By (a), we can write the given cubic as a product of $t - 1$ and a quadratic polynomial. Indeed, it is equal to $(t-1)(t^2 + t - 3)$. The solutions are: $1, [-1 + \sqrt{13}]/2, [-1 - \sqrt{13}]/2$.

**4.2.** The solutions are $x = 1, 2, 9$.

**4.4.** (d) If $u_0^3$ is the single root of the quadratic, we have $q = -2u_0^3$ and $p = -3u_0^2$, and the factorization can be checked.

(e) We can choose real $r$ and $\theta$ such that

$$u_0^3 = r^3(\cos 3\theta + i\sin 3\theta)$$

$$v_0^3 = r^3(\cos 3\theta - i\sin 3\theta).$$

The condition that $u_0 v_0$ be real requires that $u_0 = r(\cos\theta + i\sin\theta)$ and $v_0 = r(\cos\theta - i\sin\theta)$. Note that $\omega = \cos 2\pi/3 + i\sin 2\pi/3$.

**4.5.** (a) $-3, (3 + i\sqrt{3})/2, (3 - i\sqrt{3})/2$.

(b) $u^3 = 3.5690[\cos 147.30° + i\sin 147.30°]$ and $u = 1.0000 + 1.1549i$. The solutions are $2, 1, -3$.

**4.6.** $x = t + 5$ yields $t^3 - 108t + 432 = 0$, $x = -7, 11, 11$.

**4.7.** $x^3 + px + q = 0$ being the equation, set $x = a^2b + ab^2$. An argument similar to that in Exercise 4(a) leads to $3a^3b^3 = -p$ and $p(a^3 + b^3) = 3q$. Having found the determination $a$ and $b$ for the cube roots of $a^3$ and $b^3$ respectively, the pairs $(a, b)$, $(a\omega, b\omega)$, $(a\omega^2, b\omega^2)$ all yield the same values of $a^2b + ab^2$.

**4.9.** $p^3 + 8q^2 = 0$. If $y = 2x$, the latter equation becomes $y^4 = (y^2 + 6y - 18)^2$.

**4.10.** (a) $\phi = \pi/2$, $x = 2a \cos \pi/6 = \sqrt{3}a$.

(b) $\phi = 0$, $x = 2$.

**4.11.** (b) $u^6 + 2pu^4 + (p^2 - 4r)u^2 - q^2 = 0$.

(d) The equation for $u^2$ is $u^6 + 2u^4 + 13u^2 - 16 = 0$, which leads to $u = 1$ and the factorization

$$t^4 + t^2 + 4t - 3 = (t^2 + t - 1)(t^2 - t + 3).$$

The equation for $u^2$ is $u^6 - 4u^4 + 16u^2 - 64 = 0$ which leads to $u = 2$ and the factorization

$$t^4 - 2t^2 + 8t - 3 = (t^2 + 2t - 1)(t^2 - 2t + 3).$$

**4.12.** (b) Setting the discriminant of the quadratic equal to zero yields $2u^3 - qu^2 + 2(pr - s)u + (qs - p^2s - r^2) = 0$.

(c) The first equation is equivalent to $(t^2 + 1)^2 = (t - 2)^2$. For the second equation, $u$ is given by $2u^3 + 5u^2 - 4u - 7 = 0$, and we can rewrite the equation as $(t^2 - t - 1)^2 = 4(t - 1)^2$.

**4.13.** (c) A reciprocal polynomial of odd degree can be written in the form $a(x^{2k+1} + 1) + bx(x^{2k-1} + 1) + cx^2(x^{2k-3} + 1) + \cdots = 0$. Noting that when $m$ is odd we have

$$x^m + 1 = (x + 1)(x^{m-1} - x^{m-2} + x^{m-3} - \cdots - x + 1),$$

we obtain the result.

**4.14.** (b) Prove by induction on $m$. If the result is true for $m \leq r$, then

$$x^{r+1} + x^{-(r+1)} = t(x^r + x^{-r}) - (x^{r-1} + x^{-(r-1)}),$$

whence the left side is a polynomial of degree $r + 1$.

(c) Since $x = 0$ is not a solution of the equation, we can replace it by the equivalent

$$a(x^k + x^{-k}) + b(x^{k-1} + x^{-(k-1)}) + \cdots = 0.$$

A change to the variable $t$ gives an equation of degree $k$ in $t$. Having found values of $t$, we can then solve equations of the form $x^2 - tx + 1 = 0$ for $x$.

**4.15.** (a) $t = 1/2, -3$.

(b) $2x^4 + 5x^3 + x^2 + 5x + 2 = (2x^2 - x + 2)(x^2 + 3x + 1)$.

**4.17.** (c) $x = u + v$ leads to a quasi-reciprocal equation in $u$ provided that $4v^3 + v^2 - 6v + 1 = 0$. Taking $v = 1$ leads to the equation $u^4 + 4u^3 + 9u^2 + 8u + 4 = 0$. Let $t = u + 2/u$. Then $t^2 + 4t + 5 = 0$, whence $t = -2 \pm i$. The equation $u^2 + (2 \pm i)u + 2 = 0$ can be solved for $u$ (see Exercise 3.10) and then $x = u + 1$.

**5.1.** Homogeneous: (a), (b), (e), (f); Symmetric: (c).

**5.2.** The "only if" part is straightforward. Suppose the result holds. Then for each $x$, $y$, $z$, the equation can be written in the form

$$t^d f_d(x, y, z) + \sum_{k \neq d} t^k f_k(x, y, z) = t^d f(x, y, z)$$

where $f_k$ is homogeneous of degree $k$. By Exercise 1.12, $f_d = f$, $f_k = 0$, whence the result follows.

**5.3.** $cx^k$.

**5.4.** Each homogeneous polynomial can be obtained by collecting up like terms in the given polynomial.

**5.5.** Every symmetric polynomial in two variables can be expressed as a sum of polynomials of one or other of the types $cx^k y^k = cs_2^k$ or $c(x^a y^b + x^b y^a) = cs_2^a p_{b-a}$, where $a < b$ and $p_i = x^i + y^i$ for $i = 1, 2, \ldots$. Show by induction, using the expansion of $(x+y)^i$, that $p_i$ is a polynomial in $s_1$ and $s_2$.

**5.7.** The sum of the zeros is equal to $-p$. The zeros are in arithmetic progression if and only if their average is equal to one of them, i.e. if and only if $-p/3$ is a zero. This yields the condition $2p^3 + 27r = 9pq$.

**5.8.** $s_1^3 - 3s_1 s_2 + 3s_3$; $s_1 s_2^2 - s_2 s_3 - 2s_1^2 s_3$; $(s_1 - z)(s_1 - y)(s_1 - x) = s_1^3 - s_1^2 s_1 + s_1 s_2 - s_3 = s_1 s_2 - s_3$.

**5.12.** Consider the polynomial $u(x, y) = g(x, y) - g(y, x) = a_0(y) + a_1(y)x + a_2(y)x^2 + \cdots + a_n(y)x^n$. For each value of $y$, this polynomial vanishes identically in $x$, so that the polynomial coefficients $a_i(y)$ vanish identically in $y$. But this means that each $a_i$ is the zero polynomial, so that $u$ itself is the zero polynomial.

**6.1.** $\gcd(20119, 34782) = 341$; the quotients are 1, 1, 2, 1, 2, 5.

**6.2.** (b) (iv) $341 = (34782)(11) - (20119)(19)$. The pencil and paper table is

$$
\begin{array}{ccccc}
 & -2 & -1 & -2 & -1 & -1 \\
1 & -2 & 3 & -8 & 11 & -19
\end{array}
$$

**6.3.** (a) If $a$ is a multiple of $p$, then $\gcd(a, p) = p \neq 1$. On the other hand, if $g = \gcd(a, p) \neq 1$, then, since $g \mid p$, $g = p$, so $p \mid a$.

(b) Use the general result enunciated in Exercise 2.

(c) Suppose $p$ does not divide $a$. For some integers $x$, $y$, we have that $1 = ax + py$, whence $b = abx + bpy$. Since $p$ divides both terms of the right side, $p$ must divide $b$.

(d) The result holds for $n = 2$. Assume it holds for each number less than $n$. Let $p_1$ be the smallest positive divisor of $n$ exceeding 1; then $p_1$ must be prime (why?). Then either $n = p_1$, in which case this is the only representation of $n$ as a product of primes, or $n/p_1 < n$, and we can apply the induction hypothesis to obtain the desired representation. By an extension of Exercise (c), it can be shown that the only primes dividing $n$ must be the $p_i$ and thence that the representation is unique.

(e) $\quad 418 = 2.11.19 \qquad 1606 = 2.11.73$
$\qquad 20119 = 11.31.59 \quad 34782 = 2.3.11.17.31$

**6.4.** (a) If $p^a$ and $p^b$ are the highest powers of the prime $p$ dividing $m$ and $n$ respectively, then $p^{\min(a,b)}$ and $p^{\max(a,b)}$ are the highest powers of $p$ dividing $u$ and $v$ respectively.

(b) and (c) are straightforward consequences of (a).

**6.6.** (a) $x = 2 + 5k$ where $k$ is an integer.

(b) (i) no solution. (ii) $x \equiv 2 \pmod 3$.

(d) There exist $u$ and $v$ such that $g = au + mv$. Let $x = (b/g)u$.

(e) $au \equiv av \pmod m \Rightarrow m \mid a(u - v) \Rightarrow m \mid (u - v)$ (since $\gcd(a, m) = 1$) $\Rightarrow u \equiv v \pmod m$.

**7.4.** (d) $\mathbf{F}[\mathbf{t}]$ is the set of all polynomials in the $n$ variables $t_1$ with coefficients in $\mathbf{F}$. We prove that $\mathbf{F}[\mathbf{t}]$ is an integral domain by induction on the number of variables. It is true for $n = 0$ (see Exercise 5). Suppose it is true for $n - 1$ variables. Let $f$ and $g$ be two nonzero polynomials over $\mathbf{F}$. Write $f(\mathbf{t}) = a_r t_n^r + \cdots$ and $g(\mathbf{t}) = b_s t_n^s + \cdots$, where the coefficients are polynomials in the $n - 1$ variables $t_1, \ldots, t_{n-1}$ and $a_r \neq 0$, $b_s \neq 0$. Then $f(\mathbf{t})g(\mathbf{t}) = a_r b_s t_n^{r+s} + \cdots$. By the induction hypotheses, the leading coefficient is nonzero, so $fg \neq 0$.

**7.5.** (a) If $ab = 0$, $a \neq 0$, then $b = a^{-1}ab = 0$.

(b) $ac = bc \Rightarrow (a - b)c = 0 \Rightarrow a - b = 0 \Rightarrow a = b$.

**7.6.** (c) $\mathbf{Z}_m$ is a field if and only if $m$ is prime.

(d) If $m$ is composite, then $ab = m \equiv 0 \pmod m$ is possible for numbers $a$ and $b$, neither of which are divisible by $m$.

**7.7.** The following polynomials cannot be expressed as a product of polynomials of lower degree: $t, t+1, t^2+t+1, t^3+t+1, t^3+t^2+1, t^4+t^3+t^2+t+1, t^4 + t^3 + 1, t^4 + t + 1$. Since the leading coefficient is 1 and there are two choices for each other coefficient, there are $2^d$ polynomials altogether of degree $d$.

## Solutions to Problems

### Chapter 1

**8.1.** $\tan A + \tan B = -p$, $\tan A \tan B = q$. If $q = 1$, $A + B \equiv \pi/2 \pmod \pi$, which yields the result $q$. If $q \neq 1$, the expansion formula for $\tan(A + B)$ yields $-p = (1 - q)\tan(A + B)$, whence

$$\sin^2(A + B) + p \sin(A + B)\cos(A + B) + q \cos^2(A + B)$$

$$= \sin^2(A+B) + (q-1)\tan(A+B)\sin(A+B)\cos(A+B) + q\cos^2(A+B)$$
$$= \sin^2(A+B) + q\sin^2(A+B) - \sin^2(A+B) + q\cos^2(A+B) = q.$$

**8.2.**

$$\sum_{i=1}^{n}(x+i-1)(x+i) = pn \Rightarrow nx^2 + n^2x + \frac{(n+1)n(n-1)}{3} - pn = 0.$$

Hence $2r + 1 = -n$ and $r(r+1) = (n^2-1)/3 - p$. Eliminating $r$ yields $n^2 = 12p+1$. If $p = 10$, then $n = 11$ and the solutions of the quadratic are $-5$ and $-6$. In general, $n$ must have the form $6k \pm 1$, whence $p = 3k^2 \pm k$.

**8.3.** *Solution 1.* Let the roots of $ax^2 + bx + c = 0$ be $r$ and $s$. Then one is the square of the other if and only if

$$0 = (r-s^2)(s-r^2) = rs + r^2s^2 - [(r+s)^3 - 3rs(r+s)],$$

which is equivalent (using $r + s = -b/a$, $rs = c/a$) to

$$0 = a^2c + ac^2 + b^3 - 3abc.$$

*Solution 2.* Suppose that the roots of $ax^2 + bx + c = 0$ are $u$ and $u^2$. Then $u + u^2 = -b/a$, whence $au^2 + au + b = 0$. But $au^2 + bu + c = 0$, so $(b-a)u = (b-c)$. If $a = b$, then $b = c$ and the equation is a constant multiple of $x^2 + x + 1 = 0$. If $a \neq b$, then substitute $u = (b-c)/(b-a)$ into $au^2 + bu + c = 0$. In either case, we obtain

$$b^3 + a^2c + ac^2 - 3abc = 0.$$

On the other hand, let $u$ and $v$ be the roots of $ax^2 + bx + c = 0$, and suppose that $b^3 + a^2c + ac^2 - 3abc = 0$. Then $b = -a(u+v)$, $c = auv$ leads to $0 = (u-v^2)(v-u^2)$, so one of $u$ and $v$ is the square of the other.

*Solution 3.* If $u$ and $u^2$ are the roots of the given equation, then $u$ is a common root of the two equations $ax^2 + bx + c = 0$ and $ax^2 + ax + b = 0$ (since $u + u^2 = -b/a$). Hence $(b-a)u = (b-c)$. Substituting this into the equation yields the condition $a(b-c)^2 + b(b-a)(b-c) + c(b-a)^2 = 0$.

On the other hand, suppose the condition holds. If $a = b$, then $b = c$ and the equation is a constant multiple of $x^2 + x + 1 = 0$, each of whose roots is the square of the other. If $a \neq b$, then $u = (b-a)^{-1}(b-c)$ satisfies $ax^2 + bx + c = 0$ and $ax^2 + ax + b = 0$. Hence $u + u^2 = -b/a$ and $u^2$ must be the second root of $ax^2 + bx + c = 0$.

**8.4.** Let $q(t) = p(n+t) = t^2 + bt + c$. Then $p(n)p(n+1) = c(1+b+c) = q(c) = p(n+c)$. Also, $p(n)p(n+1) = p(n-c-b)$.

**8.5.** The discriminant of the second quadratic is

$$4(a+p)^2 - 12(q+ap) = (2a-p)^2 + 3(p^2-4q),$$

from which the result follows.

**8.6.** The quadratics all have the form $x^2 + ax + b$. Each alteration changes the value of $a - b$ by 1. Initially, $a - b = -10$; finally, $a - b = 10$. Hence, somewhere along the way, $a - b = 1$, and we have a quadratic with zeros $-1$, $-b$.

**8.7.** $(a + b)^2 = (c + d)^2$ implies $ab = cd$, so that $-c$ and $-d$ are the zeros of $t^2 - (a + b)t + ab$. Hence $a + d = b + c = 0$. It is easy to check that any solution of this equation satisfies the given equations.

**8.8.** Let $r$ be the real root of the given equation. Then $(r^2 + ar + c) + (br + d)i = 0$. Since the sum of the roots is nonreal, $b \neq 0$ and we have $r = -d/b$. Substituting this into $r^2 + ar + c = 0$ yields $abd = b^2c + d^2$.

On the other hand, suppose that $abd = b^2c + d^2$ and $b \neq 0$. Then $z^2 + (a + bi)z + (c + di) = (z + d/b)(z + b(c + di)/d)$ has one real and one nonreal zero.

**8.9.** A common zero of $x^2 + px + q$ and $px^2 + qx + 1$ is a zero of $x(x^2 + px + q) - (px^2 + qx + 1) = x^3 - 1 = (x - 1)(x^2 + x + 1)$. If the zero is 1, then $1 + p + q = 0$. Otherwise, the zero satisfies $x^2 + x + 1 = 0$, and it follows that $(p - 1)x + (q - 1) = 0$. Solving this equation for $x$ and substituting into $x^2 + px + q = 0$ yields $p^2 + q^2 + 1 = p + q + pq$.

**8.10.** The discriminant of the quadratic reduces to $(4q - p^2)[(1 - q)^2 + p^2]$, the second factor being nonzero by hypothesis.

**8.11.** Since $\tan(\pi \cot x) = \tan(\pi/2 - \pi \tan x)$, $\pi \cot x = \pi/2 - \pi \tan x + n\pi$ for some integer $n$. Hence

$$\tan x + 1/\tan x - (2n + 1)/2 = 0.$$

Solving this equation for $\tan x$ yields the result. The quantity under the radical is $(2n + 1)^2 - 16$, which is negative exactly when $n = -2, -1, 0, 1$.

**8.12.** The discriminant is equal to

$$\sum_{k=1}^{n}(2a_k - a_{n+1})^2 - (n - 4)a_{n+1}^2.$$

If $n \leq 4$, this is always nonnegative. If $n > 4$, take $a_i = 1$ for $1 \leq i \leq n$ and $a_{n+1} = 2$ to obtain an equation with nonreal roots.

**8.13.** *Solution 1.* Since $p(1)$, $p(-1)$, $p(i)$, $p(-i)$ belong to the unit disc, so also do $(1/2)(p(1) + p(-1)) = 1 + b$ and $(1/2)(p(i) + p(-i)) = -1 + b$. A quick sketch convinces one that $b = 0$. Since $p(1) = 1 + a$ and $p(-1) = 1 - a$ belong to the unit disc, $a = 0$.

*Solution 2.* Let $a = q + ri$, $b = s + ti$. Then

$$4 = |p(1)|^2 + |p(-1)|^2 + |p(i)|^2 + |p(-i)|^2 = 4 + 4(q^2 + r^2 + s^2 + t^2),$$

from which the result follows.

**8.14.** (a) The quadratic equation can be rewritten $(z-a)^2 = a^2 - b$. Let $u$ be such that $u^2 = a^2 - b$. Then the roots of the equation are $-a + u$ and $-a - u$. The roots of the quadratic and linear equations are collinear if and only if, for some real $t$,

$$w = (1-t)(-a+u) + t(-a-u) = -a + (1-2t)u$$

if and only if $(w+a)/u$ is real, if and only if $(w+a)\overline{u}$ is real.

   (b) Suppose the roots of $z^2 + 2cz + d = 0$ are $-c + v$ and $-c - v$. Then the four roots are collinear if and only if $u$ and $v$ are real multiples of $a - c$. (Make a sketch.)

**8.15.** Let $\cos 4\theta = (b^2 - 8ac)/b^2$. Then $\cos^2 2\theta = (b^2 - 4ac)/b^2$, from which it follows that $2\cos^2 \theta - 1 = \sqrt{b^2 - 4ac}/b$. It follows that $-(b/a)\cos^2 \theta$ is a root of the given quadratic equation.

**8.16.** The equation $z^2 - 2(\cos \phi + i \sin \phi)z + 1 = 0$ is equivalent to

$$
\begin{aligned}
[z - (\cos \phi + i \sin \phi)]^2 &= \cos^2 \phi - \sin^2 \phi - 1 + 2i \sin \phi \cos \phi \\
&= 2i \sin \phi (\cos \phi + i \sin \phi) \\
&= 2 \sin \phi [\cos(\phi + \pi/2) + i \sin(\phi + \pi/2)].
\end{aligned}
$$

Solving for $z$ yields two roots

$$u = (\cos \phi + i \sin \phi) + \sqrt{2 \sin \phi}(\cos \theta + i \sin \theta)$$

$$v = (\cos \phi + i \sin \phi) - \sqrt{2 \sin \phi}(\cos \theta + i \sin \theta)$$

where $\theta = \phi/2 + \pi/4$. (Note that $\pi/4 < \theta < 3\pi/4$.)
   Observe that

$$\cos \phi = \sin(\pi/2 + \phi) = \sin 2\theta = 2 \sin \theta \cos \theta$$

$$\sin \phi = -\cos(\pi/2 + \phi) = 2 \sin^2 \theta - 1 = 1 - 2 \cos^2 \theta.$$

Then

$$u + i = (2 \sin \theta + \sqrt{2 \sin \phi})(\cos \theta + i \sin \theta)$$

$$v + i = (2 \sin \theta - \sqrt{2 \sin \phi})(\cos \theta + i \sin \theta).$$

Since $2 \sin \theta - \sqrt{2 \sin \phi} = 2 \sin \theta - \sqrt{4 \sin^2 \theta - 2} > 0$, it follows that $\arg(u + i) = \arg(v + i) = \theta$.
   Also,

$$
\begin{aligned}
u - i &= 2 \cos \theta (\sin \theta - i \cos \theta) + \sqrt{2 \sin \phi}(\cos \theta + i \sin \theta) \\
&= [-2i \cos \theta + \sqrt{2 \sin \phi}](\cos \theta + i \sin \theta) \\
v - i &= [-2i \cos \theta - \sqrt{2 \sin \phi}](\cos \theta + i \sin \theta)
\end{aligned}
$$

and
$$|u - i| = |v - i| = \sqrt{4 \cos^2 \theta + 2 \sin \phi} = \sqrt{2}.$$

Hence $u$ and $v$ lie on a circle with center $i$ and radius $\sqrt{2}$. The line joining $u$ and $v$ passes through $-i$ and makes an angle $\theta = \phi/2 + \pi/4$ with the real axis. As $\phi$ moves from $0$ to $\pi$, $u$ traces that part of the circle in the upper half plane from $1$ to $-1$ while $v$ traces that part in the lower half plane from $1$ to $-1$.

**8.17.** The equation can be rewritten $(x - a)^2 + (x - a)(b + c) + b^2 + c^2 - bc = 0$. Solving this for $x - a$ by the quadratic formula yields the solutions $a + b\omega + c\omega^2$ and $a + b\omega^2 + c\omega$, where $\omega$ is an imaginary cube root of 1.

**8.18.** If $ax^2 + bx + c$ and $ax^2 + bx - c$ are both factorable over $\mathbf{Z}$, then $b^2 - 4ac = q^2$ and $b^2 + 4ac = p^2$ for positive integers $p$ and $q$, which must have the same parity. Let $2u = p + q$ and $2v = p - q$. Then $b^2 = u^2 + v^2$ and $2ac = uv$. One of $u$ and $v$ must be even, say $v = 2w$.

Noting that $(r + si)^2 = u + 2wi$ implies that $(r^2 + s^2)^2 = u^2 + 4w^2$, we solve $r^2 - s^2 = u$, $rs = w$ for $r$ and $s$. Thus, $r^2$ and $-s^2$ are the roots of the equation
$$t^2 - ut - w^2 = 0$$
whence $r^2 = (|b| + u)/2$ and $s^2 = (|b| - u)/2$. It remains to show that $r$ and $s$ are integers. Now, $b$ and $u$ have the same parity and satisfy
$$\left( \frac{|b| - u}{2} \right) \left( \frac{|b| + u}{2} \right) = w^2.$$

Let $d$ be a prime divisor of $w$. Then $d$ divides exactly one of the factors on the left. Otherwise, $d$ would divide their sum $b$. But $d$ divides $ac = uw$, which contradicts the coprimality of $b$ and $ac$. It follows that each factor on the left is a square, and so $r$ and $s$ are integers. It is straightforward to show that $r$ and $s$ are coprime.

Hence, $|b| = r^2 + s^2$ and $ac = rs(r^2 - s^2)$. On the other hand, if these conditions are satisfied, then
$$b^2 - 4ac = (r^2 - 2rs - s^2)^2$$
and
$$b^2 + 4ac = (r^2 + 2rs - s^2)^2$$
and the quadratic can be factored.

The condition on $a$, $b$, $c$ cannot be weakened to require merely that their greatest common divisor be 1 (as specified in the source of this problem). For a discussion, see *Amer. Math. Monthly* **47** (1940), 187–188.

**8.19.**

$$
\begin{aligned}
a \cos^2 \theta \quad &+ \quad 2h \cos \theta \sin \theta + b \sin^2 \theta \\
&= \quad (1/2)[(a + b) + (a - b) \cos 2\theta + 2h \sin 2\theta] \\
&= \quad (1/2)[(a + b) + \sqrt{(a - b)^2 + 4h^2} \sin(2\theta + \phi)]
\end{aligned}
$$

where $\tan\phi = (a-b)/2h$. The maximum (minimum) occurs when $\sin(2\theta+\phi)$ is equal to 1 ($-1$, respectively).

**8.20.** $f(x,y)$ has linear factors if and only if $f(x,y)=0$ is solvable in $x$ as a linear function of $y$. The discriminant of this quadratic equation is $4y^2 - 8by + 4a^2 + 4c$, which is a square of a linear function in $y$ exactly when $b^2 = a^2 + c$.

**8.21.** (a) Substituting $y = mx + c$ into $b^2x^2 - a^2y^2 = a^2b^2$ yields the quadratic equation

$$(a^2m^2 - b^2)x^2 + 2a^2cmx + a^2(b^2 + c^2) = 0$$

for the abscissae of the intersection point. The discriminant is $4a^2b^2(b^2 + c^2 - a^2m^2)$ and this vanishes if and only if the line is tangent to the curve.

(b) The midpoint between the points of intersection of the line $y = mx+c$ and the circle $x^2 + y^2 = r^2$ is

$$\left( \frac{-cm}{1+m^2}, \frac{c}{1+m^2} \right).$$

Let $y = mx + c$ be tangent to $b^2x^2 - a^2y^2 = a^2b^2$, so that $a^2m^2 = b^2 + c^2$ and let $(x, y)$ be on the locus. Then

$$m = -x/y \quad c = (x^2 + y^2)/y.$$

Hence $a^2x^2 = b^2y^2 + (x^2 + y^2)^2$. Note that this is independent of $r$ and contains the points $(0,0)$ and $(a,0)$. However, to take account of the fact that the points on the locus lie within the circle, we require that $x^2 + y^2 \leq r^2$.

**8.22.** $at^2 + bt + c$ has zeros $r + is$, $-r + is$ ($r \neq 0$) if and only if $at^2 - ibt - c$ has zeros $s + ir$, $s - ir$, which occurs if and only if $ib/a$, $c/a$ are real and $b^2 - 4ac \geq 0$. Also, $at^2 + bt + c$ has zeros $iu$ and $iv$ if and only if $at^2 - ibt - c$ has real zeros, which occurs if and only if $ib/a$ and $c/a$ are real and $b^2 - 4ac \leq 0$. Hence, the necessary and sufficient conditions are that $ib/a$ and $c/a$ are real.

**8.23.** If one diameter is $y = mx$, its conjugate is $ma^2y + b^2x = 0$. The line $y = mx$ intersects the ellipse in points $(u, v)$ for which $(b^2 + a^2m^2)u^2 = a^2b^2$; the line $ma^2y + b^2x = 0$ in points $(z, w)$ for which $(m^2a^2 + b^2)z^2 = m^2a^4$. We require that $u^2 + v^2 = z^2 + w^2$, which leads to $m^2a^2(b^2 - a^2) = b^2(b^2 - a^2)$. The case $a = b$ is that of a circle and all diameters are of equal length. When $a \neq b$, we must have $m = b/a$, so that the equations of the conjugate diameters are $bx + ay = 0$ and $bx - ay = 0$.

**8.24.** Invoking the given inequality for $x = 0$, $\frac{1}{2}$, 1 yields $-1 \leq c \leq 1$, $-4 \leq a + 2b + 4c \leq 4$, $-1 \leq a + b + c \leq 1$, respectively. Eliminating $b$ and

$a$ from the last two inequalities yields respectively $-6 \leq a - 2c \leq 6$ and $-5 \leq b + 3c \leq 5$. Taking account of $|c| \leq 1$ leads to $|a| \leq 8$ and $|b| \leq 8$. Hence $|a| + |b| + |c| \leq 17$. Equality holds for the polynomial $8x^2 - 8x + 1 = 2(2x - 1)^2 - 1$.

**9.1.** $a - bi$ is also a root; since the sum of the roots is 0, the third root is $-2a$. Hence the product of the roots is $-2a(a^2 + b^2) = -q$, from which the result follows.

**9.2.** Let $g(x)$ be the constant polynomial $c$. Then $f(c) = c$, so that $f(x) = x$ for each $x$.

**9.3.** Observe that $(a^2 + b^2 - 2)^2 + (c^2 + d^2 - 2)^2 + 2(ac - bd)^2 = (a^2 + c^2 - 2)^2 + (b^2 + d^2 - 2)^2 + 2(ab - cd)^2$. Each of I and II is equivalent to one side of this identity vanishing.

**9.4.** Trying $x = u + \sqrt{2}$ yields $(u^3 - 3u^2 + 5u - 6) + 3u(u - 2)\sqrt{2} = 0$, which is evidently satisfied by $u = 2$. Thus, $x = 2 + \sqrt{2}$.

**9.5.** If $a + b = 0$, i.e. $a = -b$, then both equations are identities in $x$. On the other hand, suppose $a + b \neq 0$. Two roots of the first equation are $-a$ and $-b$. Multiplying this equation by $x$ yields a nontrivial quadratic equation, so that these are the only two roots. Both of these are roots of the second equation.

**9.6.** (a) Since $(-a - \sqrt{a^2 - b})(-a + \sqrt{a^2 - b}) = b$, it follows that $b/x = -a + \sqrt{a^2 - b}$ from which the result follows.

(b) The equation for $y$ can be rewritten $y + c + \sqrt{c^2 - d} = 0$, where $c = ap - q$ and $d = bp^2 - 2apq + q^2$. Now apply (a) to $(y, c, d)$.

**9.7.** We make two initial observations:

(1) $22x - 15 - 8x^2 = -(2x - 3)(4x - 5) \geq 0$ if and only if $5/4 \leq x \leq 3/2$.

(2) $(x^2 - 2x + 1) - (22x - 15 - 8x^2) = (3x - 4)^2 \geq 0$ so that

$$1 - x + \sqrt{22x - 15 - 8x^2} < 0 \quad \text{when} \quad 5/4 \leq x \leq 3/2, \; x \neq 4/3.$$

Hence, the square roots of the given expression are

0 when $x = 4/3$

pure imaginary ($i$ times a real) when $5/4 \leq x \leq 3/2$, $x \neq 4/3$

not real, not pure imaginary otherwise.

Consider the case $x < 5/4$ or $x > 3/2$, and let the square root be $u + iv$. Then

$$u^2 - v^2 = 1 - x$$
$$4u^2v^2 = 8x^2 - 22x + 15$$

whence

$$(u^2 + v^2)^2 = (u^2 - v^2)^2 + 4u^2v^2 = (3x - 4)^2.$$

Take $u^2 + v^2 = 3x - 4$. Then $2u^2 = 2x - 3$, $2v^2 = 4x - 5$. If $x < 5/4$, the square roots are

$$\pm \left( \sqrt{\frac{3 - 2x}{2}} \, i + \sqrt{\frac{5 - 4x}{2}} \right).$$

If $x > 3/2$, the square roots are

$$\pm \left( \sqrt{\frac{2x - 3}{2}} + i\sqrt{\frac{4x - 5}{2}} \right).$$

Finally, it can be seen that, if $5/4 \leq x \leq 3/2$, the square roots are

$$\pm i \left( \sqrt{\frac{3 - 2x}{2}} - \sqrt{\frac{4x - 5}{2}} \right).$$

**9.8.** Let $p(x) = \alpha x^2 + \beta x + \gamma$ and $q(x) = \lambda x^2 + \mu x + \nu$. The condition that the given quartic is of the form $p(q(x))$ leads to, on comparison of coefficients,

$$a = \alpha \lambda^2$$
$$b = 2\alpha \lambda \mu$$
$$c = 2\alpha \lambda \nu + \alpha \mu^2 + \beta \lambda$$
$$d = 2\alpha \mu \nu + \beta \mu$$
$$e = \alpha \nu^2 + \beta \nu + \gamma.$$

The first two conditions lead to $2\mu a = \lambda b$; the second and third lead to $c\mu - d\lambda = \alpha \mu^3$. Eliminating $\mu$ from these two equations and noting that $a = \alpha \lambda^2$, we obtain the necessary condition

$$4abc - 8a^2 d = b^3.$$

On the other hand, suppose that this condition is satisfied. If $b = 0$, then $d = 0$, and we can take $p(x) = ax^2 + cx + e$, $q(x) = x^2$. Suppose that $b \neq 0$. Choose $\lambda = 1$, $\alpha = a$, $\mu = b/2a$. Any choice of $\beta$ and $\nu$ which satisfies $d = b\nu + (b/2a)\beta$ will give a correct expression for $c$. Finally, the equation for $e$ dictates the appropriate value of $\gamma$. Thus, the coefficients of $p$ and $q$ are found.

**9.9.** $xu = u + v - uv = yv$. Substituting $v/u = x/y$ into the first pair of equations yields the second pair.

**9.10.** The equation can be rewritten $(x^2 - 7x + 10)(x^2 - 7x + 12) = 360$, which leads to $(x^2 - 7x + 11)^2 = 361$. Hence, $x^2 - 7x - 8 = 0$ or $x^2 - 7x + 30 = 0$, leading to $x = -1, 8, (1/2)(7 \pm i\sqrt{71})$.

**9.11.** That $x^4 y^2 + y^4 z^2 + z^4 x^2 \geq 3x^2 y^2 z^2$ is a consequence of the arithmetic-geometric mean inequality (Exercise 1.5.9). Suppose that the polynomial is the sum of the squares of polynomials $f(x, y, z)$. Each of these must have degree 3. No polynomial has a term in $x^3$, $y^3$ or $z^3$, for such terms would result in terms in the sum in $x^6$, $y^6$ or $z^6$, respectively, with positive coefficients. No $f(x, y, z)$ can have terms in $yz^2$, for such a term would produce in the sum terms in $y^2z^4$ with positive coefficients. Since $y^2z^4$ can come *only* from squaring $yz^2$ (and not from a product of $y^2z$ and $z^3$), there would be no cancellation of such terms. Similarly, there are no terms in $zx^2$ and $xy^2$. Hence, each $f(x, y, z)$ has the form $ax^2y + by^2z + cz^2x + dxyz$. But the sum of any number of squares of such $f(x, y, z)$ would produce a term in $x^2y^2z^2$ with a positive coefficient (certainly, not $-3x^2y^2z^2$).

**9.12.** See the hints for a possible strategy. Three polynomials with the required property are

$$x^2 + (1 - \sqrt{2})y^2$$

$$uy^2 + x$$

$$-uxy + y$$

where $u$ is the positive square root of $2(\sqrt{2} - 1)$.

**9.13.** Let $n \geq 3$. Then

$$
\begin{aligned}
F_n &= (x + y)^n - x^n - y^n \\
&= (x + y)^2[(x + y)^{n-2} - (x^{n-2} + y^{n-2})] \\
&\quad + xy[xy^{n-3} + x^{n-3}y + 2x^{n-2} + 2y^{n-2}] \\
&= (x^2 + xy + y^2)F_{n-2} + xy[(x + y)^{n-2} + x^{n-2} \\
&\quad + y^{n-2} + xy^{n-3} + x^{n-3}y] \\
&= QF_{n-2} + PG_{n-3}.
\end{aligned}
$$

Similarly, $G_n = QG_{n-2} + PF_{n-3}$. These equations can be used as a basis of an induction argument that $F_n$ is a polynomial in $P$ and $Q$ when $n$ is odd and $G_n$ is a polynomial in $P$ and $Q$ when $n$ is even, once it has been checked that this is so when $n \leq 3$.

**9.14.** We have $P_0 = 1$, $P_1 = xy + xz + yz$, $P_2 = (xy + xz + yz)^2 + (x + y)(x + z)(y + z)$. Assume as an induction hypothesis that $P_m$ is symmetric in $(x, y, z)$ for $m \leq n$.

It is clear that $P_{n+1}$ is symmetric in $x$ and $y$. It suffices to show that $P_{n+1}(x, y, z) = P_{n+1}(z, y, x)$. First, observe that, for $m \leq n$,

$$P_{m+1}(x, y, z) - P_{m+1}(z, y, x) = (x + z)Q_m(x, y, z)$$

where

$$Q_m(x, y, z) = (y+z)P_m(x, y, z+1) - (x+y)P_m(x+1, y, z) + (x-z)P_m(x, y, z).$$

Since the left side vanishes for $m \leq n - 1$, so also does $Q_m(x, y, z)$. We show that $Q_n(x, y, z) = 0$.

From the recursion relation,

$$(y+z)P_n(x,y,z+1) - (x+y)P_n(x+1,y,z) + (x-z)P_n(x,y,z)$$
$$= (y+z)[(x+z+1)(y+z+1)P_{n-1}(x,y,z+2) - (z+1)^2 P_{n-1}(x,y,z+1)]$$
$$- (x+y)[(x+z+1)(y+z)P_{n-1}(x+1,y,z+1) - z^2 P_{n-1}(x+1,y,z)]$$
$$+ (x-z)[(x+z)(y+z)P_{n-1}(x,y,z+1) - z^2 P_{n-1}(x,y,z)].$$

Using $Q_{n-1}(x,y,z+1) = Q_{n-1}(x,y,z) = 0$, we have that

(i) $(y+z+1)P_{n-1}(x,y,z+2) = (x+y)P_{n-1}(x+1,y,z+1) - (x-z-1)P_{n-1}(x,y,z+1)$

(ii) $(x-z)P_{n-1}(x,y,z) = (x+y)P_{n-1}(x+1,y,z) - (y+z)P_{n-1}(x,y,z+1).$

Substituting these into $Q_n(x,y,z)$ yields a linear combination of $P_{n-1}(x+1,y,z+1)$, $P_{n-1}(x,y,z+1)$ and $P_{n-1}(x+1,y,z)$, all of whose coefficients vanish. Hence $Q_n(x,y,z) = 0$ as required.

**9.15.** Each term in the expansion is a product

$$x_1^{a_1} x_2^{a_2} x_3^{a_3} \cdots x_n^{a_n}$$

where the nonnegative integers $a_i$ satisfy

$$a_n \le 1$$
$$a_{n-1} + a_n \le 2$$

$$a_2 + a_3 + \cdots + a_n \le n - 1$$
$$a_1 + a_2 + a_3 + \cdots + a_n = n.$$

Conversely, for any choice of nonnegative integers satisfying these conditions, there is a term in the expansion with these exponents. (Each factor of the expression contributes an $x_i$ to the product; for like terms, there is no cancellation. To build a term of the required type, start with the $x_i$ of highest index $i$ and work from right to left through the product.)

Let $f(n)$ be the number of terms. Clearly, $f(1) = 1$ and $f(2) = 2$. In selecting the $a_i$, consider the possibility that the first equality holds in the $r$th line, so that we have

$$\left.\begin{array}{l} a_n = 0 \\ a_{n-1} \le 1 \\ \\ a_{n-r+2} + \cdots + a_{n-1} \le r - 2 \end{array}\right\} \quad \text{(A)}$$

$$a_{n-r+1} + \cdots + a_{n-1} = r \qquad \text{(B)}$$

$$\left.\begin{array}{l} a_{n-r} \le 1 \\ a_{n-r+1} + a_{n-r} \le 2 \\ \\ a_1 + \cdots + a_{n-r} = n - r. \end{array}\right\} \quad \text{(C)}$$

If $r = 1$, there are $f(n-1)$ ways of choosing $a_1, \ldots, a_{n-1}$ to satisfy inequalities (C). If $r = n$, there are $f(n-1)$ ways of choosing $a_1, \ldots, a_{n-1}$ to satisfy (A) (in this case, $a_n = 0$ and $a_1$ is determined by the last line). If $1 < r < n$, there are $f(r-1)$ ways of satisfying (A) and $f(n-r)$ ways of satisfying (C). Hence, with the convention that $f(0) = 1$,

$$f(n) = \sum_{r=0}^{n-1} f(r)f(n-1-r).$$

To find a closed form for $f(n)$, we make use of a *generating function*. Let $y$ be the formal power series defined by

$$\begin{aligned} y &= 1 + f(1)x + f(2)x^2 + f(3)x^3 + \cdots \\ &= 1 + x + 2x^2 + 5x^3 + 14x^4 + 42x^5 + \cdots . \end{aligned}$$

Then, with the recursion relation taken into account, it can be checked that

$$xy^2 - y + 1 = 0.$$

Solving this equation for $y$ yields

$$y = \frac{1 - (1 - 4x)^{1/2}}{2x}.$$

Expanding the right-hand side binomially, we find that

$$f(n) = \frac{1}{n} \binom{2n}{n+1} = \frac{1}{n+1} \binom{2n}{n} = \frac{1}{2n+1} \binom{2n+1}{n}.$$

This problem is E2972 posed in the *American Mathematical Monthly* **89** (1982), 698. A solution with reference appears in the *Amer. Math. Monthly* **93** (1986), 217–218.

**Note:** The numbers 1, 1, 2, 5, 14, 42, 132, 429, ..., $f(n-1)$, ... are called the Catalan numbers. An article in *Scientific American* (June, 1976; pages 120–125) draws attention to a number of interesting properties and interpretations. Euler showed that the $n$th term is the number of ways a fixed convex polygon with $n+1$ vertices can be decomposed by diagonals into nonoverlapping triangles. Catalan interpreted the $n$th term as the number of ways a chain of $n$ letters in fixed order can be equipped with $n-1$ pairs of parentheses so that two terms reside between a corresponding pair. For example, when $n = 4$, we have

$$((ab)(cd)), \ (((ab)c)d), \ (a(b(cd))), \ (a((bc)d)), \ ((a(bc))d).$$

I am indebted to W. Karpinski for the observation that

$$1 + 24 \left[ \frac{f(n-1)f(n-2)}{f(n)f(n-2) - f(n-1)^2} \right] = (2n+1)^2.$$

for $n \geq 3$. This is a straightforward manipulative exercise for the reader.

**9.16.** Let $z$ be the sum, with $x$ and $y$ its first and second terms; thus, $z = x + y$. If $u = 0$ or $u \geq 3/8$, then $x^3$ and $y^3$ are real, so that $x$ and $y$ have uniquely determined real values, as well as nonreal values. For other values of $u$, $x^3$ and $y^3$ are nonreal complex conjugates. We will assume that for each nonreal determination of $x$ and $y$, they are complex conjugates. Then $xy = 1 - 2u$ and $z$ satisfies the equation

$$z^3 = x^3 + y^3 + 3xyz = (6u - 2) + (3 - 6u)z$$

or $0 = z^3 + (6u - 3)z - (6u - 2) = (z - 1)(z^2 + z + 6u - 2)$. Hence $z = 1$ or $2z = -1 \pm \sqrt{9 - 24u}$.

When $u \geq 3/8$ and the real determinations of $x$ and $y$ are taken, then $z = 1$; test this out on your pocket calculator. The other two values of $z$ are given by the nonreal $x\omega + y\omega^2$ and $x\omega^2 + y\omega$ where $\omega$ is an imaginary cube root of unity. Observe that these numbers satisfy the quadratic equation $z^2 + z + 6u - 2 = 0$ since their sum is $-(x + y) = -1$ and their product is $x^2 + y^2 - xy = x^3 + y^3 = 6u - 2$.

When $u = 0$, then $z = -2$ corresponds to the real determination of $x$ and $y$. The nonreal determinations of $(x, y)$ are $(-\omega, -\omega^2)$ and $(-\omega^2, -\omega)$ and these both yield $z = 1$.

When $u \neq 0$, $u < 3/8$, then $z$ is always real and takes each of the three possible values corresponding to the determination of $x$ and $y$.

**9.17.** $m = (1/2)t(t-1)$, $n = (1/2)t(t+1)$ satisfy $4mn - m - n + 1 = (t^2 - 1)^2$.

**9.18.** The sum of the zeros, $u(1 + u + u^2)$, and the sum of products of pairs of the zeros, $u^3(1 + u + u^2)$ must be rational. Either $1 + u + u^2 = 0$ or $u^2$ is rational. If $u^2$ is rational, then $u(1 + u^2) = (u + u^2 + u^3) - u^2$ is rational. Since $u$ is nonrational, $1 + u^2 = 0$. Thus, the possible values of $u$ are

$$i, \quad -i, \quad (-1 + i\sqrt{3})/2, \quad (-1 - i\sqrt{3})/2.$$

**9.19.** First, observe that, if

$$u(t) = (at + b)(c_r t^r + c_{r-1} t^{r-1} + \cdots + c_1 t + c_0)$$

$$v(t) = (bt + a)(c_r t^r + c_{r-1} t^{r-1} + \cdots + c_1 t + c_0),$$

then $\Gamma(u(t)) = a^2(\Sigma c_i^2) + 2ab(\Sigma c_i c_{i+1}) + b^2(\Sigma c_i^2) = \Gamma(v(t))$.

We have $f(t) = (t + 2)(3t + 1)$. Let $g(t) = (2t + 1)(3t + 1) = 6t^2 + 5t + 1$. Then, for $n = 1, 2, \ldots,$

$$
\begin{aligned}
\Gamma(f(t)^n) &= \Gamma((t+2)^n(3t+1)^n) \\
&= \Gamma((2t+1)(t+2)^{n-1}(3t+1)^n) \\
&= \Gamma((2t+1)^2(t+2)^{n-2}(3t+1)^n) \\
&= \cdots = \Gamma((2t+1)^n(3t+1)^n) = \Gamma(g(t)^n).
\end{aligned}
$$

**9.20.** Since $(x + y)^2 = 8xy$ and $(x - y)^2 = 4xy$, the answer is $\sqrt{2}$.

# Answers to Exercises

**Chapter 2**

**1.5.** $-1062.6855$.

**1.6.** $7t^5 - 2t^3 - 3t^2 + t + 2$; $6$; $53900$.

**1.7.** $3t^4 + 5t^3 + t^2 - 2t + 6$; $2$; $94$.

**1.10.** (c) $325301 + 372391(t - 6) + 182521(t - 6)^2 + 49656(t - 6)^3 + 8099(t - 6)^4 + 792(t - 6)^5 + 43(t - 6)^6 + (t - 6)^7$.

**1.11.** $-357 + 1157(y + 2) - 1585(y + 2)^2 + 1184(y + 2)^3 - 518(y + 2)^4 + 132(y + 2)^5 - 18(y + 2)^6 + (y + 2)^7$.

**1.13.** If $f(t)$ is the given polynomial, then the polynomial sought is $f(t-3)$. Expand $f(t)$ in powers of $t + 3$ to obtain $f(t) = (t + 3)^4 - 15(t + 3)^3 + 83(t + 3)^2 - 196(t + 3) + 163$. The required polynomial is $t^4 - 15t^3 + 83t^2 - 196t + 163$.

**2.2.** By Exercise 1, $p(t) = (t - c)q(t) + p(c)$, from which the result follows.

**2.3.** If $q(s) = 0$, then $p(s) = 0$. Suppose $0 = p(s) = (s - r)q(s)$. Since an integral domain has no zero divisors and $s - r$ is nonzero, $q(s) = 0$.

**2.5.** Since $u(b) = 0$, we have that $u(t) = (t - b)w(t)$ so that $f(t) = (t - a)(t - b)w(t)$ and $v(t) = (t - a)w(t)$. Hence $u(t) - v(t) = (a - b)w(t)$ and the result follows.

**2.6.** $t^4 - 5t - 6 = (t+1)(t^3 - t^2 + t - 6) = (t-2)(t^3 + 2t^2 + 4t + 3)$. The remaining zeros of this quartic are those of $(t^3 + 2t^2 + 4t + 3) - (t^3 - t^2 + t - 6) = 3(t^2 + t + 3)$. (Observe that $t^4 - 5t - 6 = (t + 1)(t - 2)(t^2 + t + 3)$.)

**2.7.** (d) $4t^5 - 3t^4 - 7t^2 + 6 = (t^3 + 7t^2 + 3t - 2)(4t^2 - 31t + 205) + (-1341t^2 - 677t + 416)$.

**2.9.** Consider the set $S = \{f - gh : h \in \mathbf{F}[t]\}$. It contains a polynomial $r = f - gq$ of lowest degree. Suppose, if possible, $\deg r \geq \deg g$. Let $g(t) = a_m t^m + \cdots$ and $r(t) = b_{m+k} t^{m+k} + \cdots$. The polynomial $r(t) - b_{m+k} a_m^{-1} t^k g(t) = f - g(q + b_{m+k} a_m^{-1} t^k)$ belongs to $S$, but has degree less than $\deg r(t)$, contradicting the choice of $r$. Hence $\deg r < \deg g$.

**2.11.** $f(t) = q(t)(t - a)(t - b) + At + B$, for some polynomial $q(t)$ and constants $A$ and $B$. Substituting $t = a$ and $t = b$ yields

$$A = \frac{f(a) - f(b)}{a - b}, \quad B = \frac{af(b) - bf(a)}{a - b}.$$

**2.13.** (e) $f = s_1 s_2^2 - 2s_1^2 s_3 - s_2 s_3$.

**3.5.** (d) $(p_1 p_2 \cdots p_n)' = \sum_{i=1}^{n} p_1 p_2 \cdots p_i' \cdots p_n$.

(e) $(p \circ q)(t) = \Sigma a_r q(t)^r$. Differentiating yields $(p \circ q)'(t) = \Sigma r a_r q(t)^{r-1} q'(t)$ from which the result follows.

**3.15.** (a) Prove the result by induction on the degree of $p$. If $\deg p = 0$, the result holds for each $c$ with $r = 0$. Suppose the result holds for all polynomials of degree not exceeding $n - 1 \geq 0$. Let $\deg p = n$. Then, if $p(c) \neq 0$, we may take $r = 0$. If $p(c) = 0$, then $p(t) = (t - c)u(t)$ and we may apply the induction hypothesis to achieve $u(t) = (t - c)^{r-1} q(t)$ where $q(c) \neq 0$ for some positive integer $r$.

(b) Let $p(t) = (t - c)^r q(t)$ where $r \geq 1$, $q(c) \neq 0$. Then $p'(t) = (t - c)^{r-1}[rq(t) + (t - c)q'(t)]$. The quantity in square brackets does not vanish when $t = c$. Hence $p'(c) = 0 \iff r - 1 \geq 1 \iff r \geq 2$. The result follows.

(d) $\dfrac{p^{(r)}(c)}{r!}(t - c)^r$.

**4.3.** (e) The quadratic must assume either a maximum or minimum value between the two zeros. At this extremum, the derivative has a zero.

**4.4.** (f)

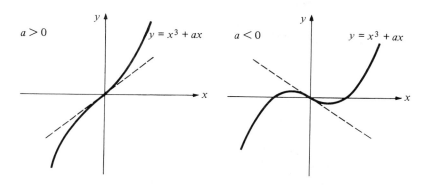

**4.5.** Set $s = x + b/3a$. The change of variables represents a horizontal shift of the origin of coordinates.

**4.6.** The graph of any cubic polynomial is a horizontal translate of the graph of a polynomial of the form $ax^3 + cx + d$, which in turn is a vertical translate of the graph $ax^3 + cx$. In the case of $ax^3 + cx$, the inflection point is at the origin and $(x, y)$ satisfies the equation $y = ax^3 + cx$ if and only if $(-x, -y)$ does. The result follows.

**4.7.** (a) Let the polynomial be $ax^3 + bx^2 + cx + d = x^3(a + b/x + c/x^2 + d/x^3)$. When $|x|$ is large enough, the quantity in parenthesis has the same sign as $a$, and the sign of the polynomial is the same as the sign of $ax^3$. Hence the polynomial takes both positive and negative values, and so must vanish at some point because of the continuity.

**4.8.** (c)

(c)

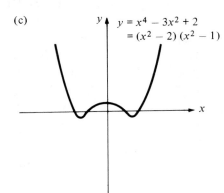

$$y = x^4 - 3x^2 + 2$$
$$= (x^2 - 2)(x^2 - 1)$$

(d)

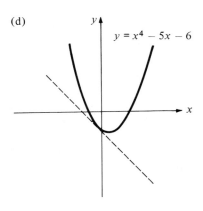

$$y = x^4 - 5x - 6$$

(e)

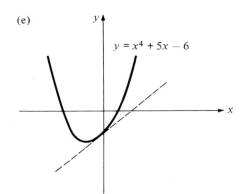

$$y = x^4 + 5x - 6$$

(f)

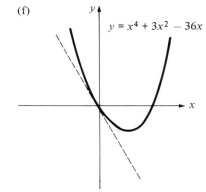

$$y = x^4 + 3x^2 - 36x$$

**4.10.** (a) Let $p(x) = 6x^5 - 15x^4 - 10x^3 + 30x^2 + 10$. Then $p'(x) = 30x^4 - 60x^3 - 30x^2 + 60x = 30(x-2)(x-1)x(x+1)$. $p(x)$ is increasing for $x \leq -1$, decreasing for $-1 \leq x \leq 0$, increasing for $0 \leq x \leq 1$, decreasing for $1 \leq x \leq 2$, increasing for $2 \leq x$. Also $p(-1) = 29 > 0$, $p(0) = 10 > 0$, $p(1) = 21 > 0$, $p(2) = 2 > 0$.

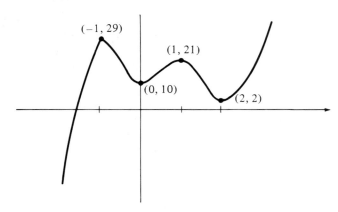

(b) Let $q(x) = 6x^5 - 15x^4 - 10x^3 + 30x^2 = x^2(6x^3 - 15x^2 - 10x + 30)$. The number of zeros of $q(x) + k$ is the number of times the line $y + k = 0$ intersects the graph of $y = q(x)$. From the graph we see that $q(x) + k$ has

exactly one simple real zero when $k < -19$ or $k > 8$;
exactly three simple real zeros when $-19 < k < -11$ or $0 < k < 8$;
exactly five simple real zeros when $-11 < k < 0$;
one double and one simple real zero when $k = -19, 8$;
one double and three simple real zeros when $k = -11, 0$.

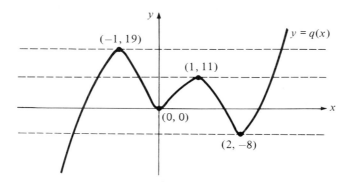

# Solutions to Problems

## Chapter 2

**5.1.** Let $f(x) = x(x - 1) \cdots (x - n + 1) - k$. Then $f'(x)$ takes values of alternate signs at $x = 0, 1, \ldots, n - 1$. Hence, $f'(x)$ has $n - 1$ distinct zeros, one between each pair $i - 1$, $i$ of integers ($1 \le i \le n - 1$). Since $\deg f'(x) = n - 1$, each of these zeros must be simple. Hence, a zero of $f(x)$ has multiplicity at most 2. We can make any zero $r$ of $f'(x)$ a zero of $f(x)$ as well by choosing $k = r(r - 1) \cdots (r - n + 1)$.

**5.2.** If all three have a common zero, the result is trivial. Otherwise, the polynomials can be written $a(x - u)(x - v)$, $b(x - u)(x - w)$, $c(x - v)(x - w)$ where $a$, $b$, $c$ are positive and $u < v < w$. Their sum is positive at $u$ and $w$, negative at $v$, and so the sum has a zero between $u$ and $v$, as well as between $v$ and $w$.

**5.3.** (a) Putting the expression over a common denominator $(1 - x)^2$ yields a numerator $n(1 - x) - (1 - x^n)$. This vanishes along with its derivative when $x = 1$, so that it is divisible by $(1 - x)^2$. We find that the given expression is equal to

$$x^{n-2} + 2x^{n-3} + 3x^{n-4} + \cdots + (n - 2)x + (n - 1).$$

(b) Setting $u = x - 1$ in the expression yields

$$u^{-2}\{-nu - [1 - (1+u)^n]\} = \sum_{k=2}^{n} \binom{n}{k} u^{k-2} = \sum_{k=2}^{n} \binom{n}{k} (x-1)^{k-2}.$$

**5.4.**

$$\begin{aligned}
(1 - x^3)^n &= [(1-x)(1+x+x^2)]^n \\
&= (1-x)^n[(1-x)^2 + 3x]^n \\
&= (1-x)^n[(1-x)^{2n} + n(3x)(1-x)^{2n-2} + \cdots]
\end{aligned}$$

which yields the result.

**5.5.** Let $u$ be a root of multiplicity exceeding 1. If $u = 0$, the condition (a) follows immediately. Let $u \neq 0$. Then $au^2 + 2bu + c = 0$ implies that

$$0 = (au^3 + 3bu^2 + 3cu + d) - u(au^2 + 2bu + c) = bu^2 + 2cu + d$$

whence

$$\begin{aligned}
0 &= a(bu^2 + 2cu + d) - b(au^2 + 2bu + c) \\
&= 2(ac - b^2)u + (ad - bc)
\end{aligned}$$

and

$$\begin{aligned}
0 &= c(bu^2 + 2cu + d) - d(au^2 + 2bu + c) \\
&= u[(bc - ad)u + 2(c^2 - bd)].
\end{aligned}$$

Condition (a) and result (b) follow immediately.
    On the other hand, assume that the condition in (a) holds. Since

$$a(ax^3 + 3bx^2 + 3cx + d) = (ax + b)(ax^2 + 2bx + c)$$

$$+ [2(ac - b^2)x - (bc - ad)],$$

any common zero of $ax^2 + 2bx + c$ and $2(ac - b^2)x - (bc - ad)$ must be at least a double zero of the cubic. But the quadratic and linear polynomials have a common zero if and only if

$$a(bc - ad)^2 + 4b(ac - b^2)(bc - ad) + 4c(ac - b^2)^2 = 0.$$

This is ensured by using condition (a) to substitute for $(bc - ad)^2$.
    **Remark.** Consider the case that $ac - b^2 = 0$. If the cubic equation has a double root, then we must have $ad - bc = 0$. On the other hand, if (a) holds, then $ad - bc = 0$ and

$$a^2(ax^3 + 3bx^2 + 3cx + d) = (ax+b)^3 + (3ax+b)(ac-b^2) + a(ad-bc) = (ax+b)^3.$$

**5.6.**

$$(r+1)S_r + \binom{r+1}{2} S_{r-1} + \cdots + (r+1)S_1 = \sum_{k=1}^{n}\sum_{j=1}^{r} \binom{r+1}{j} k^{r+1-j}$$

$$= \sum_{k=1}^{n}[(1+k)^{r+1} - k^{r+1} - 1] = (n+1)^{r+1} - 1 - n.$$

**5.7.** Let $p(p(x)) = p(x)^m$. Comparing degrees of the two sides yields $k^2 = km$, from which $k = -\infty$, 0 or $m$. If $p(x)$ is constant, then it must be 0, 1 or, in the case of odd $m$, $-1$. If $p(x)$ is nonconstant, it assumes infinitely many values, so that $p(t) - t^m$ has infinitely many zeros $t = p(x)$. But then $p(t) = t^m$.

**5.8.** The equation is satisfied if $p(t) = c$, a constant polynomial, and $q(t)$ is any polynomial with $q(0) = c$. Suppose $\deg p(t) = n \geq 1$, $\deg q(t) = r$. Then $n = r(n-1)$ from which $n = r = 2$. Hence, there exist coefficients $a$, $b$, $c$, $u$, $v$, $w$ for which

$$at^2 + bt + c = u(2at + b)^2 + v(2at + b) + w.$$

Expanding and comparing coefficients yields $1 = 4au$, $b = b + 2av$, $c = ub^2 + vb + w$, whence $v = 0$. Hence $p(t)$ can be an arbitrary quadratic $at^2 + bt + c$ while $4aq(t) = t^2 - (b^2 - 4ac)$.

**5.9.** The only constant polynomial is 0. Let $p(t) = at + b$. Then the identity is satisfied as long as $p(a) = a$, so that $p(t) = a[t + (1 - a)]$. Let $p(t) = at^2 + bt + c$. By substitution, we arrive at the requirements $a = 1/2$, $b = 0$, whence $p(t) = (1/2)t^2 + c$.

If $\deg p(t) = 3$, comparing the leading coefficients of both sides of

$$p'(p(t)) = p(p'(t)) \tag{1}$$

yields $p(t) = (1/9)t^3 + \cdots$. However, obtaining the other coefficients by substituting into (1) is an unappetizing task, and another tack is desirable. Differentiate (1) repeatedly to obtain

$$p''(p(t))p'(t) = p'(p'(t))p''(t) \tag{2}$$

$$p'''(p(t))p'(t)^2 + p''(p(t))p''(t) = p''(p'(t))p''(t)^2 + p'(p'(t))p'''(t). \tag{3}$$

If 0 is the sole zero of $p'(t)$, then $p'(t) = t^2/3$ and we must have that $p(t) = t^3/9$. Otherwise, let $r \neq 0$ and $p'(r) = 0$. Then, from (2), $p'(0)p''(r) = 0$. Either $p'(0) = 0$ or $p''(r) = 0$. In the latter case, since $p'''(t)$ is a nonzero constant, (3) gives $p'(0) = 0$. Hence

$$p'(t) = t^2/3 - rt/3$$

and

$$p(t) = t^3/9 - rt^2/6 + c.$$

Substitute $t = 0$ into (1) and (3) to obtain $3c = c^2 - rc$, whence $c = 0$ or $c = r + 3$, and

$$(2c/3 - r/3)(-r/3) = (-r/3)(r^2/9),$$

whence $6c = 3r + r^2$. Hence $(r, c) = (6, 9)$ or $(-3, 0)$. The first leads to $p(t) = t^3/9 - t^2 + 9$ which is valid, and the second to $p(t) = t^3/9 + t^2/2$ which is not (check $t = 6$). Hence $p(t) = t^3/9$ or $p(t) = t^3/9 - t^2 + 9$.

In general, if $p(t) = n^{1-n}t^n$, then $p(p'(t)) = p'(p(t))$.

**5.10.** The result holds for $n = 1$. Suppose it holds for $n = m \geq 1$. Then

$$
\begin{aligned}
(x + y)^{(m+1)} &= (x + y)(x + y - 1)^{(m)} \\
&= x \sum_{k=0}^{m} \binom{m}{k} (x - 1)^{(k)} y^{(m-k)} \\
&\quad + y \sum_{k=0}^{m} \binom{m}{k} x^{(k)} (y - 1)^{(m-k)} \\
&= \sum_{k=0}^{m} \binom{m}{k} x^{(k+1)} y^{(m-k)} + \sum_{k=0}^{m} \binom{m}{k} x^{(k)} y^{(m+1-k)} \\
&= \sum_{k=1}^{m+1} \binom{m}{k-1} x^{(k)} y^{(m+1-k)} + \sum_{k=0}^{m} \binom{m}{k} x^{(k)} y^{(m+1-k)} \\
&= x^{(m+1)} + \sum_{k=1}^{m} \left[ \binom{m}{k-1} + \binom{m}{k} \right] x^{(k)} y^{(m+1-k)} \\
&\quad + y^{(m+1)},
\end{aligned}
$$

which yields the result for $n = m + 1$. The result follows by induction.

**5.11.** If $y$ is a polynomial in $x$, the degree of the left side equals the degree of $y$, so that $y$ must be a cubic. Successive differentiations of the equation yield

$$9y''' + 4y'' + y' = 3x^2 + 10x - 2$$
$$4y''' + y'' = 6x + 10$$
$$y''' = 6.$$

Working up from the last equation, we find that $y'' = 6x - 14$, $y' = 3x^2 - 14x$ and $y = x^3 - 7x^2 + 4$. It is readily checked that this indeed satisfies the differential equation.

**5.12.** $f(x)$ has the form $ax^5 + bx^4 + cx^3$. For some polynomial $g(x)$, $ax^5 + bx^4 + cx^3 - 1 = (x - 1)^3 g(x)$. Differentiating the equation twice and setting

$x = 0$ yields $g(0) = 1$, $g'(0) = 3$, $g''(0) = 12$. Since $\deg g = 2$, use Taylor's Theorem to obtain $g(x) = 6x^2 + 3x + 1$. Hence, $f(x) = 6x^5 - 15x^4 + 10x^3$.

**5.13.** Suppose $a_3 x^3 + a_2 x^2 + a_1 x + a_0$ is the cube of a linear polynomial. Then it must have a triple zero $u$ which is a zero of its first and second derivatives. Hence $u = -a_2/3a_3$. Since $3a_3 u^2 + 2a_2 u + a_1 = 0$, we find that $a_2^2 = 3a_1 a_3$. Finally, $a_3 u^3 + a_2 u^2 + a_1 u + a_0 = 0$ leads to $9a_0 a_3 = a_1 a_2$.

On the other hand, suppose that the conditions are satisfied. Let $p = a_3^{1/3}$, $q = a_0^{1/3}$. Then, it is readily checked that the cubic is equal to $(px+q)^3$.

**5.14.** $(x - 2)(x - 1)x(x + 1)(x + 2) = x^5 - 5x^3 + 4x$ has three nonzero coefficients. If $p(x)$ has only two nonzero coefficients, it must have one of the forms $x^5 - a$, $x^5 - bx$, for otherwise 0 would be at least a double root. But neither possibility has five integer solutions. Hence $k = 3$.

**5.15.** For small values of $n$, we find that $f_0(x) = 1$, $f_1(x) = x$, $f_2(x) = x(x + 2)$, $f_3(x) = x(x + 3)^2$. As an induction hypothesis, let $n \geq 1$ and suppose that $f_n(x) = x(x + n)^{n-1}$. Then $f_n(x) = (x + n)^n - n(x + n)^{n-1}$ and

$$f'_{n+1}(x) = (n + 1)(x + n + 1)^n - (n + 1)n(x + n + 1)^{n-1}.$$

Hence

$$f_{n+1}(x) = (x + n + 1)^{n+1} - (n + 1)(x + n + 1)^n = x(x + n + 1)^n.$$

In particular, $f_{100}(1) = 101^{99}$.

**5.16.** If $f(x)$ is constant, then $f(x) = 0$ or $-1$. Otherwise, let $r$ be a zero of $f$. Then $r^2, r^4, \ldots$ must also be zeros. Since $f$ has at most finitely many zeros, $|r|$ can take only the values 0 and 1. Also, $(r - 1)^2$ is a zero of $f$, so $|r - 1| = 0$ or 1. Consulting a sketch of the complex plane assures us that the only possibilities are $r = 0$ and $r = 1$. Hence $f(x) = ax^m(x - 1)^n$. Trying this out leads to $a = -1$, $m = n$.

**5.17.** The polynomial must have the form $x^5 + ax^3 + bx$. If $u$ is a nontrivial zero, then so is $-u$, and the polynomial can be factored to $x(x - u)(x + u)(x^2 - v)$ where $u^2 v = b$, $u^2 + v = -a$. Substituting $x = 10$ yields $(10 - u)(10 + u)(100 - v) = -2967 = -3.23.43$. Now $10 - u$ and $10 + u$ are a pair of divisors of 2967 which sum to 20; the only possibilities are given by $u = 13$ and $u = 33$. Thus, there are two polynomials which satisfy the conditions: $x^5 - 226x^3 + 9633x$ and $x^5 - 1186x^3 + 105633x$.

**5.18.** The expression is equal to the Taylor expansion of $f(0)$ about the point $x$. If $f(0) = 0$, it vanishes identically and the degree is $-\infty$. If $f(0) \neq 0$, it is constant and the degree is 0.

**5.19.** (a) $p(t) = t^2 + 1$.

(b) For convenience, let $f(0) = x$, $f'(0) = y$, $f''(0) = z$; $g(0) = u$, $g'(0) = v$, $g''(0) = w$. That $cf$ is in $A$ is obvious. Suppose that $x$ and $u$ are positive. Then, since

$$(x+u)(z+w) - (y+v)^2 - r(x+u)(y+v) = x^{-1}(x+u)(xz - y^2 - rxy)$$

$$+ u^{-1}(x+u)(uw - v^2 - ruv) + (xu)^{-1}(xv - yu)^2$$

and

$$(xu)(xw + 2yv + zu) - (xv + yu)^2 - r(xu)(xv + yu)$$

$$= x^2(uw - v^2 - ruv) + u^2(xz - y^2 - rxy),$$

it follows that $f + g$ and $fg$ are in $A$ along with $f$ and $g$. If $x = 0$, then $y = 0$ and $x \geq 0$, and it is straightforward to check that $f + g$ and $fg$ are in $A$ along with $f$ and $g$.

# Answers to Exercises

### Chapter 3

**1.1.** (b) Let $x$ be any rational. For any other nonzero rational $y$, we have $x = y(y^{-1}x)$.

**1.2.** Let the coefficients of $p(t)$ and $q(t)$ be $a_i$ and $b_i$ respectively. If $p(t) = cq(t)$, then $a_i = cb_i$ for each $i$, i.e. $c|a_i$. The converse is straightforward.

**1.3.** If $at + b = f(t)g(t)$, the degrees of $f(t)$ and $g(t)$ must be 0 and 1 in some order.

**1.4.** If $f$ is a polynomial over $\mathbf{R}$ with a nontrivial factorization $f = gh$ over $\mathbf{R}$, then this is also a factorization over $\mathbf{C}$. Hence an irreducible polynomial over $\mathbf{C}$ is also irreducible over $\mathbf{R}$. An analogous argument applies for $\mathbf{R}$ and $\mathbf{Q}$.

**1.5.** Suppose, if possible, that $t^2 + 1 = (at + b)(ct + d)$. Then $ac = bd = 1$ and $ad + bc = 0$. Hence, $a$ and $c$ must be nonzero reals with the same sign. Similarly, $b$ and $d$ are nonzero with the same sign. Hence, $ad$ and $bc$ are nonzero with the same sign, and so cannot satisfy $ad + bc = 0$. Thus, $t^2 + 1$ cannot be factored over $\mathbf{R}$.

**1.6.** Note that $p(t) = f(t)g(t)$ if and only if $p(t - k) = f(t - k)g(t - k)$.

**1.7.** (a) $t^2 + c$ has nonreal zeros and so cannot be factored over $\mathbf{R}$. However, $t^2 + c = (t + i\sqrt{c})(t - i\sqrt{c})$ is a factorization over $\mathbf{C}$.
  (b) $t^2 - d^2 = (t - d)(t + d)$ is a factorization over $\mathbf{Z}, \mathbf{Q}, \mathbf{R}, \mathbf{C}$.
  (c) $t^2 + c = (t + \sqrt{-c})(t - \sqrt{-c})$ is a factorization over $\mathbf{R}, \mathbf{C}$. Suppose, if possible, that $\sqrt{-c}$ were rational; let it be $u/v$ in lowest terms. Then $-cv^2 = u^2$. Since $\gcd(u, v) = 1$, $v$ must be 1 and so $-c$ is a perfect square, a contradiction. Hence $t^2 + c$ does not have rational zeros and so is irreducible over $\mathbf{Q}$.

**1.8.** Assume that $a > 0$ and note that

$$at^2 + bt + c = a[(t - b/2a)^2 - (1/4a)(b^2 - 4ac)].$$

Apply Exercises 7 and 6. The quadratic $at^2 + bt + c$ is

(i) always reducible over $\mathbf{C}$

(ii) reducible over $\mathbf{R}$ iff $b^2 - 4ac \geq 0$

(iii) reducible over $\mathbf{Q}$ iff $b^2 - 4ac$ is a perfect square.

Reducibility over $\mathbf{Z}$ requires more careful analysis. It is necessary that the discriminant $b^2 - 4ac$ be a perfect square, say $d^2$. On the other hand, if this is true, then

$$at^2 + bt + c = (1/4a)(2at + b - d)(2at + b + d) = (u/v)(pt + q)(rt + s)$$

where $\gcd(u, v) = \gcd(p, q) = \gcd(r, s) = 1$. Comparing coefficients of both sides, we find that $v$ must divide $pr$, $ps + qr$, $qs$. Suppose $d$ is a prime divisor of $v$. Then, if (say) $d|p$, then $d\nmid q$. Hence $d|s$ and $d\nmid r$. But then $d|ps$ and $d\nmid qr$, so that $d\nmid ps + qr$, a contradiction. Hence $v$ has no prime divisors, so $v = 1$. Therefore $at^2 + bt + c$ is reducible over $\mathbf{Z}$ if $b^2 - 4ac$ is a perfect square.

**1.9.** $(5/4)t^2 - (31/45)t - (8/5) = ((5/6)t - 6/5)((3/2)t + 4/3)$, for example.

**1.10.** (a) Express the coefficients of $f$ over a least common divisor $d$. Then let $c$ be the greatest common divisor of the numerators.

(c) We can write $(c/d)g(t) = f(t) = f_1(t)f_2(t)$ where $f_1$ and $f_2$ have rational coefficients. Each $f_i$ can be written in the form $(c_i/d_i)g_i$, as in (a). The result follows with $a/b = c_1c_2d/d_1d_2c$.

(d) Let $g_1(t) = \Sigma u_i t^i$, $g_2(t) = \Sigma v_j t^j$. Let $p$, $r$, $s$ be as specified. The coefficient of $t^{r+s}$ in the product $g_1(t)g_2(t)$ is

$$\sum_{i<r} u_i v_{r+s-i} + u_r v_s + \sum_{i>r} u_i v_{r+s-i}.$$

Since the prime $p$ divides $u_i$ for $i < r$ and $v_{r+s-i}$ for $i > r$, $p$ divides both sums. But $p\nmid u_r v_s$, so that the coefficient of $t^{r+s}$ in $(a/b)g_1(t)g_2(t)$ is not an integer, yielding a contradiction.

**1.11.** Let $f(t) = t^n + a_{n-1}t^{n-1} + \cdots + a_1 t + a_0$, and suppose that $r = u/v$ (in lowest terms). Then $u^n + \sum_{i\geq 1} a_i u^{n-i}v^i = 0$. If a prime $p$ divides $v$, then $p$ must also divide $u$, contradicting the coprimality of $u$ and $v$.

**1.12.** Let $h(t) = f(t)g(t)$, where $f(t) = \Sigma a_i t^i$, $g(t) = \Sigma b_j t^j$. Then

$$c_0 = a_0 b_0$$
$$c_1 = a_0 b_1 + a_1 b_0$$
$$c_2 = a_0 b_2 + a_1 b_1 + a_2 b_0$$
$$\vdots$$

Since $p|c_0$ and $p^2 \nmid c_0$, $p$ must divide exactly one of $a_0$ and $b_0$, say $p|a_0$, $p \nmid b_0$. Then

$$p\,|\,c_1 \Longrightarrow p\,|\,a_1 b_0 \Longrightarrow p\,|\,a_1$$
$$p\,|\,c_2 \Longrightarrow p\,|\,a_2 b_0 \Longrightarrow p\,|\,a_2.$$

Continue on to find in succession that $p|a_3, \ldots, p|a_n$. But then $p|c_n = a_n b_n$, yielding a contradiction.

**1.13.** Apply the Eisenstein Criterion with prime 3.

**1.14.** $t+1$, $3t+4$.

**1.15.** $t^2 + t + 1$ has nonreal zeros and so is irreducible over $\mathbf{Z}$.

**1.16.** Assume that $h(t)$ is irreducible; use the notation of Exercise 12. If $p$ fails to divide one of $a_0$ or $b_0$, say the latter, the argument of Exercise 12 leads to a contradiction. The remaining case is that each of $a_0$ and $b_0$ are divisible by $p$ but not by $p^2$.

We show by induction that $p|a_i$, $p|b_i$ for $i = 0, 1, \ldots, m$. This is true for $i = 0$. Suppose that it has been shown for $i = 1, 2, \ldots, k-1$, where $k \leq m$. Consider

$$c_{2k} = a_0 b_{2k} + a_1 b_{2k-1} + \cdots + a_k b_k + a_{k+1} b_{k-1} + \cdots + a_{2k} b_0.$$

Since $p$ is a divisor of $c_{2k}$, $a_0, \ldots, a_{k-1}$, $b_0, \ldots, b_{k-1}$, $p$ also divides $a_k b_k$, and thus at least one of $a_k$ and $b_k$.

Now, consider

$$c_k = a_0 b_k + a_1 b_{k-1} + \cdots + a_{k-1} b_1 + a_k b_0.$$

Since $p^2$ is a divisor of $c_k$, $a_1 b_{k-1}, \ldots, a_{k-1} b_1$, $p^2$ also divides $a_0 b_k + a_k b_0$. Now, suppose $p|a_k$. Then, $p^2|a_k b_0$, so $p^2|a_0 b_k$. But $p^2 \nmid a_0$, so $p|b_k$. Similarly, if $p|b_k$, then $p|a_k$. Hence $p$ divides both $a_k$ and $b_k$.

Therefore, $p|a_i$, $p|b_i$ for $0 \leq i \leq m$, and so $p$ divides $c_{2m+1} = a_0 b_{2m+1} + \cdots + a_m b_{m+1} + a_{m+1} b_m + \cdots + a_{2m+1} b_0$, yielding a contradiction.

**1.17.** Apply the Eisenstein Criterion for $p = 2$ to show irreducibility over $\mathbf{Z}$, hence over $\mathbf{Q}$. The polynomial is reducible over $\mathbf{R}$ by the Factor Theorem since it has a real root $2^{1/r}$, when $r > 1$.

**1.18.** Construct a polynomial of the form $t^n + \cdots + 2$, where $r - 2$ terms apart from the leading and constant ones have even nonzero coefficients. Such a polynomial is irreducible by the Eisenstein Criterion with prime 2.

**1.19.** By Exercise 6, $t^{p-1} + t^{p-2} + \cdots + t + 1$ is irreducible as a polynomial in $t$ if and only if $(1+s)^{p-1} + (1+s)^{p-2} + \cdots + (1+s) + 1$ is irreducible as a polynomial in $s$. The latter polynomial can be written

$$\frac{(1+s)^p - 1}{s} = s^{p-1} + \sum_{k=2}^{p-1} \binom{p}{k} s^{i-1} + p.$$

Since $\begin{pmatrix} p \\ k \end{pmatrix}$ is divisible by $p$ for $2 \leq k \leq p-1$, the irreducibility follows from the Eisenstein Criterion with prime $p$.

**1.20.**
$$t^3 + t^2 + t + 1 = (t+1)(t^2+1)$$
$$t^5 + t^4 + t^3 + t^2 + t + 1 = (t+1)(t^2+t+1)(t^2-t+1)$$
$$t^7 + t^6 + \cdots + t + 1 = (t+1)(t^2+1)(t^4+1).$$

**1.21.** Let $u(t)$ be a polynomial of largest degree over **D** which divides both $f(t)$ and $g(t)$. If $g(t)$ does not divide $f(t)$, then $u(t)$ is a nonzero constant. By repeated application of the division algorithm (cf. Exercise 1.6.2), we can express $u(t)$ in the form $f(t)p(t) + g(t)q(t)$ for some polynomials $p(t)$ and $q(t)$ over **D**. But then $u(w) = 0$, a contradiction.

**1.22.** Suppose if possible that $g(t)$ has a nonsimple zero $w$. Then $g(w) = g'(w) = 0$. By Exercise 21, $g(t)$ divides $g'(t)$, which is a contradiction, since $0 \leq \deg g'(t) < \deg g(t)$.

**2.1.** (a) $(u, v) = (12, -10)$. $6t^2 + 2t - 20 = 6t^2 + 12t - 10t - 20 = (6t - 10)(t + 2) = 6t^2 - 10t + 12t - 20$.

(b) $at^2 + bt + c = a^{-1}[a^2t^2 + a(u+v)t + uv] = a^{-1}(at+u)(at+v)$. Let $w = \gcd(a, u)$, $a = wz$, $u = wy$. Since $a \mid vwy$, $z \mid vy$. Since $\gcd(z, y) = 1$, $z \mid v$. Thus
$$at^2 + bt + c = [zt + (u/w)][wt + (v/z)].$$

(c) $28t^2 + 57t + 14 = 28t^2 + 49t + 8t + 14 = (7t+2)(4t+7)$ is negative when $-7/4 < t < -2/7$. $20t^2 + 39t - 44 = 20t^2 + 55t - 16t - 44 = (4t+11)(5t-4)$ is negative when $-11/4 < t < 4/5$.

**2.2.** (a) $a^k - b^k = (a-b)(a^{k-1} + a^{k-2}b + \cdots + ab^{k-2} + b^{k-1})$.
(b) $a^k + b^k = (a+b)(a^{k-1} - a^{k-2}b + \cdots - ab^{k-2} + b^{k-1})$.

**2.3.** (a) $4t^2 - 20t - 11 = (2t-5)^2 - 6^2 = (2t-11)(2t+1)$.
(b) $5t^2 - 6t + 1 = (3t-1)^2 - 4t^2 = (t-1)(5t-1)$.
(c) $t^4 - 47t^2 + 1 = (t^2+1)^2 - 49t^2 = (t^2 - 7t + 1)(t^2 + 7t + 1)$.

**2.4.** (a) Any reducible cubic can be factored as the product of a linear and a quadratic. Since the leading coefficients of the factors divide that of the polynomial, the linear factor must have the form $\pm(t - k)$, where $k$ is an integer. Such $k$ is a zero.

(b) Any integer zero must divide 42. A little trial and error yields the zero 2 and the factorization $(t - 2)(t^2 - 6t + 21)$.

**2.6.** (d) $(t^4 + 4t^3 + 8t^2 - 4t + 1)(t^4 - 4t^3 + 8t^2 + 4t + 1)$.

**2.7.** (a), (b), (c), (e), (f) Irreducible.
(d) $(7t + 8)(4t - 3)$.
(g) $(t^2 + 2t - 2)(t + 3)(t - 1)$.

(h) $t^2(t+1)^2 + (t+1) = (t^3 + t^2 + 1)(t+1)$.

(i) $(3t+1)(t^3 - t^2 - 1)$.

(j) $(t-1)^3(2t+3)^2$.

(k) $(t^2+1)(t-1)^2(t+1)^2$.

(l) $(t-6)(t+2)(t+3)$.

(m) $(t-5)(t^2-2t+3)$.

(n) $(t^2+t+1)(t^3-t^2-t-1)$.

(o) $(t^2-2)(t^3-3t^2+2t-7)$.

(p) $(t^4+10t^2+1)(t^4-10t^2+1)$.

**2.8.** Suppose that $f = gh$. Write $g = u + v$ where $u$ is homogeneous and $\deg v < \deg u = \deg g$. Then $gh = uh + vh$ and $\deg vh < \deg uh = \deg f$. Since $f$ is homogeneous, we must have $vh = 0$, whence $v = 0$ and $g$ is homogeneous. Similarly, $h$ is homogeneous.

**2.9.** No. A counterexample is $x^2y^2 + 2xy - x^2 - y^2 = (xy + x - y)(xy - x + y)$.

**2.10.** (a) $(a-b)(a-5b)$.

(b) $(x-y)(x+y)(y-z)$.

(c) $(x-y)(x-z)(y-z)$.

(d) $(x+y+z)(x^2+y^2+z^2-xy-yz-zx)$.

(e) $(a-b)(a-c)(b-c)$.

(f) $(x-y)(y-z)(z-x)$.

(g) Substituting $z = ax + by$ yields $a(1+a)x^3 + b(1+b)y^3 + (a + 2ab + b^2 + 1)xy^2 + (a^2 + 2ab + b + 1)x^2y$. No choice of $a$, $b$ will make all coefficients vanish simultaneously. The polynomial is irreducible.

**2.11.** (b) From (a), $p_n(x, y, z)$ is divisible by $(x-y)(y-z)(z-x)$. Since $\deg p_n < 3$ when $n = 0, 1$, the only way $p_n$ can be divisible by a polynomial of degree 3 is for it to vanish identically.

(d) $q_2(x, y, z) = 1$; $q_3(x, y, z) = x + y + z$; $q_4(x, y, z) = x^2 + y^2 + z^2 + xy + yz + zx$.

(e) $p_n(x, y, 0) = xy(y-x)(y^{n-2} + xy^{n-3} + \cdots + x^{n-2})$.

(f) $q_n(x, y, z) = z^{n-2} + (x+y)z^{n-3} + (x^2 + xy + y^2)z^{n-4} + (x^3 + x^2y + xy^2 + y^3)z^{n-5} + \cdots + (x^{n-2} + x^{n-3}y + \cdots + y^{n-2})$.

**2.12.**
$$(x+y+z)^3 - x^3 - y^3 - z^3 = 3(x+y)(y+z)(z+x)$$

$$(x+y+z)^5 - x^5 - y^5 - z^5 = 5(x+y)(y+z)(z+x)(x^2+y^2+z^2+xy+yz+zx)$$

$$
\begin{aligned}
(x &+ y+z)^7 - x^7 - y^7 - z^7 \\
&= 7(x+y)(y+z)(z+x)(x^4 + \cdots + 2x^3y + \cdots + 3x^2y^2 + \cdots \\
&\quad + 5x^2yz + \cdots) \\
&= 7(x+y)(y+z)(z+x)[(x^2+y^2+z^2+xy+yz+zx)^2 \\
&\quad + xyz(x+y+z)].
\end{aligned}
$$

**2.13.** No. $2t^2+t-1 = (2t-1)(t+1)$ is reducible over $\mathbf{Z}$, but irreducible over $\mathbf{Z}_2$. However, if the leading coefficient of the polynomial is not divisible by $m$, then irreducibility over $\mathbf{Z}_m$ implies irreducibility over $\mathbf{Z}$. (Any factorization over $\mathbf{Z}$ would yield a factorization over $\mathbf{Z}_m$ involving polynomials of the same degree.)

**2.14.** (a) $t^2 + t + 1$ is irreducible over $\mathbf{Z}_2$, hence over $\mathbf{Z}$.

(b) $49t^2 + 35t + 11$ is irreducible over $\mathbf{Z}_2$, hence over $\mathbf{Z}$.

(c) Over $\mathbf{Z}_3$, the polynomial is equal to $t^3 + t^2 + 2t + 1$. Since it has no zero in $\mathbf{Z}_3$, it is irreducible over $\mathbf{Z}_3$. Since $3 \nmid 124$, the polynomial must be irreducible over $\mathbf{Z}$.

**2.15.** (a) Over $\mathbf{Z}_4$, the polynomial is

$$3t^4 + 2t^3 + t^2 + 2 = 3(t+1)^2(t^2+2).$$

Over $\mathbf{Z}_5$, it is

$$3t^4 + 3t^3 + t^2 + 2t = t(t+3)(3t^2 + 4t + 4).$$

Over $\mathbf{Z}_7$, it is

$$5t^3 + 5t^2 + 3t + 4 = (t+5)(5t^2 + t + 5).$$

Over $\mathbf{Z}_9$, it is

$$7t^3 + 2t^2 + 7t + 8 = (t+2)(7t^2 + 6t + 4).$$

(b) The results of (a) suggest a factorization of the form

$$(at^2 + bt + c)(ut^2 + vt + w)$$

with $(a, b, c) \equiv (7, 6, 4)$ and $(u, v, w) \equiv (0, 1, 2)$ modulo 9. Since $au = 63$, the only possibility is $a = 7$, $u = 9$. Since $cw = -10$, $c \equiv 4 \pmod 9$, $w \equiv 2 \pmod 9$, we must have $c = -5$, $w = 2$.

Since the factorization is equivalent modulo 7 to $(t+5)(5t^2 + t + 5)$ and since $9 \equiv -5 \pmod 7$, we should have $(a, b, c) \equiv (0, -1, -5)$ and $(u, v, w) \equiv (-5, -1, -5)$ modulo 7. Thus $b \equiv 6 \pmod 9$, $b \equiv 6 \pmod 7$ leads to $b \equiv 6 \pmod{63}$ and $v \equiv 1 \pmod 9$, $v \equiv -1 \pmod 7$ leads to $v \equiv -8 \pmod{63}$.

This leads to the trial $(7t^2 + 6t - 5)(9t^2 - 8t + 2)$ which works.

**2.16.** $10t^5 + 3t^4 - 38t^3 - 5t^2 - 6t + 3 = (5t^2 + 9t - 3)(2t^3 - 3t^2 - t - 1)$.

**3.1.** (b) $b$ divides every term but the first of the left side, and so divides $c_n a^n$. Since $\gcd(a^n, b) = 1$, $b$ must divide $c_n$.

**3.2.** This is a consequence of Exercise 1.

**3.3.** (a) No rational zeros. (b) 4/3. (c) 1, $-2/3$. (d) $-4$, 7.

**3.4.** The values of the polynomial at $-2, -1, 1, 2$ are, respectively, $40, 21, -11, 48$. All but $1/2$ and $4/3$ are eliminated.

**3.6.** (a) 5. (b) 3/8. (c) 2, 2/5. (d) $-1, 3, 7$. (e) $3/2, -3/2, -5/2$. (f) $-7/2, 4/3$.

**3.7.** $(3t - 4)(3t + 5)(2t^3 - 6t^2 + 9t + 3)$.

**3.8.** $-7, 2/3, -5/8, (1/2)(1 + i\sqrt{7}), (1/2)(1 - i\sqrt{7})$.

**3.9.** Newton's Table is Horner's Table for the polynomial $t^3 q(1/t)$ evaluated at $1/5$. This polynomial has zero $1/5$ iff $q(t)$ has zero 5.

**3.12.** (a) Note that $\theta/2$ and $2\theta$ are supplementary.
   (b) From (a), $\cos\theta = 2\cos^2 2\theta - 1 = 2(2\cos^2\theta - 1)^2 - 1$.
   (c) $8x^4 - 8x^2 - x + 1 = (x - 1)(2x + 1)(4x^2 + 2x - 1)$. By (b), $\cos\theta$ is a zero of this polynomial. Since $\cos\theta$ is not equal to 1 nor $-1/2$, it must be a zero of $4x^2 + 2x - 1$.
   (d) $(\sqrt{5} - 1)/4$.

**4.1.** (a) 4, 5, 12, 13, 20, 21, 28, 29, 36, 37.
   (b) 2, 7, 12, 17, 22, 27, 32, 37.
   (c) For any integer $t$, 40 divides $t^2 - 9t - 36$ iff 5 and 8 do. The congruence is satisfied by 12 and 37.

**4.4.** 22, 82.

**4.5.** 444.

**4.6.** By Exercise 1.6.6 (d), there is a number $w$ such that $a + wu \equiv b$ (mod $v$). Choose $c$ such that $0 \le c \le m - 1$ and $c \equiv a + wu$ (mod $uv$). Then $c$ is the required number.
   If $c$ and $d$ satisfy $c \equiv d \equiv a$ (mod $u$), $c \equiv d \equiv b$ (mod $v$), then $c - d$ is divisible by $u$ and $v$, hence by $uv$. Hence $c \equiv d$ (mod $m$). If both $0 \le c \le m - 1$, $0 \le d \le m - 1$, then $c = d$.

**4.8.** $504 = 7 \cdot 8 \cdot 9$. $n^8 - n^2 = n^2(n^6 - 1)$ is always divisible by 7 and 9, as well as by 8 when $n \equiv 0, 1, 3$ (mod 4). If $n \equiv 2$ (mod 4), $n^6 - 1$ is odd and $n^2$ is divisible by 4 but not by 8. Hence $n^8 - n^2$ is not divisible by 504 iff $n \equiv 2$ (mod 4).

**4.9.** (a) $t \equiv 3$ (mod 4); $t \equiv 0, 2$ (mod 9). (c) $t \equiv 11, 27$ (mod 36).

**4.10.** (a) None. (b) 11, 35. (c) 21, 45.

**4.11.** (e) 22. (f) 221. (Note that $t$ satisfies this congruence iff $-t$ does.)

**4.12.** (a) $1000 = 2^3 \cdot 5^3$. $2t^3 + t + 3 \equiv 0$ (mod 5) is satisfied for $t \equiv 3, 4$ (mod 5). The solution $t \equiv 3$ (mod 5) does not lead to a solution modulo $5^3$ (cf. Exploration E.33). However, the congruence modulo 125 is satisfied by $t \equiv 124$. The solution modulo 8 is 7. Hence, $2t^3 + t + 3 \equiv 0$ (mod 1000) has a unique solution $t \equiv 999$ (mod 1000).

(b) $83349 = 3^5 . 7^3$. For a solution, we require $t \equiv 51$, 193 or 242 (mod 243) and $t \equiv 41, 48, 90, 97, 139, 146, 156, 188, 195, 237, 244, 286, 293, 335$ or 342 (mod 343). Therefore, there are $3 \times 15 = 45$ solutions modulo 83349. For example, $t \equiv 51$ (mod 243), $t \equiv 41$ (mod 343) leads to the solution $t \equiv 25080$ (mod 83349).

**4.13.** $675 = 3^3 . 5^2$. Any solution $x$ must satisfy $x \equiv 12$ or 16 (mod 27) and $x \equiv 2, 7, 9, 12, 17, 22$ (mod 25). There are $2 \times 6 = 12$ incongruent solutions in all modulo 675. An obvious solution is $x \equiv 12$ (mod 675).

**4.14.** (a) For any zero $r$, we must have $r^5 = 4r^4 + 411r^3 + 452r^2 + 3322r + 828$. If $|r| > 2$, then

$$|r|^5 \leq 4|r|^4 + 411|r|^3 + 452|r|^2 + 3322|r| + 828 \leq 4200|r|^4 + 828$$

so that $|r| \leq 4200 + (828/|r|^4) < 5000$.

(b) Solve the congruence modulo $5000 = 2^3 . 5^4$. If $r$ is a root of the equation, then $r \equiv 1, 2$ or 3 (mod 8) and $r \equiv 7$ (mod 25) or $r \equiv 23$ (mod 625). The roots are 23 and $-18$.

**4.15.** Solving the congruence modulo $3^2 . 2^3$ yields $t \equiv 18, 63$ (mod 72). The only integer root is $-9$.

**5.3.** $t^2 - 1 = (t - 1)(t + 1)$.
$t^3 - 1 = (t - 1)(t - (-1 + i\sqrt{3})/2)(t - (-1 - i\sqrt{3})/2)$.
$t^4 - 1 = (t - 1)(t + 1)(t - i)(t + i)$.
$t^6 - 1 = (t^3 - 1)(t + 1)(t - (1 + i\sqrt{3})/2)(t - (1 - i\sqrt{3})/2)$.
$t^8 - 1 = (t^4 - 1)(t - (1+i)/\sqrt{2})(t - (1-i)/\sqrt{2})(t + (1+i)/\sqrt{2})(t + (1-i)/\sqrt{2})$.

**5.4.** Note that

$$[t - (\cos 2k\pi/n + i \sin 2k\pi/n)][t - (\cos 2k\pi/n - \sin 2k\pi/n)]$$

$$= t^2 - (2 \cos 2k\pi/n)t + 1$$

$$t^5 - 1 = (t - 1)(t^2 - (2 \cos 2\pi/5)t + 1)(t^2 - (2 \cos 4\pi/5)t + 1).$$

**5.6.** Minimum exponents and zeros of $P_{12}(t)$ $1 : 1$; $2 : -1$; $3 : (-1 \pm i\sqrt{3})/2$; $4 : \pm i$; $6 : (1 \pm i\sqrt{3})/2$; $12 :$ the remaining four zeros.

**5.7.** (b) Let $m$ be the smallest positive integer for which $w^m = 1$. By the division algorithm, we can write $n = qm + r$, where $q$ and $r$ are integers and $0 \leq r < m$. Then $w^r = w^n (w^m)^{-q} = 1$. From the minimality of $m$, it follows than $r = 0$. Hence $m | n$, and, clearly, $w$ is a primitive $m$th root of unity.

**5.8.** (a) By de Moivre's Theorem, $\cos(2k\pi/n) + i \sin(2k\pi/n) = \zeta_n^k$.
(b) Since $t^n - 1 = (t - 1)(t^{n-1} + t^{n-2} + \cdots + t + 1)$, all $n$th roots of unity except 1 itself are zeros of the second factor.
(c) Suppose $\gcd(a, n) = d$. Then, if $\zeta$ is an $n$th root of unity, $(\zeta^a)^{n/d} = (\zeta^n)^{a/d} = 1$. Hence, if $\zeta^a$ is primitive, then $d = 1$. Suppose $d = 1$ and let

$k$ be the smallest positive exponent for which $(\zeta^a)^k = 1$. It can be seen that the powers of $\zeta^a$ cycle through the set $\{1, \zeta^a, \zeta^{2a}, \ldots, \zeta^{(k-1)a}\}$. Since $(\zeta^a)^n = 1$, $n$ must be a multiple of $k$ so that $n = mk$. Since $\zeta$ is a primitive $n$th root and $\zeta^{ak} = 1$, $ak$ must be a multiple of $n$, say $ak = qn$. Hence $a = qm$, so that $m$ divides $n$ and $a$. Therefore, $m = 1$ and $\zeta^a$ is a primitive $n$th root.

(e) $\gcd(a, n) = \gcd(n - a, n)$, so that $a \longleftrightarrow n - a$ is a pairing of positive integers less than $n$ and coprime with $n$. Note that $a \neq n - a$ when $n \geq 3$.

**5.9.** The zeros of $P_n(t)$ consist of all the primitive $d$th roots of unity, where $d$ runs through all the positive divisors of $n$. For each zero, there is a corresponding linear factor. By collecting together all the linear factors for the primitive $d$th roots, we obtain the required representation.

**5.11.** $Q_9 = t^6 + t^3 + 1$;
$Q_{10} = t^4 - t^3 + t^2 - t + 1$;
$Q_{12} = t^4 - t^2 + 1$;
$Q_{14} = t^6 - t^5 + t^4 - t^3 + t^2 - t + 1$;
$Q_{15} = t^8 - t^7 + t^5 - t^4 + t^3 - t + 1$;
$Q_{16} = t^8 + 1$.

To find, for example, $Q_{15}(t)$, we look for the complementary factors of $P_{15}(t)$ whose zeros are primitive 1st, 3rd, 5th roots of unity. Thus

$$
\begin{aligned}
P_{15}(t) &= (t^5 - 1)(t^{10} + t^5 + 1) \\
&= (t^5 - 1)(t^2 + t + 1)(t^8 - t^7 + t^5 - t^4 + t^3 - t + 1).
\end{aligned}
$$

**5.12.** (a) It is straightforward to see that if $\zeta$ is a primitive $2k$th root of unity, then $\zeta^2$ is a primitive $k$th root of unity. Let $k$ be even and $\zeta^2$ be a primitive $k$th root of unity. Suppose that $\zeta$ is a primitive $r$th root of unity. Then $k \leq r \leq 2k$ and either $r = 2k$ or $r$ is odd. Since $(\zeta^k)^2 = 1$ and $\zeta^k \neq 1$, we must have $\zeta^k = -1$. Hence $\zeta^{r-k} = -1$, so that $\zeta^{2(r-k)} = 1$. Hence $r - k \geq k$, so that $r \geq 2k$. Thus, $\zeta$ is a primitive $2k$th root of unity.

The primitive $k$th roots of unity are precisely the numbers of the form $(\zeta_{2k}^a)^2$, where $\gcd(a, k) = \gcd(a, 2k) = 1$.

$$
\begin{aligned}
Q_k(t^2) &= \Pi\{(t^2 - \zeta_{2k}^{2a}) : 1 \leq a < k, \gcd(a, k) = 1\} \\
&= \Pi\{(t - \zeta_{2k}^a)(t - \zeta_{2k}^{k+a}) : 1 \leq a < k, \gcd(a, k) = 1\} \\
&= \Pi\{(t - \zeta_{2k}^a) : 1 \leq a < 2k, \gcd(a, 2k) = 1\} \\
&= Q_{2k}(t).
\end{aligned}
$$

(b) Let $k$ be odd and let $\zeta$ be a primitive $k$th root of unity. Then no power of $\zeta$ is equal to $-1$, whence it follows that no odd power of $-\zeta$ is equal to 1. Hence, if $(-\zeta)^r = 1$, then $r$ is even and $\zeta^r = 1$. But then $r$ is an even multiple of $k$, i.e. a multiple of $2k$. It follows that $-\zeta$ is a primitive $2k$th root of unity.

On the other hand, if $-\zeta$ is a primitive $2k$th root of unity, then $(-\zeta)^k = -1$, from which $\zeta^k = 1$. It is straightforward to see that $\zeta$ is a primitive $k$th root of unity. Hence $\zeta$ is a primitive $k$th root of unity iff $-\zeta$ is a primitive $2k$th root of unity. Thus $Q_{2k}(t) = Q_k(-t)$, since $\deg Q_k(t)$ is even by Exercise 8.

**5.13.** $\cos[(2k+1)\pi/n] + i\sin[(2k+1)\pi/n]$, for $k = 0, 1, \ldots, n-1$.

**5.14.** $r^{1/n}(\cos 2k\pi/n + i\sin 2k\pi/n)$ $(0 \le k \le n-1)$.

**5.15.** Let $n = kr$. Then, if $u = t^r$, we have $t^n - 1 = u^k - 1 = (u-1)(u^{k-1} + \cdots + 1)$, from which the result follows. The conjecture is false; take $k = 1$, $m = 2$.

**5.17.** Use $\zeta + \zeta^2 + \cdots + \zeta^6 = -1$ and $\zeta^7 = 1$ to check that $u + v = -1$, $uv = 2$.

**5.18.** (a) Modulo 11: 1, 3, 4, 5, 9; Modulo 13: 1, 3, 4, 9, 10, 12; Modulo 17: 1, 2, 4, 8, 9, 13, 15, 16.
(b) For any prime $p$, $u+v = \Sigma\{\zeta^a : 1 \le a \le p-1\} = -1$. When $p = 11$, 13, 17, the product $uv = 3, -3, -4$, respectively.

**5.19.** $f(x) = (x^4 + 2x^3 + 2x^2 + x)(x^8 + x^6 + x^5 + x^4 + x^3 + x)$. The faces of the dice are labeled $(4,3,3,2,2,1)$ and $(8,6,5,4,3,1)$. For more on this consult:

Duane Broline, Renumbering the faces of dice. *Math. Mag.* **52** (1979), 312–315.

J.A. Gallian & D.J. Rusin, Cyclotomic polynomials and nonstandard dice. *Discrete Math.* **27** (1979), 245–249.

Martin Gardner, Mathematical games. *Scientific American* **238** (1978), 19–32.

**6.1.** Suppose $f = u/v$. Divide $v$ into $u$ to obtain $u = pv + w$ where $\deg w < \deg v$. Then the result follows with $g = w/v$.

**6.2.** Putting the sum $A/(t-m) + B/(t-n)$ over a common denominator and equating the numerator to $at+b$ yields the condition $A(t-n) + B(t-m) = at + b$ for $t \ne m, n$. This can be interpreted as saying that the polynomial $A(t-n) + B(t-m) - (at+b)$ vanishes for infinitely many values of $t$ (all but $t = m, n$). But this implies that it must be the zero polynomial (by Exercise 2.2.4) and so vanishes for $t = m$ and $t = n$.

$$A = (am+b)/(m-n); \quad B = (an+b)/(n-m).$$

**6.3.** $A = 3$, $B = -2$.

**6.4.** (a) The result can be proved by induction on $k$. It is clearly true when $k = 1$. Suppose it has been established when the degree of the denominator is less than $k$.

Consider the difference

$$\frac{p(t)}{q(t)} - \frac{c_1}{t - a_1} = \frac{p(t) - c_1 q_1(t)}{q(t)}$$

where $q(t) = (t - a_1)q_1(t)$ and $c_1$ is a constant to be determined. The degree of the numerator is strictly less than the degree of $q(t)$. Since $q_1(a_1) \neq 0$, we can choose $c_1$ so that $p(a_1) - c_1 q_1(a_1) = 0$. Thus $p(t) = c_1 q_1(t) + (t - a_1)p_1(t)$, for some polynomial $p_1(t)$.

Hence

$$\frac{p(t)}{q(t)} = \frac{c_1}{t - a_1} + \frac{p_1(t)}{q_1(t)}.$$

Since $\deg p_1(t) < \deg q_1(t)$, the induction hypothesis can be applied to $p_1(t)/q_1(t)$.

(b) If the sum is put over a common denominator, the numerator is

$$p(t) = \Sigma c_i (t - a_1) \cdots (\widehat{t - a_i}) \cdots (t - a_n). \qquad (*)$$

(The hat denotes a deleted term.) Note that

$$q'(t) = \Sigma (t - a_1) \cdots (\widehat{t - a_j}) \cdots (t - a_n),$$

so that $p(a_i) = c_i q'(a_i)$.

(c)

$$f(t) = \sum_{i=1}^{n} [b_i(t - a_1) \cdots (\widehat{t - a_i}) \cdots (t - a_n)/q'(a_i)].$$

**6.5.** (a)

$$\sum_{k=2}^{n} \frac{1}{k(k-1)} = \sum_{k=2}^{n} \frac{1}{k-1} - \frac{1}{k} = 1 - \frac{1}{n} = \frac{n-1}{n}.$$

(b) When $n = 2$, the sum is $1/6$. For $n \geq 3$, we have

$$\sum_{k=2}^{n} \frac{1}{k^3 - k} = \frac{1}{2} \sum_{k=2}^{n} \left[ \frac{1}{k-1} - \frac{2}{k} + \frac{1}{k+1} \right]$$

$$= \frac{1}{2} \left[ \sum_{k=1}^{n-1} \frac{1}{k} - 2 \sum_{k=2}^{n} \frac{1}{k} + \sum_{k=3}^{n+1} \frac{1}{k} \right]$$

$$= \frac{1}{2} [1 - 1/2 - 1/n + 1/(n+1)]$$

$$= \frac{(n+2)(n-1)}{4n(n+1)}.$$

(c) If $x = 0$, the sum is equal to $n - 1$. If $x = -1, -1/2, \ldots, -1/n$, the sum is not defined.

$$\sum_{k=2}^{n} \frac{1}{(kx+1)((k-1)x+1)} = \frac{1}{x} \sum_{k=2}^{n} \left[ \frac{1}{(k-1)x+1} - \frac{1}{kx+1} \right]$$

$$= \frac{1}{x}\left[\frac{1}{x+1} - \frac{1}{nx+1}\right] = \frac{n-1}{(x+1)(nx+1)}.$$

**6.6.** We have $t^3+t^2+15t-27 = [A(t-3)+B](t^2+6t+27)+(Ct+D)(t-3)^2$. Setting $t=3$ yields $B=1$. Comparison of coefficients leads to $A=2/3$, $B=1$, $C=1/3$, $D=0$.

**6.7.** $t^4 - 3t^3 + t^2 - 3t = t(t-3)(t^2+1)$. The rational function has the representation
$$-\frac{1}{t} + \frac{2}{t-3} + \frac{-t+1}{t^2+1}.$$

**6.8.**
$$\frac{1}{t^2-1} = \frac{1}{2}\left[\frac{1}{t-1} - \frac{1}{t+1}\right]$$

$$\frac{1}{t^3-1} = \frac{1}{3}\left[\frac{1}{t-1} - \frac{t+2}{t^2+t+1}\right]$$
$$= \frac{1}{3}\left[\frac{1}{t-1} + \frac{\omega}{t-\omega} + \frac{\omega^2}{t-\omega^2}\right]$$

where $\omega = (-1+i\sqrt{3})/2$.

$$\frac{1}{t^4-1} = \frac{1}{4}\left[\frac{1}{t-1} - \frac{1}{t+1} - \frac{2}{t^2+1}\right]$$
$$= \frac{1}{4}\left[\frac{1}{t-1} - \frac{1}{t+1} + \frac{i}{t-i} - \frac{i}{t+i}\right]$$

$$\frac{1}{t^5-1} = \frac{1}{5}\left[\frac{1}{t-1} - \frac{t^3+2t^2+3t+4}{t^4+t^3+t^2+t+1}\right]$$
$$= \frac{1}{5}\left[\frac{1}{t-1} + \frac{\zeta}{t-\zeta} + \frac{\zeta^2}{t-\zeta^2} + \frac{\zeta^3}{t-\zeta^3} + \frac{\zeta^4}{t-\zeta^4}\right].$$

If $(t^n-1)^{-1} = \sum_{i=0}^{n-1} a_i(t-\zeta_n^i)^{-1}$, then $a_i$ must satisfy

$$a_i \prod_{j\neq i}(\zeta_n^i - \zeta_n^j) = 1 \quad \text{or} \quad a_i\zeta_n^i \prod_{i=1}^{n-1}(1-\zeta_n^i) = 1.$$

Since the product is equal to $t^{n-1} + \cdots + t + 1$ evaluated at $t=1$, we find that $a_i = \zeta_n^i/n$.

## Solutions to Problems

### Chapter 3

**7.1.** (a) $(2x^2 + 1)^2 - 4x^2 = (2x^2 + 2x + 1)(2x^2 - 2x + 1)$.
(b) $(x^2 - 2)^2 - 16x^2 = (x^2 + 4x - 2)(x^2 - 4x - 2)$.
(c) $(xy - x - y + 1)(xy - 1) = (x - 1)(y - 1)(xy - 1)$.
(d) $(xz - 1)(yz - 1)(xy - 1)$.
(e) $3(a - b)(b - c)(c - a)$.
(f) $(a + c)(b + c)(a + b - c)$.
(g) $(a + b)(b + c)(c + a)$.
(h) $(a + b)(b + c)(c - a)$.
(i) $(x^2 + x + 1)(x^8 - x^7 + x^5 - x^4 + x^3 - x + 1)$.
(j) $(2x + z)(x^2 - 2xz + z^2 + 3y^2)$.
(k) $(x + y + z - xyz - 2)(x + y + z - xyz + 2)$.
(l) $(x - y)(y - z)(z - x)(xy + yz + zx)$.
(m) $12abc(a + b + c)$.
(n) $3abc(a + b)(b + c)(c + a)$.
(o) $2abc(a + b + c)$.
(p) $(x + y + z)^2[(x + y + z)^3 - 5(x + y + z)(xy + yz + zx) + 15xyz] = (x+y+z)^2[x^3+y^3+z^3 - 2x^2y - 2xy^2 - 2y^2z - 2yz^2 - 2z^2x - 2zx^2 + 6xyz]$.
(q) $(x + y + z + w)(x + y - z - w)(x - y + w - z)(x - y - z + w)$.
(r) $(xy - z^2)(yz - x^2)(zx - y^2)$.
(s) $(x^2 + y^2 - z^2)(y^2 + z^2 - x^2)(z^2 + x^2 - y^2)$.
(t) $(x + y)(y + z)(z + x)$.
(u) $[(a^2 + b^2) + (2ab - c^2)][(a^2 + b^2) - (2ab - c^2)] = [(a + b)^2 - c^2][(a - b)^2 + c^2] = (a + b + c)(a + b - c)(a^2 + b^2 + c^2 - 2ab)$.

**7.2.** $(x^4 - x^3 + x^2 + 2x - 6) = (x^2 - x + 3)(x^2 - 2)$.
$(x^4 + x^3 + 3x^2 + 4x + 6) = (x^2 - x + 3)(x^2 + 2x + 2)$.

**7.3.** Since $[p(x)]^5 - x$ should vanish when $x = 1, 2, 3$, we require that $p(1) = 1$, $p(2) = 2^{1/5}$, $p(3) = 3^{1/5}$. A possible polynomial $p$ is the quadratic $ax^2 + bx + c$, where

$$2a = 3^{1/5} - 2^{6/5} + 1, \quad b = 2^{1/5} - 1 - 3a, \quad c = 1 - a - b.$$

**7.4.** Since $i$ should be a zero of the polynomial, we have that $(-a+b-1) + (a + b - 5)i = 0$. Hence $a = 2$, $b = 3$, and

$$(2x + 3)(x^5 + 1) - (5x + 1) = 2(x^6 + 1) + 3x(x^4 - 1)$$

$$= (x^2 + 1)[2(x^4 - x^2 + 1) + 3x(x^2 - 1)].$$

**7.5.** (a) The zeros of $x^2 + px + 1$ are reciprocals, say $r$ and $1/r$. These are also zeros of $ax^3 + bx + c$. Since the sum of the zeros of the cubic is 0, its third zero is $-(r + 1/r)$. Since $-c/a$ is the product of the zeros and $b/a$ is the sum of all products of pairs of the zeros,

$$\frac{a^2 - c^2}{ab} = \frac{1 - (c/a)^2}{b/a} = 1.$$

(b) The zeros of $cx^3 + bx^2 + a$ are $r$, $1/r$, $-(r+1/r)^{-1}$. Hence both cubics are divisible by $x^2 + px + 1 = (x - r)(x - 1/r)$.

**7.6.** If the polynomial is equal to $(ux + vy + w)(qx + ry + s)$, then $uq = 3$, $ur + vq = 2p$, $vr = 2$, $us + wq = 2a$, $vs + wr = -4$, $ws = 1$. By multiplying each factor by a suitable constant (the first by $s$, the second by $w$) if necessary, we may suppose that $w = s = 1$. Hence $uq = 3$, $u + q = 2a$, so that $u$ and $q$ are the zeros of $t^2 - 2at + 3 = (t - a)^2 - (a^2 - 3)$. Also $vr = 2$, $v + r = -4$, so that $v$ and $r$ are zeros of $t^2 + 4t + 2$. We may assume that $r = -2 + \sqrt{2}$ and $v = -2 - \sqrt{2}$.

Hence,

$$
\begin{aligned}
2p &= ur + vq = u(r - v) + (u + q)v = u(2\sqrt{2}) - 2(2 + \sqrt{2})a \\
&\Rightarrow p + 2a = (u - a)\sqrt{2} \\
&\Rightarrow p^2 + 4ap + 4a^2 = 2(u - a)^2 = 2(a^2 - 3) \\
&\Rightarrow p^2 + 4ap + 2a^2 + 6 = 0.
\end{aligned}
$$

**7.7.** Let the polynomial be $t^3 + at^2 + bt + c$ and its zeros be $r$, $s$ and $rs$. Then $(1 + r)(1 + s) = 1 - a$ and $rs(1 + r)(1 + s) = b - c$. If $a \neq 1$, then $rs = (b - c)/(1 - a)$ is rational and $t - rs$ is a factor of the cubic. (Observe that, in fact, $1 - a$ is a divisor of $b - c$, since $rs$ must be an integer.) If $a = 1$, then, say, $r = -1$ and $b = c$. In this case, $t + 1$ is a factor of the cubic.

**7.8.** Setting $x = 0$ and $x = 1$, we find that $a$ must divide both 90 and 92. Hence $a = -2, -1, 1$ or 2. Since $x^2 - x - 2 = (x - 2)(x + 1)$, and 2 and $-1$ are not zeros of $x^{13} + x + 90$, $a \neq -2$. Similarly, $a \neq 1$. Thus, $a = -1$ or $a = 2$.

If $u$ is a zero of $x^2 - x - 1$, then $u^2 = u + 1$, whence $u^4 = 3u + 2$, $u^8 = 21u + 13$, $u^{12} = 144u + 89$, $u^{13} = 233u + 144 \neq -u - 90$. Hence $u$ is not a zero of $x^{13} + x + 1$. Thus $a \neq -1$.

If $v$ is a zero of $x^2 - x + 2$, then $v^2 = v - 2$, $v^4 = -3v + 2$, $v^8 = -3v - 14$, $v^{12} = 45v - 46$, $v^{13} = -v - 90$. Hence both zeros of $x^2 - x + 2$ are zeros of $x^{13} + x + 90$. Thus, $a = 2$.

Checking, we find that

$$x^{13} + x + 90 = (x^2 - x + 2)(x^{11} + x^{10} - x^9 - 3x^8 - x^7 + 5x^6 + 7x^5$$

$$- 3x^4 - 17x^3 - 11x^2 + 23x + 45).$$

**7.9.** It is the cube of $b(b - a - c)$.

**7.10.** Let $u = r + s$, $v = rs$. Then $q(x) = x^2 - ux + v$ and $p(x) = (x - r)(x - s) + (r + s - x) = x^2 - (u + 1)x + (u + v)$. Expressing the given polynomial as the product of $p(x)$ and $q(x)$ and comparing coefficients leads to

$$2a + 1 = -(2u + 1) \implies a + u + 1 = 0$$

$$(a - 1)^2 = u^2 + 2(u + v) \implies 2(u + v) = (a - 1)^2 - u^2 = -4a$$

$$4 = v(u + v) = -2av.$$

Since $u = -a - 1$ and $u + v = -2a$, it follows that $v = -a + 1$ and that $4 = 2a^2 - 2a$. Hence $a = 2$ or $a = -1$.

Since $b = -u(u+v) - v(u+1) = -(3a^2+a)$, we have that $(a, b) = (2, -14)$ or $(-1, -2)$. Thus, the possibilities are

$$x^4 + 5x^3 + x^2 - 14x + 4 = (x^2 + 3x - 1)(x^2 + 2x - 4)$$

$$x^4 - x^3 + 4x^2 - 2x + 4 = (x^2 + 2)(x^2 - x + 2).$$

**7.11.** $320n^2 + 144n - 243 = (an + b)(cn + d)$ implies that $b$ and $d$ are powers of 3, $\gcd(ac, 3) = 1$ and $ad + bc = 144 \equiv 0 \pmod 9$. Hence, without loss of generality, we can take $b = 9$, $c = -27$. We find that $320n^2 + 144n - 243 = (8n + 9)(40n - 27)$.

Hence

$$3(81^{n+1}) + (16n - 54)9^{n+1} - (320n^2 + 144n - 243)$$

$$= [3(9^{n+1}) + (40n - 27)][9^{n+1} - (8n + 9)]$$

$$= [27(9^n - 1) + 40n][9(9^n - 1) - 8n]$$

$$= 8^2[27(9^{n-1}+9^{n-2}+9^{n-3}+\cdots+1)+5n][9(9^{n-1}+9^{n-2}+9^{n-3}+\cdots+1)-n].$$

It is straightforward to check that, modulo 8, for each $n$, each factor of the last member in square brackets is congruent to 0. Hence the given quantity is divisible by $8^4 = 2^{12}$.

**7.12.** If $n = 2m$, $a = 2b$, then $x^n + x^a + 1 = (x^m + x^b + 1)^2$. If $f(x) = x^n + x^a + 1$ with exactly one of $n$ and $a$ even, then $f'(x) = x^k$ where $k = n - 1$ or $a - 1$, and $\gcd(f, f') = 1$. If $n$ and $a$ are both odd, then $f'(x) = x^{n-1} + x^{a-1}$, so that $f(x) = xf'(x) + 1$ and $\gcd(f, f') = 1$. Since any repeated factor of $f(x)$ must also divide $f'(x)$, the result follows.

**7.13.** If $at^2 + bt + c$ and $at^2 + bt + c + 1$ are both reducible over $\mathbf{Z}$, then for some integers $m$ and $n$,

$$b^2 - 4ac = m^2 \quad \text{and} \quad b^2 - 4ac - 4a = n^2.$$

Hence $4a = m^2 - n^2$.

(a) In the case that $a = 1$, this yields $4 = m^2 - n^2 = (m - n)(m + n)$. Since $m$ and $n$ have the same parity, we must have $m = 2$, $n = 0$, whence

$b^2 = 4(c+1)$. Thus, $c+1$ is a square, say $u^2$, from which $b = 2u$. Conversely, if $b = 2u$ and $c = u^2 - 1$, both $t^2 + bt + c$ and $t^2 + bt + c + 1$ are evidently reducible over $\mathbf{Z}$.

(b) If $a = 3$, we are led to $b^2 = 4(4 + 3c)$. Choosing $c = 7$ makes $4 + 3c$ a square, and we obtain the examples

$$3t^2 + 10t + 7 = (3t + 7)(t + 1)$$

$$3t^2 + 10t + 8 = (3t + 4)(t + 2).$$

**7.14.** Modulo $x^q - x - 1$, it can be shown by induction that

$$x^{q^r} \equiv x + r, \quad (r = 1, 2, \ldots).$$

This is clearly true for $r = 1$. If it is true for $r = k$, then

$$x^{q^{k+1}} \equiv (x + 1)^{q^k} \equiv x^{q^k} + 1 \equiv (x + k) + 1 = x + (k + 1).$$

Hence,

$$x^m - 1 \equiv x \cdot x^q \cdots x^{q^{p-1}} - 1 \equiv x(x + 1)(x + 2) \cdots (x + p - 1) - 1$$

$$\equiv x^p - x - 1.$$

If $n = 1$, then $p = q$ and $x^p - x - 1 \equiv 0$. If $n > 1$, then $q > p$ and $x^p - x - 1 \not\equiv 0$.

**7.15.** $(x^4 - 1)^4 - x - 1 = (x^4 - 1)^4 - x^4 + (x^4 - x - 1) = (x^4 - x - 1)$ $[(x^4 - 1)^3 + (x^4 - 1)^2 x + (x^4 - 1)x^2 + x^3 + 1]$. The second factor vanishes for $x = -1$ and $x = 0$, and so it is equal to

$$x(x + 1)(x^{10} - x^9 + x^8 - 3x^6 + 3x^5 - 2x^4 + 3x^2 - 2x + 1).$$

**7.16.** Let $\omega$ be an imaginary cube root of unity so that $\omega^2 + \omega + 1 = 0$. Then, when $t = \omega$,

$$(t + 1)^n - t^n - 1 = [(\omega + 1)^n - 1 - \omega^n] = -(\omega^{2n} + 1 + \omega^n) = 0$$

when $n \equiv 1, 5 \pmod 6$. Hence $t - \omega$ divides the given polynomial. Since $t^2 + t + 1$ is the product of two factors of the type $t - \omega$, this too divides the given polynomial. The derivative of $(t + 1)^n - t^n - 1$ is $n[(t + 1)^{n-1} - t^{n-1}]$. When $n \equiv 1 \pmod 6$, this too vanishes for $t = \omega$ and hence is divisible by $t^2 + t + 1$. From these facts, the results can be obtained by setting $x = ty$.

**7.17.** The given polynomial can be written as a rational function with numerator

$$(t^4 - 1)^{k+1} + (t + 1)^{k+1} t^{4k-1}$$

and denominator $(t+1)^k$. It suffices to show that the numerator is divisible by $t^5 + 1$, or, equivalently, vanishes when we set $t^5 = -1$. This can be seen by expanding the numerator binomially and pairing the terms:

$$t^{4k+4} - \binom{k+1}{1} t^{4k} + \binom{k+1}{2} t^{4k-4} - \binom{k+1}{3} t^{4k-8} + \cdots + (-1)^{k+1}$$

$$+ t^{4k-1} + \binom{k+1}{1} t^{4k} + \binom{k+1}{2} t^{4k+1} + \binom{k+1}{3} t^{4k+2} + \cdots + t^{5k}.$$

**7.18.** Let $p$ be any prime exceeding 3. We show that $x^2 - x + 1$ divides $x^{2p} - x^p + 1$. Let $w$ be a zero of the quadratic. Then $w$ is a primitive 6th root of unity. Since $p \equiv 1$ or 5 (mod 6), $w^p$ is also a primitive 6th root of unity, so that $w$ is a zero of $x^{2p} - x^p + 1$, from which the divisibility follows.

Suppose that $m = kp$, with $p$ as above and $x = 2^k$. Then it follows that $4^k - 2^k + 1$ divides $4^m - 2^m + 1$.

**7.19.** $n$ is clearly odd. Let $n + 1 = 2r$. Then

$$4^n + n^4 = (2^n + n^2)^2 - n^2 2^{n+1} = (2^n + n2^r + n^2)(2^n - n2^r + n^2).$$

If $r = 2$ or 3, the number is easily cnecked to be composite. If $r \geq 4$, then $2^n + n^2 > 2^n = 2^{r-1}2^r > (2r-1)2^r$, so that both factors of the number exceed 1 and the number is composite. If $r = n = 1$, the number is prime.

**7.20.** Let $2k = m + 1$. Then

$$2^{2m} + 1 = (2^m + 1)^2 - 2^{2k} = (2^m + 2^k + 1)(2^m - 2^k + 1).$$

There are four cases to be considered. If $(k, m) \equiv (2,3)$ or $(3,1)$ (mod 4), then $2^m - 2^k + 1 \equiv 0$ (mod 5); if $(k, m) \equiv (1,1)$ or $(0,3)$ (mod 4), then $2^m + 2^k + 1 \equiv 0$ (mod 5). In any case, one of the factors is divisible by 5 and, when $m > 3$, it is straightforward to check that both factors exceed 5.

**7.21.** $f(x,y) = (x - y)q(x,y) \Rightarrow f(y,x) = (y - x)q(y,x) \Rightarrow q(x,y) = -q(y,x)$. Since $q(x,y)$ is a polynomial, this last equation persists for $y = x$, so that $q(x,x) = 0$. By the Factor Theorem, $q(x,y) = (x - y)r(x,y)$ and the result follows.

**7.22.** The equation can be rewritten

$$\begin{aligned} 0 &= nx^{n-1} + (n-1)x^{n-2} + \cdots + 2x - (n-1)(n+1) \\ &= [nx - (n+1)][x^{n-2} + 2x^{n-3} + 3x^{n-4} + \cdots + (n-1)], \end{aligned}$$

whence $x = (n+1)/n$ is a solution.

**8.1.** $yx = x^2 + 1 \Rightarrow x^2 - x^3 - x = x^2 - yx^2 = (1-y)x^2 = y^2x^2 = x^4 + 2x^2 + 1 \Rightarrow x^4 + x^3 + x^2 + x + 1 = 0$, whence the result follows.

**8.2.** If $b = 2a$, both sides of the equation are undefined. Let $b \neq 2a$. Now $a^3 + b^3 = (a + b)(a^2 - ab + b^2)$ and $a^3 + (a - b)^3 = (2a - b)[a^2 - a(a - b) +$

$(a-b)^2] = (2a-b)(a^2 - ab + b^2)$. If $a^2 + b^2 = ab$ (i.e. $a/b$ is a primitive 6th root of unity), the left side of the equation is undefined and the equation is not valid. Hence, the equation is valid as long as $b \neq 2a$ and $a^2 + b^2 \neq ab$.

**8.3.** An obvious solution is $x = 0$. Since, for every nonzero integer $x$,

$$\left(x^2 + \frac{x}{2}\right)^2 < x^4 + x^3 + x^2 + x + 1 < \left(x^2 + \frac{x}{2} + 1\right)^2,$$

$x$ cannot be even. If $x$ is odd, we must have that

$$x^4 + x^3 + x^2 + x + 1 = \left(x^2 + \frac{x}{2} + \frac{1}{2}\right)^2,$$

which reduces to $x^2 - 2x - 3 = 0$. The only other solutions are $x = -1$, 3.

**8.4.** *First solution.* Suppose $z$ is a common zero of the two polynomials. Then

$$z + 1 = z^5 = z(az + b)^2 = (a^4 - 3a^2 b + b^2)z + (a^3 b - 2ab^2).$$

Since $z$ is nonrational (why?),

$$a^4 - 3a^2 b + b^2 = 1 \qquad\qquad (1)$$

$$a^3 b - 2ab^2 = 1. \qquad\qquad (2)$$

Eliminate $b^2$ from (1) and (2) to get

$$5a^3 b = 2a^5 - 2a - 1. \qquad\qquad (3)$$

Squaring (3) and using (2) to eliminate $b^2$ gives

$$25a^5(a^3 b - 1) = 8a^{10} - 16a^6 - 8a^5 + 8a^2 + 8a + 2. \qquad\qquad (4)$$

Use (3) to eliminate $b$ and obtain

$$a^{10} + 3a^6 - 11a^5 - 4a^2 - 4a - 1 = 0. \qquad\qquad (5)$$

But (5) has no rational roots. Hence the assumption of a common zero leads to a contradiction.

*Second solution.* If $z$ is a common zero, then $z$ is a zero of the remainder $(a^4 - 3a^2 b + b^2 - 1)z - (a^3 b - 2ab^2 - 1)$ when the quintic is divided by the quadratic. This leads to (1), (2), which when solved as a linear system for $b$ and $b^2$ yields

$$5a^3 b = 2a^5 - 2a - 1 \qquad 5ab^2 = a^5 - a - 3.$$

Elimination of $b$ yields (5) and the argument proceeds as before.

**8.5.** *First solution.* Let $u$, $v$, $w$ be the zeros of $x^3 + ax^2 + 11x + 6$ and $u$, $v$, $z$ the zeros of $x^3 + bx^2 + 14x + 8$. Then

$$u + v + w = -a \quad uv + uw + vw = 11 \quad uwv = -6$$
$$u + v + z = -b \quad uv + uz + vz = 14 \quad uvz = -8$$

$$\Rightarrow \; w(u + v) = 11 - uv \quad z(u + v) = 14 - uv \quad 6z = 8w$$
$$\Rightarrow \; 6(14 - uv) = 8(11 - uv) \Rightarrow uv = 2$$
$$\Rightarrow \; w = -3, \; z = -4 \Rightarrow u + v = -3.$$

Hence $a = 6$, $b = 7$.

*Second solution.* The common zeros of two polynomials are zeros of their difference $(a - b)x^2 - 3x - 2$. Now

$$6 + 11x + ax^2 + x^3 = (2 + 3x + (b - a)x^2)(3 + x)$$
$$+ (4a - 3b - 3)x^2 + (1 + a - b)x^3.$$

Since $2 + 3x + (b-a)x^2$ divides $6 + 11x + ax^2 + x^3$, it follows that $4a - 3b - 3 = 1 + a - b = 0$, or $a = 6$, $b = 7$.

**8.6.** Since the $r - a_i$ are distinct nonzero integers and at most two can have the same absolute value, their absolute values arranged in increasing order are respectively at least equal to $1, 1, 2, 2, 3, 3, \ldots, n, n$. Hence

$$\Pi |r - a_i| \geq 1^2 \cdot 2^2 \cdots \cdot n^2 = (n!)^2.$$

However, since $\Pi(r - a_i) = (-1)^n(n!)^2$, equality actually occurs. Thus, the numbers $r - a_i$ are $+1, -1, +2, -2, \ldots, +n, -n$ in some order and $r - a_1 + r - a_2 + \cdots + r - a_{2n} = 1 - 1 + 2 - 2 + \cdots + n - n = 0$. The result follows.

**8.7.** Let $t^3 - mt^2 - mt - (m^2 + 1)$ have an integer zero $t$. Then $t$ must be such that the quadratic equation

$$m^2 + (t^2 + t)m - (t^3 - 1) = 0$$

has an integer solution $m$. The discriminant of this quadratic is

$$t^4 + 6t^3 + t^2 - 4 = (t^2 + 3t - 4)^2 + (24t - 20) = (t + 4)^2(t - 1)^2 + 4(6t - 5).$$

If $t \geq 1$, then

$$(t^2 + 3t - 4)^2 + (24t - 20) \geq (t^2 + 3t - 3)^2,$$

so that $t \leq 8$. If $t \leq -5$, then

$$(t^2 + 3t - 4)^2 + (24t - 20) \leq (t^2 + 3t - 5)^2,$$

so that $t \geq -15$. Thus, $-15 \leq t \leq 8$. Checking possibilities leads to the following polynomials:

| $m$ | $t^3 - mt^2 - mt - (m^2 + 1)$ |
|---|---|
| $-77$ | $t^3 + 77t^2 + 77t - 5930 = (t + 10)(t^2 + 67t - 593)$ |
| $-13$ | $t^3 + 13t^2 + 13t - 170 = (t + 10)(t^2 + 3t - 17)$ |
| $-7$ | $t^3 + 7t^2 + 7t - 50 = (t - 2)(t^2 + 9t + 25)$ |
| $-2$ | $t^3 + 2t^2 + 2t - 5 = (t - 1)(t^2 + 3t + 5)$ |
| $0$ | $t^3 - 1 = (t - 1)(t^2 + t + 1)$ |
| $1$ | $t^3 - t^2 - t - 2 = (t - 2)(t^2 + t + 1)$ |

**8.8.** Suppose, if possible, that $m$ is an integer zero of $f$. Let $k$ be any positive integer. Determine an integer $c$ so that $r = m - kc$ is one of $\{1, 2, 3, \ldots, k\}$. Then $f(r) = f(m - kc) \equiv f(m) = 0 \pmod{k}$. The result follows by a contradiction argument.

**8.9.** For some polynomial $g(t)$,

$$f(t) - 12 = (t - a)(t - b)(t - c)(t - d)g(t).$$

If $f(k) = 25$, then substituting $k$ into this equation yields a representation of 13 as the product of at least four distinct integers, an impossibility.

**8.10.** Since $m + n + k = 0$, 1 is a zero of the quadratic. The other zero must be $k/m = -1 - n/m$.

**8.11.** We must have

$$(ad - bc)(x - p)^2 = (5d - b)x^2 + (8d - 10b)x + (14d - 7b).$$

The discriminant of the right side must vanish, so that

$$(4d - 5b)^2 = 7(5d - b)(2d - b),$$

which simplifies to $(3d - 2b)(2d + b) = 0$. Similarly,

$$(ad - bc)(x - q)^2 = (a - 5c)x^2 + (10a - 8c)x + (7a - 14c)$$

yields $(3c - 2a)(2c + a) = 0$.

Further information is obtained by comparing quadratic and constant coefficients in the two equations:

$$\begin{array}{ll} a + b = 5 & c + d = 1 \\ ap^2 + bq^2 = 14 & cp^2 + dq^2 = 7. \end{array}$$

Experimenting with $|p| = 1$, $|q| = 2$ leads to the possibility

$$(a, b, c, d, p, q) = (2, 3, -1, 2, 1, -2).$$

**8.12.** Observe that

$$
\begin{aligned}
0 &= -w^7 - x^7 - y^7 - z^7 = (x + y + z)^7 - x^7 - y^7 - z^7 \\
&= 7(x + y)(y + z)(z + x)[(x^2 + y^2 + z^2 + xy + yz + zx)^2 \\
&\quad + xyz(x + y + z)].
\end{aligned}
$$

Now,

$$
\begin{aligned}
4(x^2 &+ y^2 + z^2 + xy + yz + zx)^2 + 4xyz(x + y + z) \\
&= [(x + y)^2 + (y + z)^2 + (z + x)^2]^2 - 4xyzw \\
&= [(w + z)^2 + (w + x)^2 + (w + y)^2]^2 - 4xyzw \\
&= [3w^2 + 2w(x + y + z) + x^2 + y^2 + z^2]^2 - 4xyzw \\
&= [w^2 + x^2 + y^2 + z^2]^2 - 4xyzw \\
&\geq [2|wx| + 2|yz|]^2 - 4xyzw \\
&\geq [4(|wxyz|)^{1/2}]^2 - 4xyzw \\
&\geq 12|xyzw| \geq 0
\end{aligned}
$$

by a double application of the Arithmetic–Geometric Mean Inequality (Exercise 1.2.17), with equality iff $|x| = |y| = |z| = |w| = 0$. Otherwise, at least one of $x + y$, $y + z$, $z + x$ vanishes. In any case, $w(w + x)(w + y)(w + z)$ assumes only the value 0.

**8.13.** This follows directly from Exercise 2.12. A generalization to four variables appears in *Crux Mathematicorum* **2** (1976), 180.

**8.14.** (a) Using the third equation to eliminate $c$ from the other two yields

$$
a(x^2z^2 - xy^2z) + b(xy^3 + y^2z^2) = 0
$$

$$
a(x^2z^2 + x^3y) + b(y^2z^2 - x^2yz) = 0.
$$

Eliminating $a$ and taking account of $bxy \neq 0$ leads to

$$
x^3z^3 + y^3z^3 + x^3y^3 + x^2y^2z^2 = 0.
$$

(b) Determining $abc$ by multiplying the right sides gives

$$
abc = 2abc + a^2bx^2z^{-2} + a^2cx^2y^{-2} + b^2cx^{-2}y^2 + ab^2y^2z^{-2}
$$

$$
+ ac^2z^2y^{-2} + bc^2x^{-2}z^2.
$$

Multiplying by $xyz$ gives

$$
\begin{aligned}
0 &= abcxyz + a^2x^3(byz^{-1} + czy^{-1}) + b^2y^3(czx^{-1} + axz^{-1}) \\
&\quad + c^2z^3(axy^{-1} + byx^{-1}) \\
&= abcxyz + a^3x^3 + b^3y^3 + c^3z^3.
\end{aligned}
$$

(c) Cubing the three equations and adding (taking account of (a) and (b)) yields

$$
\begin{aligned}
a^3 + b^3 + c^3 &= b^3 y^3 (z^{-3} + x^{-3}) + c^3 z^3 (x^{-3} + y^{-3}) \\
&\quad + a^3 x^3 (y^{-3} + z^{-3}) + 9abc \\
&= b^3 y^3 (-x^{-1} y^{-1} z^{-1} - y^{-3}) + c^3 z^3 (-x^{-1} y^{-1} z^{-1} - z^{-3}) \\
&\quad + a^3 x^3 (-x^{-1} y^{-1} z^{-1} - y^{-3}) + 9abc \\
&= -(a^3 + b^3 + c^3) - x^{-1} y^{-1} z^{-1} (a^3 x^3 + b^3 y^3 + c^3 z^3) + 9abc,
\end{aligned}
$$

whence the result.

**8.15.** Multiplying out the left side yields

$$
3 + \frac{xz(z-x) + xy(x-y)}{yz(y-z)} + \frac{zy(y-z) + xy(x-y)}{xz(z-x)} + \frac{xz(z-x) + yz(y-z)}{xy(x-y)}
$$

$$
= 3 + [(x/yz)(x - y - z) + (y/xz)(-x + y - z) + (z/xy)(-x - y + z)]
$$

$$
= 3 + 2(x^3 + y^3 + z^3)/(xyz) = 9
$$

by Exercise 1.5.9(a).

**8.16.** Let each quantity be equal to $u$. Then

$$
x^3 - myz = ux^2 \quad \text{and} \quad y^3 - mxz = uy^2.
$$

Subtract and divide by $x - y$ to get

$$
x^2 + xy + y^2 + mz = u(x + y).
$$

Similarly, $z^2 + zy + y^2 + mx = u(y + z)$. Subtract and divide by $x - z$ to get $x + y + z - m = u$.

**8.17.** Since $0 = (r + 1)^3 + c(r + 1)^2 + d(r + 1) + 1 = r^3 + (3 + c)r^2 + (3 + 2c + d)r + (2 + c + d)$, $r$ is a zero of $x^3 + (3 + c)x^2 + (3 + 2c + d)x + (2 + c + d)$. Since $P(x)$ is irreducible, this cubic must coincide with $P(x)$ and so $a = 3 + c$, $b = 3 + 2c + d$, $-1 = 2 + c + d$, whence $b = c$ and $a = 3 + b$. Let $s$ be a second zero of $P(x)$; the third zero is $1/rs$, and we obtain

$$
-3 = b - a = rs + r^{-1} + s^{-1} + r + s + r^{-1} s^{-1}.
$$

Hence

$$
\begin{aligned}
0 &= (r^2 + r)s^2 + (r^2 + 3r + 1)s + (r + 1) \\
&= [(r + 1)s + 1][rs + (r + 1)],
\end{aligned}
$$

from which it follows that $s = -(r + 1)^{-1}$ or $-(r + 1)r^{-1}$.

**8.18.** $\theta$ satisfies the equation

$$
0 = 1 + \theta^m = (1 + \theta)(1 - \theta + \theta^2 - \theta^3 + \cdots - \theta^{m-2} + \theta^{m-1}).
$$

Since $\theta \neq 1$, it is the second factor which vanishes and we obtain

$$
\begin{aligned}
1 &= 1 - \theta + \theta^2 - \theta^3 + \cdots + \theta^{m-1} - \theta^m \\
&= (1 - \theta)(1 + \theta^2 + \cdots + \theta^{m-1}).
\end{aligned}
$$

Hence,

$$(1 - \theta)^{-1} = 1 + \theta^2 + \theta^4 + \cdots + \theta^{m-1} = \theta + \theta^3 + \theta^5 + \cdots + \theta^{m-2}.$$

**8.19.** (a) If $p(x) = x^2 + bx + c$, $q(x) = x^2 + dx + e$, choose $h$ and $k$ such that $2h + b = d$ and $h^2 + bh + c - k = e$.

(b) The reducible monic quadratics are of the form $(x - r)(x - s)$, where $r, s \in \mathbf{F}$; there are $m(m+1)/2$ of these. The total number of monic quadratics is $m^2$, so there are $m(m - 1)/2$ irreducible quadratics. From (a), it can be seen that for each of the $m$ choices of linear coefficient, the number of irreducible quadratics is the same. Hence, for a given choice, there are $(m - 1)/2$ irreducible quadratics for appropriate values of $k$.

**8.20.** If the four terms are $a - d$, $a$, $a + d$, $a + 2d$, we obtain

$$(a - d)a(a + d)(a + 2d) + d^4 = (a^2 + ad - d^2)^2.$$

If $a^2 + ad - d^2 = b^2$, then $(2a + d)^2 = 4b^2 + 5d^2$. For example, $b = 1$, $d = 3$ leads to a fourth power. A more general solution appears in *Amer. Math. Monthly* **57** (1950), 186.

**8.21.** (b) Suppose $f = u_1^{a_2} u_2^{a_2} \cdots u_k^{a_k}$, where the $u_i$ are irreducible with any pair coprime and $a_i \geq 1$, and $g = u_1^{b_1} u_2^{b_2} \cdots u_k^{b_k}$, where $b_i \geq 0$. Since $f = g^m h$ with $g$ not dividing $h$, it follows that $a_i \geq mb_i$ for each $i$ and that, for some irreducible factor, say $u_1$, $a_1 < (m+1)b_1$. Now $u_1^{a_1-1}$ but not $u_1^{a_1}$ divides $f'$. Since $a_1 - 1 < (m + 1)b_1 - 1 < (m + 1)b_1$, it follows that $g^{m+1}$ does not divide $f'$.

**8.22.** By de Moivre's Theorem (Exercise 1.3.8), $\cos n\theta + i \sin n\theta = u(\cos \theta, \sin^2 \theta) + i \sin \theta v(\cos \theta, \sin^2 \theta)$, for some polynomials $u$ and $v$. Let $x = \cos \theta$. Then the result holds with $f(x) = T_n(x) = \cos n\theta$ and $g(x) = v(x, 1 - x^2)$.

**8.23.** Since $x^k - 1 = \Pi\{Q_d(x) : d|k\}$, $\{m\}!$ is a product of factors $Q_d(x)$, where each $Q_d(x)$ occurs as often as $d$ divides a number in the set $\{1, 2, \ldots, m\}$; the exponent of $Q_d(x)$ is $\lfloor m/d \rfloor$. It suffices to show that, for each positive integer $d$,

$$\lfloor (m + n)/d \rfloor + \lfloor m/d \rfloor + \lfloor n/d \rfloor \leq \lfloor 2m/d \rfloor + \lfloor 2n/d \rfloor.$$

Let $m = ud + r$, $n = vd + s$ where $0 \leq r, s < d$. Then the difference between the right and left sides is

$$\lfloor 2r/d \rfloor + \lfloor 2s/d \rfloor - \lfloor (r + s)/d \rfloor.$$

This is clearly nonnegative if $0 \leq r+s < d$. If $d \leq r+s < 2d$, then at least one of $2r$ and $2s$ is no less than $d$ and the expression is again nonnegative.

**8.24.** *First solution.* $(x, y, z$ real$)$. Let $x = \tan u$, $y = \tan v$, $z = \tan w$. Then $(1 - xy - yz - zx)\tan(u + v + w) = x + y + z - xyz$. Observe first that $xy + yz + zx - 1$ and $x + y + z - xyz$ cannot vanish simultaneously. Otherwise, $x + y + z = z(1 - xz - yz)$ so that $0 = (x + y)(1 + z^2)$. Thus, we can see that the sum of any two variables is 0, which is impossible. Hence $x + y + z = xyz$ iff $\tan(u + v + w) = 0$ iff $u + v + w$ is a multiple of $\pi$.

Suppose $x + y + z = xyz$. Then $u + v + w$ is a multiple of $\pi$, so that $2u + 2v + 2w$ is a multiple of $\pi$. Then $\tan 2u + \tan 2v + \tan 2w = (\tan 2u)(\tan 2v)(\tan 2w)$, and the result follows.

*Second solution.* Putting the left side over a common denominator yields the numerator

$$2(x + y + z) + 2xyz(xy + yz + zx) - 2[(x + y + z)(xy + yz + zx) - 3xyz]$$

which, under the stated conditions, reduces to $8xyz$.

**8.25.**

$$d_1 = (x_2 y_3 - x_3 y_2)(x_3 x_1 + y_3 y_1)(x_1 x_2 + y_1 y_2)$$

$$= x_1 x_2 x_3(x_1 x_2 y_3 - x_1 y_2 x_3) + y_1 y_2 y_3(y_1 x_2 y_3 - y_1 y_2 x_3)$$

$$+ x_1 y_1(x_2^2 y_3^2 - x_3^2 y_2^2).$$

Obtain a similar expression for $d_2$ and $d_3$, and check the required identity directly.

**8.26.** (a) $a = b + 1$ iff 1 is a zero of the quadratic; $a + b + 1 = 0$ iff $-1$ is a zero. Suppose the quadratic has a zero $u$ which is a nonreal root of unity. Then $\bar{u} = u^{-1}$ is also a zero of the quadratic. Since the product of the zeros is 1, $b = 1$. Since $|u + u^{-1}| < 2$, it follows that $(a, b) = (-1, 1), (0, 1), (1, 1)$; these possibilities yield quadratics whose zeros are primitive cube, 4th and 6th roots of unity respectively.

(b) Let $f(t)$ be the first quadratic and $g(t)$ be the second. If $f(1) = 0$, then $(b + 1)^2 = a^2$ and either $g(1)$ or $g(-1)$ vanishes. If $f(-1) = 0$, then $a^2 + (1 - b)^2 = 0$ so that $a = 0$, $b = 1$ and $g(i) = 0$. Otherwise $b^2 = 1$ and $a^2 - 2b = -1, 0, 1$. We must have $(a, b) = (-1, 1), (1, 1)$ and the result follows.

**8.27.** We have that $au = -bv - c$ and $a\bar{u} = -b\bar{v} - c$, whence $a^2 = b^2 + c^2 + 2bc \cos \theta$, where $\theta = 2k\pi/n$ for some $k, n \in \mathbf{N}$ and $v = \cos \theta + i \sin \theta$. Thus, $\cos \theta$ is a rational number; write it as $p/q$ in lowest terms with $q$ positive.

Suppose, if possible, that $q \geq 3$. Since $\cos 2\theta = (2p^2 - q^2)/q^2$ and the greatest common divisor of $2p^2 - q^2$ and $q^2$ is either 1 or 2, $\cos 2\theta$ has a denominator in lowest terms at least equal to $\frac{1}{2}q^2 > q$. We find that $\cos \theta$, $\cos 2\theta$, $\cos 4\theta$, $\cos 8\theta$, ... are all rationals, each of whose denominators in lowest terms exceeds that of its predecessor. Hence, these cosines are all unequal. However, let $n = 2^s m$ with $m$ odd. Since the powers of 2 cannot be all incongruent modulo $m$, there exist $i, j \in \mathbf{N}$ with $i > j$ and $2^i \equiv 2^j \pmod{m}$, whence $2^j(2^{i-j} - 1) \equiv 0 \pmod{m}$. Let $r = i - j$. Then $2^{r+s}\theta - 2^s\theta = 2^s(2^r - 1)2k\pi/m$ is a multiple of $2\pi$ and $\cos 2^{r+s}\theta = \cos 2^s\theta$, yielding a contradiction.

Therefore, the only possible values of $\cos\theta$ are $0$, $\pm\frac{1}{2}$ and $\pm 1$. Since $v$ is not real, $v$ must be a nonreal fourth or sixth root of unity. Similarly, so is $u$. Suppose, if possible, that $v = i$. Then $a^2 u^2 = c^2 - b^2 + 2bci$ and $a^3 u^3 = (3b^2 c - c^3) + (b^3 - 3bc^2)i$. Since $bc(b^2 - 3c^2) \neq 0$, $u^2$ and $u^3$ are nonreal and so $u$ is neither a fourth nor sixth root of unity. Hence $v \neq i$; similarly, $v \neq -i$ and $u \neq \pm i$. Hence $u$ and $v$ are both nonreal sixth roots of unity and we have one of $u = v$, $u = -v$, $u = v^2$, $u = -v^2$. Since the first two possibilities would imply that $u$ was rational, we must have $\pm av^2 + bv + c = 0$. Since the monic irreducible polynomial with zero $v$ is $t^2 \pm t + 1$, the required result follows.

**8.28.** *First solution.* Let $u = \cos(\pi/14) + i\sin(\pi/14)$, so that $2ix = u - u^{-1}$. Since $iu = \cos(4\pi/7) + i\sin(4\pi/7)$, $iu$ is a zero of $t^6 + t^5 + t^4 + t^3 + t^2 + t + 1$, whence

$$(-u^6 + u^4 - u^2 + 1) + i(u^5 - u^3 + 1) = 0$$

or

$$-(u^3 - u^{-3}) + i(u^2 + u^{-2}) + (u - u^{-1}) - i = 0.$$

Since $-8ix^3 = (u^3 - u^{-3}) - 3(u - u^{-1})$ and $-4x^2 = (u^2 + u^{-2}) - 2$, we can substitute for $x$, divide by $i$ and obtain the required equation.

*Second solution.* Applying de Moivre's Theorem and using $\cos^2(\pi/14) = 1 - x^2$ yields

$$
\begin{aligned}
1 = \sin\pi/2 &= 7(1 - x^2)^3 x - 35(1 - x^2)^2 x^3 + 21(1 - x^2)x^5 - x^7 \\
&= 7x - 56x^3 + 112x^5 - 64x^7.
\end{aligned}
$$

This can be manipulated to

$$0 = (x + 1)(8x^3 - 4x^2 - 4x + 1)^2.$$

Since $x \neq -1$, the result follows.

**8.29.** Let $v = \sin\pi/7$. The length of the side of the heptagon is $2v$. We have, by de Moivre's Theorem,

$$0 = \sin\pi = 7v - 56v^3 + 112v^5 - 64v^7.$$

Setting $x = 2v$ yields the result. Other roots of the equation are $v = \pm\sin(2\pi/7)$, $\pm\sin(3\pi/7)$, so that the roots of the equation in $x$ are the lengths of the sides and diagonals and their negatives.

A similar problem posed for a regular undecagon (11-gon) is Problem 2864, *Amer. Math. Monthly* **27** (1920), 482; **28** (1922), 91.

**8.30.** We have the following table of solutions:

| Modulo | the congruence is satisfied by $n$ congruent to | |
|---|---|---|
| $3^3 = 27$ | 1 | $(\bmod 3)$ |
| $3^4 = 81$ | 1, 4 | $(\bmod 9)$ |
| $3^5 = 243$ | 19 | $(\bmod 27)$ |
| $3^6 = 729$ | 46 | $(\bmod 81)$ |
| $3^7 = 2187$ | 208 | $(\bmod 243)$ |

Suppose $m \geq 5$ and $n \equiv a \pmod{3^{m-2}}$ satisfies $f(n) \equiv 0 \pmod{3^m}$. Then $a \equiv 1 \pmod 9$ and $f(a) = 3^m k$. Let $n = a + 3^{m-2}b$, so that $n^2 \equiv a^2 + 3^{m-2}2ab$ and $n^3 \equiv a^3 + 3^{m-1}a^2 b \pmod{3^{m+1}}$; then

$$f(n) \equiv f(a) + 3^{m-1}a^2 b + 3^{m-1} \cdot 264ab - 3^{m-1} \cdot 37b \pmod{3^{m+1}}.$$

Hence $f(n) \equiv 0 \pmod{3^{m+1}} \iff 3k + (a^2 + 264a - 37)b \equiv 0 \pmod 9$ $\iff 3k + (a - 4)(a - 2)b \equiv 0 \pmod 9$. Since $a \equiv 1 \pmod 9$, $a - 4 = 3u$, where $u$ is not divisible by 3, so that

$$f(n) \equiv 0 \pmod{3^{m+1}} \iff k + u(a - 2)b \equiv 0 \pmod 3.$$

The last congruence is uniquely solvable for $b$ modulo 3.

Since $f(n) \equiv 0 \pmod{3^m}$ has a unique solution $a$ modulo $3^{m-2}$ (and therefore nine solutions $a + c3^{m-2}$, where $0 \leq c \leq 8$, modulo $3^m$), it follows that $f(n) \equiv 0 \pmod{3^{m+1}}$ has a unique solution $a + 3^{m-2}b$ modulo $3^{m-1}$, and the required result follows by induction.

**8.31.** Suppose such a representation were possible. Then $g(t)$ cannot be constant and $tg(t)f(t + 1) - tg(t + 1)f(t) = g(t)g(t + 1)$. Since $f(t)$ and $g(t)$ have no common divisor of positive degree, $g(t) \mid tg(t + 1)$, so that $tg(t + 1) = u(t)g(t)$. By considering degrees and leading coefficients, and by noting that $g(t)$ is not constant, we see that $u(t) = t - c$ where $c \neq 0$. Hence $g(t)$ has zeros $c + k$ for $k = 1, 2, \ldots$, which contradicts $g(t)$ having positive degree.

# Answers to Exercises

## Chapter 4

**1.1.** $(x, y, z) = (7, 9, 13)$.

**1.3.** $(x, y, z) = (10, 4, -6)$, $(-10, -4, 6)$.

**1.4.** If $(x, y, z)$ satisfies

$$ax + by + cz = px + qy + rz = 0, \qquad (*)$$

then by Exercise 4.2, $x : y : z = (br - cq) : (cp - ar) : (aq - bp)$. If there is a solution of the form $(x, y, z) = (u^2, u, 1)$, then (a) follows, and $y^2 = xz$ shows that the condition (b) is necessary.

If (b) holds, then for every solution of system $(*)$, $y^2 = xz$. If there is a solution with nonzero $z$, then there is a solution with $z = 1$; if $y = u$, then $x = u^2$ and the two quadratic equations have a common root. On the other hand, if it turns out that $z = 0$, then $y = 0$. Since $ap \neq 0$, $x = 0$, so 0 is a common root of the two quadratic equations.

**1.5.** (a) $12(b - 4a) = (a - b)^2$.

**1.6.** (b) Since $x^{-1} : y^{-1} : z^{-1} = (qr - p^2) : (pr - q^2) : (pq - r^2)$, we obtain $a(qr - p^2) + b(pr - q^2) + c(pq - r^2) = 0$.

**1.8.** $x$, $y$, $z$ are the zeros of $t^3 - 12t^2 + 41t - 42$ and are therefore 2, 3, 7 in some order.

**1.9.** (a) 3, 5, 7. (b) 1, $-2$, 3.

**1.10.** $(x, y) = (u, u)$ or $(-4v, v)$ where $u = 1$ or $(-2 \pm i\sqrt{3})/7$ and $v = 2$ or $(-1 \pm i\sqrt{55})/14$.

**1.11.** $(x, y, z) = (6, 8, 2)$.

**2.1.** (g) $x = 7$ satisfies (d), but not (a), (b), (c).

(h) Consider an equation of the type $U = V$. This implies $U^2 = V^2$. However, if $U^2 = V^2$, then either $U = V$ or $U = -V$. Hence, $U^2 = V^2$ is valid under a wider range of circumstances than is $U = V$. We have that (a) is equivalent to the given equation; (a) $\Rightarrow$ (b); (b) $\Longleftrightarrow$ (c); (c) $\Rightarrow$ (d).

(i) $\sqrt{x + 2} - \sqrt{x - 3} = 1$.

**2.2.** The equation becomes $cy^2 + aky + (ad - bc) = 0$. If this equation has a nonnegative solution $y$, then there is exactly one corresponding solution $x$ to the given equation. Any negative or nonreal solution $y$ yields no solution $x$.

If $c = 0$, there is a unique solution $y = -d/k$ and a solution $x$ when $d/k < 0$.

If $c \neq 0$, there are no solutions $x$ if $a^2 k^2 < 4c(ad - bc)$ or if $a^2 k^2 \geq 4c(ad - bc) \geq 0$ and $ak/c > 0$. If $a^2 k^2 > 4c(ad - bc) \geq 0$ and $ak/c < 0$, then there are two nonnegative solutions $y$ and two solutions $x$. If $a^2 k^2 = 4c(ad - bc)$ and $ak/c \leq 0$, or if $a^2 k^2 > 4c(ad - bc)$ and $4c(ad - bc) < 0$, then there is one nonnegative solution $y$ and one solution $x$.

Discussion of problems of this type can be found in Goro Nagase, Existence of real roots of a radical equation, *Math. Teacher* **80** (1987), 369–370.

**2.3.** (a) $y = \sqrt{x + 4}$ leads to $y^2 - y - 6 = 0$. The solution $y = 3$ corresponds to $x = 5$; $y = -2$ is extraneous.

(b) $y = \sqrt{x + 1}$ leads to $y^2 - 4y + 3 = 0$, and $(x, y) = (8, 3), (0, 1)$.

(c) $y = \sqrt{x + 1}$ leads to $y^2 + 4y + 3 = 0$. There are no solutions $x$.

**2.4.** Both equations lead to $0 = (x - 2)(4x - 7)$.
 (a) $2, 7/4$; (b) no solutions.

**2.5.** (b) Let $u = 14 + x$, $v = 14 - x$. Apply (a) to obtain $w = 4$ and $64 = 28 + 12\sqrt[3]{196 - x^2}$. This simplifies to $27 = 196 - x^2$, whence $x = \pm 13$. Both are valid solutions.

**2.6.** Set $y = x^2 + 18x + 45$ and obtain $y - 15 = 2\sqrt{y}$. There are two roots $y = 9$ and $y = 25$. The first is extraneous; the second leads to a quadratic equation which can be solved for $x$.

**2.7.** (a) Since $\sqrt{x^2 + 9} > x$ for all real $x$, the left side of the equation is always strictly positive.
 (b) Suppose that the equation is solvable. Then

$$x^2 + b = (1 - x)^2 = 1 - 2x + x^2$$

so that $2x = 1 - b$. Testing this out, we find that

$$1 - x + \sqrt{x^2 + b} = (1/2)[1 + b + \sqrt{(1 + b)^2}]$$

$$= \begin{cases} (1/2)[1 + b + (1 + b)] = 1 + b \neq 0 & \text{when } b > -1 \\ (1/2)[1 + b - (1 + b)] = 0 & \text{when } b \leq -1. \end{cases}$$

Thus, there is no solution when $b > -1$ and $x = (1 - b)/2$ is a solution when $b \leq -1$.

**3.2.** (a) See Exercises 1.2.4, 1.4.4, 1.4.11, 1.4.12, 1.4.13.
 (b) Let the reciprocal polynomial have the variable $t$. If it has even degree $2k$, the substitution $x = t + t^{-1}$ leads to a polynomial equation of degree $k$ in $x$. If $k \leq 4$, then, by (a), the number of solutions, counting multiplicity is, $k$. Each of the solutions gives rise to two values of $t$, namely the roots of $t^2 - xt + 1 = 0$. If the reciprocal polynomial has odd degree, it is the product of $t + 1$ and a polynomial of even degree, and the result follows.

**3.3.** (a) The solutions for various $n$ are: $2 : 2, -2$; $3 : -1, -1, 2$; $4 : -2, 0, 0, 2$; $6 : -2, -1, -1, 1, 1, 2$; $8 : -2, -\sqrt{2}, -\sqrt{2}, 0, 0, \sqrt{2}, \sqrt{2}, 2$.
 (c) If $t_1$ and $t_2$ are the roots of $t^2 - yt + 1 = 0$, then, for each $n \geq 3$, $t_i^n = yt_i^{n-1} - t_i^{n-2}$, from which the result follows.
 (f) The result is true for $n = 1, 2$. Suppose it holds up to $n = k \geq 2$. Then

$$\begin{aligned} 2\cos(k + 1)\theta &= y(2\cos k\theta) - (2\cos(k - 1)\theta) \\ &= yf_k(2\cos\theta) - f_{k-1}(2\cos\theta) = f_{k+1}(2\cos\theta). \end{aligned}$$

**3.5.** The four equations are $a = w + v - u^2$; $b = z + uw - uv$; $c = uz + vw$; $d = vz$. Eliminating $w$ yields the system $b = z + au + u^3 - 2uv$; $c = uz + av + u^2v - v^2$; $d = vz$. Eliminating $z$ yields the system $c = bu + 3u^2v -$

$au^2 - u^4 + av - v^2$; $d = bv + 2uv^2 - auv - u^3v$. Rearranging terms, we find
that

$$v^2 - (3u^2 + a)v + (u^4 + au^2 - bu + c) = 0$$

$$2uv^2 + (b - au - u^3)v - d = 0$$

whence $(b + ua + 5u^3)v - (2u^5 + 2au^3 - 2bu^2 + 2cu + d) = 0$.

Eliminating $v$, we find that

$$(2u^5 + 2au^3 - 2bu^2 + 2cu + d)^2 - (3u^2 + a)(2u^5 + 2au^3 - 2bu^2$$

$$+ 2cu + d)(5u^3 + au + b) + (u^4 + au^2 - bu - c)(5u^3 + au + b)^2 = 0,$$

an equation in $u$ of degree 10 which cannot in general be reduced to an
equation of lower degree.

**3.6.** (a) Suppose, if possible, that $\sqrt{2} = a + b\sqrt{3}$. Then $2 = (a^2 + 3b^2) + 2ab\sqrt{3}$. Since $\sqrt{3}$ is not rational, $ab = 0$. But this cannot occur. If $3^{1/3} = a + b\sqrt{3}$, then $3 = (a^3 + 9ab^2) + 3b(a^2 + b^2)\sqrt{3}$. Hence $b = 0$, which is impossible. Since $\mathbf{Q}(\sqrt{3}) \subseteq \mathbf{R}$, $i \notin \mathbf{Q}(\sqrt{3})$.

(b) Suppose $i = a + b\sqrt{-3}$. Then $-1 = (a^2 - 3b^2) + 2abi\sqrt{3}$, so $ab = 0$. But either $a = 0$ or $b = 0$ is impossible.

(c) $\sqrt{2} \in \mathbf{C}$ but $\sqrt{2} \notin \mathbf{Q}(i)$.

(e) $(a + b\sqrt{d})^{-1} = (a^2 - b^2d)^{-1}(a - b\sqrt{d})$.

(f) Any field which contains $\mathbf{Q}$ and $\sqrt{d}$ must contain all the elements of the form $a + b\sqrt{d}$, where $a, b \in \mathbf{Q}$, hence must contain $\mathbf{Q}(\sqrt{d})$. On the other hand, $\mathbf{Q}(\sqrt{d})$ is itself a field, and so the result follows.

(h) It suffices to choose $b$, $c$ such that $b^2 - 4c$ is equal to $d$ multiplied by a square, say $b^2 - 4c = 4d$. This can be arranged if $c + d = 1$, $b = 2$. An example is $t^2 + 2t + (1 - d)$.

**3.7.** (b)

$$(a + b\sqrt{2} + c\sqrt{3} + d\sqrt{6})^{-1} = [(a^2 + 2b^2 - 3c^2 - 6d^2)^2 - 8(ab - 3cd)^2]^{-1}$$

$$[(a + b/\sqrt{2}) - \sqrt{3}(c + d\sqrt{2})][(a^2 + 2b^2 - 3c^2 - 6d^2) - 2\sqrt{2}(ab - 3cd)].$$

**3.8.** (a)

$$(a + 2^{1/4}b + 2^{1/2}c + 2^{3/4}d)^{-1} = [(a^2 - 4bd + 2c^2)^2 - 2(2ac - b^2 - 2d^2)^2]^{-1}$$

$$[(a + 2^{1/2}c) - 2^{1/4}(b + 2^{1/2}d)][(a^2 - 4bd + 2c^2) - 2^{1/2}(2ac - b^2 - 2d^2)].$$

(b)

$$(1 + 2^{1/4} + 2^{1/2} + 2^{3/4})^{-1} = -(1 - 2^{1/4})$$

$$(1 + 2^{3/4})^{-1} = (1 - 4.2^{1/4} + 2.2^{1/2} - 2^{3/4})/(-7).$$

**3.9.** (a) Let $f(t) = at^3 + bt^2 + ct + k$ and suppose that $u \in \mathbf{F}(\sqrt{d})$ is a zero of $f(t)$. If $u \in \mathbf{F}$, then the result follows. Otherwise, $u = v + w\sqrt{d}$, with $v, w \in \mathbf{F}$, $w \neq 0$, and

$$0 = [a(v^3 + 3vw^2d) + b(v^2 + w^2d) + cv + k]$$

$$+ \left[a(3v^2 w + w^3 d) + 2bvw + cw\right]\sqrt{d}.$$

Since $\sqrt{d} \notin \mathbf{F}$, it follows that both expressions in square brackets must vanish, resulting in $0 = f(v - w\sqrt{d})$. The third zero of the cubic is $-(b/a) - 2v$, a member of $\mathbf{F}$.

(b) By (a), we see that $g(t)$ has a zero in $\mathbf{F}_{n-1}$, hence in $\mathbf{F}_{n-2}$, etc.

**3.10.** (b) Using $\zeta^6 = -\zeta^3 - 1$, it is straightforward to show that $\mathbf{Q}(\zeta)$ is closed under addition, subtraction and multiplication. As for the inverse of $\zeta^3 + \zeta$, by the Euclidean algorithm, we obtain

$$(t^6 + t^3 + 1) = (t^3 - t + 1)(t^3 + t) + (t^2 - t + 1)$$

$$(t^3 + t) = (t + 1)(t^2 - t + 1) + (t - 1)$$

$$(t^2 - t + 1) = t(t - 1) + 1.$$

Hence

$$
\begin{aligned}
1 &= (t^2 - t + 1) - t[(t^3 + t) - (t + 1)(t^2 - t + 1)] \\
&= (t^2 + t + 1)(t^2 - t + 1) - t(t^3 + t) \\
&= (t^2 + t + 1)[(t^6 + t^3 + 1) - (t^3 - t + 1)(t^3 + t)] - t(t^3 + t) \\
&= (t^2 + t + 1)(t^6 + t^3 + 1) - (t^5 + t^4 + t + 1)(t^3 + t).
\end{aligned}
$$

Setting $t = \zeta$ yields

$$1 = -(\zeta^5 + \zeta^4 + \zeta + 1)(\zeta^3 + \zeta),$$

whence $(\zeta^3 + \zeta)^{-1} = -(\zeta^5 + \zeta^4 + \zeta + 1)$.
Similarly, $\zeta^{-1} = -(\zeta^5 + \zeta^2)$.

(e) There are many ways of verifying the zeros. For example,

$$f(u^2 - 2) = u^6 - 6u^4 + 9u^2 - 1 = (3u - 1)^2 - 6u^4 + 9u^2 - 1$$

$$= -6u(u^3 - 3u + 1) = 0.$$

Since the sum of the three zeros is 0, the third zero must be $2 - u - u^2$. The reciprocal of any element in $\mathbf{Q}(u)$ can be found by using the Euclidean algorithm as outlined in (b).

**4.2.** (b) The corresponding real system is $2x - y - 3 = 0$, $x + 2y + 4 = 0$.

**4.3.** The corresponding real system is $px - qy + r = 0$, $qx + py + s = 0$. The lines are perpendicular and hence intersect in exactly one point.

**4.5.** (e) The corresponding real system is

$$(x + 1/2)^2 - (y + 1)^2 = 1/4$$

$$(2x + 1)(y + 1) = 0$$

and the points of intersection are $(-1, -1)$ and $(0, -1)$.

**4.8.** (a) The equivalent real system is

$$x^3 - 3xy^2 - 7x + 6 = 0$$
$$y(3x^2 - y^2 - 7) = 0.$$

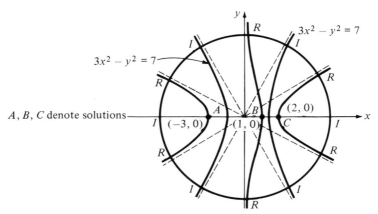

(b) The equivalent real system is

$$x^3 - 3xy^2 = x + 1$$
$$3x^2y - y^3 = y$$

which can be written

$$y^2 = (1/3)(x^2 - 1 - 1/x)$$
$$y(3x^2 - y^2 - 1) = 0.$$

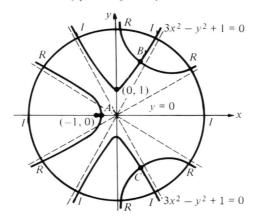

**4.9.** (b) Note that $(3x^2 - 1)^2 - (8x^4 - 4x^2 - 3x + 3) = x^4 - 2x^2 + 3x - 2 = (x - 1)(x + 2)(x^2 - x + 1)$, from which the sign of this difference is easily determined.

When $x < -2$,

$$0 < 4x^2(2x^2 - 1) < 8x^4 - 4x^2 - 3x + 3 < (3x^2 - 1)^2;$$

when $-2 < x < 1$,

$$0 \le (3x^2 - 1)^2 < 8x^4 - 4x^2 - 3x + 3;$$

when $1 < x$,

$$0 < x^2 + 3 = 8x^2 - 4x^2 - 3x^2 + 3 < 8x^4 - 4x^2 - 3x + 3 < (3x^2 - 1)^2.$$

From these inequalities follow the desired conclusions.

(c) $\tan \theta = \sqrt{2} - 1$; $\tan^2 \theta = 3 - 2\sqrt{2}$; $\tan 3\theta = \sqrt{2} + 1$; $\tan^2 3\theta = 3 + 2\sqrt{2}$. The asymptotes of (C) are the straight lines $y^2 = (3 \pm 2\sqrt{2})x^2$ or $y = \pm(\tan \theta)x$, $y = \pm(\tan 3\theta)x$. Note that the locus of (D) is the union of the loci $y = 0$ and $y^2 = x^2 - 1 + (3/4x)$; its asymptotes are the straight lines $y = 0$, $x = 0$ and $y = \pm x$.

(d)

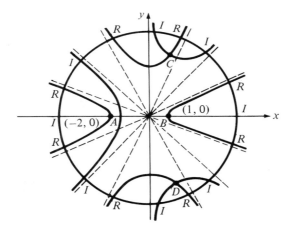

**5.1.** (b)

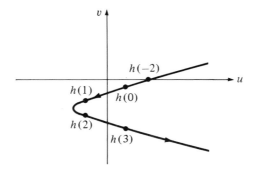

**5.2.** (f)

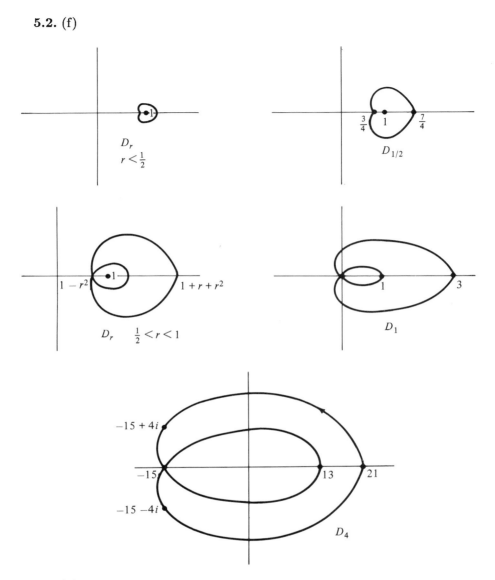

$D_r$
$r < \frac{1}{2}$

$\frac{3}{4}$   $1$   $\frac{7}{4}$

$D_{1/2}$

$1 - r^2$   $1$   $1 + r + r^2$

$D_r$   $\frac{1}{2} < r < 1$

$1$   $3$

$D_1$

$-15 + 4i$

$-15$   $13$   $21$

$-15 - 4i$

$D_4$

**5.3.** (b) The quadratic has a single zero of multiplicity 2. If the quadratic has two distinct zeros on the curve $|z| = 1$, the image $D_1$ of this curve would pass through the origin twice and therefore have a double loop. For $r$ close to 1, $D_r$ would also display the double loop, and the loop would disappear for some value of $r$ strictly less than 1. In the present case, the double loop occurs when $r > 1$, but disappears precisely at the point when it passes through the origin.

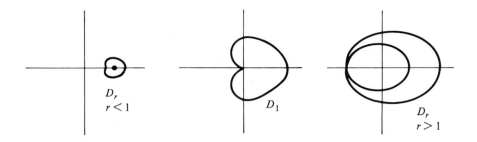

**5.4.**

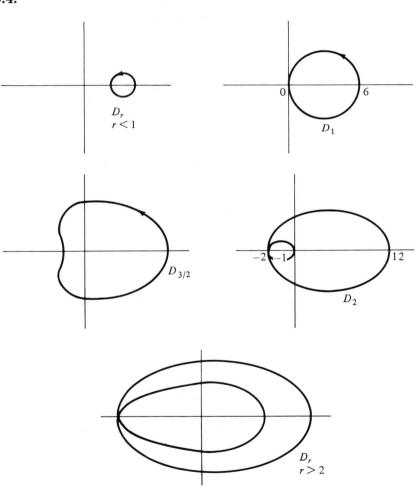

**6.1.** Let $m_i$ be the multiplicity of the zero $t_i$. We have that $p(t) = (t - t_1)^{m_1} p_1(t)$, where $p_1(t_1) \neq 0$. Since $t_2 - t_1 \neq 0$, $t_2$ must be a zero of $p_1(t)$ of multiplicity $m_2$ so that $p(t) = (t - t_1)^{m_1}(t - t_2)^{m_2} p_2(t)$, where $p_2(t)$ vanishes at neither $t_1$ nor $t_2$. Continuing in this way, we find that $p(t) = q(t)(t - t_1)^{m_1} \cdots (t - t_k)^{m_k}$, where $q(t)$ is a polynomial over **C** which does not vanish at any $t_i$. If $\deg q(t) \geq 1$, then $q(t)$ must have a zero, by the Fundamental Theorem. But such a zero would be an additional zero of $p(t)$, yielding a contradiction. Hence $q(t)$ must be a constant.

**6.3.** (a) If $s_i$ is a zero of $p(t)$, so is $\bar{s}_i$. Therefore, $p(t)$ is divisible by $(t - s_i)(t - \bar{s}_i)$.

(b) $p(t)$ can be written as the product of linear polynomials over **C**. Linear factors corresponding to nonreal zeros can be paired off and combined into real quadratic factors, as in (a).

**6.4.** For any complex number $k$, $p(t) - k$ is a polynomial over **C** with exactly $n$ zeros counting multiplicity. The only values of $k$ for which the zeros are not distinct are those for which $p(t) - k$ and its derivative $p'(t)$ have a zero in common. Thus, $p(t) - k$ has a multiple zero only when $k = p(r)$ for some zero $r$ of $p'(t)$. Since $p'(t)$ has only finitely many zeros, the result follows.

**6.6.** We can write

$$a_n t^n + a_{n-1} t^{n-1} + \cdots + a_0 = a_n(t - r_1) \cdots (t - r_n).$$

The result follows from expanding the right side and comparing coefficients.

**6.7.** Since $0 = p(a_m) = \Pi(a_m - b_j)$, it follows for some $j$, $a_m = b_j$. Let $j = n$, say. Then

$$\prod_{i=1}^{m-1}(t - a_i) = \prod_{j=1}^{n-1}(t - b_j).$$

Continue the argument, pairing off and cancelling factors $t - a_i$ and $t - b_j$.

**6.8.** Suppose $p(t)$ and $q(t)$ have degree not exceeding $n$ and $p(a_i) = q(a_i)$. Then $(p - q)(t)$ has degree not exceeding $n$ and at least $n + 1$ zeros, whence it is the zero polynomial. The result follows.

**6.9.**

$$p(t) = C(t^2 - 2t + 1)^m \prod_{i=1}^{k}(t^2 - (r_i + 1/r_i)t + 1)^{m_i}$$

for complex zeros $r_i$ with multiplicity $m_i$ within the closed unit disc. Each quadratic factor is a reciprocal polynomial and it is straightforward to prove that their product is also a reciprocal polynomial.

**6.10.** Let $f(a_i)$ be prime for $0 \leq i \leq 2n$, and suppose, if possible that $f(t) = g(t)h(t)$ where $\deg g(t) = m \geq 1$, $\deg h(t) = n - m \geq 1$. Then either $|g(a_i)| = 1$ for more than $2m$ of the $a_i$ or $|h(a_i)| = 1$ for more than

$2(n-m)$ of the $a_i$. Suppose the former occurs. Then either $g(t)-1$ or $g(t)+1$ must vanish for at least $m+1$ of the $a_i$, contradicting the fact that $\deg(g(t)\pm1)=m$. Hence, the factorization $f=gh$ is not possible and $f(t)$ is irreducible.

## Solutions to Problems

### Chapter 4

**7.1.** The equation can be rewritten $(x-a)^4-6a^2x^2=0$. Factoring the left side as a difference of squares leads to two quadratic equations.

**7.2.** Let $u=x^2+x-1$, $v=x^2-x-2$, so that $u-v=2x+1$. Let

$$\begin{aligned}
g(x) &= (x^2+x-1)^4-(x^2-x-2)^4-(2x+1)^4 \\
&= u^4-v^4-(u-v)^4=2v(u-v)(2u^2-uv+v^2).
\end{aligned}$$

Then $g(x)=0\Longrightarrow(1)\ v=0\Longrightarrow x=-1$ or $x=2$

$$\begin{aligned}
\text{or (2)}\quad &u-v=0\Longrightarrow x=-1/2 \\
\text{or (3)}\quad &0=2u^2-uv+v^2=2x^4+2x^3-x^2+x+4.
\end{aligned}$$

In case (3), we have $4u=(1\pm i\sqrt7)v$, which yields the quadratic equations

$$(3-i\sqrt7)x^2+(5+i\sqrt7)x+(-2+2i\sqrt7)=0$$

$$(3+i\sqrt7)x^2+(5-i\sqrt7)x+(-2-2i\sqrt7)=0$$

for the remaining solutions. [As a check, note that the product of the two quadratic polynomials is equal to

$$(3x^2+5x-2)^2+7(x^2-x-2)^2=8(2x^4+2x^3-x^2+x+4).]$$

**7.3.** The substitution $x=1-t$ yields the equation

$$t^4-2t^3-4t^2+2t+1=0$$

or $(t^2+t^{-2})-2(t-t^{-1})-4=0$. Set $u=t-t^{-1}$, whence $u^2=(t^2+t^{-2})-2$. Hence $u^2-2u-2=0$. We solve successively for $u$, $t$ and $x$.

**7.4.** The equation can be rewritten

$$(x^2+5ax+5a^2)^2-a^4=b^4,$$

whence $x^2+5ax+5a^2=\pm\sqrt{a^4+b^4}$ (cf. Problem 3.8.20).

**7.5.** The equation implies $x+a\sqrt x+b=x+c^2-2c\sqrt x$, whence $(a+2c)^2x=(c^2-b)^2$.

For there to be infinitely many solutions, it is necessary that $a = -2c$ and $b = c^2$. Suppose these conditions hold. Then the equation reduces to

$$\sqrt{(\sqrt{x} - c)^2} = c - \sqrt{x}.$$

If $c < 0$, the equation has no solution. If $c \geq 0$, then it is satisfied by all $x$ for which $0 \leq x \leq c^2$. Hence there are infinitely many solutions iff $c > 0$ and $a = -2c$, $b = c$.

**7.6.** Square both sides:

$$2x + 2\sqrt{x^2 - 2x + 1} = k^2.$$

There are two possible cases to consider: (1) $x > 1$; (2) $x \leq 1$. If $x > 1$, then the equation reduces to $4x = k^2 + 2$. A necessary condition for a solution $x > 1$ is that $k > \sqrt{2}$. If $\sqrt{k} > 2$, then it can be checked that $x = (k^2 + 2)/4$ indeed satisfies the equation. If $x \leq 1$, then the equation reduces to $2x + 2(1-x) = k^2$, or $2 = k^2$. Hence, for a solution, it is necessary that $k = \sqrt{2}$. On the other hand, let $x$ be any number with $1/2 \leq x \leq 1$. Then it is readily checked that the square of the left side is 2.

Thus, if $k^2 > 2$, then $x = (k^2+2)/4$ satisfies the equation, while if $k^2 = 2$, any $x$ with $1 \leq 2x \leq 2$ satisfies the equation.

**7.7.** Squaring both sides of the equation and collecting terms gives:

$$2ab - y = 2\sqrt{aby + a^2b^2},$$

where $y = x[x - (a + b)]$. Squaring again leads to $y^2 = 8aby$, whence $y = 0$ or $y = 8ab$. The solution $y = 8ab$ is extraneous (since $2ab - y$ would be negative). Hence $x = 0$ or $x = a + b$. The first of these is extraneous, but $x = a + b$ is a valid solution.

**7.8.** Let $y = x^2 + 5x$. The equation becomes

$$(y + 4)(y + 6) + (y - 6) = 0,$$

whose solutions are $y = -2$ and $y = -9$. We can now form two quadratic equations for the four values of $x$.

**7.9.** (a) $p(t) = t + a(t - u)(t - v)$ (by the Factor Theorem)
$\Longrightarrow p(t) - u = (t - u)[a(t - v) + 1];$
$\qquad p(t) - v = (t - v)[a(t - u) + 1]$
$\Longrightarrow p(p(t)) - t = p(p(t)) - p(t) + p(t) - t$

$$
\begin{aligned}
&= a[p(t) - u)(p(t) - v) + (t - u)(t - v)] \\
&= a(t - u)(t - v)[a^2(t - u)(t - v) + a(2t - u - v) + 1 + 1] \\
&= a(t - u)(t - v)[a(p(t) - t) + a(2t) + (b - 1) + 2] \\
&= a(t - u)(t - v)[a^2 t^2 + a(b + 1)t + ac + b + 1].
\end{aligned}
$$

(Note that $a(u + v) = 1 - b$, being the sum of the zeros of $p(t) - t$.)

Hence, two zeros of the quartic $p(p(t)) - t$ are $u$ and $v$ and the remaining zeros are zeros of $at^2 + a(b + 1)t + ac + b + 1$.

(b) Here, $a = 1$, $b = -3$, $c = 2$, $u = 2 + \sqrt{2}$, $v = 2 - \sqrt{2}$ and the quadratic equation for the remaining two zeros is $t^2 - 2t = 0$. Hence $t = 2 \pm \sqrt{2}$, 0 or 2.

**7.10.** For the radical $\sqrt{x^2 - 1}$ to be defined, we require $|x| \geq 1$. If $x < -1$, then $x + \sqrt{x^2 - 1} = \sqrt{x^2 - 1} - |x| < 0$ and the left side of the equation is not defined. Hence $x \geq 1$.

*First solution.* Let $x = (1/2)(u + u^{-1})$, where $u \geq 1$. Then $\sqrt{x^2 - 1} = (1/2)(u - u^{-1})$. Since $x(x + 1) = (u^2 + 1)(u + 1)^2/4u^2$ and $x(x - 1) = (u^2 + 1)(u - 1)^2/4u^2$, squaring the equation and simplifying yields $u^4 = 3$, whence $x = \frac{1}{2}(3^{1/4} + 3^{-1/4})$. This checks out.

*Second solution.* Squaring the equation yields

$$(x - \sqrt{x^2 - 1})^3 = 2x\sqrt{x^2 - 1}[x - \sqrt{x^2 - 1}]$$

$$\Longrightarrow (x - \sqrt{x^2 - 1})^2 = 2x\sqrt{x^2 - 1}$$

$$\Longrightarrow 2x^2 - 1 = 4x\sqrt{x^2 - 1} \Longrightarrow 12x^4 - 12x^2 - 1 = 0,$$

from which $x^2$ and eventually $x$ can be determined.

**7.11.** Let $v = \cos(\pi/14)$. By de Moivre's Theorem, $\cos(\pi/2) = 0 = 64v^7 - 112v^5 + 56v^3 - 7v = v[(4v^2)^3 - 7(4v^2)^2 + 14(4v^2) - 7]$. The same equation is valid for $v = \cos(3\pi/14)$ and $v = \cos(5\pi/14)$, so the three roots of the given equation must be $4\cos^2(\pi/14)$, $4\cos^2(3\pi/14)$ and $4\cos^2(5\pi/14)$, of which the first is the largest (cf. Problem 3.8.29).

**7.12.** If $1 - x^2 = -x^3$, then $(1 - x)(1 - x^2)^2 = (1 - x)x^6 = (x^2 - x^3)x^4 = x^4$. If $1 - x^2 = x$, then $(1 - x)(1 - x^2)^2 = (1 - x)x^2 = x^2 x^2 = x^4$. The zeros $u$, $v$, $w$ of $x^3 - x^2 + 1$ are distinct and each is a zero of $f(x) = x^4 - (1 - x)(1 - x^2)^2$. By the Factor Theorem, $f(x)$ is divisible by $x^3 - x^2 + 1 = (x - u)(x - v)(x - w)$. Similarly, $x^2 + x - 1$ divides $f(x)$. Since the cubic and quadratic have no zeros in common, $f(x) = k(x^3 - x^2 + 1)(x^2 + x - 1)$. Checking leading coefficients of both sides reveals that $k = 1$.

**7.13.** The equation can be rewritten

$$(x^2 - 9x - 1)^{10} - 10x^9(x^2 - 9x - 1) + 9x^{10} = 0.$$

The quadratic equation $x^2 - 9x - 1 = x$ has real roots, and each of these roots satisfies the given equation.

**8.1.** Multiply the first equation by $y$, the second by $x$, and take the difference: $0 = (x^3 - y^3)(2xy - 1)$. Either $x = y$ or $2xy = 1$. If $x = y$, then $x = y = 0$ or $9/8$. If $y = 1/2x$, then the first equation becomes

$8x^6 - 9x^3 + 1 = 0$, whence $x^3 = 1$ or $1/8$. The solutions $(x, y)$ are $(0, 0)$, $(9/8, 9/8)$, $(1, 1/2)$, $(1/2, 1)$.

**8.2.** First, note that, if $u + v = u^{-1} + v^{-1}$, then $(u + v)(1 - uv) = 0$, whence $u = -v$ or $uv = 1$. Applying this to $u = x/a$, $v = b/y$, we have that, either $x/a + b/y = 0$ or $x/a = y/b$. Similar options occur for $(x, z)$ and $(y, z)$. Suppose $x/a + b/y = 0$. Then $c = z$, and $a/x + b/y = 0$, so that $x/a = a/x$. Thus, $x = \pm a$, $y = \mp b$. Similarly, we can handle $y/b + c/z = 0$ and $x/a + c/z = 0$.

The only remaining possibility is $x/a = y/b = z/c$. If the common value is $t$, then $t^2 - t + 2 = 0$.

The complete set of solutions $(x, y, z)$ consists of $(a, b, -c)$, $(a, -b, c)$, $(-a, b, c)$, $(ta, tb, tc)$ where $2t = 1 \pm i\sqrt{7}$.

**8.3.** $xy + xz + yz = -3ab$, $xyz = a^3 + b^3$, so that $x$, $y$, $z$ are the zeros of

$$t^3 - 3abt - (a^3 + b^3) = [t - (a + b)][t^2 + (a + b)t + a^2 - ab + b^2].$$

**8.4.** $x + y = z + 2$, $(x + y)^2 = z^2 + 8 \Longleftrightarrow z = 1$, $x + y = 3$.

$x^3 + y^3 - 1 = 86 - 3xy \Longleftrightarrow (x + y)^3 - 3xy(x + y) = 87 - 3xy \Longleftrightarrow xy = -10$.

$(x, y, z) = (5, -2, 1)$ or $(-2, 5, 1)$.

**8.5.** Let $u = a^2 - x^2 = b^2 - y^2 = c^2 - z^2$. Then

$$
\begin{aligned}
u &- \sqrt{a^2 - u}\sqrt{b^2 - u} = \sqrt{c^2 - u}(\sqrt{a^2 - u} + \sqrt{b^2 - u}) \\
\Rightarrow &\quad [2u^2 - (a^2 + b^2)u + a^2b^2] - 2u\sqrt{a^2 - u}\sqrt{b^2 - u} \\
&= (c^2 - u)(a^2 + b^2 - 2u + 2\sqrt{a^2 - u}\sqrt{b^2 - u}) \\
\Rightarrow &\quad 2c^2u + a^2b^2 - c^2(a^2 + b^2) = 2c^2\sqrt{a^2b^2 - (a^2 + b^2)u + u^2} \\
\Rightarrow &\quad 4a^2b^2c^2u + a^4b^4 + a^4c^4 + b^4c^4 \\
&\quad - 2a^2b^2c^2(a^2 + b^2 + c^2) = 0 \\
\Rightarrow &\quad u = \frac{a^2 + b^2 + c^2}{2} - \frac{1}{4}\left[\frac{a^2b^2}{c^2} + \frac{b^2c^2}{a^2} + \frac{c^2a^2}{b^2}\right] \\
&\quad (x, y, z) = \left(\pm\frac{1}{2}(bca^{-1} - abc^{-1} - acb^{-1}),\right. \\
\end{aligned}
$$

$$\pm\frac{1}{2}(acb^{-1} - abc^{-1} - bca^{-1}), \; \pm\frac{1}{2}(abc^{-1} - acb^{-1} - bca^{-1})).$$

The task of checking these solutions is left to the reader.

**8.6.** Multiplying the second equation by $14xy$ and adding the result to the first yields

$$
\begin{aligned}
-7y^4 &+ 14x^3y - 67x^2y^2 + 56xy^3 + 64 = 0 \\
\Rightarrow &\quad -7y^4 + 56xy^3 - 67x^2y^2 + 14x^3y + (x^2 - 7xy + 4y^2)^2 = 0 \\
\Rightarrow &\quad x^4 - 10x^2y^2 + 9y^4 = 0 \\
\Rightarrow &\quad (x - 3y)(x + 3y)(x - y)(x + y) = 0.
\end{aligned}
$$

Checking the possibilities leads to $(x, y) = (2,2), (-2,-2), (3,1), (-3,-1)$, and four other solutions given by $3x^2 + 2 = 0$, $x + y = 0$ and $17y^2 + 4 = 0$, $x + 3y = 0$.

**8.7.** Let $u = -x+y+z$, $v = x-y+z$, $w = x+y-z$. The equations become $vw = a^2$, $uw = b^2$, $uv = c^2$. Then $b^2v = a^2u$, $uv = c^2$ lead to $u = \pm bca^{-1}$, $v = \pm acb^{-1}$. Also $w = \pm abc^{-1}$. Then $2x = v + w$, $2y = u + w$, $2z = u + v$.

**8.8.** From the first two equations $x : y : z = (b-c) : (c-a) : (a-b)$. Using this in the last equation leads to $(x, y, z) = (b - c, c - a, a - b)$.

**8.9.** Since $xy - xz = b - c$, we find that $2xy = a+b-c$ and $2xz = a-b+c$. Hence $z(a + b - c) = y(a - b + c)$. This, with $2yz = -a + b + c$ yields

$$2(a + b - c)z^2 = (a - b + c)(-a + b + c).$$

Similarly, $x^2$ and $y^2$ can be found.

**8.10.** (a) $w = u + v - u^{-1}$, $w^{-1} = u + v - v^{-1}$

$$\Rightarrow \quad 1 = (u + v)^2 - (u + v)(u^{-1} + v^{-1}) + u^{-1}v^{-1}$$
$$\Rightarrow \quad 1 - u^{-1}v^{-1} = (u + v)^2(1 - u^{-1}v^{-1})$$
$$\Rightarrow \quad uv = 1, u + v = 1 \quad \text{or} \quad u + v = -1.$$

Hence $(u, v, w) = (z, z^{-1}, z)$, $(z, 1 - z, (z - 1)z^{-1})$ or $(z, -(1 + z), -(1 + z)z^{-1})$, for arbitrary nonzero $z$.

(b) Equating first and second, first and third, second and fourth, third and fourth members, respectively, yields

$$uv(w - 1)(w - u) = w(v - 1)(uv - 1) \qquad \ldots(1)$$
$$uv(w - 1)(vw - 1) = w(u - 1)(uv - 1) \qquad \ldots(2)$$
$$u(uv - 1)(vw - 1) = vw(w - 1)(u - 1) \qquad \ldots(3)$$
$$vw(uv - 1)(w - u) = u(v - 1)(w - 1) \qquad \ldots(4)$$

The solutions in which any one of $u = 1$, $v = 1$, $w = 1$, $u = w$, $uv = 1$, $vw = 1$ occur are covered by the following: $(u, v, w) = (1, z, z^{-1})$, $(z, 1, z)$, $(z, z^{-1}, 1)$ where $z$ is nonzero.

If none of the foregoing possibilities occur, then we can cancel freely in manipulating the above four equations. From (1) and (2), we have

$$(u - 1)(w - u) = (v - 1)(vw - 1),$$

and from (3) and (4),

$$(vw)^2(u - 1)(w - u) = u^2(v - 1)(vw - 1).$$

Hence, $u = vw$ or $u = -vw$.

Suppose that $u = vw$. Substituting for $w$ in (2) leads to $v^2 = 1$, and we get the solutions $(u, v, w) = (-1, 1, -1)$, $(-1, -1, 1)$ already noted.

Suppose that $u = -vw$. Substituting for $w$ in $u^{-1} + w + uw^{-1} = v^{-1} + w^{-1} + vw$ leads to

$$(v - u)(1 + v + u - uv) = 0.$$

If $v = u$, then $w = -1$. Expressing everything in terms of $u$, we find that the given system is satisfied iff

$$0 = u^3 + u^2 + u - 3 = (u - 1)(u^2 + 2u + 3)$$

and

$$0 = 3u^3 + u^2 - u - 1 = (3u - 5)(u^2 + 2u + 3) + 14.$$

But these are inconsistent and there is no solution with $v = u$. If $uv = 1 + u + v$ (and $u = -vw$), then the first member of the given system is equal to the fourth and the second is equal to the third. Thus, the system is satisfied iff (4) holds:

$$-u(u + v)(-uv^{-1} - u) = u(v - 1)(-uv^{-1} - 1)$$

or

$$u^2(u + v)(1 + v) = -u(v - 1)(u + v)$$

or $u(1 + v) = -(v - 1)$ or $u + v + uv = 1$ or $u + v = 0$. Hence $(u, v, w) = (i, -i, 1)$ or $(-i, i, 1)$, which has already been noted.

**8.11.** Let $u = x + y$, $v = xy$. Then the two equations become $u^2 - 2v = 13$, $u(13 - v) = 35$, whence $0 = u^3 - 39u + 70 = (u - 5)(u - 2)(u + 7)$. Hence $x$ and $y$ are the zeros of any one of the quadratics $t^2 - 5t + 6$, $2t^2 - 4t - 9$, $t^2 + 7t + 18$.

**8.12.** $x$, $y$, $z$ are the zeros of the cubic

$$t^3 - 3at^2 + (2a^2 - a - 7)t + (4a^2 + 10a - 6) = (t + 2)[t - (2a - 1)][t - (a + 3)].$$

**8.13.** Suppose $a$, $b$, $c$ are all distinct. From the first equation and $(y - z) + (z - x) + (x - y) = 0$, it follows that $x - y = t(a - b)$, $y - z = t(b - c)$, $z - x = t(c - a)$ for some $t$. The second equation then determines $t$. Finally, writing two of the variables in terms of a third, we find from the third equation that

$$3x = e + t(2a - b - c), \quad 3y = e + t(2b - a - c), \quad 3z = e + t(2c - a - b).$$

If exactly two of $a$, $b$, $c$ are equal, say $a = b \neq c$, then $x = y$ and the second equation is consistent iff $d = 0$. If $d = 0$, any value of $t$ will work and the solution can be completed using the third equation as before.

If $a = b = c$, then the first equation imposes no restriction. We can substitute $z = e - (x + y)$ into the second equation, choose an arbitrary value of $x$ and solve a cubic equation in $y$ and thence determine $z$.

**8.14.** From the last two equations

$$(y+z)^2 \overset{+}{_-} (y-z)^2 + 4yz = 9/4 - 3x + x^2 + 3/4 = x^2 - 3x + 3.$$

Hence $a^3 = x(x^2 - 3x + 3) - 1 = (x-1)^3$, so that $x = 1 + a$, $1 + \omega a$ or $1 + \omega^2 a$, where $\omega$ is an imaginary cube root of unity. Substituting this into the last two equations yields quadratic equations for $y$ and $z$.

**8.15.** Adding and subtracting the two equations yields

$$2(x^2 - y^2) = (a+b)x - (a-b)y$$

$$4xy = (a-b)x + (a+b)y.$$

Hence

$$2(x+yi)^2 = [(a+b) + (a-b)i](x+yi)$$

and

$$2(x-yi)^2 = [(a+b) - (a-b)i](x-yi).$$

If real solutions $x$, $y$ are sought, these equations are equivalent. We obtain $x + yi = 0$ whence $(x, y) = (0, 0)$ or $2(x + yi) = (a+b) + (a-b)i$, whence $(x, y) = ((a+b)/2, (a-b)/2)$.

However $x$ and $y$ may be nonreal and a more careful analysis is needed. If $x + yi$ and $x - yi$ are both nonzero, then we are led to the second solution above. The remaining cases are as follows:

(1) $x + yi = 0$ and $2(x - yi) = (a+b) - (a-b)i$

$$\Rightarrow (x, y) = \left( \frac{(a+b) - (a-b)i}{4}, \frac{(a-b) + (a+b)i}{4} \right);$$

(2) $x - yi = 0$ and $2(x + yi) = (a+b) + (a-b)i$

$$\Rightarrow (x, y) = \left( \frac{(a+b) + (a-b)i}{4}, \frac{(a-b) - (a+b)i}{4} \right).$$

**8.16.** Equating two expressions for $b^2 c^2$ yields

$$\begin{aligned} 0 &= (v^2 + u^2 + z^2)(w^2 + y^2 + u^2) - [vw + u(y+z)]^2 \\ &= (uw - vy)^2 + (uv - wz)^2 + (u^2 - yz)^2. \end{aligned}$$

Hence, $uw = vy$, $uv = wz$, $u^2 = yz$, and similarly, $vw = ux$, $v^2 = xz$, $w^2 = xy$.

By substitution, we obtain $x(x+y+z) = a^2$, $y(x+y+z) = b^2$, $z(x+y+z) = c^2$, whence $(x+y+z)^2 = a^2 + b^2 + c^2 = d^2$. Therefore $(x, y, z, u, v, w) = \pm(a^2/d, b^2/d, c^2/d, bc/d, ca/d, ab/d)$.

**8.17.** Eliminating $u$ from each adjacent pair of equations yields

$$v(y-x) = b - ax$$

$$vy(y - x) = c - bx$$
$$vy^2(y - x) = d - cx$$

whence

$$(c - bx)^2 = (b - ax)(d - cx).$$

The solutions for which $y = x$ can occur only if $x = b/a = c/b = d/c$. In this case, $u$ and $v$ are restricted only by $u + v = a$. Henceforth, we exclude this possibility.

Consider the possibility that $y = 0$. This is feasible only if $bd = c^2$, in which case $x = c/b = d/c$, $u = b/x$ and $v = a - u$.

Now $(ac - b^2)x^2 - (ad - bc)x + (bd - c^2) = 0$. Suppose that $ac - b^2 \neq 0$. Then we can solve for $x$ and determine the remaining variables from $y(b - ax) = c - bx$, $v(y - x) = b - ax$, $u = a - v$. If $ac - b^2 = 0$, $ad - bc \neq 0$, then the quadratic equation collapses to a linear equation with a single root $x = b/a = c/b$. But then $0 \neq d - cx = y^2(b - ax) = 0$, a contradiction. The remaining possibility is that all coefficients of the quadratic vanish. In this case, $l/a = c/b = d/c$. From the three equations at the head of the solution, we find that

(i) $v = 0$, $x = b/a$ so that $u = a$ and $y$ is arbitrary;

(ii) $y = b/a = c/b = d/c$ so that either $u = 0$, $v = a$ and $x$ is arbitrary or else $u \neq 0$ and $x = y = b/a$.

A method for dealing with general equations of this type is given in S. Ramanujan, Note on a set of simultaneous equations, *J. Ind. Math. Soc.* 4 (1912), 94–96 = *Collected papers* (#3), 18–19 (Chelsea, 1927, 1962).

**8.18.** From the given system, we can derive

$$(xy + yz + zx) + 3u + 10v = 0 \quad \text{or} \quad 8 + 3u + 10v = 0$$

$$(v + y)(z - x) = 0$$
$$(v + z)(y - x) = 0$$
$$(v + x)(z - y) = 0.$$

The first two equations of the given system show that $x$, $y$, $z$ cannot all be equal. Also, $x$, $y$, $z$ cannot all be distinct, since then $v + y$, $v + z$, $v + x$ could not vanish simultaneously. Hence, exactly two of $x$, $y$, $z$ are equal, say $x = y \neq z$. Then $2x + z = 5$, $2x^2 + z^2 = 9$, $v + x = 0$ and $3u + 10v = -8$. Hence $(x, y, z, u, v) = (2, 2, 1, 4, -2)$ or $(4/3, 4/3, 7/3, 16/9, -4/3)$.

**8.19.** If any of $x$, $y$, $z$ vanish, then $(x, y, z)$ is one of $(0, 0, 0)$, $(a, 0, 0)$, $(0, b, 0)$, $(0, 0, c)$. Suppose, if possible, that $xyz \neq 0$. Adding these three equations yields

$$2(ax + by + cz) = (y - z)^2 + (z - x)^2 + (x - y)^2$$

whence $ax = (x-y)(x-z)$, $by = (y-x)(y-z)$, $cz = (z-x)(z-y)$. Hence

$$ax(z-y) = by(x-z) \Rightarrow z(ax+by) = (a+b)xy$$

$$\Rightarrow bx^{-1} + ay^{-1} - (a+b)z^{-1} = 0.$$

Similarly

$$cx^{-1} - (a+c)y^{-1} + az^{-1} = 0$$

$$-(b+c)x^{-1} + cy^{-1} + bz^{-1} = 0.$$

From any two of these equations, $x : y : z = 1 : 1 : 1$, so that $by + cz = cz + ax = ax + by = 0$, which is feasible only if $a = b = c = 0$.

**8.20.** If $a = b$, it is straightforward to verify that the equation is satisfied iff $x = y$. Suppose $a \neq b$ and let $x = (1+u)/(1-u)$, $y = (1+v)/(1-v)$. The equations become

$$(ab+1)(u^2+1) = (a^2+1)(uv+1)$$

$$(ab+1)(v^2+1) = (b^2+1)(uv+1).$$

(At this point, some obvious solutions can be noted: $(u,v) = (a,b)$, $(-a,-b)$, $(i,i)$, $(-i,-i)$.) These equations lead to

$$(b^2+1)(u^2+1) = (a^2+1)(v^2+1)$$

or

$$(a^2+1)v^2 = (b^2+1)u^2 + (b+a)(b-a)$$

and

$$(a^2+1)v = (ab+1)u + a(b-a)u^{-1}.$$

Hence

$$[(a^2+1)(b^2+1) - (ab+1)^2]u^2 + [(a^2+1)(b+a)(b-a) - 2(ab+1)a(b-a)]$$
$$- a^2(b-a)^2 u^{-2} = 0$$
$$\Rightarrow 0 = u^4 + (1-a^2)u^2 - a^2 = (u^2 - a^2)(u^2 + 1).$$

Therefore, the four obvious solutions are the only solutions and we can now determine $x$ and $y$.

**8.21.** *First solution.* Let $u = x+y$, $v = xy$. The three equations become $u^3 - 3uv + z^3 = 8$, $u^2 - 2v + z^2 = 22$, $z^2 + uz + v = 0$. Hence $(u+z)^3 = 8$ and $(u+z)^2 + 2z^2 = 22$, whence $(u+z, z^2) = (2,9)$, $(2\omega, 11 - 2\omega^2)$, $(2\omega^2, 11 - 2\omega)$. This leads to the solutions $(x,y,z) = (-3,2,3)$, $(2,-3,3)$, $(3,2,-3)$, $(2,3,-3)$ and a number of nonreal solutions such as $(2\omega, -\sqrt{11 - 2\omega^2}, \sqrt{11 - 2\omega^2})$. Here, $\omega$ is an imaginary cube root of unity.

*Second solution.* The last equation is $(z+x)(z+y) = 0$. Let $z = -x$; the case $z = -y$ can be similarly handled. Then $y^3 = 8$, so that $y = 2, 2\omega, 2\omega^2$. Then $2x^2 = 22 - y^2$ permits $x$ to be determined.

**8.22.** *First solution.* Clearly $0 < x < y$. If $x \leq 1$, then $(x+y)^2 \geq 9$, so that $y \geq 2$. Hence

$$9(y^3 - 1) \leq 9(y^3 - x^3) = 7(x+y)^2 \leq 7(y+1)^2$$

$$\Rightarrow \ (y-2)(9y^2 + 11y + 8) \leq 0.$$

Since $9y^2 + 11y + 8$ has a negative discriminant, it is always positive. Hence $y = 2$, $x = 1$. On the other hand, if $x \geq 1$, then $(x+y)^2 \leq 9$, so $y \leq 2$. Then $(y-2)(9y^2 + 11y + 8) \geq 0$ and $y = 2$, $x = 1$.

    *Second solution.* Since $x > 0$, let $x = z^2$. Then $7z^{-2} + z^6 = y^3 = (3z^{-1} - z^2)^3$, whence

$$\begin{aligned}
0 &= 2z^9 - 9z^6 + 27z^3 + 7z - 27 \\
&= (z-1)(2z^8 + 2z^7 + 2z^6 - 7z^5 - 7z^4 - 7z^3 + 20z^2 + 20z + 27) \\
&= (z-1)[(z^2 + z + 1)(2z^6 - 7z^3 + 20) + 7].
\end{aligned}$$

The polynomial $2z^6 - 7z^3 + 20$, considered as a polynomial in $z^3$, has negative discriminant, and so is always positive. Since $z^2 + z + 1$ is also positive, the term in square brackets never vanishes, and so $z = 1$. Thus, $(x, y) = (2, 1)$ is the only solution.

**8.23.** Since $x^2 = (y+z)^2 - 2(1+a)yz = (y-z)^2 + 2(1-a)yz$, it follows that

$$2(1+a)yz = (x+y+z)(y+z-x)$$

$$2(1-a)yz = (x+y-z)(x+z-y),$$

whence

$$4(1-a^2)x^2y^2z^2 = (x+y+z)(y+z-x)(x+y-z)(x+z-y)x^2.$$

    Similar equations hold for $1 - b^2$ and $1 - c^2$. From these it can be seen that, when $|a|, |b|, |c|$ are distinct from 1,

$$x^2(1-a^2)^{-1} = y^2(1-b^2)^{-1} = z^2(1-c^2)^{-1}.$$

If $a = 1$, then $x^2 = (y-z)^2$, whence $y = x+z$ or $z = x+y$. If $y = x+z$, then $xz = -bxz$ and $xy = cxy$, whence $x = 0$, or $xz \neq 0$, $y = 0$, $b = -1$, or $xy \neq 0$, $z = 0$, $c = 1$, or $xyz \neq 0$, $b = -1$, $c = 1$. The cases $z = x+y$ and $a = -1$ can be handled similarly.

**8.24.** Adding the first two equations yields

$$x^2 + y^2 + z^2 = 30 \qquad\qquad\qquad\qquad (*)$$

whence $(x+y+z)^2 = 30 + 2(-7) = 16$.

    Taking the difference of the first two equations yields $x^2 - z^2 = 24$.

    Case (i): From $x + z = 4 - y$, $x - z = 24(4 - y)^{-1}$, we obtain $(x, y, z) = (u + 6u^{-1}, \ 4 - 2u, \ u - 6u^{-1})$ where $2u = 4 - y$. Substituting this into

$(*)$ yields $(u - 3)(3u^3 + u^2 - 4u - 12) = 0$. Conversely, any solution of this equation leads to a solution of $x + y + z = 4$, $x^2 - z^2 = 24$, $x^2 + y^2 + z^2 = 30$, and hence of the given system.

Case (ii): From $x + z = -4 - y$, $x - z = 24(-4 - y)^{-1}$, we obtain $(x, y, z) = (v + 6v^{-1}, -4 - 2v, v - 6v^{-1})$ where $2v = -4 - y$. Substituting this into $(*)$ yields $(v + 3)(3v^3 - v^2 - 4v + 12) = 0$. Any solution of this equation yields a solution of the original system.

Hence $(x, y, z) = (5, -2, 1)$, $(-5, 2, -1)$, $(u + 6u^{-1}, 4 - 2u, u - 6u^{-1})$ where $u(u - 1)(3u + 4) = 12$ or $(v + 6v^{-1}, -4 - 2v, v - 6v^{-1})$ where $v(v + 1)(3v - 4) = -12$.

**8.25.** For any quadratic polynomial $f(t)$, it is easy to verify that

$$\Sigma k f(k^2) x_k = a f(a).$$

In particular, if $f(t) = (t - a)^2$, we obtain that

$$\Sigma k(k^2 - a)^2 x_k = 0$$

whence each term of the sum on the left must vanish. This implies either that each $x_k$ vanishes, in which case $a = 0$, or that $a = m^2$ for $m$ equal to one of 1, 2, 3, 4, 5, in which case $x_m = m$ and the remaining $x_k$ vanish.

**8.26.** Suppose we have a real solution $(x, y, z)$. Then $x + y = 2 - z$ and $xy = (z - 1)^2$, so that

$$0 \le (x - y)^2 = (2 - z)^2 - 4(z - 1)^2 = z(4 - 3z).$$

Hence, we must have $0 \le z \le 4/3$. Then $p = xyz = z(z - 1)^2$ assumes all values between 0 and 4/27 inclusive (sketch a graph of the function $z(z - 1)^2$).

Conversely, if $0 \le p \le 4/27$, then the equation $z(z - 1)^2 = p$ is solvable for real $z$ with $0 \le z \le 4/3$. For such $z$, the quadratic $t^2 - (2 - z)t + (z - 1)^2$ has real zeros $x$, $y$, and we have a real solution $(x, y, z)$ of the given system.

**8.27.** The second equation implies that

$$\frac{x + y}{x - y} + \frac{x - y}{x + y} = \pm \left[ 3 + \frac{1}{3} \right]$$

whence one of $x = \pm 2y$, $y = \pm 2x$ holds. The solutions $(x, y)$ to the equations are $(2, 1)$, $(-2, -1)$, $(1, 2)$, $(-1, -2)$, $(2i, -i)$, $(-2i, i)$, $(i, -2i)$, $(-i, 2i)$.

**8.28.** The first two equations represent spheres in Cartesian 3-space with radii 7 and centers $(5, 2, 6)$ and $(11, 7, 2)$. Since the centers are less than 14 units apart, the spheres intersect in a circle lying on the plane $12x + 10y - 8z = 109$. Since the plane $38x - 56y - 13z = 0$ passes through the point $(8, 9/2, 4)$ which is at the same time midway between the centers of the sphere and at the center of the circle of intersection, it must intersect this circle in exactly two points, $(x_1, y_1, z_1)$ and $(x_2, y_2, z_2)$. It is easily verified that $(16 - x_1, 9 - y_1, 8 - z_1)$ satisfies the three equations, and so this point can be none other than $(x_2, y_2, z_2)$.

**8.29.** Suppose if possible that at least two of $x$, $y$, $z$ are equal. Then $x = y = z = t$ where $t^3 + (a - 1)t + b = 0$. By Exercise 1.4.4, there is a unique real value of $t$ which satisfies this equation.

If $x$, $y$, $z$ are all distinct, then from

$$z - y = (y - x)(y^2 + yz + x^2 + a)$$
$$x - z = (z - y)(z^2 + zy + y^2 + a)$$
$$y - x = (x - z)(x^2 + xz + z^2 + a),$$

we have that

$$1 = (y^2 + yx + x^2 + a)(z^2 + zy + y^2 + a)(x^2 + xy + y^2 + a) > a^3,$$

which contradicts the condition on $a$.

**8.30.** Since $az + cx = bz + cy = ay + bx$, it follows that

$$x : y : z = a(b + c - a) : b(c + a - b) : c(a + b - c).$$

Taking a suitable constant of proportionality, values of $x$, $y$, $z$ satisfying the given equations can be found.

**8.31.**

$$z^2 - abz + (a^2 + b^2 - 4) = z^2 - (xy + xy^{-1} + x^{-1}y + x^{-1}y^{-1})z$$
$$+ (x^2 + x^{-2} + y^2 + y^{-2}) = [z - (xy^{-1} + x^{-1}y)][z - (xy + x^{-1}y^{-1})].$$

**8.32.** Let $t = ux/a^2 = vy/b^2 = wz/c^2$. Then $x/a = t(a/u)$, etc. so that $t^2(a^2/u^2 + b^2/v^2 + c^2/w^2) = 1$. Hence

$$(x/u + y/v + z/w)^2 = (ta^2/u^2 + tb^2/v^2 + tc^2/w^2)^2$$
$$= a^2/u^2 + b^2/v^2 + c^2/w^2.$$

**8.33.** A complete set of solutions of the equation $\sqrt{y - a} + \sqrt{z - a} = 1$ can be described geometrically as follows. Let $BC$ be a segment of length 1. Each solution $(y, z)$ of the equation corresponds to a point $P$ on the segment of length 1 parallel to and distant $\sqrt{a}$ from $BC$.

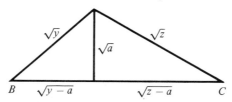

For the system of three equations, form a triangle $ABC$ whose sides are three segments of length 1. The solution of the system is given by the squares of the lengths of $AP$, $BP$ and $CP$, where $P$ is any point distant $\sqrt{a}$, $\sqrt{b}$, $\sqrt{c}$ from $BC$, $CA$, $AB$ respectively. Since there is exactly one such point, the system has exactly one solution.

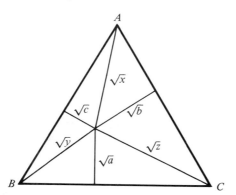

**8.34.** Suppose $abc \neq 0$. Then $xyz \neq 0$ and the system becomes $y^{-1}+z^{-1} = a^{-1}$, etc., whence $2x^{-1} = b^{-1} + c^{-1} - a^{-1}$, etc. If, say $a = 0$ and $bc \neq 0$, then $yz = 0 \Rightarrow x = 0 \Rightarrow y = z = 0$. If, say $a = b = 0$ and $c \neq 0$, then either $z = 0$ and $x$, $y$ are arbitrary solutions of the third equation or else $x = y = 0$ and $z$ is arbitrary. If $a = b = c = 0$, then two of $x$, $y$, $z$ must vanish and the third is arbitrary.

**9.1.** For each complex $z$, $z$ and $-z - 1/a$ have the same image under the mapping. Hence the mapping is one-one on the closed unit disc iff $|z| \leq 1 \Rightarrow |z + 1/a| > 1$. Taking $z = 0$ yields the necessary condition $|a| < 1$. If further, $|a| \neq 1/2$, then taking $z = -|a|/a \neq -1/2a$ yields the condition $|a| < 1/2$. Hence, it is necessary that $|a| \leq 1/2$.

But this condition is also sufficient. The intersection of the closed discs with radii 1 and centers 0 and $-1/a$ is empty if $|a| < 1/2$ and contains a single point if $|a| = 1/2$. Hence, for $|a| \leq 1/2$, the conditions $|z| \leq 1$, $|z + 1/a| \leq 1$ are simultaneously fulfilled iff $|a| = 1/2$ and $z = -z - 1/a = -1/2a$.

**9.2.** If the straight line passes through the origin, its image under $z \longrightarrow z^2$ is a ray emanating from the origin and not a parabola. So we exclude this case. Let $w$ be the point on the line closest to the origin. Then a typical point on the line is represented by $w + iwr = w(1 + ir)$, where $r$ is real. The image of this point is $w^2(1 - r^2 + 2ir)$. The locus of $1 - r^2 + 2ir$ is a parabola whose axis coincides with the real axis and whose vertex is at 1. The locus of $w^2(1 - r^2 + 2ir)$ is the image of this parabola under a dilatation followed by a rotation, and so is a parabola whose vertex is at $w^2$.

**9.3.** Three points in $\mathbf{C}$ are vertices of an equilateral triangle iff they can be obtained from the points 1, $\omega$, $\omega^2$ by a dilatation followed by a translation. The center of the triangle is represented by one third of the sum of the numbers corresponding to its vertices.

The roots of $x^3 + px^2 + qx + r = 0$ are vertices of an equilateral triangle iff they have the form $-p/3 + u$, $-p/3 + u\omega$, $-p/3 + u\omega^2$ for some nonzero $u$. This happens iff the cubic equation can be written in the form $(x + p/3)^3 = u^3$ for some nonzero $u$, which occurs iff $p^2 = 3q$, $r = p^3/27 - u^3$, i.e. $p^3 \neq 27r$.

A generalization can be found in *Crux Mathematicorum* 9 (1983), 218–221.

**9.4.** First, we determine the area of a triangle with vertices $u$, $v$, $w$ in the complex plane. This area remains unchanged if we subject the points to a translation (subtraction of $u$) followed by a rotation (multiplication by $|v - u|^{-1}(\bar{v} - \bar{u})$) to obtain the vertices $0$, $|v - u|$, $|v - u|^{-1}(w - u)(\bar{v} - \bar{u})$. Using the half-base-times-height formula for area yields

$$\frac{1}{2}|v - u| \cdot \mathrm{Im}[|v - u|^{-1}(w - u)(\bar{v} - \bar{u})]$$

$$= \frac{1}{4}|u\bar{v} + v\bar{w} + w\bar{u} - \bar{u}v - \bar{v}w - \bar{w}u|.$$

Refer to Exercises 1.4.2–4. A change of variables $x = y + a/3$ converts the equation to one of the form $y^3 + py + q = 0$ whose zeros are obtained from those of the given equation by a translation and whose coefficients $p$, $q$ are expressible in terms of $a$, $b$, $c$. If $u$, $v$, $w$ are the zeros of the polynomial in $y$, then $w = -(u + v)$ and the formula for the area reduces to $(3/4)|u\bar{v} - \bar{u}v|$. Now $u = r + s$, $v = r\omega + s\omega^2$, where $18r^3 = -9q + \sqrt{81q^2 + 12p^3}$, $s = -p/3r$ and $\omega$ is an imaginary cube root of unity. The area is thus

$$|\omega^2 - \omega|(|r|^2 - |s|^2) = \sqrt{3}(|r|^2 - |s|^2).$$

**9.5.** If the zeros are $u$, $iv$, $-iv$ ($u$, $v$ real), then $b = -au$, $c = av^2$ and $d = -auv^2$, from which the necessity of the conditions follow.

On the other hand, if $ac > 0$ and $bc = ad$, we have that

$$at^3 + bt^2 + ct + d = (at + b)(t^2 + a^{-1}c),$$

from which the sufficiency of the conditions can be deduced.

**9.6.** Taking note of the fact that $|a| = |b| = |c| = |d| = 1$, we find that, when $|z| = 1$,

$$|p(z)|^2 = p(z)\overline{p(z)} = 4 + (a\bar{d})z^3 + (\bar{a}d)z^{-3} + (a\bar{c} + b\bar{d})z^2$$

$$+ (\bar{a}c + \bar{b}d)z^{-2} + (a\bar{b} + b\bar{c} + c\bar{d})z + (\bar{a}b + \bar{b}c + \bar{c}d)z^{-1}.$$

Let $\omega$ be an imaginary cube root of unity. Since $1 + \omega + \omega^2 = 0$,

$$|p(z)|^2 + |p(\omega z)|^2 + |p(\omega^2 z)|^2 = 12 + 3a\bar{d}z^3 + 3\bar{a}dz^{-3}.$$

Choosing $z$ to be a cube root of $\bar{a}d$ gives the right side the value 18, whence at least one of the three terms on the left side is at least 6. The result follows.

The result can be generalized to polynomials of degree $n$. See Problem 4426 in *Amer. Math. Monthly* **58** (1951), 113; **59** (1952), 419.

**9.7.** Let $n = 1$ and $p(x) = ax + b$. For $m = 0$, the condition is that $c_1 ax + (c_0 + c_1 b) = 0$ identically; for $m = 1$, the condition is that $(c_0 + c_1)ax + (c_0 + c_1)b + c_1 a = 0$ identically. In either case, it can be seen that $c_0 = c_1 = 0$.

Suppose that the result holds for polynomials of degrees up to $n - 1$ inclusive. Let $\deg p(x) = n$ and $0 \leq m \leq n - 1$. Then, differentiating the condition leads to a similar condition for the derivative $p'(x)$ and the result follows in a straightforward way from the induction hypothesis.

The only case left to consider is that

$$c_0 p(x) + c_1 p(x + 1) + \cdots + c_n p(x + n) = 0$$

identically. Since the leading coefficient must vanish, $c_0 + c_1 + \cdots + c_n = 0$, whence

$$0 = c_1[p(x+1) - p(x)] + c_2[p(x+2) - p(x)] + \cdots + c_n[p(x+n) - p(x)]$$

$$= (c_1 + c_2 + \cdots + c_n)q(x) + (c_2 + \cdots + c_n)q(x+1)$$
$$+ (c_3 + \cdots + c_n)q(x+2) + \cdots + c_n q(x+n-1)$$

where $q(x) = p(x+1) - p(x)$, a polynomial of degree $n - 1$. The desired result now follows from the induction hypothesis.

**9.8.** Let

$$f(t) = c \prod_{i=1}^{k}(t - r_i)^{m_i},$$

where $m_1 + m_2 + \cdots + m_k = n$. Then

$$f'(t) = c \prod_{i=1}^{k}(t - r_i)^{m_i - 1} g(t)$$

for some polynomial $g(t)$ which does not vanish at any $r_i$. The degree of $g(t)$ is

$$(n - 1) - \sum_{i=1}^{k}(m_i - 1) = k - 1.$$

But since a power of $f'(t)$ divides a power of $f(t)$, each zero of $g(t)$ is a zero of $f(t)$. But, since $g(t)$ and $f(t)$ do not share any zero, $g(t)$ is constant, so that $k = 1$, $m_1 = n$. It is easily checked that any polynomial of the form $c(t - r)^n$ satisfies the conditions of the problem.

**9.9.** Suppose that $n = \deg p(t) \geq \deg q(t)$ and that, if possible, $(p - q)(t)$ is not identically zero. Let the distinct zeros of $p(t)$ and $q(t)$ be $u_1, \ldots, u_r$

and those of $p(t) + 1$ and $q(t) + 1$ be $v_1, \ldots, v_s$. Clearly, the $u_i$ are distinct from the $v_j$. Since each $u_i$ and each $v_j$ is a zero of $(p - q)(t)$, it follows that

$$n = \max(\deg p(t), \deg q(t)) \geq \deg(p - q)(t) \geq r + s.$$

Consider the derivative $p'(t)$, which is also the derivative of $p(t) + 1$. Since the multiplicity of each zero of $p(t)$ and of $p(t) + 1$ exceeds its multiplicity as a zero of $p'(t)$ by 1, $p'(t)$ has at least $n - r$ zeros, counting multiplicity, from among the $u_i$ and $n - s$ zeros, counting multiplicity, from among the $v_j$. Hence

$$(n - r) + (n - s) \leq \deg p'(t) = n - 1 \Rightarrow n \leq r + s - 1 < r + s.$$

But this contradicts the earlier inequality, and so $(p - q)(t)$ must be the zero polynomial.

**9.10.** Suppose the polynomial is $t^3 + at^2 + bt + c$. Let $u = 1 - 2^{1/3} + 2^{2/3}$. Then $u^2 = -3 + 3 \cdot 2^{2/3}$ and $u^3 = -9 + 9 \cdot 2^{1/3}$. Since $u$ is a zero of the polynomial,

$$(-9 - 3a + b + c) + (9 - b)2^{1/3} + (3a + b)2^{2/3} = 0.$$

The coefficients in parentheses vanish when $(a, b, c) = (-3, 9, -9)$ and we obtain the polynomial $t^3 - 3t^2 + 9t - 9$.

# Answers to Exercises

### Chapter 5

**1.2.** (a) $-0.77$, $1.93$; (b) $-0.43$; (c) $0.15$, $3.47$.

**1.3.** (c) For example, the final polynomial of Exercise 2 has $p(3) = -257$ and $p(4) = 1133$. Taking $a = 3$, $b = 4$ yields the next approximation 3.293. With $a = 3.293$, $b = 4$, we are led to 3.377.

**1.5.** (c) 0.148.
    (e) (ii)

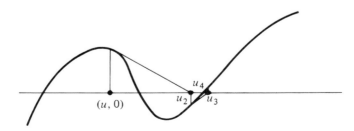

**1.7.** (b) $a_{n+1}^2 - c = (a_n^2 - c)^2/(4a_n^2)$. It follows that for $n \geq 2$, $a_n^2 \geq c$, and, since $a_n - a_{n+1} = (a_n^2 - c)/(2a_n)$, $\{a_n\}$ eventually decreases. Since

$a_n^2 - c < a_n^2 + c < 2a_n^2$ (for $n \geq 2$), we have that $a_{n+1}^2 - c < (a_n^2 - c)/2$, so that $a_{n+1}^2 - c < 2^{-(n-1)}(a_2^2 - c)$. From this estimate, the result follows.

**1.8.** Let the polynomial be $p(t)$. When $t \leq -1$, $p(t) > 3t^4 - t^2 > 0$ and when $-1 < t < 0$, $p(t) > -t^2 - 3t = -t(t + 3) > 0$. Hence all of the real zeros of $p(t)$ are positive. Two zeros are 0.295 and 1.336. As can be seen by the methods of Section 2, there are only two real zeros.

**1.9.** (b) Each successive $u_n$ seems to be farther from $r$ than its predecessor.

(c)

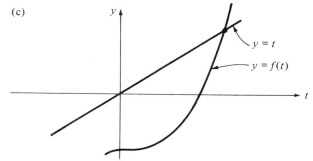

(d, e) Beginning with 3, the sequence of successive approximations is: $\{3, 3.10723, 3.121968, 3.123983, 3.124258, 3.124297, 3.124302, \ldots\}$. Beginning with 4, the sequence is: $\{4, 3.239609, 3.139972, 3.126440, 3.124595, 3.124342, 3.124307, 3.124303\}$. The zero, to five decimal places, is 3.12430.

**2.6.** (c) The proof is by induction on the degree of $p(t)$. To give a feel for the argument, examine the case that $\deg p(t) = 1$. Since $p(t) = at$ is trivial to analyze, let $p(t) = at + b$, $b \neq 0$. Then $(t - r)p(t) = at^2 + (b - ra)t - rb$. If $ab > 0$, then $a$ and $-rb$ differ in sign; if $ab < 0$, then $a$ and $-rb$ have one sign and $b - ra$ the opposite sign. In any case, the result holds.

Suppose the result holds for polynomials of degree up to $n - 1 \geq 0$. Let $p(t) = a_n t^n + q(t)$, where $q(t) = a_{n-1}t^{n-1} + \cdots + a_1 t + a_0$. Then

$$
\begin{aligned}
(t - r)p(t) &= a_n t^{n+1} - ra_n t^n + (t - r)q(t) \\
&= a_n t^{n+1} + (a_{n-1} - ra_n)t^n + \cdots.
\end{aligned}
$$

Suppose $p(t)$ has $k$ sign changes.

Case (i): $a_n$ *has the same sign as the next nonzero coefficient.* Then $q(t)$ has $k$ sign changes. The leading coefficients of $(t - r)p(t)$ and $(t - r)q(t)$ have the same sign and all other corresponding coefficients except that of $t^n$ have the same value. Hence $(t - r)p(t)$ has at least as many sign changes as $(t - r)q(t)$, and the result follows.

Case (ii): $a_n$ *differs in sign from the next nonzero coefficient.* Then $q(t)$ has $k - 1$ sign changes. Whether or not $a_{n-1}$ vanishes, the coefficients of $t^{n+1}$ and $t^n$ in $(t - r)p(t)$ have opposite signs, so that this polynomial has one more sign change than $(t - r)q(t)$, i.e. at least $k + 1$ sign changes.

**2.7.** Let $r_1, r_2, \ldots, r_k$ be all the positive zeros of $p(t)$, each taken as often as its multiplicity. Then $p(t) = (t - r_1)(t - r_2) \cdots (t - r_k)q(t)$ where $q(t)$ has no positive zeros and, say, $m$ sign changes. Applying Exercise 2.6, we find that $p(t)$ has $k + m$ sign changes. The second part follows from Exercise 5.

**2.10.** The two adjacent vanishing coefficients correspond to consecutive powers of $t$, one even and one odd. Let the signs $+$, $-$, $0$ of the *remaining* $n - 1$ coefficients be written out in order. Denoting the polynomial by $p(t)$, we see that there is a sign change between two consecutive coefficients in a given position which are nonzero for exactly one of $p(t)$ and $p(-t)$. If two nonzero coefficients are separated by a zero coefficient (apart from $a_k$ and $a_{k+1}$), then there are at most two sign changes involved with $p(t)$ and $p(-t)$ together. As a consequence, there are at most $n - 2$ changes of sign for $p(t)$ and $p(-t)$ together, and so at most $n - 2$ real zeros. Since $p(0) \neq 0$, the result follows.

**2.11.** Multiplying the given polynomial by $t - 1$ yields

$$a_n t^n + (a_{n-1} - a_n)t^n + \cdots + (1 - a_3)t^3 - 1.$$

By Exercise 10, this has at most $n - 1$ real zeros.

A similar argument shows that, if any three consecutive coefficients of a real polynomial are equal and nonzero, the zeros of the polynomial are not all real. See problem E1283 in *Amer. Math. Monthly* **64** (1957), 592; **65** (1958), 286.

**2.12.** (a)

$$f(r) = a_n r^n + (a_{n-1}r^{n-1} + \cdots + a_{p+1}r^{p+1}) + a_p r^p$$

$$+ a_{p-1}r^{p-1} + \cdots + a_1 r + a_0.$$

For $0 \leq i \leq p$, if $a_i < 0$, then $Mr^i \geq 0 > a_i r^i$, while if $a_i > 0$, $Mr^i \geq a_i r^i$. For $p < i < n$, $a_i > 0$. Hence

$$f(r) \geq a_n r^n + M(r^p + r^{p-1} + \cdots + r + 1).$$

(b) The first inequality is obvious. Since $r > r - 1$,

$$a_n r^{n-p-1}(r - 1) > a_n(r - 1)^{n-p-1}(r - 1) > M.$$

(c) From (b),

$$a_n r^n > Mr^{p+1}(r - 1)^{-1} > M(r^{p+1} - 1)/(r - 1).$$

Now apply (a).

**2.14.** (a) $1 + \sqrt[6]{3} < 2.201$; $1 + 3/2 = 2.5$.
  (b) $1 + 3 = 4$; $1 + 1 = 2$.

**2.15.** (a) Let $q(a) = q(b) = k$. Then $a$ and $b$ are zeros of the polynomial $q(t) - k$ whose derivative is $p(t)$.

(b) Suppose there are $k$ distinct real zeros, $r_1, r_2, \ldots, r_k$ of $q(t)$. Since the multiplicity of each $r_i$ as a zero of $p(t)$ is one less than its multiplicity as a zero of $q(t)$, the total multiplicity of all the $r_i$ as zeros of $p(t)$ is $m - k$. However, Rolle's theorem provides an additional zero of $p(t)$ between adjacent pairs of the $r_i$ for an additional $k - 1$ zeros.

(c) We can take $2q(t) = t^4 - 2t^3 - t^2 + 2t = (t+1)t(t-1)(t-2)$. By Rolle's theorem, $p(t)$ has a zero in each of the intervals $(-1, 0)$, $(0, 1)$, $(1, 2)$.

(d) Let $f(t) = a_0 t + a_1 t^2/2 + \cdots + a_n t^{n+1}/(n+1)$. Since $f(0) = f(1) = 0$, by Rolle's theorem, $f'(t) = a_0 + a_1 t + \cdots + a_n t^n$ has at least one zero in the interval $(0, 1)$.

**2.16.** It is straightforward to see the result by examining the graph of $f(t)$. Here is an analytic argument. If not all the zeros of $f(t)$ are real, we may take $k = 0$. Suppose all the zeros $u_i$ of $f(t)$ are real. Then the zeros $v_i$ of $f'(t)$ are all real, and we can label them so that

$$u_1 \geq v_1 \geq u_2 \geq v_2 \geq \cdots \geq u_{n-1} \geq v_{n-1} \geq u_n \quad (\text{where } n = \deg f(t)).$$

With no loss of generality, we can assume that the leading coefficient of $f(t)$ is 1, so that $f(t) = (t - u_1)(t - u_2) \cdots (t - u_n)$. For $t \geq u_2$, we have that

$$(t - u_1)(t - u_2) = \left[t - \frac{1}{2}(u_1 + u_2)\right]^2 - \frac{1}{4}(u_1 - u_2)^2,$$

so that

$$f(t) \geq -\frac{1}{4}(u_1 - u_2)^2 (u_2 - u_3)(u_2 - u_4) \cdots (u_2 - u_n).$$

Let $k$ be any positive real exceeding the absolute value of the right side of this inequality. Then $f(t) + k$ does not vanish for $t \geq u_2$. If $f(t) + k$ has $m$ real zeros, then all are less than $u_2$. The derivative $f'(t)$ of $f(t) + k$ has at least $m - 1$ real zeros less than $u_2$. But then $m - 1 \leq n - 2$, so that $m \leq n - 1$ and we have the required result.

**2.17.** (c) Referring to the Taylor Expansion of $p(t)$ about 0, we see that the number of sign changes of the coefficients is equal to the number of sign changes in the sequence $(p(0), p'(0), \ldots)$. On the other hand, for sufficiently large positive $v$, $p^{(k)}(v)$ has the same sign as the leading coefficient of $p(t)$ for each $k$. Take this value of $v$ with $u = 0$ in the Fourier–Budan criterion; $A$ is the number of sign changes in the coefficients and $B = 0$.

(d) For $t = -2$, the sequence of derivative values has signs $+, -, +, -, +$; for $t = 0$, the signs are $-, +, +, 0, +$. The Fourier–Budan Theorem provides for 1 or 3 zeros in the interval $[-2, 0]$. However, since $t^4 + t^2 - 4t - 3$ has one sign change, the Descartes' Rule of Signs puts an upper bound of 1 on the number of zeros.

$$[t^4 + t^2 + 4t - 3 = (t^2 + t - 1)(t^2 - t + 3).]$$

**2.21.** (a) The real zeros are $-\sqrt{2}$, $\sqrt{2}$, approx. 3.087.

(b) Zeros are $-7$, $2/3$, $-5/8$, $(1/2)(1 \pm i\sqrt{7})$.

(c) $2t^4 + 5t^3 + t^2 + 5t + 2 = (2t^2 - t + 2)(t^2 + 3t + 1)$.

(d) Polynomial is $(4t - 7)(4t^5 + 7t^4 + 13t^3 + 22t^2 + 3t + 3)$. Real zeros are $7/4$ and approx. $-1.6896$.

(e) There are three real and four nonreal zeros. The polynomial factors as $(4t + 3)(t^2 + t + 2)(4t^4 - 7t^3 - 11t + 6)$. The nonrational zeros are approximately 0.4912 and 2.1829.

**3.1.** (b) If $|w| \leq 1$, then the inequality is satisfied. Otherwise $|w|^{-1} < 1$ and (a) can be applied.

**3.2.** (b)
$$|z_0| = |(z_0 + z_1 + \cdots + z_n) - (z_1 + \cdots + z_n)|$$
$$\leq |z_0 + \cdots + z_n| + |z_1 + z_2 + \cdots + z_n|.$$

Applying (a) and transferring terms yields the result. Strict equality holds iff all the $z_i$ have the same argument.

**3.3.** (b) If $|w| \leq 1$, the inequality holds. Otherwise, $|w|^{-1} < 1$ and, from (a),
$$0 \geq |a_n w^n|[1 - K(|w|^{-1} + |w|^{-2} + \cdots + |w|^{-n}]$$
$$> |a_n w^n|(|w| - 1)^{-1}(|w| - 1 - K).$$

**3.5.** (a) By Descartes' Rule of Signs, there is at most one positive zero. Since the polynomial is negative at $t = 0$ and positive for large $t$, there is at least one positive zero.

(b) Since the polynomial in (a) is positive for $t > r$, it follows that for $|w| > r$,
$$|w^n + a_{n-1}w^{n-1} + \cdots + a_1 w + a_0| \geq |w^n| - |a_{n-1}w^{n-1}| - \cdots - |a_0| > 0.$$

[A generalization of Exercise 2 and this Exercise appears in Emeric Deutsch, Bounds for the zeros of polynomial, *Amer. Math. Monthly* **88** (1981), 205–206.]

**3.6.** $(1 - w)g(w) = b_0 + (b_1 - b_0)w + \cdots + (b_n - b_{n-1})w^n - b_n w^{n+1}$. For $|w| \leq 1$, $w \neq 1$, we have from Exercise 2(b) that

$$|(1 - w)g(w)| > b_0 - |b_1 - b_0| - |b_2 - b_1| - \cdots - |b_n - b_{n-1}| - |b_n|$$
$$= b_0 + (b_1 - b_0) + (b_2 - b_1) + \cdots + (b_n - b_{n-1}) - b_n = 0.$$

Since, also, 1 is not a zero of $g(t)$, the required result follows.

(b) $w$ is a zero of $g(t)$ iff $w^{-1}$ is a zero of $b_0 t^n + b_1 t^{n-1} + \cdots + b_{n-1}t + b_n$, whence the required result follows from (a).

(c) It is straightforward to see that $0 < a_n u^n \leq a_{n-1}u^{n-1} \leq \cdots \leq a_1 u \leq a_0$. If $b_i = a_i u^i$, then $w$ is a zero of $f(t)$ iff $w/u$ is a zero of $g(t)$. By (a),

$|w/u| > 1$, so that $|w| > u$. A similar argument using (b) and the fact that $a_n v^n \geq a_{n-1} v^{n-1} \geq \cdots \geq a_1 v \geq a_0$ yields $|w| < v$.

**3.8.** (a) Case 1: *Both zeros are nonreal.* If the zeros are $u + vi$ and $u - vi$, then $(1 + c) + b = (1 - u)^2 + v^2$ and $(1 + c) - b = (1 + u)^2 + v^2$ are both positive, so that $|b| < 1 + c$ always holds. Hence $c < 1 \iff u^2 + v^2 < 1 \iff$ both zeros lie in the interior of the complex unit disc.

Case 2: *Both zeros are real.* If the zeros are $r$ and $s$, then $(1 + c) - b = (1 + r)(1 + s)$ and $(1 + c) + b = (1 - r)(1 - s)$. Suppose that $|r| < 1$, $|s| < 1$. Then $-1 < c = rs < 1 \Rightarrow 0 < 1 + c < 2$. Also, $1 \pm r$ and $1 \pm s$ are positive, so that $(1 + c) \pm b > 0 \Rightarrow b < 1 + c$. On the other hand, if $|b| < 1 + c < 2$, then $(1 \pm r)(1 \pm s) > 0$, so that $1 + r$ and $1 + s$ have the same sign, as do $1 - r$ and $1 - s$. Since $|rs| = |c| < 1$, this forces $-1 < r$, $s < 1$, as required.

For other treatments of this criterion, consult problems E1313 in *Amer. Math. Monthly* **65** (1958), 284, 776 and 1029 in *Crux Math.* **11** (1985), 83; **12** (1986), 191–193.

(b) Case 1: *The zeros are nonreal.* Let the zeros be $r$, $u + vi$, $u - vi$. Then

$$1 - b + c - d = (1 + r)[(1 + u)^2 + v^2]$$

$$1 + b + c + d = (1 - r)[(1 - u)^2 + v^2]$$

$$1 - c + bd - d^2 = [1 - (u^2 + v^2)][(1 - ru)^2 + r^2 v^2].$$

Hence $-1 < r < 1$, $u^2 + v^2 < 1 \iff |b + d| < 1 + c$ and $c - bd < 1 - d^2$.

Let all the zeros lie within the open disc defined by $|z| < 1$. Then the product $-d$ of the zeros satisfies $|d| < 1$. Hence $1 - d^2 > 0$ and $d(b + d) \leq |d(b + d)| < |b + d| < 1 + c$ so that $bd - c < 1 - d^2$. Hence $|c - bd| < 1 - d^2$ and $|b + d| < 1 + c = |1 + c|$.

On the other hand, suppose $|bd - c| < 1 - d^2$ and $|b + d| < |1 + c|$. Clearly $c - bd < 1 - d^2$. We show that $1 - b + c - d$ and $1 + b + c + d$ must both be positive. Suppose, if possible, that $1 - b + c - d \leq 0$ and $1 + b + c + d \geq 0$. Then $1 + c$ and $-(1 + c)$ do not exceed $b + d$ so that $|1 + c| \leq |b + d|$, a contradiction. On the other hand, if $1 - b + c - d \geq 0$ and $1 + b + c + d \leq 0$, then $1 + c$ and $-(1 + c)$ do not exceed $-(b + d)$, so that $|1 + c| \leq -(b + d) = |b + d|$, a contradiction. Since $1 - b + c - d$ and $1 + b + c + d$ also cannot both be negative, we must have $|b + d| < 1 + c$ and all zeros lie inside the open unit disc.

Case 2: *All zeros are real.* Let the zeros be $p$, $q$, $r$ with $p \leq q \leq r$. Then

$$1 - b + c - d = (1 + p)(1 + q)(1 + r)$$

$$1 + b + c + d = (1 - p)(1 - q)(1 - r)$$

$$1 - c + bd - d^2 = (1 - pq)(1 - pr)(1 - qr).$$

Let all the zeros lie within the open unit disc. From the above equations, $|b + d| < 1 + c = |1 + c|$ and $c - bd < 1 - d^2$. As in Case 1, it can be shown that $bd - c < 1 - d^2$, so that $|bd - c| < 1 - d^2$.

On the other hand, suppose that $|bd - c| < 1 - d^2$ and $|b + d| < |1 + c|$. As in Case 1, it can be argued that $1 - b + c - d$ and $1 + b + c + d$ have the same sign. Suppose $1 - b + c - d < 0$ and $1 + b + c + d < 0$. Then $|b + d| < -(1 + c)$. Since $0 < 1 - d^2$ and $bd - c < 1 - d^2$, it follows that $d(b + d) < 1 + c < 0$, so that $|1 + c| < |d| \, |b + d| < |b + d|$, a contradiction.

Hence $1 - b + c - d > 0$ and $1 + b + c + d > 0$. Noting that $-1 + b - c + d$ and $1 + b + c + d$ are the respective values of the polynomial at $-1$ and $1$, we see that one of the following possibilities obtains:

(i) $-1 < p < 1 < q \leq r$

(ii) $p \leq q < -1 < r < 1$

(iii) $-1 < p \leq q \leq r < 1$.

Ad (i), $1 - qr < 0$ and $(1 - pq)(1 - pr)(1 - qr) > 0$ implies that $p > 0$ and $pq < 1 < pr < pqr = -d$, which contradicts $1 - d^2 > 0$. Ad (ii), $1 - pq < 0$ implies that $r < 0$ and $d = |pqr| > |pr| > 1 > |qr|$, which contradicts $1 - d^2 > 0$. Hence (iii) must hold.

**3.9.** (d) If the zeros of a monic cube are $r$, $u + iv$, $u - iv$, with $r$, $u$, $v$ real, then the coefficients are positive $\iff r + 2u < 0$, $2ru + u^2 + v^2 > 0$ and $r(u^2 + v^2) < 0$. The last inequality forces $r$ to be negative. However, we can choose $u$ to be a small positive value and $v$ a large value to achieve the other two inequalities. For example, taking $r = -3$, $u = 1$, $v = 3$ leads to the polynomial $t^3 + t^2 + 4t + 30$.

**3.10.** (d) Let $p(t) = 8a^2t^3 + 8abt^2 + 2(b^2 + ac)t + (bc - ad)$. We are given that $p(t)$ has a real zero $u$. If $bc > ad$, then $p(t) > 0$ for $t \geq 0$, so that $u < 0$. If $bc \leq ad$, then $p(t) = 2t(2at + b)^2 + 2act + (bc - ad)$ shows that $p(t)$ has no negative zero. If $bc = ad$, then $u$ must be zero and it can be checked directly that $f(\pm\sqrt{c/a}) = 0$. If $bc < ad$, then $u > 0$.

**3.11.** (a) The reasoning, analogous to that of Section 4.5, can be outlined as follows. If the Nyquist diagram does make at least one circuit of the origin, then by shrinking the semi-circle in the right side of the $z$-plane down to a point, we find that its image curve must at some stage pass through the origin. Suppose the Nyquist diagram does not make a circuit of the origin. Let $D$ be the image of the circle of radius $M$. Since the disc of radius $M$ contains all the zeros of the polynomial, $D$ must wind around the origin $n$ times ($n$ being the degree of the polynomial). Consider three curves:

$D_1$: the image of the semi-circular arc $\{z : |z| = M,\ \mathrm{Re}\,z \geq 0\}$,
$D_2$: the image of the semi-circular arc $\{z : |z| = M,\ \mathrm{Re}\,z \leq 0\}$,
$D_3$: the image of the line segment $\{z : z = yi,\ y \text{ real},\ |y| \leq M\}$.

$D = D_1 \cup D_2$. A point tracking around $D$ traces around $D_1$ and $D_2$ in succession. $D_1 \cup D_3$ does not make any circuit of the origin, so that all the winding around the origin is done by the $D_2$ part of the curve. By making a few sketches, we can see that $D_2 \cup D_3$ (the image of the left semi-circle) must wind around the origin $n$ times. But then the left semi-circle must surround all $n$ zeros of the polynomial, leaving none for the right semi-circle to surround.

(b)

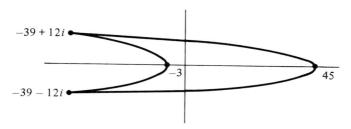

**3.12.** (a) The Nyquist diagram indicates that the polynomial is unstable.

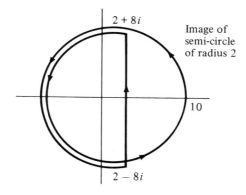

Image of semi-circle of radius 2

(b) Every real zero must be negative. Suppose $u + iv$ is a nonreal zero. Then $(u^3 + 2u^2 + 3u + 1) = v^2(3u + 2)$ and $3u^2 + 4u + 3 = v^2$. Eliminating $v^2$ leads to $8u^3 + 16u^2 + 14u + 5 = 0$, whence $u$ cannot be nonnegative. Hence any nonreal zero must have negative real part.

(c) Every real zero must be negative. If $u + vi$ is a nonreal zero, then

$$(u^4 + 3u^3 + u^2 + u + 8) - (6u^2 + 9u + 1)v^2 + v^4 = 0 \qquad \ldots (1)$$

$$(4u^3 + 9u^2 + 2u + 1) = (4u + 3)v^2 \qquad \ldots (2)$$

Eliminating $v^2$ leads to

$$(u^4 + 3u^2 + u^2 + u + 8)(4u + 3)^2 - (6u^2 + 9u + 1)(4u + 3)(4u^3 + 9u^2 + 2u + 1)$$
$$+ (4u^3 + 9u^2 + 2u + 1)^2 = 0.$$

The leading coefficient of the left side is negative and the constant coefficient is positive. Hence the equation is satisfied for at least one positive value of $u$. Such a positive value of $u$ leads to a pair of real values of $v$, so that the system has a solution $u + vi$ with $u > 0$. Hence the given equation is not stable.

## Solutions to Problems

**Chapter 5**

**4.1.** Let $u$, $v$ be any two distinct real zeros, if such exist of $t^3 - 3t + k$. Then $u^3 - v^3 = 3(u - v)$, whence $u^2 + uv + v^2 = 3$. If both $u$ and $v$ are positive and less than 1, then the left side is less than 3, yielding a contradiction. Hence at most one real zero lies in $[0, 1]$.

An alternative solution can be based on the graph of the polynomial, using the fact that there is a unique maximum when $t = -1$ and a unique minimum when $t = 1$.

**4.2.** By Descartes' Rule of Signs, there is at most one positive root. The left side can be rewritten as

$$n!x(1 + x)(1 + x/2)(1 + x/3) \cdots (1 + x/n),$$

which for $x = 1/n!$ exceeds 1. Since the left side is less than 1 for $x = 0$, it must equal 1 for some $x$ between 0 and $1/n!$.

**4.3.** Clearly, $x = 1$ satisfies the equation and $x = 0$ does not. Let $n \geq 2$. The equation can be rewritten

$$n = x^{-n} + x^{-n+1} + \cdots + x^{-1}.$$

If $|x| \geq 1$, then $|x^{-n} + \cdots + x^{-1}| \leq n$ with equality iff both $|x| < 1$ and all powers of $x^{-1}$ have the same argument iff $x = 1$. Therefore, for all $x$ satisfying the equation, either $x = 1$ or $|x| < 1$.

A number of other solutions, some using Exercise 3.6, can be found in *Amer. Math. Monthly* **66** (1959), 143–144.

**4.4.** With the quartic written as $(x^2 - 2)^2 - x(5x^2 + 7)$, it is evident that it is positive when $x$ is negative.

**4.5.** We prove by induction that, if $|k| < 2$, then $f_n(t) = k$ has $2^n$ distinct real roots. The result is clearly true when $n = 0$. Suppose it holds for $n - 1$. Then $f_n(t) = k \iff f_{n-1}(t) = \sqrt{2 + k}$ or $f_{n-1}(t) = -\sqrt{2 + k}$. By the induction hypothesis, each of the alternative equations has $2^{n-1}$ distinct real roots, and any root of one is not a root of the other. The result follows.

**4.6.** The equation can be rewritten

$$t^{n+1} + at^n + \cdots + a^{n+1} = t^{n+1} + bt^n + \cdots + b^{n+1}.$$

After multiplication by $(t - a)(t - b)$, this becomes $f(t) = 0$ where

$$
\begin{aligned}
f(t) &= (t^{n+2} - a^{n+2})(t - b) - (t^{n+2} - b^{n+2})(t - a) \\
&= (a - b)t^{n+2} - (a^{n+2} - b^{n+2})t + ab(a^{n+1} - b^{n+1}).
\end{aligned}
$$

We have that $f(a) = f(b) = 0$ and $f'(t) = (n+2)(a-b)t^{n+1} - (a^{n+2} - b^{n+2})$.

If $n$ is even, $f'(t)$ has exactly one real zero, counting multiplicity, so that $f(t)$ has at most two. These are already accounted for by the extraneous roots $a$ and $b$ of $f(t) = 0$, so that the original equation has no root.

If $n$ is odd, then $f'(t)$ has exactly two zeros counting multiplicity, so $f(t)$ has at most three. Since two of these are accounted for by $a$ and $b$, then given equation has at most one root.

**4.7.** Let

$$f(t) = 2t^3 - pt^2 + qt - r = (t - a_1)(t - a_3)(t - a_5)$$

$$+ (t - a_2)(t - a_4)(t - a_6).$$

Since $f(a_i)$ is positive for $i = 1, 4, 5$ and negative for $i = 2, 3, 6$, there must be a real zero in each of the intervals $(a_{2j-1}, a_{2j})$ for $j = 1, 2, 3$.

**4.8.** By Descartes' Rule of Signs, there is at most one positive root. Denoting the left side by $f(x)$, we have that

$$f(a + b + c) = 2\Sigma a^2 b + 4abc > 0$$

and

$$27f(2(a + b + c)/3) = -[10\Sigma a^3 - 6\Sigma a^2 b + 6abc] = -\{[5(a + b) - c][a - b]^2$$

$$+ [5(b + c) - a][b - c]^2 + [5(c + a) - b][c - a]^2\}.$$

Without loss of generality, suppose $a \geq b \geq c$. If $a \leq 5(b + c)$, then $f(2(a + b + c)/3) \leq 0$. On the other hand, if $a > 5(b + c)$, then

$$10\Sigma a^3 - 6\Sigma a^2 b + 6abc > [2a^3 - 6a^2(b + c)] + [2a^3 - 6a(b^2 + c^2)]$$

$$+ [2a^3 - 6bc(b + c)] > 4a^2(b + c) + 44a(b^2 + c^2) + 4bc(b + c) > 0,$$

so again $f(2(a + b + c)/3) \leq 0$. The required result follows.

**Remark.** Another solution can be found in *Crux Math.* **11** (1985), 129. See also the discussion to problem 787 in *Crux Math.* **10** (1984), 56–58, in which it is pointed out that, if $a$, $b$, $c$ are three sides of a given quadrilateral, the length $x$ of the fourth side in order to achieve maximum area is given by the equation of the problem. In fact, the points can be arranged in a semi-circle whose diameter is the fourth side.

**4.9.** By the Fundamental Theorem of Algebra, we have that

$$(x - a_1)(x - a_2) \cdots (x - a_n) - 1 = (x - r_1)(x - r_2) \cdots (x - r_n),$$

whence it can be seen that the second equation has $n$ zeros $a_i$.

**4.10.** Let $p(x) = k(x - a_1)(x - a_2) \cdots (x - a_n)$. Then $p'(a_1) = k(a_2 - a_1)(a_3 - a_1) \cdots (a_n - a_1)$. For $i = 2, 3, \ldots, n$,

$$|a_i - a_1| \le |a_i - b| + |b - a_1| < 2|b - a_i|.$$

Hence

$$|p(b)| = k \prod_{i=1}^{n} |b - a_i| \ge k2^{-n+1}|b - a_1| \prod_{i=2}^{n} |a_i - a_1|$$
$$= 2^{-n+1}|p'(a_1)(b - a_1)|.$$

**4.11.** Let the left side be $f(t)$ and suppose that $a \le b \le c$. We have that

$$f(t) = (t - b)[(t - a)(t - c) - e^2] - (t - a)d^2 - (t - c)f^2 - 2def.$$

The quadratic $(t - a)(t - c) - e^2$ is nonpositive for $t = a$ and $t = c$ and so has real zeros $u$ and $v$ for which $u \le a \le c \le v$. Thus $(a - u)(c - u) = (v - a)(v - c) = e^2$, whence

$$f(u) = (\sqrt{a - u}\,d \pm \sqrt{c - u}\,f)^2 \ge 0$$

and

$$f(v) = -(\sqrt{v - a}\,d \pm \sqrt{v - c}\,f)^2 \le 0.$$

Since the leading coefficient of $f(t)$ is positive, $f(t)$ has a zero not exceeding $u$, a zero not less than $v$ and a zero in the interval $[u, v]$. The only way in which $f(t)$ might fail to have three real zeros is that $u = a = b = c = v$. But, in this case, $e = 0$ and $f(t) = (t - a)[(t - a)^2 - d^2 - f^2]$ evidently has three real roots.

**Remark.** The roots of this equation are the eigenvalues of the real symmetric matrix $(a\,f\,e\,/\,f\,b\,d\,/\,e\,d\,c)$, which must be real from matrix theory.

**4.12.** By Descartes' Rule of Signs, there are at most two positive roots. If $f(x)$ denotes the left side, then

$$f(x) = (x^2 - 10^{10})^2 - (x^2 - 10^{10}) - (x + 1)$$
$$= (x^2 - 10^{10} - 0.5)^2 - (x + 1.25).$$

Since $f(10^5) < 0$ and $f(10^5 \pm 10^{-2}) = [10^{-4} \pm 2.10^3 - 0.5]^2 - [10^5 \pm 10^{-3} + 1.25] > 10^6 - 10^5 - 2 > 0$, the two roots lie in the open intervals (99999.99, 100000) and (100000, 100000.01). If $x < 0$, it is straightforward to see that $f(x) > 0$, so that there are no negative roots.

The positive roots are solutions of the two equations

$$x = [10^{10} + 0.5 + (x + 1.25)^{1/2}]^{1/2} \tag{1}$$

$$x = [10^{10} + 0.5 - (x + 1.25)^{1/2}]^{1/2}. \tag{2}$$

We solve by successive approximation, by evaluating the right side at $x = 10^5$. With the aid of the binomial theorem, this yields

$$
\begin{aligned}
(1) \quad x &= 10^5[1 + (316.73)10^{-10}]^{1/2} = 10^5 + (158.36)10^{-5} \\
&= 100000.0016
\end{aligned}
$$

$$
\begin{aligned}
(2) \quad x &= 10^5[1 - (315.73)10^{-10}]^{1/2} = 10^5 - (157.86)10^{-5} \\
&= 99999.9984.
\end{aligned}
$$

These solutions check out.

**4.13.** Suppose that $p(x) = ax^2 - bx + c$ has two distinct zeros in $(0,1)$, and that $a$, $b$, $c$ are integers with $a$ positive. Then $b > 0$, $c = p(0) > 0$, $a - b + c = p(1) > 0$, $b^2 - 4ac > 0$ and the product $c/a$ of the zeros is less than 1. Thus, it is necessary that
  (1) $0 < c < a$,
  (2) $b < a + c$,
  (3) $4ac < b^2$.
Since $a$ and $c$ are positive integers, $(a-1)(c-1) \geq 0$, whence $b^2 > 4ac \geq 4(a + c - 1) \geq 4b \Rightarrow b \geq 5 \Rightarrow 2a > a + c \geq 6 \Rightarrow a \geq 4$. Now $a = 4$ forces $c = 2$ or $c = 3$, either of which make (2) and (3) incompatible. Hence $a \geq 5$.
   Consider the possibilities for $a = 5$. If $c \geq 2$, then (2) and (3) are incompatible. However, $c = 1$ forces $b = 5$ and yields the polynomial $5x^2 - 5x + 1$, which has the stated property.

**4.14.** Since each $q_n(x)$ has nonnegative coefficients, any real zero cannot be positive. It is readily checked that $q_1(x)$ and $q_2(x)$ each have a negative zero. Suppose, for some $n \geq 2$, it has been established that $q_i(0) > 0$ and that each $q_i(x)$ has at least one negative zero and that the greatest such zero $x_i$ satisfies $x_{i-1} < x_i$ ($2 \leq i \leq n$). Then, for $1 \leq i \leq j \leq n$, $q_i(x) > 0$ for $x_j < x \leq 0$.
   From the recursion relations, it follows that $q_{n+1}(0) > 0$ and $q_{n+1}(x_n) < 0$, so that $q_{n+1}(x)$ has at least one zero in the interval $(x_n, 0)$. Thus, there is a largest real zero $x_{n+1}$ and it satisfies $x_n < x_{n+1} < 0$.
   In fact, it can be shown that $x_{2m+2} > -1/(m + 1)$ for $m \geq 1$. For, from the recursion relations, we find that

$$
q_{2m+2}(-1/(m + 1)) = -q_{2m-1}(-1/(m + 1)).
$$

If $q_{2m+2}(-1/(m + 1)) < 0$, then $x_{2m+2} > -1/(m + 1)$. On the other hand, if $q_{2m+2}(-1/(m + 1)) > 0$, then $q_{2m-1}(-1/(m + 1)) < 0$, so that $x_{2m+2} > x_{2m-1} > -1/(m + 1)$.
   Therefore, if $r$ is any positive number, no matter how small, we can choose a positive integer $m$ so that $-r < -1/(m + 1) < 0$. Then, for $n \geq 2m + 2$, $-r < x_n < 0$, so that the $x_n$ get closer and closer to 0.

**Remark.** These polynomials occur as the denominators of a sequence of convergents in a continued fraction representation of the infinite series $1 - x + 2x^2 - 6x^3 + 24x^4 - 120x^5 + \cdots$ studied by L. Euler around 1760. See E.J. Barbeau, Euler subdues a very obstreperous series *Amer. Math. Monthly* **86** (1979), 356–372.

**4.15.** *First solution.* Let the three zeros be $u$, $v$, $w$ with $u \geq v \geq w$. Let $u - w = r$ and $u - v = s$, so that $v - w = r - s$. Hence

$$
\begin{aligned}
2(a^2 - 3b) &= 2(u + v + w)^2 - 6(uv + vw + wu) \\
&= (u - w)^2 + (u - v)^2 + (v - w)^2 \\
&= r^2 + s^2 + (r - s)^2 \\
&= 2(s - r/2)^2 + 3r^2/2 \geq 0.
\end{aligned}
$$

Under the condition $0 \leq s \leq r$, the quadratic (in $r$ and $s$) is minimized when $s = r/2$ and maximized when $s = 0$ or $s = r$. Hence

$$(3/2)r^2 \leq 2(a^2 - 3b) \leq 2r^2,$$

so that

$$(a^2 - 3b)^{1/2} \leq u - w \leq (2/\sqrt{3})(a^2 - 3b)^{1/2} < 2(a^2 - 3b)^{1/2}.$$

(Solution due to David Ash.)

*Second solution.* By a change of variable $x = t + a/3$, we can write $t^3 + at^2 + bt + c = x^3 + px + q$ where $p = b - a^2/3$. Consider the graph of $y = x^3 + px + q$. It is centrally symmetric about the point $(0, q)$ and has parallel tangents at $(x_1, q)$ and $(x_2, q)$.

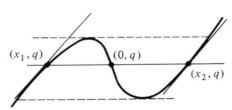

The distance between the roots of $x^3 + px + q = 0$ is the distance between the outer points of intersection of the curve and the line $y = 0$. From the shape of the curve, it can be seen that this distance lies somewhere between the values $x_2 - x_1$ and $x_4 - x_3$, where $x_1$, $0$, $x_2$ are the roots of the equation $x^3 + px + q = q$ and $x_3$, $x_4$ are the roots of the equation $x^3 + px + q = r$, where $r$ is chosen so that the equation has a double root.

Clearly, $x_2 - x_1 = 2(-p)^{1/2}$ (so that, in particular $a^2 - 3b = -3p > 0$). Now, $x^3 + px + (q - r)$ has a double zero when it shares a zero with its derivative $3x^2 + p$, i.e. when its zeros are either $(-p/3)^{1/2}$, $(-p/3)^{1/2}$, $-2(-p/3)^{1/2}$ or the negative of these. Hence $x_4 - x_3 = 3(-p/3)^{1/2}$.

The distance between the outer zeros of $x^3+px+q$ is equal to the distance between the outer zeros of the given polynomial, and it lies between

$$x_4 - x_3 = 3(-p/3)^{1/2} = (a^2 - 3b)^{1/2}$$

and

$$x_2 - x_1 = 2(-p)^{1/2} = (2/\sqrt{3})(a^2 - 3b)^{1/2}.$$

**4.16.** It is clear that $z^6 + 6z + 10 = 0$ has no pure imaginary roots. Since $6z + 10 \geq 0$ when $z \geq -5/3$ and $z^6 + 6z \geq 0$ when $z \leq -5/3$, there are no real roots. Since the coefficients are real, the number of zeros in the first and fourth quadrants are equal as are the number of zeros in the second and third quadrants. This it suffices to consider zeros $z = r(\cos\theta + i\sin\theta)$ for which $r > 0$ and $0 < \theta < \pi$, i.e. $\sin\theta > 0$.

For such a root

$$r^6 \cos 6\theta + 6r \cos\theta + 10 = 0 \qquad \ldots(1)$$

$$r^6 \sin 6\theta + 6r \sin\theta = 0. \qquad \ldots(2)$$

From (2), $r^5 = -6\sin\theta/\sin 6\theta \Rightarrow \sin 6\theta < 0 \Rightarrow \pi/6 < \theta < \pi/3, \pi/2 < \theta < 2\pi/3$ or $5\pi/6 < \theta < \pi$. Substituting for $r^5$ into (1) yields

$$0 = r(r^5 \cos 6\theta + 6\cos\theta) + 10 = (6r\sin 5\theta)/(\sin 6\theta) + 10$$

so that $r = -(5\sin 6\theta)/(3\sin 5\theta)$.

Hence $\theta$ must satisfy the equation $f(\theta) = 0$ where

$$f(\theta) = 6.3^5 \sin\theta \sin^5 5\theta - 5^5 \sin^6 6\theta.$$

Note that $f(\theta)$ is positive for $\theta = \pi/6, \pi/2$ and $5\pi/6$, and negative for $\theta = \pi/3$ and $2\pi/3$. Since $f(\pi) = 0$, we look at $f(\pi - \phi) = f(\phi)$, where $\phi$ is very small. In this case, $\sin\phi \doteq \phi$, $\sin 5\phi \doteq 5\phi$ and $\sin 6\phi \doteq 6\phi$, so

$$f(\phi) \doteq 6.3^5 \cdot 5^5 \phi^6 - 5^5 \cdot 6^6 \phi^6 < 0.$$

Thus $f(\phi)$ must vanish in each of the open intervals

$$(0, \pi/6), \ (\pi/6, \pi/3), \ (\pi/3, \pi/2), \ (\pi/2, 2\pi/3), \ (2\pi/3, 5\pi/6), \ (5\pi/6, \pi).$$

Note that the vanishing of $f(\phi)$ entails that $\sin 5\theta$ must be positive. Since $\sin 6\theta < 0$ for $\theta$ in the second, fourth and sixth intervals, we obtain a positive value of $r$ and a viable solution $(r, \theta)$ to the equations (1) and (2). (The zeros of $f(\theta)$ in the other three intervals lead to inadmissible solutions with $r < 0$.)

Thus there are six zeros of $z^6 + 6z + 10$, one in the first quadrant whose argument is between $\pi/6$ and $\pi/3$, two others in each of the second and third quadrants, and one in the fourth quadrant.

**4.17.** Let $f(x)$ denote the left side. We make use of Rolle's Theorem,

$$f'(x) = u_1(x)x^{15}(x+1)^{63} + v_1(x)x^8(x+1)^{26} + w_1(x)x^3(x+1)^7,$$

where $\deg u_1(x) = \deg v_1(x) = \deg w_1(x) = 1$. The positive zeros of $f'(x)$ are the same as those of

$$g(x) = u_1(x)x^{12}(x+1)^{56} + v_1(x)x^5(x+1)^{19} + w_1(x).$$

Differentiating twice yields

$$g''(x) = u_3(x)x^{10}(x+1)^{54} + v_3(x)x^3(x+1)^{17}$$

where $\deg u_3(x) = \deg v_3(x) = 3$. The positive zeros of $g''(x)$ are the same as those of

$$h(x) = u_3(x)x^7(x+1)^{37} + v_3(x).$$

Differentiating four times yields

$$h^{(4)}(x) = u_7(x)x^3(x+1)^{34}$$

where $\deg u_7(x) = 7$.

The number of positive zeros of $h^{(4)}(x)$ does not exceed $7 = \deg u_7(x)$. Hence the number of positive zeros of $h(x)$ and $g''(x)$ does not exceed 11, the number of positive zeros of $g(x)$ and $f'(x)$ does not exceed 13, and the number of positive zeros of $f(x)$ does not exceed 14.

**4.18.** This problem could be solved by using Exercise 3.8 and the delineation of the cases for real and nonreal zeros made through Exercise 6.1.9. However, let us take a more elementary approach.

Let $f(t) = t^3 - t^2 + a$. For $f(t)$ not to have a real zero outside of the closed interval $[-1, 1]$, it is necessary that $a = f(1) \geq 0$ and $-2 + a = f(-1) \leq 0$, i.e. $0 \leq a \leq 2$. Since the zeros of $f'(t)$ are 0 and 2/3, $f(t)$ has a maximum at $t = 0$ and a minimum at $t = 2/3$. All zeros are real iff $a = f(0) \geq 0$ and $-4/27 + a = f(2/3) \leq 0$ iff $0 \leq a \leq 4/27$. For all $a$ in this range, all three zeros lie in $[-1, 1]$.

(To get an intuitive idea of what is happening, we imagine how the zeros of $f(t)$ vary with $a$ near $a = 0$ and $a = 4/27$. Let $a = 0$; $f(t)$ has a double zero at $t = 0$ and a simple zero at $t = 1$. As $a$ decreases through zero, two real zeros of $f(t)$ converge to 0 and then split into a complex conjugate pair moving away from 0 while one real zero increases along the real axis through 1. Thus we can see that for $a < 0$, the real zero will lie outside the unit disc. Now let $a = 4/27$. Then $f(t)$ has a double zero at $t = 2/3$ and a simple zero at $t = -1/3$. As $a$ increases through 4/27, two real zeros of $f(t)$ converge to 2/3 and then split into a complex conjugate pair moving away from 2/3 while one real zero moves to the left through $-1/3$. It will be possible for $a$ to increase further above 4/27 before the zeros leave the disc.

We know the real zero does not leave until $a = 2$. When will the nonreal zero leave?)

Let $a > 4/27$ and suppose that the three zeros are $r$, $u + iv$ and $u - iv$. Then

$$r + 2u = 1 \quad 2ru + u^2 + v^2 = 0 \quad r(u^2 + v^2) = -a.$$

We eliminate $u$ and $r$ from these three equations. From the first and third

$$-a = (1 - 2u)(u^2 + v^2).$$

From the second and third

$$-2au + (u^2 + v^2)^2 = 0.$$

Hence

$$-a^2 = [a - (u^2 + v^2)^2](u^2 + v^2).$$

Thus, $u^2 + v^2$ is the unique positive zero of the polynomial $g(t) = t^3 - at - a^2$. This zero lies in the interval

$$(0, 1] \iff 0 \le g(1) \iff 0 \le 1 - a - a^2 \iff a \le (1/2)(\sqrt{5} - 1).$$

[Note that $4/27 < (1/2)(\sqrt{5} - 1) < 2$.]

Hence, all three zeros of $f(t)$ lie in the closed unit disc iff $0 \le a \le (1/2)(\sqrt{5} - 1)$.

**4.19.** We can make a change of variable $t = s + (b + c)/2$ to obtain a polynomial $g(s) = s^3 + us^2 - v^2 s - uv^2$ whose zeros are $-u$, $-v$, $v$ where $u \ge v \ge 0$. The zeros of $g(s)$ and $g'(s)$, respectively, are equal to the zeros of $f(t)$ and $f'(t)$ translated by $-(b + c)/2$. Hence it suffices to show the result for $g(s)$, i.e. that $g'(s) = 3s^2 + 2us - v^2$ has a zero in the closed interval $[0, v/3]$.

Clearly $g'(0) \le 0$, $g'(v/3) = 2v(u - v)/3 \ge 0$. If $g'(0) = 0$, then $v = 0$ and $g(s)$ has 0 as a double zero, so that $f(t)$ has $b = c$ as a double zero. If $g'(v/3) = 0$, then $g(s)$ has $-v = -u$ as a double zero, so that $f(t)$ has $a = b$ as a double zero.

**4.20.** Observe that $u = (1/2)(\sqrt{5} - 1)$ is the unique positive zero of the polynomial $1 - t - t^2$, so that, if $|z| < u$, then $1 - |z| - |z|^2 > 0$.

First, suppose that $n_1 \ge 2$. Then, for $|z| < u < 1$,

$$
\begin{aligned}
|1 + z^{n_1} + z^{n_2} + \cdots + z^{n_k}| &\ge 1 - |z|^{n_1} - |z|^{n_2} - \cdots - |z|^{n_k} \\
&\ge 1 - |z|^2 - |z|^3 - \cdots > 1 - |z|^2(1 - |z|)^{-1} \\
&= (1 - |z| - |z|^2)(1 - |z|)^{-1} > 0.
\end{aligned}
$$

On the other hand, let $n_1 = 1$. Observe that

$$(1 - z)(1 + z + z^{n_2} + \cdots + z^{n_k}) = 1 - z^2 + g(z),$$

where $g(z)$ is a sum of the form $\Sigma \pm z^k$ with $k \geq 2$. By an estimate similar to that above,

$$|(1 - z)(1 + z + z^{n_2} + \cdots)| > (1 - |z| - |z|^2)(1 - |z|)^{-1} > 0.$$

Thus, the result follows.

**4.21.** Suppose that all the zeros lie on a straight line. Since the sum of the zeros is 0, the line must pass through 0, so that the zeros are $-w$, $uw$, $(1 - u)w$ for some complex $w$ and real $u$ with $0 \leq u \leq 1$. We have that

$$[-u - (1 - u) + u(1 - u)]w^2 = 12(1 + i\sqrt{3})$$

$$\Rightarrow (u^2 - u + 1)w^2 = -24(\cos \pi/3 + i \sin \pi/3)$$

$$= 24(\cos 4\pi/3 + i \sin 4\pi/3).$$

Hence $w = \pm r(\cos 2\pi/3 + i \sin 2\pi/3)$, where $(u^2 - u + 1)r^2 = 24$.

By considering the product of the zeros, we find that

$$a = w^3(u - u^2) = \pm r^3(u - u^2) = \pm 24^{3/2}(u - u^2)(u^2 - u + 1)^{-3/2}$$

$$= \pm 24^{3/2}(1/4 - v)(3/4 + v)^{-3/2}$$

where $0 \leq v = (1/2 - u)^2 \leq 1/4$. Hence $a$ must be real.

(**Remark.** One has the suspicion that $|a|$ is minimized when the equation has a double root. This will occur when $u = 1 - u = 1/2$. In fact, by finding the zeros $\pm\sqrt{2}(1 - i\sqrt{3})$ of the derivative $3[z^2 + 4(1 + i\sqrt{3})]$, we can identify the roots of the equation in the critical situation as $\pm\sqrt{2}(1 - i\sqrt{3})$ with multiplicity 2 and $\mp 2\sqrt{2}(1 - i\sqrt{3})$. This will occur when $a = \mp 32\sqrt{2}$. In view of this, we would expect $24^{3/2}[1/4 - v][3/4 + v]^{-3/2}$ to assume its maximum value of $32\sqrt{2}$ when $v = 0$. Thus, we look at the difference of the squares of these two quantities.)

Now

$$2^{11} \quad - \quad 24^3[1/4 - v]^2[3/4 + v]^{-3}$$

$$= \quad 2^9[3/4 + v]^{-3}[4(3/4 + v)^3 - 27(1/4 - v)^2]$$

$$= \quad 2^7[3/4 + v]^{-3}v(4v - 9)^2 \geq 0.$$

Hence

$$|a| = 24^{3/2}[1/4 - v][3/4 + v]^{3/2} \leq \sqrt{2}^{11} = 32\sqrt{2}.$$

On the other hand, suppose $a$ is real and satisfies $|a| \leq 32\sqrt{2}$. Then we can find $u$ such that $\pm 24^{3/2}u(1 - u)(1 - u + u^2)^{-3/2}$ assumes the value $a$. Then, if $r$ and $w$ are determined by the equations above, we find that the symmetric functions in $-w$, $uw$, $(1 - u)w$ yields the coefficients of the equation and therefore must be its roots.

This result is generalized in *El. Math.* **12** (1957), 12.

**4.22.** Denote the left side by $p(t)$. Since $p(t)$ is positive for large $|t|$, $p(t)$ assumes its minimum value on $\mathbf{R}$ at some point $c$; we must have that $p'(c) = 0$. But then $p(c) = p'(c) + c^{2n}/(2n)! > 0$, so that $p(t)$ has a positive minimum and is therefore everywhere positive.

**4.23.** Suppose $p(t)$ is real iff $t$ is real. Then for all real values of $k$, $p(t) + k$ has the same property. Hence all the zeros of $p(t) + k$ are real for each value of $k$. By Exercise 5.2.16, this can occur only if $\deg p(t) = 1$.

**4.24.** Let $q(t) = a_0 t + a_1 t^2/2 + \cdots + a_n t^{n+1}/(n+1)$. Then $q(0) = 0$, $q'(t) = p(t)$ and $q'(0) = a_0 > 0$, so that $q(t)$ is increasing at $t = 0$. However, $q(1) - q(-1) = 2(a_0 + a_2/3 + a_4/5 + \cdots) < 0$, so that $q(1) < q(-1)$. Hence $q(t)$ cannot be increasing on the whole of the interval $[-1, 1]$, so that $q'(t) = p(t)$ must assume both positive and negative values there. Hence, for some $r \in [-1, 1]$, $p(r) = 0$.

**4.25.** Let the zeros of $p(z)$ be $r_i$ ($1 \le i \le n$). Then, for $z \ne r_i$,

$$n - kp'(z)/p(z) = n - k\Sigma(z - r_i)^{-1}.$$

Suppose that $|z| > R + |k|$, then

$$|z - r_i| > |z| - |r_i| > (R + |k|) - R = |k|.$$

Hence $|k\Sigma(z - r_i)^{-1}| \le |k|\Sigma(|z - r_i|)^{-1} < n$, so that $np(z) - kp'(z) \ne 0$. The result follows.

**4.26.** Let $d$ be the greatest common divisor of $p$ and $q$, and let $z^d = 1$. Then $z^{p+q} = z^p = 1$ and so $z$ is a zero of the polynomial. On the other hand, if $|z| = 1$ and $bz^p = az^{p+q} + b - a$, then

$$b = |bz^p| = |az^{p+q} + b - a| \le a + (b - a) = b,$$

so that $|az^{p+q} + b - a| = a + (b - a)$. Hence $z^{p+q} = 1$ and so $z^p = 1$. Hence $z^d = 1$.

**4.27.** Suppose there exists a root $z$ with $|z| \ne 1$. Since the reciprocal of any zero is also a zero, we may suppose that $z = r(\cos\theta + i\sin\theta)$ with $r > 1$. Since $z^n(z - u) = (1 - uz)$, we have that $|z|^{2n}|z - u|^2 = |1 - uz|^2$. Thus

$$r^{2n}(r^2 - 2ru\cos\theta + u^2) = (1 - 2ru\cos\theta + u^2 r^2).$$

Since $r > 1$ and $r^2 - 2ru\cos\theta + u^2 \ge (r - |u|)^2 > 0$,

$$r^2 - 2ru\cos\theta + u^2 < 1 - 2ru\cos\theta + u^2 r^2 \Rightarrow (r^2 - 1)(1 - u^2) < 0,$$

which contradicts $|u| \le 1$ and $r > 1$. Hence $|z| = 1$ for each zero $z$.

A number of solutions to this problem is provided in *Amer. Math. Monthly* **72** (1965), 1143–1144. One of them yields the result as a special case of the following: if $f(t) = \Sigma a_r t^r$ and $g(t) = \Sigma \bar{a}_{n-r} t^r$ satisfy $a_n g(t) = \bar{a}_0 f(t)$, then all of the zeros of $f(t)$ lie on the unit circle.

**4.28.** If $k = 0$, the equation is $x^n + x^{n-1} + \cdots + x + 1 = 0$ and the result holds. Suppose the result has been established for $k \leq r$. Let

$$f(x) = (n+1)^{-(r+1)} x^n + n^{-(r+1)} x^{n-1} + \cdots + 2^{-(r+1)} x + 1$$

$$g(x) = x f(x).$$

Then

$$g'(x) = (n+1)^{-r} x^n + n^{-r} x^{n-1} + \cdots + 2^{-r} x + 1.$$

Suppose $n$ is odd. Then, by the induction hypothesis, $g'(x)$ has exactly one real zero. By Rolle's Theorem, $g(x)$ has at most two real zeros. Since $g(0) = 0$ and $f(0) \neq 0$, $f(x)$ has at most one real zero. But, since $\deg f(x)$ is odd, $f(x)$ has a real zero.

If $n$ is even, $g'(x)$ has no real zero, so $g(x)$ has at most one real zero. Since $g(0) = 0$ and $f(0) \neq 0$, $f(x)$ can have no real zero.

**4.29.** The equation can be rewritten as $f(x) = 0$ where

$$f(x) = n(a_1 - x)(a_2 - x) \cdots (a_n - x)$$

$$- \sum_{i=1}^{n} a_i (a_1 - x) \cdots (\widehat{a_i - x}) \cdots (a_n - x).$$

Now $f(a_i) = (-1)^i a_i p_i$ where $p_i > 0$. If all the $a_i$ have the same sign, then the signs of the $f(a_i)$ alternate and each interval $(a_i, a_{i+1})$ $(1 \leq i \leq n-1)$ contains a root of the equation. Since 0 is an additional root, the equation has $n$ distinct real roots.

Suppose $a_k < 0 < a_{k+1}$. Then the above argument provides for a root of the equation in each interval $(a_i, a_{i+1})$ $(1 < i < k - 1, k + 1 \leq i \leq n)$. Now $f(a_k)$ and $f(a_{k+1})$ have the same sign and $f(0) = 0$. Hence, either $f(x)$ has at least two distinct roots in $(a_k, a_{k+1})$ or else a double root at 0. Hence the equation in this case has at least $n - 1$ distinct real roots, or $n$ roots if we count multiplicity. The case $n = 2$, $a_1 = -1$, $a_2 = 1$ yields an equation with a double root at 0.

**4.30.** Let $|z| < 1/(k+1)$. Then

$$
\begin{aligned}
|1 &+ a_1 z + a_2 z^2 + \cdots + a_n z^n| \\
&\geq 1 - |a_1| \, |z| - |a_2| \, |z|^2 - \cdots - |a_n| \, |z|^n \\
&\geq 1 - k[(k+1)^{-1} + (k+1)^{-2} + \cdots + (k+1)^{-n}] \\
&\geq 1 - k(1/k) = 0.
\end{aligned}
$$

**4.31.** Let $\text{Re } z > a$. Then

$$
\begin{aligned}
|z + a_{n-1}| &> \text{Re}(z + a_{n+1}) > a + \text{Re}(a_{n-1}) \\
&= |a_{n-2}|a^{-1} + |a_{n-3}|a^{-2} + \cdots + |a_0|a^{-(n-1)} \\
&> |a_{n-2}| |z|^{-1} + |a_{n-3}| |z|^{-2} + \cdots + |a_0| |z|^{-(n-1)} \\
\Rightarrow |f(z)| &\geq |z^n + a_{n-1}z^{n-1}| - |a_{n-2}| |z|^{n-2} - \cdots - |a_0| > 0.
\end{aligned}
$$

Let $\text{Re } z < -b$. Then

$$
\begin{aligned}
-|z + a_{n-1}| &\leq \text{Re}(z + a_{n-1}) < -b + \text{Re}(a_{n-1}) \\
&= -[|a_{n-2}|b^{-1} + |a_{n-3}|b^{-2} + \cdots + |a_0| |b|^{-(n-1)}] \\
&< -|a_{n-2}| |z|^{-1} - |a_{n-3}| |z|^{-2} - \cdots - |a_0| |z|^{-(n-1)} \\
\Rightarrow |f(z)| &\geq |z^n + a_{n-1}z^{n-1}| - |a_{n-2}| |z|^{n-2} - \cdots > 0.
\end{aligned}
$$

**4.32.** $q(t) = f(t)g(t)$ where $f(t) = tp(t) + p'(t)$ and $g(t) = tp'(t) + p(t)$. The function $h(t) = tp(t)$ has at least $n + 1$ distinct zeros (including 0), so that $h'(t) = g(t)$ has at least $n$ distinct real zeros, all distinct from the zeros of $p(t)$ exceeding 1.

Let $\{u_1, \ldots, u_r\}$ be the set of simple zeros of $p(t)$ exceeding 1, and $\{v_1, \ldots, v_s\}$ be the set of multiple zeros. Note that $r + s = n$. Each $v_i$ is a zero of both $p(t)$ and $p'(t)$ and so is a zero of $f(t)$. From a sketch of a graph, it can be seen that the sign of $f(u_i) = p'(u_i)$ alternates with $i$ and that $f(t)$ has at least $r - 1$ distinct zeros apart from the $v_i$. Hence $f(t)$ has at least $r + s - 1 = n - 1$ zeros exceeding 1.

It remains to check that these zeros are distinct from the zeros of $g(t)$ identified above. Suppose $f(w) = g(w) = 0$, $|w| \neq 1$. Then $(w - 1)(p(w) - p'(w)) = 0$, so that $p(w) = p'(w) = (1 + w)^{-1}g(w) = 0$ so $w$ is not one of the zeros of $g(t)$ distinct from those of $p(t)$.

**4.33.** The result is clear if $\deg p(x) = 1$. If $n = 2$ and $p(x) = ax^2 + bx + c$, then the left side is the discriminant $b^2 - 4ac$ which is nonnegative. Suppose the result has been established for polynomials of degree up to $n - 1$. Let $p(x)$ be as specified and write it as $(x - r)q(x)$, where $r$ is a real zero of $p(x)$ and $\deg q(x) = n - 1$. Then

$$
(n - 1)^2(p'(x))^2 - n(n - 1)p(x)p''(x) = n(x - r)^2[(n - 2)q'(x)^2
$$

$$
- (n - 1)q(x)q''(x)] + [q'(x)(x - r) - (n - 1)q(x)]^2 \geq 0.
$$

**4.34.** Let $(u, v)$ be a given open interval and let $a_i/b_i$ $(i = 1, 2, \ldots, n - 1)$ be any $n - 1$ rationals in the interval. The polynomial $g(t) = \Pi(b_i t - a_i)$ is a polynomial of degree $n - 1$ all of whose zeros lie within the interval $(u, v)$. The derivative $g'(t)$ has $n - 2$ zeros $r_j$ with $u < r_1 < r_2 < \cdots < r_{n-2} < v$. The $n$ numbers $g(u)$, $g(r_1)$, $g(r_2)$, $\ldots$, $g(r_{n-2})$, $g(v)$ are all nonzero and alternate in sign.

Let $c = \min\{|g(u)|, |g(v)|, |g(r_i)|\}$ and let $k$ be a large positive integer such that $(|u| + |v|)^n/k < c$. Define $f(t) = t^n/k + g(t)$. Then the signs of $f(t)$ agree with those of $g(t)$ at $u$, $r_j$, $v$ and so the signs of $f(t)$ at these $n$ points alternate. Hence $f(t)$ has $n - 1$ zeros in the interval $(u, v)$. The polynomial $kf(t)$ fulfils the requirements.

**4.35.** Since $(1+ix)^m = f(x)+ig(x)$, it follows that $(1-ix)^m = f(x)-ig(x)$. Hence

$$2[af(x) + bg(x)] = (a + bi)(1 - ix)^m + (a - bi)(1 + ix)^m.$$

Suppose that $z$ is a zero of $af(x) + bg(x)$. Then, clearly $z \neq \pm i$, and

$$\frac{|1 + iz|^m}{|1 - iz|^m} = \frac{|a + bi|}{|a - bi|} = 1 \Rightarrow |1 + iz| = |1 - iz|$$

$\Rightarrow z$ is real (Exercise 1.3.13).

# Answers to Exercises

### Chapter 6

**1.1.** Let the roots of the given equation be $u$, $v$, $w$.

*First solution.* $u^2 + v^2 + w^2 = (u + v + w)^2 - 2(uv + uw + vw) = -5$; $u^2v^2 + u^2w^2 + v^2w^2 = (uv + uw + vw)^2 - 2uvw(u + v + w) = -11$; $u^2v^2w^2 = (uvw)^2 = 100$. The equation $x^3 + 5x^2 - 11x - 100 = 0$ has the required roots.

*Second solution.* The roots of $x^3 + x^2 + 3x + 10 = 0$ are $-u$, $-v$, $-w$. Since

$$\begin{aligned}
(x^2 &- u^2)(x^2 - v^2)(x^2 - w^2) \\
&= (x - u)(x - v)(x - w)(x + u)(x + v)(x + w) \\
&= (x^3 - x^2 + 3x - 10)(x^3 + x^2 + 3x + 10) \\
&= (x^3 + 3x)^2 - (x^2 + 10)^2 = x^6 + 5x^4 - 11x^2 - 100.
\end{aligned}$$

If $y = x^2$, then

$$(y - u^2)(y - v^2)(y - w^2) = y^3 + 5y^2 - 11y - 100$$

is a polynomial with the required zeros.

**1.2.** Since $uvw = -1/2$, the polynomial sought has zeros $3u$, $3v$, $3w$ and so is $4t^3 - 21t^2 - 27t + 54$.

**1.3.** With $s_i$ representing the symmetric functions of $m$, $n$, $p$, $q$, we find that the required polynomial is

$$(t - mn)(t - mp)(t - mq)(t - np)(t - nq)(t - pq)$$

$$
\begin{aligned}
= \;& t^6 - (\Sigma mn)t^5 + (\Sigma m^2 np + 3mnpq)t^4 \\
& - (\Sigma m^3 npq + 2\Sigma m^2 n^2 pq + \Sigma m^2 n^2 p^2)t^3 \\
& + (\Sigma m^3 n^2 p^2 q + 3m^2 n^2 p^2 q^2)t^2 - (\Sigma m^3 n^3 p^2 q^2)t \\
& + m^3 n^3 p^3 q^3 \\
= \;& t_6 - s_2 t^5 + (s_1 s_3 - s_4)t^4 - (s_1^2 s_4 + s_3^2 - 2s_2 s_4)t^3 \\
& + (s_1 s_3 - s_4)s_4 t^2 - s_2 s_4^2 t + s_4^3 \\
= \;& t^6 - 2t^5 - 2t^4 + 4t^3 + 2t^2 - 2t - 1.
\end{aligned}
$$

The given quartic has 1 as a double zero; indeed,

$$
t^4 - 3t^3 + 2t^2 + t - 1 = (t-1)^2(t^2 - t - 1).
$$

The sextic will have two double zeros, and we find that

$$
t^6 - 2t^5 - 2t^4 + 4t^3 + 2t^2 - 2t - 1 = (t-1)(t+1)(t^2 - t - 1)^2.
$$

**1.4.** $(1/5)(5t^3 + 6t^2 - 2t + 1)$.

**1.5.** If $s_r$ is the $r$th symmetric function of the zeros $t_i$ of $p(t)$ and $u_r$ is the $r$th symmetric function of $t_i^{-1}$, then

$$
u_r = s_{n-r}/s_n = (-1)^r a_{n-r}/a_n,
$$

whence the monic polynomial with zeros $t_i^{-1}$ is

$$
(a_0)^{-1} \sum_{r=0}^{n} (-1)^{2r} a_{n-r} t^r.
$$

**1.6.** $a_n t^n + k a_{n-1} t^{n-1} + \cdots + k^{n-1} a_1 t + k^n a_0$.

**1.7.** (a) The polynomial must have the form $a_n t^n + a_{n-2} t^{n-2} + \cdots$ so that its derivative is $n a_n t^{n-1} + (n-2)a_{n-2} t^{n-3} + \cdots$. Since the next to leading coefficient of the derivative vanishes, the result follows.

(b)

$$
w_1^2 + \cdots w_n^2 = (w_1 + \cdots + w_n)^2 - 2\Sigma w_i w_j = -2(n-2)a_{n-2}/n
$$

$$
= (n-2)[(z_1 + \cdots + z_n)^2 - 2\Sigma z_i z_j]/n.
$$

It is conjectured by I.J. Schoenberg in *Amer. Math. Monthly* **93** (1986), 8–11 that in fact $\Sigma|w_i|^2 \le (n-2)(\Sigma|z_i|^2/n)$. The reader may wish to establish this when $\deg p = n = 3$. See also problem E3115, *Amer. Math. Monthly* **92** (1985), 666; **94** (1987), 689.

**1.8.** (a) The sum of the squares of the zeros is $(a_{n-1}/a_n)^2 - (2a_{n-2}/a_n)$. If the zeros are real, then the sum of their squares is positive and the first inequality follows. The second inequality follows from the fact that, if

$a_0 \neq 0$, the sum of the reciprocals of the squares of the zeros is $(a_1/a_0)^2 - (2a_2/a_0)$.

For a quadratic, a necessary and sufficient condition for real zeros is that $a_1^2 \geq 4a_0a_2$. Thus to find a counterexample, we choose the $a_i$ so that $2a_0a_2 \leq a_1^2 < 4a_0a_2$. For example, the coefficients of $2t^2 + 3t + 2$ satisfy both conditions, but the zeros are nonreal.

(b) In this case, $a_5^2 = 4 < 2a_4a_6 = 6$, so that not all the zeros are real.

(c) By Rolle's Theorem, if all the zeros of $p(t) = \Sigma a_i t^i$ are real, then its $(n-2)$th derivative has at least two real zeros. Hence the discriminant of

$$(n-2)(n-3)\cdots(3)[n(n-1)a_nt^2 + 2(n-1)a_{n-1}t + 2a_{n-2}]$$

is positive. This yields the desired necessary condition.

For $n = 3$, the condition becomes $a_2^2 \geq 3a_1a_3$ (cf. Exercises 9 (b)). A cubic polynomial with not all its zeros real which satisfies the condition is $t^3 + 5t^2 + 7t$. A counterexample of degree $n$ is $t^n + 5t^{n-1} + 7t^{n-2}$.

**Remarks.** The problem for the case $n = 5$ was posed in the 1983 *USA Mathematical Olympiad*. Observe that it would be unreasonable for a condition which does not involve the constant coefficient to be sufficient for the zeros to be real; for, if $p(t)$ is any polynomial over **R**, $k$ could be chosen sufficiently great that $p(t)+k$ has at most one real zero. See Exercise 5.2.16.

**1.9.** (a) If $z = 0$, then $xy = b$. If $z \neq 0$, then $xyz = -c = z^3 + az^2 + bz$ yields the result. Also

$$\begin{aligned}(x-y)^2 &= (x+y)^2 - 4xy = (-a-z)^2 - 4(z^2 + az + b)\\ &= -[3z^2 + 2az - (a^2 - 4b)].\end{aligned}$$

(b) $(x-y)^2 \geq 0 \Rightarrow 3z^2 + 2az - (a^2 - 4b)$ must be negative $\Rightarrow 3t^2 + 2at - (a^2 - 4b)$ must have real zeros $u$ and $v$ $(u \leq v)$ and $z$ must satisfy $u \leq z \leq v \Rightarrow$ the discriminant $16(a^2 - 3b)$ must be positive.

(c) Consider the graph of $p(t)$.

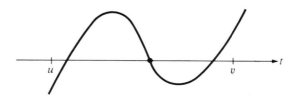

A change in the value of $c$ results in a vertical translation of the graph. For any $c$, $u$ is no greater than the smallest real zero and $v$ is no less than the largest. Hence $p(u) \leq 0 \leq p(v)$.

This can be written as

$$u^3 + au^2 + bu \leq -c \leq v^3 + av^2 + bv.$$

Substituting for $u$ and $v$ yields

$$(2a^3 - 9ab) - (2a^2 - 6b)\sqrt{a^2 - 3b} \le -27c$$

$$\le (2a^3 - 9ab) + (2a^2 - 6b)\sqrt{a^2 - 3b}$$

and this yields the first inequality. The second inequality can be found by squaring.

(d) The inequality can be rewritten as

$$(2a^3 - 9ab + 27c)^2 \le 4(a^2 - 3b)^3$$

from which it follows that the discriminant $4(a^2 - 3b)$ of the quadratic $p'(t)$ is positive. Let $r$ and $s$ be the real zeros of $p'(t)$. If $r < s$, then $3r = -a - (a^2 - 3b)^{1/2}$ and $3s = -a + (a^2 - 3b)^{1/2}$. We have, upon dividing $p(t)$ by $p'(t)$, that

$$\begin{aligned} 27p(r) &= 3(3r + a)p'(r) - 6(a^2 - 3b)r + 3(9c - ab) \\ &= 0 + 2(a^2 - 3b)^{3/2} + (2a^3 - 9ab + 27c) > 0 \end{aligned}$$

while

$$\begin{aligned} 27p(s) &= 3(3s + a)p'(s) - 6(a^2 - 3b)s + 3(9c - ab) \\ &= 0 - 2(a^2 - 3b)^{3/2} + (2a^3 - 9ab + 27c) < 0. \end{aligned}$$

Hence $p(t)$ has a real zero less than $r$, between $r$ and $s$ and greater than $s$. If $r = s$, then $a^2 - 3b = 2a^3 - 9ab + 27c = 0$ and $p(r) = 0$. In this case, $p$ has a triple zero at $r$. Thus, all zeros of $p(t)$ are real.

**1.10.** Let the roots be $a$, $b$, $c$, $d$ with $ab = -5$. Then $cd = 4$. We have that

$$-1 + (a + b)(c + d) = -7$$

$$-5(c + d) + 4(a + b) = -23$$

$$(a + b) + (c + d) = 1.$$

Hence $a + b = -2$ and $c + d = 3$, so that the equation can be rewritten

$$(t^2 + 2t - 5)(t^2 - 3t + 4) = 0$$

and thence solved.

**1.11.** (a) Let $a$, $b$, $c$, $d$ be the roots and suppose that $(ab)^2 = (cd)^2 = s$. Then $-p = (a + b) + (c + d) = -r/ab$, which yields the result. On the other hand, if $r^2 = p^2 s$, let $t = x + r/px$. The equation can be rewritten as $t^2 + pt + (q - 2r/p) = 0$. Each solution $t$ of this quadratic leads to an equation $px^2 - ptx + r = 0$ the product of whose roots for either value of $t$ is $r/p$ (cf. Problem 1.4.17).

(b) *First solution.* Suppose the zeros are $a$, $b$, $c$, $d$ with $a + b = c + d = -p/2$. Then $p^2/4 + ab + cd = q$ and $(-p/2)(ab + cd) = -r$. For these to be consistent, we require $2r/p = q - p^2/4$.

*Second solution.* Make the change of variables $x = -\frac{1}{2}p - y$. Substitution into the original equation yields

$$y^4 + py^3 + qy^2 + (qp - r - \frac{1}{4}p^3)y + (-p^4/16 + p^2q/4 - rp/2 + s) = 0.$$

This equation has exactly the same roots as the original equation and so its coefficients must be the same. In particular, $qp - r - p^3/4 = r$, which yields the result.

(c) Yes. The change of variable $x = -p/2 - y$ as in (b) leads to exactly the same equation for $y$. Hence the roots $a$, $b$, $c$, $d$ of the equation in $x$ are the same as the roots $-p/2 - a$, $-p/2 - b$, $-p/2 - c$, $-p/2 - d$ of the equation in $y$.

If, say, $a = -p/2 - b$ and $c = -p/2 - d$, then $a + b = c + d = -p/2$. If, say, $a = -p/2 - a$, then $-p/4$ is a root of the given equation. But, in any case, $-p/4$ is a zero of the derivative $4x^3 + 3px^2 + 2qx + r$, so $-p/4$ is a double root, and again pairs of roots have the same sum.

**2.3.** (b) If a cubic over **R** does not have all its zeros real, they must be of the form $r$, $u + vi$, $u - vi$ ($r$, $u$, $v$ real, $v \neq 0$). The discriminant is then

$$[(r - u)^2 + v^2]^2 (2vi)^2 = -4v^2[(r - u)^2 + v^2]^2 < 0.$$

An example of a quartic with nonreal zeros and positive discriminant is $t^4 + 4$ whose zeros are $1 + i$, $1 - i$, $-1 + i$, $-1 - i$ and whose discriminant is $2^{14}$.

**2.4.** Suppose $p(t)$ has a nonreal zero $t_1$. Then its complex conjugate $t_2$ is also a zero. Let $t_1 = u + vi$, $t_2 = u - vi$. Then one of the factors in $D$ is the square $(t_1 - t_2)^2 = -4v^2$. If $t_3$ is another nonreal zero with complex conjugate $t_4$, then $D$ contains the product

$$(t_1 - t_3)^2(t_2 - t_4)^2 = |t_1 - t_3|^4.$$

On the other hand, if $t_5$ is a real zero, then $D$ contains the product

$$(t_1 - t_5)^2(t_2 - t_5)^2 = |t_1 - t_5|^4.$$

Thus, one of the square factors involving $t_1$ and $t_2$ is negative while the remaining factors involving either of these zeros can be combined into a positive product.

Now look at factors of $D$ not involving $t_1$ and $t_2$. If there is another nonreal complex conjugate pair of zeros, the terms involving them can be combined to give a negative product. This argument can be continued on to show that the sign of the discriminant is $(-1)^k$ where $k$ is the number of pairs of complex conjugate zeros.

**2.5.** Let the zeros of $t^3 + at^2 + bt + c$ be $x$, $y$, $z$. Then

$$
\begin{aligned}
D &= (x-y)^2(x-z)^2(y-z)^2 \\
&= \Sigma x^4 y^2 + 2\Sigma x^3 y^2 z - 2\Sigma x^4 yz - 2\Sigma x^3 y^3 - 6x^2 y^2 z^2 \\
&= (a^2 b^2 - 2b^3 - 3c^2 - 2a^3 c + 4abc) + 2(abc - 3c^2) \\
&\quad - 2(a^3 c - 3abc + 3c^2) - 2(b^3 - 3abc + 3c^2) - 6c^2 \\
&= a^2 b^2 - 4a^3 c + 18abc - 4b^3 - 27c^2.
\end{aligned}
$$

**2.6.** (a) $1$, $-1$, $i$, $-i$; $-256$.
   (b) $i$, $-i$, $2i$, $-2i$; $5184$.

**2.7.** The discriminant is nonzero $\Longleftrightarrow$ all the zeros of the polynomial are distinct, i.e. simple $\Longleftrightarrow$ the polynomial and its derivative do not have any zero in common $\Longleftrightarrow$ the only common divisors of the polynomial and its derivative are constant.

**3.1.** *First solution.* $a + b + c = 2$ and $a^2 + b^2 + c^2 = (a + b + c)^2 - 2(ab + ac + bc) = 2$. Since $a^3 = 2a^2 - a - 5$, etc.,

$$a^3 + b^3 + c^3 = 2(a^2 + b^2 + c^2) - (a + b + c) - 15 = -13.$$

Since $a^4 = 2a^3 - a^2 - 5a$, etc.,

$$a^4 + b^4 + c^4 = 2(a^3 + b^3 + c^3) - (a^2 + b^2 + c^2) - 5(a + b + c) = -38.$$

*Second solution.* $a^4 + b^4 + c^4 = s_1^4 - 4s_1^2 s_2 + 2s_2^2 + 4s_3 s_1 = -38$.

**3.3.** For each $i$,

$$t_i^n + a_{n-1} t_i^{n-1} + \cdots + a_1 t_i + a_0 = 0.$$

Multiplying by $t^r$ and summing over $1 \le i \le n$ yields the result.

**3.7.**

$$
\begin{array}{ll}
p_1 + c_1 = 0 & p_1 = -c_1 \\
p_2 + c_1 p_1 + 2c_0 = 0 & p_2 = c_1^2 - 2c_0 \\
p_3 + c_1 p_2 + c_0 p_1 = 0 & p_3 = -c_1^3 + 3c_0 c_1 \\
p_4 + c_1 p_3 + c_0 p_2 = 0 & p_4 = c_1^4 - 4c_0 c_1^2 + 2c_0^2 \\
p_5 + c_1 p_4 + c_0 p_3 = 0 & p_5 = -c_1^5 + 5c_0 c_1^3 - 5c_0^2 c_1.
\end{array}
$$

For $t^2 - 3t + 2$, $p_1 = 3$, $p_2 = 5$, $p_3 = 9$, $p_4 = 17$, $p_5 = 33$. For $t^2 + t + 1$, $p_1 = p_2 = p_4 = p_5 = -1$, $p_3 = 2$.

**3.8.** $p_1 = -7$; $p_2 = 61$; $p_3 = -466$; $p_4 = 3621$; $p_5 = -28082$.

**3.9.** Let $z_1, z_2, \ldots, z_n$ be zeros of the polynomial

$$f(t) = t^n + c_{n-1} t^{n-1} + \cdots + c_1 t + c_0,$$

and let $p_k$ be the sum of their $k$th powers. Then, by exercise 6, $c_{n-k} = 0$ for $k = 1, 2, \ldots, n$, so that $f(t) = t^n$. Hence, each $z_i$ vanishes.

## Solutions to Problems

### Chapter 6

**4.1.** (a) If the roots are $r - s$, $r$, $r + s$, then $a = -3r$, $b = 3r^2 - s^2$ and $c = -r(r^2 - s^2)$. The result follows by direct substitution.

(b) If the roots are $rs^{-1}$, $r$, $rs$, then $a = -r(1 + u)$, $b = r^2(1 + u)$ and $c = -r^3$, where $u = s + s^{-1}$. The result follows by direct substitution.

**4.2.** Let $u = \tan A$, $v = \tan B$, $w = \tan C$ where $A$, $B$, $C$ are the angles of a triangle. Since

$$0 = \tan(A + B + C) = \frac{u + v + w - uvw}{1 - (uv + vw + wu)},$$

$$u + v + w = uvw = p/q.$$

Also,

$$
\begin{aligned}
uw + vw + uv &= (u + v)w + uv = (uv - 1)w^2 + uv \\
&= 1 + (1 + w^2)(uv - 1) \\
&= 1 + \frac{(\sec^2 C)(\sin A \sin B - \cos A \cos B)}{\cos A \cos B} \\
&= 1 - \frac{\cos(A + B)}{\cos A \cos B \cos^2 C} = 1 + \frac{1}{\cos A \cos B \cos C} \\
&= 1 + 1/q.
\end{aligned}
$$

The result follows.

**4.3.** (a) Suppose, if possible, that $a$ is a double zero of $x^3 - x^2 - x - 1$. Then $a$ is a zero of the derivative $3x^2 - 2x - 1$, and hence a zero of

$$8x + 10 = (3x^2 - 2x - 1)(3x - 1) - 9(x^3 - x^2 - x - 1).$$

But then $a = -5/4$, which cannot be a zero of the cubic.

(b) The result is readily checked using symmetric functions of the roots for $n = 1$ and $n = 2$. Use induction, based on relations such as the following for $k \geq 0$

$$\frac{b^{k+3} - c^{k+3}}{b - c} = \frac{b^{k+2} - c^{k+2}}{b - c} + \frac{b^{k+1} - c^{k+1}}{b - c} + \frac{b^k - c^k}{b - c}.$$

**4.4.** The zeros of $17x^4 + 36x^3 - 14x^2 - 4x + 1$ are in harmonic progression iff the zeros of $x^4 - 4x^3 - 14x^2 + 36x + 17$ are in arithmetic progression. Suppose that the zeros of the latter polynomial are $a - 3d$, $a - d$, $a + d$, $a + 3d$. Then $a = 1$ and $(a^2 - 9d^2)(a^2 - d^2) = 17$, whence $(d^2 - 2)(9d^2 + 8) = 0$. Checking the coefficient $-14$ reveals that $9d^2 + 8 \neq 0$. Hence $d = \sqrt{2}$ and the roots of the given equation are the reciprocals of $1 - 3\sqrt{2}$, $1 - \sqrt{2}$, $1 + \sqrt{2}$, $1 + 3\sqrt{2}$.

**4.5.** $k = 86$.

**4.6.** Let $u = \tan A$, $v = \tan B$, $w = \tan C$, and let the fourth root be $y$. We have $p - y = u + v + w = uvw = s/y$, $uv + uw + vw + y(u + v + w) = q$ and $(uv + uw + vw)y + uvw = r$. Hence $uv + uw + vw = q - s$, so $(q - s)y + (p - y) = r$. Hence $(q - s - 1)y = r - p$.

Suppose, if possible, that $q = s + 1$. Then $uv + uw + vw = 1$. Together with $u + v + w = uvw$, this implies that

$$0 = 1 - (u + v)w - uv = (1 + w^2)(uv - 1)$$

so $uv = 1$. Similarly $uw = vw = 1$ and we obtain a contradiction. Hence $q - s - 1 \neq 0$ and $y = (r - p)/(q - s - 1)$.

**Remark.** It can also be seen that $y^2 - py + s = 0$. Eliminating $y$ from the two equations gives a necessary constraint on the coefficients, which reflects the fact that not every set of three numbers can be represented as the tangents of the angles of a triangle.

**4.7.** $x + y = 5 - z$ and

$$(x + y)z + xy = 3 \Rightarrow xy = z^2 - 5z + 3 \Rightarrow 0 \leq (x - y)^2 = (x + y)^2 - 4xy$$

$$= -(3z^2 - 10z - 13) = -(z + 1)(3z - 13),$$

from which the result follows.

**4.8.** Suppose $a$ is as required. Then $u + v + w = 6$

$$\Longleftrightarrow \quad (u - 1) + (v - 2) = -(w - 3)$$
$$\Longleftrightarrow \quad (u - 1)^3 + (v - 2)^3 + 3(u - 1)(v - 2)(u + v - 3)$$
$$= -(w - 3)^3$$
$$\Longleftrightarrow \quad (u - 1)(v - 2)(u + v - 3) = 0.$$

Similarly

$$(u - 1)(w - 3)(u + w - 4) = 0$$

and

$$(v - 2)(w - 3)(v + w - 5) = 0.$$

**Case 1.** $u = 1$, $v = 2$, $w = 3$ are the zeros of the polynomial $x^3 - 6x^2 + 11x - 6$, which is not of the required form.

**Case 2.** $u = 1$, $v + w = 5$, so that $a = u(v + w) + vw = 5 - a \Rightarrow a = 5/2$. In this case, $v$ and $w$ are the zeros of $x^2 - 5x - 5/2$ and the cubic is

$$(x - 1)(x^2 - 5x - 5/2) = x^2 - 6x^2 + 5x/2 + 5/2.$$

**Case 3.** $v = 2$, $u + w = 4$, so that $a = 8 - a/2 \Rightarrow a = 16/3$. The cubic is

$$(x - 2)(x^2 - 4x - 8/3) = x^3 - 6x^2 + 16x/3 + 16/3.$$

**Case 4.** $w = 3$, $u + v = 3$, so that $a = 9 - a/3 \Rightarrow a = 27/4$. The cubic is

$$(x - 3)(x^2 - 3x - 9/4) = x^3 - 6x^2 + 27x/4 + 27/4.$$

**4.9.** Let $y_k = x_k - 1$ for each $k$. Then the $y_k$ are the zeros of

$$(y + 1)^n + a(y + 1)^{n-1} + a^{n-1}(y + 1) + 1 = y^n + (n + a)y^{n-1} + \cdots$$

$$+ [n + a(n - 1) + a^{n-1}]y + (2 + a + a^{n-1})$$

and

$$n - 3 = \Sigma(x_k + 2)(x_k - 1)^{-1} = \Sigma(1 + 3y_k^{-1}) = n + 3\Sigma y_k^{-1}.$$

Hence

$$-1 = \Sigma y_k^{-1} = -[n + a(n - 1) + a^{n-1}](2 + a + a^{n-1})^{-1}$$

$$\Rightarrow a^{n-1} + a + 2 = a^{n-1} + a(n - 1) + n$$

$$\Rightarrow a(2 - n) = n - 2 \Rightarrow n = 2 \text{ or } a = -1.$$

The case $a = -1$ yields the polynomial $x^n - x^{n-1} - x + 1 = (x - 1)(x^{n-1} - 1)$. But in this case, one of the zeros is 1 and the left side of the given equation is undefined. Hence $a \neq -1$. The case $n = 2$ yields the polynomial $x^2 + 2ax + 1$, whose zeros can be verified to satisfy the condition, provided $a \neq -1$.

**4.10.** $u$, $v$, $w$ are zeros of $t^3 - pt - q$, where

$$p = -(uv + vw + wu) = (1/2)(u^2 + v^2 + w^2)$$

and $q = uvw$. The result follows from adding the equation $u^{n+3} = pu^{n+1} + qu^n$ to the corresponding equations for $v$ and $w$.

**4.11.** Let $u = r + s$, $v = rs$, $w = p + q$, $z = pq$. Then $u + w = a$, $vz = d$, $uw + v + z = b$, $uz + vw = c$ and the zeros of the required quartic are $pu + v$, $qu + v$, $rw + z$ and $sw + z$. Note that

$$(pu + v) + (qu + v) = uw + 2v = b + v - z$$

$$(rw + z) + (sw + z) = uw + 2z = b + z - v$$

$$(pu + v)(qu + v) = (uz + vw)u + v^2 = cu + v^2$$

$$(rw + z)(sw + z) = (uz + vw)w + z^2 = cw + z^2.$$

The sum of the zeros is $2(uw + v + z) = 2b$. The sum of the products of pairs of zeros is

$$(uw + 2v)(uw + 2z) + c(u + w) + v^2 + z^2 = (uw + v + z)^2 + 2vz + c(u + w)$$

$$= b^2 + 2d + ac.$$

The sum of all products of three of the zeros is

$$(uw + 2v)(cw + z^2) + (uw + 2z)(cu + v^2)$$
$$= acuw + uw(v^2 + z^2) + 2c(vw + zu) + 2vz(z + v)$$
$$= abc - c(u + w)(v + z) + uw(v^2 + z^2) + c^2 + c(vw + zu)$$
$$\quad + 2db - 2uvzw$$
$$= abc + c^2 + 2bd - (uz + vw)(uv + wz) + uw(v^2 + z^2) - 2uvzw$$
$$= abc + c^2 + 2bd - vz(u^2 + w^2 + 2uw)$$
$$= abc + c^2 + 2bd - a^2d.$$

The product of the zeros is

$$(cu + v^2)(cw + z^2) = c^2uw + c(uz^2 + wv^2) + v^2z^2$$
$$= c^2(uw + v + z) - c^2(v + z)$$
$$\quad + c[c(z + v) - vz(u + w)] + v^2z^2$$
$$= bc^2 - acd + d^2.$$

Hence, the required quartic is

$$x^4 - 2bx^3 + (2d + b^2 + ac)x^2 - (2bd - a^2d + c^2 + abc)x$$
$$+ (d^2 + bc^2 - acd) = 0.$$

**Remark.** Problem 4860 in the *Amer. Math. Monthly* **66** (1959), 596; **67** (1960), 598 generalizes this problem to polynomials of degree $n$ in which symmetric functions of degree $q$ of any $p$ of the zeros are taken. Here, $n = 4$, $p = 3$, $q = 2$.

**4.12.** $a$ must be a value assumed by $-(t^4 - 14t^3 + 61t^2 - 84t)$ for four integer values of $t$ (possibly counting repetitions). Now

$$t^4 - 14t^3 + 61t^2 - 84t = t(t - 3)(t - 4)(t - 7)$$

so that the left side assumes the same value for $k$ and $7 - k$. Suppose that $a$ is such that $k$ and $7 - k$ are zeros for some integer $k$. Then

$$t^4 - 14t^3 + 61t^2 - 84t + a = [t^2 - 7t + k(7 - k)][t^2 - 7t + (k - 3)(k - 4)].$$

The discriminant of the second factor is $50 - (2k - 7)^2$, which is square only for $k = 0, 1, 3, 4, 6, 7$. Hence $a = 0$ or $a = 36$. Indeed, $t^4 - 14t^3 + 61t^2 - 84t + 36 = (t - 1)^2(t - 6)^2$.

**4.13.** Suppose that $p(z)$ is of degree $n \geq 2$. Then $p(z)p(-z) - p(z)$ is of degree $2n$ and the coefficient of $z^{2n-1}$ is 0. Hence the sum of the roots of the equation is 0. It follows that if there are zeros in one half plane, then there must be zeros in the other, in order for the imaginary parts to cancel in the sum. Any zero of $p(z)$ satisfies the equation, so that if $p(z)$ has any nonreal zeros, the result follows. Suppose $p(z)$ has only real zeros. Then $p(z) = c\Pi(z - r_i)$ where all the $r_i$ are real and $c$ is nonreal. The polynomial $p(-z) - 1 = c\Pi(-z - r_i) + 1$ assumes a nonreal value when $z$ is real and not equal to $-r_i$ for any $i$ and assumes the value 1 for $z = -r_i$. Hence all the zeros of $p(-z) - 1$, which also satisfy the equation, are nonreal and the result follows.

If $p(z) = iz + 1/2$, then the equation becomes $(iz + 1/2)^2 = 0$, both of whose roots are in the upper half plane.

**4.14.** The polynomials $\pm(t \pm 1)$ satisfy the requirements, so we may assume that the degree $n$ exceeds 1. Suppose that the leading coefficient is 1, so that the polynomial $p(t)$ has the form

$$t^n + c_{n-1}t^{n-1} + c_{n-2}t^{n-2} + \cdots + c_1 t + c_0$$

with zeros $r_i$. Note that all the $r_i$ fail to vanish. Then

$$0 \le \Sigma r_i^2 = c_{n-1}^2 - 2c_{n-2} \Rightarrow c_{n-2} = -1 \Rightarrow \Sigma r_i^2 = 3.$$

Similarly,

$$0 \le \Sigma r_i^{-2} = (c_1/c_0)^2 - 2(c_2/c_0) \Rightarrow \Sigma r_i^{-2} = 3.$$

Hence

$$\sum_{i=1}^{n}(r_i^2 + r_i^{-2}) = 6.$$

Since $r_i^2 + r_i^{-2} \ge 2$ for each $i$, it follows that $n \le 3$. If $n = 3$, then $|r_i| = 1$ for each $i$ and the only possibilities are $t^3 + t^2 - t - 1 = (t+1)^2(t-1)$ and $t^3 - t^2 - t + 1 = (t-1)^2(t+1)$. If $n = 2$, then $t^2 - t - 1$ and $t^2 + t - 1$ are the only possibilities and both satisfy the requirements. If the leading coefficient is to be $-1$, then the only possibilities are the negatives of these polynomials.

**4.15.** Let the zeros of $f(x)$ be $r_i$ and those of $f'(x)$ be $s_j$. Then, since $f'(x) = f(x)\Sigma(x - r_i)^{-1}$. For each $j$, $0 = f'(s_j) = f(s_j)\Sigma(s_j - r_i)^{-1}$. Since $f(s_j) \ne 0$, $\Sigma(s_j - r_i)^{-1} = 0$.

**4.16.** The purported roots are of the form $w = v^3(4v^3 - 3v)^{-1} = (4 - 3v^{-2})^{-1}$ where $2v = u + u^{-1}$ and $u$ is a primitive 7th root of unity. Since $4v^2 = u^2 + u^{-2} + 2$ and $8v^3 = u^3 + u^{-3} + 6v$, it follows that

$$8v^3 + 4v^2 - 4v - 1 = 0.$$

Thus, the three zeros of $8t^3 + 4t^2 - 4t - 1$ are $v_1 = \cos(2\pi/7)$, $v_2 = \cos(4\pi/7)$ and $v_3 = \cos(6\pi/7)$, and so the three zeros of $t^3 + 4t^2 - 4t - 8$ are $v_1^{-1}$, $v_2^{-1}$, $v_3^{-1}$. We first determine the cubic whose zeros are $w_i^{-1}$ where $w_i = (4 - 3v_i^{-2})^{-1}$.

We have that

$$\Sigma w_i^{-1} = 12 - 3\Sigma v_i^{-2} = 12 - 3(24) = -60,$$

$$\Sigma w_i^{-1} w_j^{-1} = 48 - 24\Sigma v_i^{-2} + 9(\Sigma v_i^{-2} v_j^{-2})$$

$$= 48 - 24(24) + 9(80) = 192$$

and

$$w_1^{-1} w_2^{-2} w_3^{-1} = 64 - 48(\Sigma v_i^{-2}) + 36(\Sigma v_i^{-2} v_j^{-2}) - 27v_1^{-2} v_2^{-2} v_3^{-2}$$

$$= 64 - 48(24) + 36(80) - 27(64) = 64.$$

Hence $w_i^{-1}$ are the zeros of $t^3 + 60t^2 + 192t - 64$ and so the $w_i$ are the zeros of $64t^3 - 192t^2 - 60t - 1$, as required.

**4.17.** $a$, $b$, $c$, $d$ are the zeros of the quartic $t^4 - wt^3 - zt^2 - yt - x$, from which $w = a+b+c+d$, $z = -(ab+ac+ad+bc+bd+cd)$, $y = abc+abd+acd+bcd$, $x = -abcd$.

**4.18.** First, note that $\Sigma u^3 = -\Sigma u^2 v = 3uvw$. Let the polynomial with zeros $u$, $v$, $w$ be $t^3 + pt + q$. Then

$$x + y + z = \Sigma uv^2 - 6uvw = 9q$$

$$xy + yz + zx = \Sigma u^3 v^3 + \Sigma u^4 vw - 3\Sigma u^3 v^2 w + 12u^2 v^2 w^2$$

$$= (uv + vw + wu)^3 + uvw[\Sigma u^3 - 6\Sigma u^2 v + 6uvw] = p^3 + 27q^2$$

$$xyz = (-q)(-4p^3 - 27q^2) = 4p^3 q + 27q^3.$$

Hence $x$, $y$, $z$ are the zeros of the polynomial

$$t^3 - 9qt^2 + (p^3 + 27q^2)t - (4p^3 q + 27q^3).$$

Now,

$$\Sigma x^3 + a\Sigma x^2 y + bxyz = [(9q)^3 - 27q(p^3 + 27q^2) + 3(4p^3 q + 27q^3)]$$

$$+ a[9q(p^3 + 27q^2) - 3(4p^3 q + 27q^3)] + b[4p^3 q + 27q^3]$$

$$= 27(3 + 6a + b)q^3 + (-15 - 3a + 4b)p^3 q.$$

This expression vanishes when $(a, b) = (-1, 3)$.

**4.19.** The cubic polynomial $f(t) = (t + a)(t + b)(t + c) - x(t + b)(t + c) - y(t + a)(t + c) - z(t + a)(t + b)$ has zeros $u$, $v$, $w$, and so $f(t) = (t - u)(t - v)(t - w)$. Hence

$$f(-a) = -x(b - a)(c - a) = (-a - u)(-a - v)(-a - w),$$

so that $x(a - b)(a - c) = (a + u)(a + v)(a + w)$. The variables $y$ and $z$ can similarly be isolated.

## Answers to Exercises

**Chapter 7**

**1.5.**
$$p_i(t) = \frac{(t - a_1) \cdots (t \widehat{- a_i}) \cdots (t - a_n)}{(a_i - a_1) \cdots (a_i - a_n)}.$$

**1.6.**

$$\begin{aligned} q(t) &= (7/15)(t - 1)(t - 3) - (1/3)(t + 2)(t - 3) + (1/10)(t + 2)(t - 1) \\ &= (7/30)t^2 - (43/30)t + (16/5). \end{aligned}$$

**1.7.** 45, 55, 66, 78, . . ..

**1.8.** $n$th terms from 6th column back to 1st column: $0, 3, 3n+8, (3/2)n^2 + (13/2)n + 7, (1/2)n^3 + (5/2)n^2 + 4n + 2, f(n) = (1/8)n^4 + (7/12)n^3 + (7/8)n^2 + (5/12)n$.

**1.10.** (e)
$$\Delta^k f(t) = \sum_{r=0}^{k} (-1)^{k-r} \binom{k}{r} f(t + r).$$

(g)(ii)

$$\begin{aligned} f(n) &= 2 + 9(n - 1) + (15/2)(n - 1)(n - 2) + (11/6)(n - 1)(n - 2)(n - 3) \\ &\quad + (1/8)(n - 1)(n - 2)(n - 3)(n - 4) \\ &= 2 + 9(n - 1) + (15/2)(n^2 - 3n + 2) + (11/6)(n^3 - 6n^2 + 11n - 6) \\ &\quad + (1/8)(n^4 - 10n^3 + 35n^2 - 50n + 24) \end{aligned}$$

which is the same as the answer to Exercise 1.8.

**1.11.** $6 + 44(n - 1) + (131/2)(n - 1)^{(2)} + 34(n - 1)^{(3)} + (22/3)(n - 1)^{(4)} + (2/3)(n - 1)^{(5)}$.

**1.12.** (a) $-0.00107, -0.00003, -0.00031$.
  (b) $1.32827$.
  (c) $1.32832$.

**1.13.** $\log 1.25 = 0.22314$; $\log 0.75 = -0.28768$; $\log 2.1 = 0.74194$; $\log 2.71828 = 1.00000$.

**1.15.** Yes. Both polynomials have degree 6 and are each the uniquely determined polynomial taking the assigned values.

**1.16.** (b) $t^4 = t^{(4)} + 6t^{(3)} + 7t^{(2)} + t$, $t^5 = t^{(5)} + 10t^{(4)} + 25t^{(3)} + 15t^{(2)} + t$.
  (c) The proof is by induction. Suppose all powers of $t$ up to $t^k$ can be written as a linear combination of factorial powers. Since $t^{k+1} - t^{(k+1)}$

is a polynomial of degree $k$, each of its terms can be written as a linear combination of factorial powers and the result follows.

**1.17.** (a) Use Exercise 16. We can write

$$f(t) = b_n t^{(n)} + b_{n-1} t^{(n-1)} + \cdots + b_1 t + b_0.$$

By repeated application of the $\Delta$-operator, we find that $\Delta^n f(t) = n! b_n$ and $\Delta^{n+1} f(t) = 0$.

The converse is not true. If $f(t) = \sin 2\pi t$, then $\Delta f(t) = f(t+1) - f(t) = 0$ for each $t$ and all differences vanish.

(b) $\Delta b^t = (b-1) b^t$, so that $\Delta^n b^t = (b-1)^n b^t$ is nonzero. Hence $b^t$ is not a polynomial.

**1.18.** (a) The Lagrange Polynomial which assumes the value $f(j)$ at $t = j$ $(j = 0, 1, \ldots, k)$ is given by

$$\sum_{i=1}^{k} \frac{f(i) t(t-1) \cdots (t-i+1)(t-i-1) \cdots (t-k)}{i(i-1) \cdots (2)(1)(-1)(-2) \cdots (-k+i)}.$$

Since this is the unique polynomial of degree not exceeding $k$ with the assigned values, the result follows.

(b) The expression is the Lagrange polynomial with the value $f(a_j)$ when $t = a_j$.

**1.19.** (a) $(1/2) t(t+1)$, or, more generally,

$$(1/n!) t(t+1) \cdots (t+n-1).$$

(b) Express $f(t)$ as in Exercise 18(a). If $n > k \geq i \geq 0$, then

$$\binom{n}{i} \binom{n-i-1}{k-i}$$

is an integer, while if $n < 0 \leq i \leq k$,

$$\binom{n}{i} \binom{n-i-1}{k-i} = (-1)^i \binom{|n|+i-1}{i} \cdot (-1)^{k-i} \binom{|n|+k}{k-i}$$

is an integer. Hence, for all $n$, the coefficient of $f(i)$ in the expansion for $f(n)$ is an integer and the result follows.

**1.20.**

$$h(t) = \sum_{r=0}^{n} \binom{t}{r} = \sum_{r=0}^{n} t^{(r)}/r!$$

$$h(n+1) = 2^{n+1} - 1.$$

**2.2.** (b) (i) $1/2$; (ii) $1/2$; (iii) $3/4$.

**2.4.** (a) Sketch graphs. $p_1(t) = 0$. $p_2(t) = 1/2$, since the difference $t^2 - 1/2$ achieves its maximum absolute values with alternating signs 3 times at $t = -1, 0, 1$. $p_3(t)$ is of the form $mt$ for some value of $m$, for it can be seen from a sketch that there is a value of $m$ strictly between 0 and 1 for which $|t^3 - mt| \leq |1 - m|$ for $-1 \leq t \leq 1$, and that $t^3 - mt$ assumes the value $\pm(1-m)$ with alternating signs at the four points $-1, -b, b, 1$ where $0 < b < 1$. The function $(t^3 - mt) + (1 - m) = (t + 1)(t^2 - t + 1 - m)$ is nonnegative for $-1 \leq t \leq 1$ and assumes its minimum value 0 when $t = b$. Hence $t^2 - t - 1 - m = (t-b)^2$, so that the discriminant $1 - 4(1-m) = 4m - 3$ vanishes. Hence $m = 3/4$ and $p_3(t) = 3t/4$.

(b) $C_1 = t$, $C_2 = (2t^2 - 1)/2$, $C_3 = (4t^3 - 3t)/4$.

(c) $C_k = T_k/2^{k-1}$. The maximum absolute value of $C_k$ and $T_k$ on $[-1, 1]$ is assumed at the $k + 1$ points $t = \cos i\pi/k$ where $0 \leq i \leq k$.

**2.5.** (b)

$$B(f, n; t) = \sum_{k=0}^{n} \binom{n}{k} t^k (1 - t)^{n-k} = [(1 - t) + t]^n = 1$$

$$B(g, n, t) = \sum_{k=0}^{n} \left(\frac{k}{n}\right)\binom{n}{k} t^k (1 - t)^{n-k}$$

$$= t \sum_{k=1}^{n} \binom{n-1}{k-1} t^{k-1}(1 - t)^{n-k} = t.$$

(c) $B(t^2, n; t) = [(n - 1)t^2 + t]/n$. (See Remark after Answer 2.6.)

**2.6.** (b) Since $B(f, n; 0) = f(0)$ and $B(f, n; 1) = f(1)$, if $k \neq 1$, the eigenfunction $f(t)$ must satisfy $f(0) = f(1) = 0$.

$n = 1$: $k = 1$ and $f(t) = at + b$

$n = 2$: $k = 1$ and $f(t) = at + b$
$k = 1/2$ and $f(t) = a(t^2 - t) = at(t - 1)$

$n = 3$: $k = 1$ and $f(t) = at + b$
$k = 2/3$ and $f(t) = at(t - 1)$
$k = 2/9$ and $f(t) = a(2t^3 - 3t^2 + t) = at(t - 1)(2t - 1)$

$n = 4$: $k = 1$ and $f(t) = at + b$
$k = 3/4$ and $f(t) = at(t - 1)$
$k = 3/8$ and $f(t) = at(t - 1)(2t - 1)$
$k = 3/32$ and $f(t) = a(14t^4 - 28t^3 + 17t^2 - 3t)$
$\qquad = at(t - 1)(14t^2 - 14t + 3)$.

**Remark.**

$$
\begin{aligned}
B(f,n;t) &= \sum_{k=0}^{n}\sum_{r=k}^{n} f\left(\frac{k}{n}\right)\binom{n}{k}(-1)^{r-k}\binom{n-k}{r-k}t^r \\
&= \sum_{r=0}^{n}\sum_{k=0}^{r}(-1)^{r-k}f\left(\frac{k}{n}\right)\binom{n}{r}\binom{r}{k}t^r \\
&= \sum_{r=0}^{n}\binom{n}{r}\left[\sum_{k=0}^{r}(-1)^{r-k}\binom{r}{k}f\left(\frac{k}{n}\right)\right]t^r \\
&= \sum_{r=0}^{n}\binom{n}{r}\Delta^r f(0)t^r = (1+t\Delta)^n f(0)
\end{aligned}
$$

where $\Delta f(t) = f(t+1/n) - f(t)$ in this situation. This gives a handy way of computing the Bernstein polynomials in ascending powers of $t$. It can be readily seen that $\deg B(f,n;t) \le \deg f(t)$ when $f(t)$ is itself a polynomial. In particular, when $f(t) = t^2$, then $f(0) = 0$, $\Delta f(0) = 1/n^2$ and $\Delta^2 f(0) = 2/n^2$ so that

$$
B(t^2,n;t) = n(1/n^2)t + \binom{n}{2}(2/n^2)t^2.
$$

Suppose $B(f,n;t) = kf(t)$ for a polynomial $f(t) = at^r + bt^{r-1} + \cdots$ of degree $r$. Differentiating both sides of the equation $r$ times leads to

$$
r!\binom{n}{r}\Delta^r f(0) = kr!a.
$$

Now, $f(t) = at^r + \cdots$ can be written in terms of factorial powers as $at^{(r)} + \cdots$ where, here, $t^{(r)} = t(t-1/n)(t-2/n)\cdots(t-(r-1)/n)$ (to take account of the changed differencing interval). Then $t^{(r)} = (r/n)t^{(r-1)}$, so that

$$
\Delta^s t^{(r)} = r(r-1)(r-2)\cdots(r-s+1)n^{-s}t^{(r-s)}.
$$

Hence $\Delta^r f(0) = ar!n^{-r}$ and $k = n(n-1)\cdots(n-r+1)n^{-r}$. Every eigenvalue must have this form.

**3.1.** (a) $(x^2 - y^2)^2 + (z^2 - w^2)^2 + 2(xy - zw)^2$.

(b) If $a$, $b$, $c$, $d > 0$, then we can find real $x$, $y$, $z$, $w$ such that $a = x^2$, etc. From (a) $(a+b+c+d)/4 \ge xyzw = (abcd)^{1/4}$ as required.

**3.3.** Consider the case $n = 3k$. We have that

$$
(y_1^{3k} + y_2^{3k} + y_3^{3k}) + (y_4^{3k} + \cdots) + \cdots \ge 3[(y_1y_2y_3)^k + (y_4y_5y_6)^k + \cdots]
$$

$$
\ge 3ky_1y_2y_3\cdots.
$$

**3.4.** The inequality holds for $n = 2$ and $n = 3$. For $n = 2^r3^s$, we can use Exercise 2 repeatedly.

**3.5.** (a) $w_i = 1/n$.

(c)

$$a_1^{w_1} a_2^{w_2} \cdots a_n^{w_n} = a_1^{w_1} \cdots a_{n-2}^{w_{n-2}} (a_{n-1}^u a_n^v)^{w_{n-1}+w_n}$$

$$\leq w_1 a_1 + \cdots + w_{n-2} a_{n-2} + (w_{n-1} + w_{n-2}) a_{n-1}^u a_n^v$$

$$\leq w_1 a_1 + \cdots + w_{n-2} a_{n-2} + (w_{n-1} + w_{n-2})(u a_{n-1} + v a_n),$$

which yields the result by (b).

**3.7.** (a) Apply the AGM inequality to $x$ and $x^{-1}$.

(b) $1 - 4x(1 - x) = (2x - 1)^2 \geq 0$.

**3.8.** This is a consequence of the CSB inequality applied to $a_1^{1/2}$, $a_2^{1/2}$, ... and $a_1^{-1/2}$, $a_2^{-1/2}$, ....

**3.9.** (a) When $n = 1$, equality holds without any restriction on $x_1$. Let $n = 2$ and suppose only that both variables have the same sign. If either vanishes, we have essentially the $n = 1$ case. Otherwise,

$$(1 + x_1)(1 + x_2) = 1 + x_1 + x_2 + x_1 x_2 > 1 + x_1 + x_2.$$

For $n \geq 3$, there is a simple induction argument for the case that all $x_i \geq -1$. Assuming the result holds for $n - 1$, we find that the left side exceeds $(1 + x_1 + x_2 + \cdots + x_{n-1})(1 + x_n)$ which in turn exceeds the right side by the $n = 2$ case. (Where is the condition $x_i \geq -1$ used?)

However, with the stronger hypothesis on the $x_i$, a more delicate argument is needed. Again, assume the result for the $n-1$ case. Then (assuming for convenience all $x_i$ are nonzero),

$$(1 + x_1)(1 + x_2) \cdots (1 + x_{n-1})(1 + x_n) - (1 + x_1 + \cdots + x_{n-1} + x_n)$$

$$= [(1 + x_1) \cdots (1 + x_{n-1}) - (1 + x_1 + \cdots + x_{n-1})]$$

$$+ x_n [(1 + x_1) \cdots (1 + x_{n-1}) - 1].$$

The first term on the right is positive by the induction hypothesis. If all the $x_i$ are positive, then the second term is clearly positive. If all the $x_i$ are negative, then $-1 \leq 1 + x_i \leq 1$, and both factors of the second term are negative. If any of the $x_i$ vanishes, we essentially have the case of a lower $n$. Equality holds iff at most one of the $x_i$ is nonzero.

## Solutions to Problems

### Chapter 7

**4.1.** By the CSB inequality applied to the quartuples $(1 - x)$, $(x - y)$, $(y - z)$, $z$ and $1, 1, 1, 1$, we have that

$$1 = (1 - x) + (x - y) + (y - z) + z$$

$$\leq 4^{1/2}[(1 - x)^2 + (x - y)^2 + (y - z)^2 + z^2]^{1/2} = 1.$$

Equality occurs only when $1 - x = x - y = y - z = z = 1/4$, so that the only solution is $(x, y, z) = (3/4, 1/2, 1/4)$.

**4.2.** $z^2 + (1/2)(x^2 + y^2) - (x + y)z = (1/2)[(z - x)^2 + (z - y)^2] \geq 0$.

**4.3.** Let the zeros be $r_1, r_2, \ldots, r_n$. Then $r_1 + r_2 + \cdots + r_n = 1$ and $r_1^{-1} + r_2^{-1} + \cdots + r_n^{-1} = n^2$. By the CSB inequality,

$$n = \Sigma r_i^{1/2} r_i^{-1/2} \leq \sqrt{\Sigma r_i}\sqrt{\Sigma r_i^{-1}} = n.$$

Thus, equality occurs between the middle members and $r_i^2 = r_i/r_i^{-1}$ must be the same for each $i$.

**4.4.** Let the zeros be $r_1, r_2, \ldots, r_n$. If all the zeros were real, then, by the AGM inequality,

$$r_1^2 + r_2^2 + \cdots + r_n^2 \geq n(r_1^2 r_2^2 \cdots r_n^2)^{1/n}.$$

Expressing the symmetric functions in terms of the coefficients yields $a^2 - 2b \geq nk^{2/n}$, which contradicts the hypothesis. Hence, not all zeros are real.

**4.5.** $x^3 + y^3 + z^3 - x^2 y - y^2 z - z^2 x$

$$
\begin{aligned}
&= x^2(x - y) + y^2(y - z) - z^2(x - z) \\
&= (x^2 - z^2)(x - y) + (y^2 - z^2)(y - z) \\
&= (x - y)(x - z)(x + z) + (y - z)^2(y + z).
\end{aligned}
$$

If either $x \leq y \leq z$ or $x \geq y \geq z$, this expression is clearly seen to be nonnegative and the inequality follows. For the other possible relations among $x$, $y$, $z$, a corresponding expression for the difference between the two sides occurs which makes the desired inequality plain.

**4.6.** The difference of the two sides is

$$z[x^2 + y^2 + z^2 - xy - yz - zx] = (z/2)[(x - y)^2 + (y - z)^2 + (z - x)^2].$$

**4.7.** First, consider the case that $a \neq 0$. Let $p$, $q$, $r$, $s$ be the (positive) zeros of $at^4 - bt^3 + ct^2 - t + 1$, and let $u = 1/p + 1/q$ and $v = 1/r + 1/s$. Then $u + v = 1$, $p + q = pqu$ and $r + s = rsv$. It is required to show that $c/a - b/a \geq 80$.

Now $c/a - b/a = (p + q)(r + s) + pq + rs - (p + q) - (r + s) = pqrsuv + pq(1 - u) + rs(1 - v) = pqrsuv + pqv + rsu$. By the AGM inequality, $(\sqrt{pq})^{-1} \leq (p^{-1} + q^{-1})/2$, whence $pq \geq 4u^{-2}$. Similarly, $rs \geq 4v^{-2}$. Hence

$$
\begin{aligned}
c/a - b/a &\geq 16(uv)^{-1} + 4(uv)^{-2}(u^3 + v^3) \\
&= 16(uv)^{-1} + 4(uv)^{-2}(u^2 - uv + v^2) \quad \text{(since } u + v = 1) \\
&= 4[3(uv)^{-1} + (v^{-2} + u^{-2})] \\
&= 4[(uv)^{-1} + (u^{-1} + v^{-1})^2] = 4[(uv)^{-1} + (uv)^{-2}] \\
&= 4(1/2 + 1/uv)^2 - 1.
\end{aligned}
$$

Since, subject to $u + v = 1$, $uv \leq 1/4$, it follows that $c/a - b/a \geq 4(9/2)^2 - 1 = 80$. Equality occurs everywhere in the above inequalities iff $p = q = r = s = 4$.

If $a = 0$, $b \neq 0$, then it is required to show that $c/b \geq 1$, i.e. $p + q + r \geq 1$ where $p$, $q$, $r$ are the zeros of the now cubic polynomial. But

$$p + q + r = (p + q + r)(1/p + 1/q + 1/r) \geq 3.$$

Finally, if $a = b = 0$, it must be shown that $c \geq 0$, which is obvious.

**4.8.** For $n = 1$, the inequality is $(x - 1)^2 \geq 0$. Suppose the inequality holds for $n = k - 1 \geq 1$. Then

$$x^{k+1} - (k + 1)x + k = x[x^k - kx + (k - 1)] + k(x - 1)^2 \geq 0$$

and the result follows by induction.

**4.9.** Let $c = ua^3$ and $d = vb^3$. Then the given condition can be rewritten

$$u^2 a^6 + v^2 b^6 = (a^2 + b^2)^3$$

which reduces to $u^2 w^3 + v^2(1 - w)^3 = 1$, where $w = a^2(a^2 + b^2)^{-1}$. We have to show that $u^{-1} + v^{-1} \geq 1$, or equivalently $u + v \geq uv$.

To get a purchase on the situation, let us examine the special case in which $ad = bc$. This is equivalent to $vb^2 = ua^2$, so that $w = v(u + v)^{-1}$ and the given condition reduces to

$$u^2 v^3 + v^2 u^3 = (u + v)^3 \iff u^2 v^2 = (u + v)^2 \iff 1 = (u^{-1} + v^{-1})^2.$$

This suggests that we try to show that

$$1 = u^2 w^3 + v^2(1 - w)^3 \geq u^2[v/(u + v)]^3 + v^2[(u/(u + v)]^3$$
$$= u^2 v^2/(u + v)^2 = (1/u + 1/v)^{-2}. \tag{*}$$

To this end, observe that

$$
\begin{aligned}
(u &+ v)^3 u^2 w^3 + (u + v)^3 v^2 (1 - w)^3 - u^2 v^3 - u^3 v^2 \\
&= u^2[(u + v)^3 w^3 - v^3] + v^2[(u + v)^3(1 - w)^3 - u^3] \\
&= u^2[(u + v)w - v][(u + v)^2 w^2 + (u + v)vw + v^2] \\
&\quad + v^2[(u + v)(1 - w) - u][(u + v)^2(1 - w)^2 \\
&\quad + (u + v)u(1 - w) + u^2] \\
&= [(u + v)w - v]\{u^2[(u + v)^2 w^2 + (u + v)vw + v^2] \\
&\quad - v^2[(u + v)^2(1 - 2w + w^2) + (u + v)u(1 - w) + u^2]\} \\
&= [(u + v)w - v](u + v)[(u^2 - v^2)(u + v)w^2 \\
&\quad + v(u + v)(u + 2v)w - v^2(2u + v)] \\
&= (u + v)[(u + v)w - v]^2[(u^2 - v^2)w + v(2u + v)] \\
&= (u + v)[(u + v)w - v]^2[u^2 w + 2uv + (1 - w)v^2] \geq 0,
\end{aligned}
$$

from which the inequality $(*)$ follows, since $0 < w < 1$. Equality occurs $\iff (u+v)w = v \iff (u+v)(1-w) = u \iff w/(1-w) = v/u \iff a^2 u = b^2 v \iff ad = bc$.

**4.10.** The proof follows a modified induction argument on $x$. The result clearly holds for $x = 0$. Suppose $1 \le x \le 5$. Then $N(x) = x$, while

$$x - (x+1)(x+5)/12 = (5-x)(x-1)/12 \ge 0$$

and

$$(x^2 + 6x + 12)/12 - x = [(x-3)^2 + 3]/12 > 0.$$

Suppose $x \ge 6$ and that the result has been established for integers up to $x - 6$, i.e. that

$$(x^2 - 6x + 5)/12 \le N(x-6) \le (x^2 - 6x + 12)/12.$$

Since $N(x) = N(x-6) + x$, the result follows immediately.

**4.11.** Suppose that $A$ and $B$ are acute. If $u = \tan A/2$ and $v = \tan B/2$, then $0 \le u, v \le 1$ and $\tan C/2 = \cot(A+B)/2 = (1-uv)/(u+v)$. Denoting the left side of the required inequality by $S$, we have that

$$\begin{aligned} S &= u^2 + v^2 + (1-uv)^2(u+v)^{-2} \\ &= [(u+v) + (1-uv)(u+v)^{-1}]^2 - 2. \end{aligned}$$

Since $4uv \le (u+v)^2$, $(uv)(u+v)^{-1} \le (u+v)/4$. Hence

$$S \ge [3(u+v)/4 + (u+v)^{-1}]^2 - 2 \ge 3 - 2 = 1$$

by application of the AGM inequality to $3(u+v)/4$ and $(u+v)^{-1}$.

**4.12.** The zeros of the quadratic are real iff $u^2 + 18u + 9 \ge 0$, i.e. when $u \le -9 - 6\sqrt{2}$ or $u \ge -9 + 6\sqrt{2}$. The sum of the squares of the zeros is $5u^2 + 12u + 5 = 5(u + 6/5)^2 - 11/5$. The overall minimum of this quantity occurs at $u = -6/5$, and the minimum subject to the constraint on $u$ occurs for the value of $u$ closest to $-6/5$, namely $u = -9 + 6\sqrt{2}$.

**4.13.**

$$x + y = m(x/m) + n(y/n) = x/m + \cdots + x/m + y/n + \cdots + y/n$$

$$\ge (m+n)[(x/m)^m(y/n)^n]^{1/(m+n)}$$

by the AGM inequality.

**Remark.** An elegant, albeit contrived, variant on the idea in the solution is the problem of maximizing $x^2 y$ subject to $x, y \ge 0$ and

$$x + y + \sqrt{2x^2 + 2xy + 3y^2} = k.$$

See Problem 358, *Crux Mathematicorum* **4** (1978); 161; **5** (1979), 84.

**4.14.** *First solution.* The inequality clearly holds when $y = 0$, so we need consider only nonnegative values of $y$. The difference

$$\frac{1}{1+x^2} - \frac{y-x}{y+x} = \frac{x[(2x-y)^2 + (8-y^2)]}{4(1+x^2)(y+x)}$$

has the numerator $(y/2)(8-y^2)$ when $x = y/2$, and so its minimum is nonnegative if and only if $y \leq 2\sqrt{2}$. The answer is $2\sqrt{2}$.

*Second solution.* As above, we suppose that $y \geq 0$. Since $y + x > 0$, the inequality is equivalent to

$$y \leq \min\{x + 2/x : x > 0\}.$$

By the AGM inequality, $x + 2/x \geq 2\sqrt{2}$, with equality iff $x = 2/x = \sqrt{2}$, so that the largest $y$ is $2\sqrt{2}$.

**4.15.** $f(t) = (t-r)g(t)$ where

$$g(t) = b_{n-1}t^{n-1} + \cdots + b_1 t + b_0.$$

Since $f(t)$ is irreducible, it has no double zeros and so $g(r) \neq 0$. Let $M = \Sigma|b_k|(|r|+1)^k$, so that $|g(t)| \leq M$ for $|t - r| \leq 1$. Since $f(t)$ is irreducible of degree exceeding 1, $f(t)$ has no rational zero, so that for any rational $p/q$, $|f(p/q)| = s/q^n \geq 1/q^n$ for some integer $s$.
If $|p/q - r| \leq 1$, then

$$1/q^n \leq |f(p/q)| = |p/q - r|\,|g(p/q)| \leq M|p/q - r|.$$

If $|p/q-r| \geq 1$, then $|p/q-r| \geq 1/q^n$. Let $k = \min(1, 1/M)$. Then $|p/q-r| \geq k/q^n$ as required.

**Remark.** This result asserts that a rational approximation of a zero of a polynomial over **Z** is, in some sense, not very close to the zero. To look at the matter in another way, note that from this result follows that, if for each $m$, a nonrational number $w$ satisfies $|w - p/q| < 1/q^m$ for infinitely many distinct rational numbers $p/q$, then $w$ is not the solution of a polynomial equation with integer coefficients. (For suppose it were the solution of an irreducible equation of degree $n$; then we would have $k/q^n \leq |p/q-r| < 1/q^{n+1}$ for infinitely many rationals $p/q$. Since for any denominator $q$, $|p/q - r| < 1/q^{n+1} \leq 1$ can occur at most finitely often, we must have that $q < 1/k$ for infinitely many positive integers $q$ — an impossibility.) Other results on approximation of nonrationals by rationals can be found in Chapter 11 of G.H. Hardy & E.M. Wright, *An Introduction to the Theory of Numbers* (4th ed., Oxford, 1960).

**4.16.** Multiply the numerator and denominator of the second (resp. third) member by $x$ (resp. $xy$) and simplify to obtain the constant value 2.

**5.1.** The Lagrange polynomial of degree not exceeding $n-1$ which assumes the values $f(a_i)$ at $a_i$ is

$$\sum_{i=1}^{n} \frac{f(a_i)(t-a_1)\cdots(\widehat{t-a_i})\cdots(t-a_n)}{p'(a_i)}$$

and this must be the polynomial $f(t)$. Hence the coefficient of $t^{n-1}$ must vanish and the result follows.

**5.2.** (a) The Lagrange polynomial of degree not exceeding 3 which assumes the value $t^4$ at $t = a,b,c,d$ is

$$\Sigma \frac{a^4(t-b)(t-c)(t-d)}{(a-b)(a-c)(a-d)}.$$

This must be equal to $t^4 - (t-a)(t-b)(t-c)(t-d)$. Comparing the coefficient of $t^3$ yields the result.

(b) The quadratic polynomial which assumes the values $a^5$, $b^5$, $c^5$ at $a$, $b$, $c$ respectively is

$$\frac{a^5(t-b)(t-c)}{(a-b)(a-c)} + \frac{b^5(t-a)(t-c)}{(b-a)(b-c)} + \frac{c^5(t-a)(t-b)}{(c-a)(c-b)}.$$

We can determine this polynomial in another way as

$$t^5 - (t^3 - ut^2 + vt - w)(t^2 + rt + s)$$

where $u = a+b+c$, $v = ab+bc+ca$, $w = abc$ and $r$ and $s$ are chosen to make the coefficients of $t^4$ and $t^3$ vanish, i.e.

$$r = u$$

$$s = ru - v = u^2 - v.$$

Then the left side of the required identity is equal to the coefficient of $t^2$, namely

$$w - rv + su = w - 2uv + u^3$$

as required.

**5.3.** The proof is by double induction. The result holds for $k = 1$ and any $n$, as well as for $n = 1$ and any $k$. Suppose it has been established for $k = 1, 2, \ldots, r-1$ and any $n$ as well as for $k = r$ and $1 \leq n \leq m-1$. Then

$$
\begin{aligned}
S_r(m) &= S_r(m-1) + S_{r-1}(m) \\
&= \frac{m(m+1)\cdots(m-1+r)[(m-1)(2m-2+r)+(r+2)(2m+r-1)]}{(r+2)!} \\
&= \frac{m(m+1)\cdots(m-1+r)(2m+r)(m+r)}{(r+2)!}
\end{aligned}
$$

as required.

**5.4.** Let $c, m, n$ be fixed, and define $g(x) = (x+c)^n$. Then $f(m) = \Delta^m g(0)$, which yields the result immediately.

**5.5.** We have that

$$g(x) = \frac{g(-1)x(x-1)}{2} + g(0)(1-x^2) + \frac{g(1)x(x+1)}{2}$$

so that

$$\begin{aligned} g'(x) &= [g(-1) - 2g(0) + g(1)]x + (1/2)[g(1) - g(-1)] \\ &= g(1)(x+1/2) + g(-1)(x-1/2) - 2g(0)x. \end{aligned}$$

Suppose that $1/2 \le |u| \le 1$. Then $u + 1/2$ and $u - 1/2$ have the same sign so that $|u + 1/2| + |u - 1/2| = 2|u|$. Thus

$$\begin{aligned} |g'(u)| &\le |g(1)|\,|u+1/2| + |g(-1)|\,|u-1/2| + 2|g(0)|\,|u| \\ &\le |u+1/2| + |u-1/2| + 2|u| = 4|u|. \end{aligned}$$

Equality occurs if $g(x) = 2x^2 - 1$, so that $K_u = 4|u|$.

Suppose $0 \le |u| \le 1/2$. Since the bound for $u$ is the same as that for $-u$, with no loss of generality, we can take $0 \le u \le 1/2$. There are several types of quadratic functions to consider:

(1) $g(x)$ is increasing on the interval $[-1, 1]$;

(2) $g(x)$ is decreasing on the interval $[-1, 1]$;

(3) $g(x)$ has a minimum at a point $c$ in $[-1, 0]$;

(4) $g(x)$ has a minimum at a point $c$ in $[0, 1]$;

(5) $g(x)$ has a maximum at a point $c$ in $[-1, 1]$.

Cases (2) and (5) need not be treated directly since $-g(x)$ falls under one of the other cases.

If $g(x)$ satisfies (1), then

$$h(x) = 2\left(\frac{g(x) - g(-1)}{g(1) - g(-1)}\right) - 1$$

is a quadratic such that $|h(x)| \le 1$ on $[-1, 1]$ and $|g'(u)| \le |h'(u)|$ for all $u$. (The graph of $h(x)$ is obtained from that of $g(x)$ by expanding the vertical scale.) Since $h(1) = 1$ and $h(-1) = -1$, the graphs of two such functions $h(x)$ cross only when $x = \pm 1$. Hence $h(0)$ is as small as possible when $h'(-1) = 0$ (i.e. when $h(x) = -1 + (x+1)^2/2$) and as large as possible when $h'(1) = 0$ (i.e. when $h(x) = 1 - (x-1)^2/2$). Thus $-1/2 \le h(0) \le 1/2$. Since

$$h'(u) = (u + 1/2) - (u - 1/2) - 2h(0)u = 1 - 2h(0)u,$$

$$|g'(u)| \le |h'(u)| \le 1 + u.$$

For case (3), it suffices to consider functions $g(x)$ whose minimum value is $-1$ and maximum is $+1$. Then $g(x) = -1 + 2(1-c)^{-2}(x-c)^2$, so that $g'(u) = 4(1-c)^{-2}(u-c)$. We wish to maximize $|g'(u)|$ over all possible values of $c$. By using calculus or the fact that the equation

$$\frac{4(u-t)}{(1-t)^2} = \frac{4(u-c)}{(1-c)^2}$$

has $c$ as a double root, one finds that the maximum value is $(1-u)^{-1}$. Indeed

$$\frac{1}{1-u} - \frac{4(u-c)}{(1-c)^2} = \frac{(1+c-2u)^2}{(1-u)(1-c)^2} \ge 0$$

with equality iff $c = 2u - 1$. Hence $|g'(u)| \le (1-u)^{-1}$ with equality iff $g(x) = -1 + [(x+1-2u)^2]/[2(1-u)^2]$.

For case (4), we can consider functions of the form $g(x) = -1 + 2(1+c)^{-2}(x-c)^2$ where $g'(u) = 4(1+c)^{-2}(u-c)$. Now, checking the extreme case $c = 0$ and $c = 1$, we find that

$$4u - \frac{4(u-c)}{(1+c)^2} = \frac{4c(uc+u+1)}{(1+c)^2} \ge 0$$

$$\frac{4(u-c)}{(1+c)^2} - (u-1) = \frac{(1-c)[(3u+1) - (u+1)c]}{(1+c)^2} \ge 0.$$

Hence $|g'(u)| \le \max(4u, 1-u)$.
Thus, for $|u| \le 1/2$,

$$|g'(u)| \le \max(1 + |u|, (1 - |u|)^{-1}, 4|u|, 1 - |u|) = (1 - |u|)^{-1}.$$

In conclusion, we have

| Condition on $u$ | $K_u$ | Function yielding equality |
|---|---|---|
| $1/2 < |u| \le 1$ | $4|u|$ | $2x^2 - 1$ |
| $0 \le u \le 1/2$ | $(1-u)^{-1}$ | $-1 + (x+1-2u)^2/[2(1-u)^2]$ |
| $-1/2 \le u \le 0$ | $(1+u)^{-1}$ | $-1 + (x-1-2u)^2/[2(1+u)^2]$. |

**Remark.** Similar problems were posed in the Putnam Competition (6 A1 and 29 A5). For some perspective on the situation, see A.M. Gleason, R.E. Greenwood, L.M. Kelly, *The William Lowell Putnam Mathematical Competition Problems and Solutions: 1938-1964* p. 207. The Tchebychef polynomials occur yet again: if $p(x)$ is a polynomial of degree $n$ for which $-1 \le p(x) \le 1$ for $-1 \le x \le 1$, then $|p'(x)| \le n^2$ with equality occurring for some $x$ when $p(x)$ is a Tchebychef polynomial.

**5.6.** We deal first with the quadratic situation. Let

$$f(x) = (x - v)(x - u) = (v - x)(u - x) = (x - (v + u)/2)^2 - ((v - u)/2)^2$$

where $-1 \leq u \leq v \leq 1$. When $u \leq x \leq v$, then $-1 \leq f(x) \leq 0$ with equality on the left iff $v = -u = 1$, $x = 0$. Also

$$f(1)f(-1) = (1 - u^2)(1 - v^2) \leq 1$$

with equality iff $u = v = 0$. Hence, if $f(x) \neq x^2$, at least one of $f(1)$ and $f(-1)$ is less than 1. If $u + v < 0$, then $0 < f(-1) < 1$, and if $u + v > 0$, then $0 < f(1) < 1$.

There are certain cases in which the result is clear:

(1) All of the $x_i$ have the same sign.

(2) The degree of $p(x)$ is 1.

(3) The degree of $p(x)$ is 2, from the above analysis.

Assume the result holds for polynomials of degree less than $n \geq 3$. Let $-1 < a < 0 < b < 1$ and $p(x) = (x - x_1)(x - x_2) \cdots (x - x_n)$ where $x_1 < 0 < x_n$ and $-1 \leq x_1 \leq x_2 \leq \cdots \leq x_n \leq 1$. With no loss of generality, we can take $-a \leq b$. Suppose that $x_1 \leq a < 0 < b \leq x_n$. Then, if $q(x) = (x - x_2) \cdots (x - x_{n-1})$, then

$$|p(x)| = |(x - x_1)(x - x_n)| \, |q(x)| \leq |q(x)| \quad \text{for} \quad x_1 \leq x \leq x_n.$$

By the induction hypothesis applied to $q(x)$, $|p(a)| \geq 1$ and $|p(b)| \geq 1$ cannot both occur.

Now

$$p(a)p(b) = \prod_{i=1}^{n} g(x_i) \quad \text{where} \quad g(x) = (a - x)(b - x).$$

Since $|g(x)| \leq 1$ for $a \leq x \leq b$, if also $g(-1) < 1$ or $a \leq x_1$, it follows that $|g(x_i)| \leq 1$ for each $i$ and that $|p(a)| \geq 1$ cannot occur.

This leaves the situation that $-1 \leq x_1 < a < 0 < x_n < b < 1$ and $g(-1) = (1 + a)(1 + b) > 1$. Now $2(1 + a) > g(-1) > 1$, so that $a > -1/2$. Suppose if possible that $|p(a)| \geq 1$ and $|p(b)| \geq 1$. Then we must have that $x_n - a > 1$, so that $x_n > 1/2$. Then

$$|q(a)| = \frac{|p(a)|}{(a - x_1)(x_n - a)} \geq \frac{|p(a)|}{(a + 1)(1 - a)} > |p(a)| > 1$$

and

$$|q(b)| = \frac{|p(b)|}{(b - x_1)(b - x_n)} \geq \frac{|p(b)|}{(b + 1)(b - 1/2)}$$

$$> \frac{|p(b)|}{2(1/2)} = |p(b)| > 1$$

which contradicts the induction hypothesis.

**5.7.** Since only differences of the $n_i$ are involved, it suffices to obtain the result when all the $n_i$ are nonnegative. Let

$$G(\mathbf{x}) = G(x_0, x_1, \ldots, x_k) = \prod \frac{(x_j - x_i)}{(j - i)}$$

both products taken over $0 \leq i < j \leq k$. We have to show that $G$ takes integer values whenever $\mathbf{x} = \mathbf{n}$ is a vector with integer entries. This will be proved by induction on $n = \max(n_0, n_1, \ldots, n_k)$.

If $n \leq k - 1$, since all the $n_i$ are nonnegative and there are $k + 1$ of them, two of the $n_i$ must be equal and $G(\mathbf{n}) = 0$. If $n = k$, then either two $n_i$ are equal or else $\{n_0, n_1, \ldots, n_k\} = \{0, 1, 2, \ldots, k\}$. In either case, $G(\mathbf{n})$ is an integer.

Suppose it has been shown that $G(\mathbf{n})$ is an integer when $n \leq r$ for some $r \geq k$. Let $n = r + 1$. Without loss of generality, we may suppose that $n_0 = r + 1$. If, for some $i > 0$, $n_i = r + 1$, then $G(n) = 0$. Suppose that $n_i \leq r$ $(1 \leq i \leq r)$. Let

$$f(t) = G(t, n_1, n_2, \ldots, n_k).$$

By the induction hypothesis, $f(t)$ is an integer for $t = 0, 1, 2, \ldots, k, \ldots, r$. Since $\deg f(t) = k$, it follows from Exercise 1.19, that $f(r+1)$ is an integer. Hence $G(\mathbf{n})$ is an integer if $n = r + 1$.

**5.8.** Let $n$ be a positive integer to be determined later, and let $p(x) = 0.5[1 + (2x - 1)^n]$. Since all the coefficients of $1 + (2x - 1)^n$ are even, $p(x)$ has integer coefficients. Also

$$|p(x) - 0.5| = 0.5|2x - 1|^n.$$

If $0.19 \leq x \leq 0.81$, then $|p(x) - 0.5| \leq 0.5(0.62)^n$. Now choose $n$ sufficiently large that $(0.62)^n < 2/1981$.

**5.9.** *First solution.* In a diagram, let the graph of the function $\sqrt{1 - x^2}$ be given. Let us choose $p$ and $q$ so that the maximum value of the left side over all $x$ in $[0, 1]$ is made as small as possible. From the diagram, we see that the line with equation $y = px + q$ should go through the points $(0, 1 + u)$ and $(1, u)$ for some positive value of $u$. Hence, slope $p$ of the line must be $-1$ and it must also go through $(1/\sqrt{2}, 1/\sqrt{2} - u)$. This is possible only if $q = (1/2)(1 + \sqrt{2})$ and $u = (1/2)(\sqrt{2} - 1)$. From this, we see that the inequality of the problem is always satisfied if and only if $p = 1$ and $q = (1/2)(1 + \sqrt{2})$.

*Second solution.* Let $f(x) = \sqrt{1 - x^2} - px - q$ $(0 \leq x \leq 1)$. We first establish that $p$ has to be negative. For,

$$f(0) = 1 - q \leq (1/2)(\sqrt{2} - 1) \Rightarrow -q \leq (1/2)(\sqrt{2} - 1) - 1$$

and

$$f(1) = -p-q \geq -(1/2)(\sqrt{2}-1) \Rightarrow p \leq -q+(1/2)(\sqrt{2}-1) \leq (\sqrt{2}-1)-1 < 0.$$

Now, write $x = \cos\theta$, so that

$$f(\cos\theta) = \sin\theta - p\cos\theta - q = \sqrt{1+p^2}\sin(\theta+\phi) - q,$$

where $\sqrt{1+p^2}\sin\phi = -p$ and $\sqrt{1+p^2}\cos\phi = 1$. $f$ assumes its maximum value when $\theta + \phi = \pi/2$, i.e. when

$$\cos\theta = \sin\phi = -p/\sqrt{1+p^2},$$

and we have that

$$f(-p/\sqrt{1+p^2}) = \sqrt{1+p^2} - q.$$

Let $0 \leq u, v \leq 1$. Then $|f(u) - f(v)| \leq \sqrt{2} - 1$. In particular,

$$\sqrt{1+p^2} - 1 = f(-p/\sqrt{1+p^2}) - f(0) \leq \sqrt{2} - 1 \Rightarrow \sqrt{1+p^2} \leq \sqrt{2}$$

$$\Rightarrow p^2 \leq 1 \Rightarrow p \geq -1$$

and

$$\begin{aligned}
\sqrt{1+p^2} + p &= f(-p/\sqrt{1+p^2}) - f(1) \leq \sqrt{2} - 1 \\
&\Rightarrow \sqrt{1+p^2} \leq (\sqrt{2}-1) - p \\
&\Rightarrow 1+p^2 \leq (3-2\sqrt{2}) - 2(\sqrt{2}-1)p + p^2 \\
&\Rightarrow 2(\sqrt{2}-1)p \leq 2 - 2\sqrt{2} \Rightarrow p \leq -1.
\end{aligned}$$

Putting these facts together yields $p = -1$. Now,

$$f(0) \geq -(1/2)(\sqrt{2}-1) \Rightarrow q \leq (1/2)(\sqrt{2}+1)$$

and

$$f(-p/\sqrt{1+p^2}) = f(1/\sqrt{2}) = \sqrt{2}-q \leq (1/2)(\sqrt{2}-1) \Rightarrow q \geq (1/2)(\sqrt{2}+1),$$

so that $q = (1/2)(\sqrt{2}+1)$.

So far, we have established what $p$ and $q$ must be if the inequality is valid for each $x$; we must now show that this choice works, i.e. that

$$|\sqrt{1-x^2} + x - (1/2)(\sqrt{2}+1)| \leq (1/2)|\sqrt{2}-1| \quad \text{for} \quad 0 \leq x \leq 1.$$

Since $0 \leq (1 - \sqrt{2}x)^2 = 1 - 2\sqrt{2}x + 2x^2$, it follows that

$$1 - x^2 < 2 - 2\sqrt{2}x + x^2 \Rightarrow \sqrt{1-x^2} \leq \sqrt{2} - x.$$

Since $x^2 \leq x$, it follows that $1 - x \leq \sqrt{1 - x^2}$. Hence

$$-x + (1/2)(\sqrt{2} + 1) - (1/2)(\sqrt{2} - 1) \leq \sqrt{1 - x^2}$$

$$\leq -x + (1/2)(\sqrt{2} + 1) + (1/2)(\sqrt{2} - 1),$$

which yields the result.

**5.10.** By Lagrange's Formula,

$$p(x) = \sum_{k=0}^{n} y_k \frac{(x - x_0) \cdots (x - x_{k-1})(x - x_{k+1}) \cdots (x - x_n)}{(x_k - x_0) \cdots \cdots (x_k - x_n)}.$$

Now, $T_{n+1}(x) = \cos(n+1)u = \cos(n+1) \arccos x$ vanishes if and only if $x = \cos u_k$ for some $k$. Hence $T_{n+1}(x) = c_{n+1}(x - x_1) \cdots (x - x_n)$. Thus, when $x \neq x_k$ for any $k$, we can write

$$p(x) = T_{n+1}(x) \sum_{k=0}^{n} \frac{y_k}{(x - x_k)T'_{n+1}(x_k)}.$$

In the special case that each $y_k \doteq 1$, we find that

$$1 = T_{n+1}(x) \sum_{k=0}^{n} 1/[(x - x_k)T'_{n+1}(x_k)].$$

Differentiating $T_{n+1}(x) = \cos(n+1)u$ with respect to $u$ yields

$$T'_{n+1}(x)(-\sin u) = -(n+1)\sin(n+1)u.$$

In particular, when $x = x_k$, we have that $u = u_k$ and

$$T'_{n+1}(x_k) = (n+1)(-1)^k/(\sin u_k)$$

and the result follows.

# Solutions to Problems

## Chapter 8

**1.** The equation is equivalent to $(3y - 5)(9x - 3y - 5) = 34$. Try out all the divisors $3y - 5$ of 34 to obtain the solutions $(x, y) = (-1, -4)$, $(-1, 1)$, $(5, 2)$, $(5, 13)$.

**2.** $0 = a^3 - b^3 - c^3 - 3abc = (a - b - c)(a^2 + b^2 + c^2 + ab + ac - bc)$. Since the second factor never vanishes for $a$, $b$, $c$ not all zero, $a = b + c$. This yields $(a, b, c) = (2, 1, 1)$.

**3.** $24 = (x + y + z)^3 - (x^3 + y^3 + z^3) = 3\Sigma x^2 y + 6xyz \Rightarrow 8 = x(y + z)^2 +$
$(x^2 + yz)(y + z) = (y + z)(x + z)(x + y) = (3 - x)(3 - y)(x + y)$. Hence
$(x, y, z) = (1, 1, 1)$, $(4, 4, -5)$, $(4, -5, 4)$, $(-5, 4, 4)$.

**4.** The equation can be rewritten $4(2a + 1)^2 = (2b + 1)^2 + 3$. Since 4 and 1
are the only squares which differ by 3, the result follows.

**5.** The equation is equivalent to

$$
\begin{aligned}
0 &= x^2 y^2 + 3x^2 y + xy - 3xy^2 + 2y^3 \\
&= y[2y^2 + (x^2 - 3x)y + (x + 3x^2)].
\end{aligned}
$$

Either $y = 0$ and $x$ is arbitrary or the quadratic in $y$ has integer zeros. For
the latter case, it is necessary that the discriminant

$$(x^2 - 3x)^2 - 8(x + 3x^2) = x(x - 8)(x + 1)^2$$

be a square. Hence, either $x = -1$, $0$, $8$, or else $x^2 - 8x = (x - 4)^2 - 16$
is a perfect square, i.e. $x = 9$. Thus, the solutions $(x, y)$ with $y \neq 0$ are
$(-1, -1)$, $(8, -10)$, $(9, -6)$, $(9, -21)$.

**6.** $(u + vi)^3 = y + 2i$ leads to $v(3u^2 - v^2) = 2$, whence $v$ divides 2. Trying
the possibilities yields $(u, v) = (\pm 1, 1)$, $(\pm 1, -2)$. Since $(u - vi)^3 = y - 2i$,
we can take $x = (u + vi)(u - vi)$, and obtain the solutions

$$(x, y) = (u^2 + v^2, u(u^2 - 3v^2)) = (2, \mp 2), (5, \mp 11).$$

**7.** $a^2 xy + abx + acy + ad = 0 \Rightarrow (ax + c)(ay + b) = bc - ad$. If $bc = ad$, then
there are infinitely many solutions if and only if either $c$ or $b$ is a multiple
of $a$ (in which case one of the left factors can be made to vanish and the
other can be arbitrary). If $bc \neq ad$, then $ax + c$ and $ay + b$ both divide a
nonzero integer and there are at most finitely many possibilities for both
$x$ and $y$.

**8.** *First solution.* Suppose $xy > 0$. Then, since

$$(x - y)(x^2 + xy + y^2) = 2(xy + 4),$$

we have that $x > y$. Since at least one of $|x|$ and $|y|$ differs from 1, $x^2 +$
$xy + y^2 > xy + 4$, so that $x - y < 2$. Since $x \neq y$, we have that $x = y + 1$,
which leads to $y^2 + y - 7 = 0$ with no solution in integer $y$.
   Suppose $x < 0$ and $y > 0$. Then

$$8 = x^3 - 2xy - y^3 \leq -x^2 - 2xy - y^2 = -(x + y)^2,$$

which is impossible. Suppose $x > 0$ and $y < 0$. Then $8 = x^3 - 2xy - y^3 > x^3$,
so that $x$ must be 1. This does not work.
   Hence $xy = 0$ and we have the solution $(x, y) = (2, 0)$, $(0, -2)$.

*Second solution.* Let $y = x + u$. Then

$$(3u + 2)x^2 + (3u^2 + 2u)x + (u^3 + 8) = 0.$$

The discriminant of this quadratic equation is

$$-3u^4 + 4u^3 + 4u^2 - 96u - 64 = -(3u + 2)(u^3 - 2u^2 + 32)$$

which is a positive square only for $u = -2$. This leads to the solutions $(x, y) = (0, -2), (2, 0)$.

**9.** Since $\sqrt{2}$ is nonrational and $x$, $y$, $z$, $t$ are to be rational, the equation is equivalent to the system

$$x^2 + 2y^2 + z^2 + 2t^2 = 27$$

$$xy + zt = 5.$$

Hence

$$(5x^2 - 27xy + 10y^2) + (5z^2 - 27zt + 10t^2) = 0$$

$$\Rightarrow (5x - 2y)(x - 5y) + (5z - 2t)(z - 5t) = 0.$$

Trying $x - 5y = z - 5t = 0$ reduces both equations of the system to $y^2 + t^2 = 1$ and we obtain, for example, the solution

$$(x, y, z, t) = \left( \frac{10s}{1 + s^2}, \frac{2s}{1 + s^2}, \frac{5 - 5s^2}{1 + s^2}, \frac{1 - s^2}{1 + s^2} \right)$$

where $s$ is any rational. [Trying $5x - 2y = z - 5t = 0$ reduces both equations to $2y^2 + 25t^2 = 25$. If we write $y = 5u$, then this becomes $2u^2 + t^2 = 1$ and we obtain, for example, the solution $(x, y, z, t) = (4/3, 10/3, 5/3, 1/3)$. Trying $5x - 2y = 5z - 2t = 0$ reduces both equations to $2(y^2 + t^2) = 25$ and we obtain for example the solution $(x, y, z, t) = (1, 5/2, 1, 5/2)$.]

**10.** Let $u = \sqrt[3]{x + \sqrt{y}}$ and $v = \sqrt[3]{x - \sqrt{y}}$. Observe that

(1) $u + v = z$.

(2) $u^3 + v^3 + 3uv(u + v) = z^3 \Rightarrow 3uvz = z^3 - u^3 - v^3 \Rightarrow uv$ is a rational number. Suppose $uv = p/q$ in lowest terms.

(3) $u^3 v^3 = (x + \sqrt{y})(x - \sqrt{y}) = x^2 - y$, an integer. Then $p^3 = (x^2 - y)q^3$, so that any prime which divides $q$ must divide $p$. Since $\gcd(p, q) = 1$, we must have $q = 1$. Hence $uv = p$, an integer.

Thus, $u$ and $v$ are zeros of the quadratic $t^2 - zt + p$, with the result that $2u = z + \sqrt{w}$ and $2v = z - \sqrt{w}$, where $w = z^2 - 4p$. The value of $w$ can be either positive or negative. If positive, $u$ and $v$ are real and we decide arbitrarily that $u \geq v$. If $w$ is negative, then $\sqrt{w}$ is consistently taken to be one of the square roots of $w$.

We can now solve for $x$ and $y$.

$$8(x + \sqrt{y}) = 8u^3 = (z^3 + 3zw) + (3z^2 + w)\sqrt{w}$$

$$8(x - \sqrt{y}) = 8v^3 = (z^3 + 3zw) - (3z^2 + w)\sqrt{w}$$

so that

$$8x = z^3 + 3zw = z(z^2 + 3w)$$

$$8\sqrt{y} = (3z^2 + w)\sqrt{w} \Rightarrow 64y = (3z^2 + w)^2 w.$$

Hence, if there is a solution, we must have

$$(x, y, z) = (z(z^2 + 3w)/8, \ w(3z^2 + w)^2/64, z) \qquad (*)$$

for suitable $z$ and $w$. On the other hand, $(*)$ yields solutions since

$$\sqrt[3]{x + \sqrt{y}} + \sqrt[3]{x - \sqrt{y}}$$

$$\begin{aligned}
&= (1/2)[\sqrt[3]{(z^3 + 3wz) + (3z^2 + w)\sqrt{w}} \\
&\quad + \sqrt[3]{(z^3 + 3wz) - (3z^2 + w)\sqrt{w}}] \\
&= (1/2)\sqrt[3]{(z + \sqrt{w})^3} + \sqrt[3]{(z - \sqrt{w})^3} \\
&= (1/2)[(z + \sqrt{w}) + (z - \sqrt{w})] \\
&= z.
\end{aligned}$$

We need conditions to ensure that $x$, $y$, $z$ are integers. Suppose that $z = 2s$ is even. Then $0 \equiv (3z^2 + w)^2 w = (12s^2 + w)^2 w \pmod{64} \Rightarrow w = 2r$ is even $\Rightarrow (6s^2 + r)^2 r \equiv 0 \pmod 8 \Rightarrow r$ is even $\Rightarrow 4|w$. Suppose $z$ is odd. Then

$$z(z^2 + 3w) \equiv 0 \pmod 8 \Rightarrow w \equiv 5 \pmod 8.$$

Hence, the integer solutions of the given equation are given by $(*)$ where either $(z, w) = (2s, 4t)$ or $(z, w) = (2k + 1, 8m + 5)$.

**Examples:**

(i) $z = 1$, $w = 5$ yields $(x, y, z) = (2, 5, 1)$. In this case, $u = (1/2)(1 + \sqrt{5})$ and $u^3 = 2 + \sqrt{5}$.

(ii) $w = 0$ yields the obvious solution $x = s^3$, $y = 0$, $z = 2s$.

(iii) $z = 6$, $w = -4$ yields $u = 3 + i$, $v = 3 - i$, $(x, y, z) = (18, -676, 6)$.

**11.** Let $u = v/w$ be a given rational value for the polynomial. The equation $3x^2 - 5x + (4 - u^2) = 0$ has discriminant $(12v^2 - 23w^2)/w^2$ and so will have rational solutions when $12v^2 - 23w^2$ is a perfect square.

Try $w = 1$. Then we wish to find $v$ and $z$ such that $12v^2 - 23 = z^2$. Two obvious solutions are $(v, z) = (2, 5)$, $(4, 13)$. We can find as many more as we wish by the following device:

If $t^2 - 12s^2 = 1$ and $z^2 - 12v^2 = -23$, then

$$(t + s\sqrt{12})(t - s\sqrt{12}) = 1$$

$$(z + v\sqrt{12})(z - v\sqrt{12}) = -23$$

$$\Rightarrow (t + s\sqrt{12})(z + v\sqrt{12})(t - s\sqrt{12})(z - v\sqrt{12}) = -23$$

$$\Rightarrow [(tz + 12sv) + (tv + sz)\sqrt{12}][(tz + 12sv)$$
$$- (tv + sz)\sqrt{12}] = -23$$

$$\Rightarrow (tz + 12sv)^2 - 12(tv + sz)^2 = -23.$$

In particular, $t^2 - 12s^2 = 1$ is satisfied by $(t, s) = (7, 2)$, so that if $(v, z)$ satisfies $12v^2 - 23 = z^2$, then we can obtain another solution $(7v+2z, 7z+24v)$. Hence $(v, z) = (2, 5)$ gives rise, successively to $(24, 83)$, $(334, 1157)$, ..., and $(v, z) = (4, 13)$ gives rise successively to $(54, 187)$, $(752, 2605)$, .....

We can try other values of $w$ to obtain solutions. For example, $w = 2$ requires making $4(3v^2 - 23)$ a square, which will occur if $v = 3$. Here are some possibilities:

| $u$ | $x$ such that $3x^2 - 5x + 4 = u^2$ |
|---|---|
| 2 | $5/3, 0$ |
| 4 | $3, -4/3$ |
| 24 | $44/3, -13$ |
| 54 | $32, -91/3$ |
| 334 | $581/3, -192$ |
| 752 | $435, -1300/3$ |
| 3/2 | $1/2, 7/6$ |

12. Suppose, if possible, that there is a nontrivial rational solution. Because of the homogeneity of the left side, there must be a nontrivial integer solution. For such a solution, $xyz \neq 0$ (otherwise, either 3 or 9 must be the cube of a rational number). Let $(x, y, z) = (u, v, w)$ be a nontrivial solution which minimizes $|x| + |y| + |z|$. Clearly, $u = 3t$ for some integer $t$. Then $(x, y, z) = (v, w, t)$ is also a solution and $|v| + |w| + |t| < |u| + |v| + |w|$, which contradicts the minimal property of $(u, v, w)$. The result follows.

13. (a) The equation is equivalent to $x + y = 4$ and the general solution is given by $(x, y) = (2 - t, 2 + t)$ for $t \in \mathbf{Z}$.

(b) The equation is equivalent to $x + (y - 6)x + y(y - 6) = 0$. The discriminant of the quadratic in $x$ is equal to $-3(y - 6)(y + 2)$, and this is square only if $y = -2, 0, 4, 6$. Hence the solutions $(x, y) = (4, -2)$, $(6, 0)$, $(-2, 4)$, $(0, 6)$.

(c) *First solution.* The equation is equivalent to

$$8(x^2 + xy + y^2 + 1) = (x + y)(x^2 + y^2).$$

From this, it is clear that $x$ and $y$ have the same parity and that $x + y > 0$. If $x$ and $y$ are both positive, then $(x, y) \neq (1, 1)$ and

$$x^2 + y^2 < x^2 + xy + y^2 + 1 < 2(x^2 + y^2)$$

so that $8 < x + y < 16$. Trying out in turn $y = 10 - x$, $y = 12 - x$, $y = 14 - x$ yields only the solutions $(x, y) = (2, 8)$, $(8, 2)$.

If $x$ and $y$ have opposite sign, then

$$x^2 + xy + y^2 + 1 < x^2 + y^2,$$

so that $x + y < 8$. Trying out in turn $y = 2 - x$, $y = 4 - x$, $y = 6 - x$ yields no further solutions.

*Second solution.* Since $x$ and $y$ have the same parity, let $2u = x + y$ and $v = xy$. Then $x^2 + y^2 = 4u^2 - 2v$ and the equation becomes

$$8(4u^2 - v + 1) = 2u(4u^2 - 2v)$$

$$\Rightarrow (u - 2)v = 2u^3 - 8u^2 - 2 = 2(u - 2)(u^2 - 2u - 4) - 18.$$

We have that $u \neq 2$ and

$$
\begin{aligned}
0 \;\leq\; (x - y)^2 &= 4(u^2 - v) \\
&= 4(u - 2)^{-1}[u^3 - 2u^2 - 2u^3 + 8u^2 + 2] \\
&= 4(u - 2)^{-1}[-u^2(u - 6) + 2] = 4(2 - u)^{-1}[u^2(u - 6) - 2].
\end{aligned}
$$

Hence, $u - 2 \,|\, 18$ and $3 \leq u \leq 6$. The only possibility is that $(u, v) = (5, 16)$, so that $(x, y) = (2, 8)$, $(8, 2)$.

**14.** All the zeros must be negative, since $f(x) > 0$ for $x \geq 0$. Denote them by $-r_i$ $(1 \leq i \leq n)$. Since $r_1 r_2 \cdots r_n = 1$, the AGM inequality yields

$$a_{n-k} = \Sigma r_1 r_2 \cdots r_k \geq \binom{n}{k},$$

whence $f(x) \geq (x + 1)^n$ for $x > 0$. The result follows.

**15.** Adding the equations yields $(x+y+z)^2 = (x+y+z)$, whence $x+y+z = 0$ or $x+y+z = 1$. From the difference of the last two equations, we find that $(y - z)(y + z - 2x + 1) = 0$, whence $y = z$ or $y + z = 2x - 1$. There are four cases:

(i) $x + y + z = 0$; $y = z$. then $-2y = x = x^2 + 2y^2 = 6y^2$, so $y = 0$ or $y = -1/3$. Hence, $(x, y, z) = (0, 0, 0)$ or $(2/3, -1/3, -1/3)$.

(ii) $x + y + z = 0$; $y + z = 2x - 1$. Then $x = 1/3$ and we find that $y$ and $z$ are the zeros of the quadratic $9t^2 + 3t + 1$. Hence, $(x, y, z) = (1/3, \omega/3, \omega^2/3)$ or $(1/3, \omega^2/3, \omega/3)$.

(iii) $x + y + z = 1$; $y = z$. Then $y = z = \frac{1}{2}(1-x)$ and $x^2 + 2y^2 = x$ imply $0 = 3x^2 - 4x + 1 = (3x-1)(x-1)$. Hence, $(x,y,z) = (1,0,0)$ or $(1/3, 1/3, 1/3)$.

(iv) $x + y + z = 1$; $y + z = 2x - 1$. Then $x = 2/3$, and $y$ and $z$ are the zeros of $9t^2 - 3t + 1$. Hence $(x,y,z) = (2/3, -\omega/3, -\omega^2/3)$ or $(2/3, -\omega^2/3, -\omega/3)$.

**Remark.** An alternative approach starts with multiplying the second equation by $y$ and the third by $z$, and subtracting to get $0 = y^3 - z^3 = (y-z)(y^2 + yz + z^2)$ whence $y = z$, $y = \omega z$ or $y = \omega^2 z$. The reader should follow up these three possibilities.

**16.** $Q(x) - P(x) = (x-u)F(x)$ where $\deg F \le 2$, $F(x) \ge 0$ for $x \ge u$ and $F(x) \le 0$ for $x \le u$. Hence $F(x) = (x-u)G(x)$ where $\deg G(x) \le 1$ and $G(x) \ge 0$ for each $x$. Thus $G(x)$ must be zero or constant. Since the case that $G(x) = 0$ is straightforward, we can suppose that $G(x) = a$ and

$$Q(x) = P(x) + (x-u)^2 a \quad \text{where } a > 0.$$

Similarly

$$Q(x) = R(x) - (x-u)^2 b \quad \text{where } b > 0.$$

Thus the result follows with $k = b/(a+b)$.

The result does not hold when $P$, $Q$, $R$ are of degree 4. For example, let $P(x) = x^4 - x^2$, $Q(x) = x^4$, $R(x) = 2x^4 + x^2$ (so $u = 0$). If $Q = kP + (1-k)R$, then taking $x = 1$ yields $3k = 2$ while taking $x = 2$ yields $6k = 5$, which are inconsistent.

**17.** A straight line can intersect the graph of the quartic curve in exactly four points if and only if the curve is convex somewhere between each of the outer pair of intersection points and concave somewhere in between the inner pair (sketch a diagram). This occurs exactly when there are two inflection points, i.e. the second derivative of the quartic polynomial has two distinct real zeros. Thus $P(a,b) = 3a^2 - 8b$.

**18.**

$$2b = a + c \quad \Rightarrow \quad 2b^3 = b(ab + bc) = b(3 - ac) = 3b - 10$$
$$\Rightarrow \quad 0 = 2b^3 - 3b + 10 = (b+2)(2b^2 - 4b + 5).$$

Hence $b = -2$, $p = -(a + b + c) = -3b = 6$ and the equation is

$$0 = x^3 + 6x^2 + 3x - 10 = (x-1)(x+2)(x+5).$$

**19.** *First solution.* Suppose, if possible, that $p(p(x)) = q(q(x))$ has a real solution $x = w$. Let $u = p(w)$, $v = q(w)$. By hypothesis, $u \ne v$. We have that $p(u) = q(v)$, while $p(v) = p(q(w)) = q(p(w)) = q(u)$. Hence the polynomial $p(x) - q(x)$ has opposite signs for $x = u$ and $x = v$, and so must vanish between $u$ and $v$. But this contradicts the hypothesis.

*Second solution.* Since $p(x)$ and $q(x)$ are continuous, then $p(x) > q(x)$ for all $x$ or $p(x) < q(x)$ for all $x$. Suppose the former. Then $p(p(x)) > q(p(x)) = p(q(x)) > q(q(x))$ for all $x$ and the result follows.

**20.** Suppose, first, that $AB$ and $AC$ lie along the lines $x = b$ and $y = c$. Then the polynomial $P(b, y)$ has three distinct zeros and so vanishes identically. Similarly, $P(x, c)$ must be the zero polynomial. Hence $P(x, y) = k(x - b)(y - c)$. Since $A'$ lies on neither $AB$ nor $AC$, $P(x, y)$ vanishes at the coordinates of $a'$ only if $k = 0$.

Otherwise, we may suppose that the lines $AB$, $AC$ and $BC$ lie along the respective lines $y = m_i x + k_i$ ($i = 1, 2, 3$). Since for each $i$, $P(x, m_i x + k_i) = 0$ has three distinct roots, the left side must be the zero polynomial. But then, by the Factor Theorem, $P(x, y)$ is divisible by the three distinct factors $y - m_i x - k_i$. Since $\deg P(x, y) \le 2$, this is possible only if $P(x, y) = 0$.

**21.** By the Factor Theorem, $(a - b)(b - c)(c - a)$ divides both the numerator and the denominator. Arguing from degree and coefficients, the denominator must be equal to this and the numerator must have an additional symmetric linear factor, so that the result follows.

**22.** Let $u = (3/5)^{1/7}$, $v = u + u^{-1}$. Then $u^3 + u^{-3} = v^3 - 3v$,

$$u^5 + u^{-5} = v^5 - 5(v^3 - 3v) - 10v = v^5 - 5v^3 + 5v$$

and

$$u^7 + u^{-7} = v^7 - 7(v^5 - 5v^3 + 5v) - 21(v^3 - 3v) - 35v$$
$$= v^7 - 7v^5 + 14v^3 - 7v,$$

whence $v$ is a zero of the polynomial

$$t^7 - 7t^5 + 14t^3 - 7t - (3/5 + 5/3).$$

Multiplying this by 15 yields the required polynomial.

**23.** (a) $p(x) = 1 - \prod(1 - x^{a_i}/n_i)$, from which the result follows immediately.
(b) Let $r = \gcd(m, n)$ and $d$ be the number of prime factors counting repetitions of $r$. Then $m = ru$, $n = rv$, $k = ruv$, $c = a + b - d$ and

$$p(x) = \frac{x^d}{r}\left[1 - \left(1 - \frac{x^{a-d}}{u}\right)\left(1 - \frac{x^{b-d}}{v}\right)\right].$$

If, say, $m = 2^a$, then $r = 2^d$ and $p(2) = 1$. On the other hand, if $p(2) = 1$, then $2^d/r$ must equal 1 and either $2^{a-d} = u$ or $2^{b-d} = v$.

**Remark.** A generalized version of this problem in which there are $s$ $n_i$ not pairwise relatively prime is found in the *Canadian Mathematical Bulletin* **24** (1981), 507, P292.

**24.** *First solution.* Let

$$f(y) = (y - \sqrt{2} - \sqrt{3})(y - \sqrt{2} + \sqrt{3})(y + \sqrt{2} - \sqrt{3})(y + \sqrt{2} + \sqrt{3})$$
$$= y^4 - 10y^2 + 1.$$

Then

$$f(x - \sqrt{5}) = x^4 - 4\sqrt{5}x^3 + 20x^2 - 24$$
$$= -[(3x^4 - 20x^2 + 24) - (x - \sqrt{5})4x^3].$$

Since $f(\sqrt{2}+\sqrt{3}+\sqrt{5}) = 0$, we can take $p(x) = 3x^4 - 20x^2 + 24$, $q(x) = 4x^3$.
  *Second solution.* Let $u = \sqrt{2}+\sqrt{3}+\sqrt{5}$, $v = \sqrt{2}+\sqrt{3}$. Then $u^2 - 2uv + v^2 = 5$ and $v^2 = 5 + 2\sqrt{6} \Rightarrow u^2 - 2uv = -2\sqrt{6} \Rightarrow u^4 - 4u^3v + 4u^2v^2 = 24 \Rightarrow u^4 - 4u^3v + 4u^2(5 - u^2 + 2uv) = 24 \Rightarrow 4u^3v = 3u^4 - 20u^2 + 24$, which yields the result. [Solution by Jeff Higham.]

  **Remarks.** Other possibilities are $p(x) = -5x^7 + 194x^5 - 1520x^3 + 3120x$, $q(x) = 576$ (Colin Springer) and $p(x) = x^5 - 24x$, $q(x) = x^4 - 20x^2 - 24$ (Graham Denham). From this result, it can be deduced that $\sqrt{2}+\sqrt{3}+\sqrt{5}$ must be nonrational. The problem is due to Gregg Petruno, who obtained a more general result; consult Gregg N. Petruno, Sums of irrational square roots are irrational. *Math. Magazine* **61** (1988), 44–45.

**25.** Square both sides and rearrange terms to obtain

$$\sqrt{2p + 1 - x^2}\sqrt{3x + p + 4} = x^2 + 3x + 2 \qquad \ldots (1)$$

Square again, rearrange and factor to obtain

$$(x^2 + x - p)(x^2 + 8x + 2p + 9) = 0 \qquad \ldots (2)$$

There are four roots to equation (2):

$$a = (-1 + \sqrt{4p + 1})/2 \qquad b = (-1 - \sqrt{4p + 1})/2$$
$$c = -4 + \sqrt{7 - 2p} \qquad d = -4 - \sqrt{7 - 2p}$$

For a viable solution $x$ to the given equation, it is necessary that all radicals and quantities under radicals be positive:

(a) $x^2 + 3x + 2 = (x + 2)(x + 1) \geq 0$, i.e. $x \leq -2$ or $x \geq -1$;

(b) $x^2 \leq 2p + 1$;

(c) $3x + p + 4 \geq 0$;

(d) $x^2 + 9x + 9 + 3p \geq 0$.

Thus, from (b), if $p < -1/2$, there is no solution. If $p < -1/4$, then $a$ and $b$ are not solutions. If $p > 7/2$, $c$ and $d$ are not solutions. If $p \leq 7/2$, then $d \leq -4$, so that $3d + p + 4 < 0$ and $x = d$ denies condition (c). Hence, $d$ is never a solution of the given equation.

Suppose that $c$ is a solution. Then, since $-1/2 \leq p$, $c \leq -4 + 2\sqrt{2} < -1$, so that, by condition (a), $c \leq -2$, and, by condition (c), $p \geq -4 + 6 = 2$. Hence, if $p < 2$, then $c$ is not a solution. Thus, there are no solutions if $p < -1/4$.

Let $p \geq -1/4$. Then

$$2p + 1 - a^2 = (\sqrt{p + 1/4} + 1/2)^2$$

$$3a + p + 4 = (\sqrt{p + 1/4} + 3/2)^2$$

$$a^2 + 9a + 3p + 9 = p - a + 9a + 3p + 9 = 8a + 4p + 9$$
$$= (\sqrt{4p + 1} + 2)^2$$

and $a$ satisfies the equation.

Also

$$2p + 1 - b^2 = (\sqrt{p + 1/4} - 1/2)^2$$

$$3b + p + 4 = (\sqrt{p + 1/4} - 3/2)^2$$

$$b^2 + 9b + 3p + 9 = (\sqrt{4p + 1} - 2)^2.$$

If $p \geq 2$, then

$$\sqrt{2p + 1 - b^2} + \sqrt{3b + p + 4} = (\sqrt{p + 1/4} - 1/2) + (\sqrt{p + 1/4} - 3/2)$$
$$= \sqrt{4p + 1} - 2 = \sqrt{b^2 + 9b + 3p + 9}.$$

If $0 < p < 2$, then the left side is

$$(\sqrt{p + 1/4} - 1/2) + (3/2 - \sqrt{p + 1/4}) = 1,$$

which is not equal to

$$\sqrt{b^2 + 9b + 3p + 9} = |\sqrt{4p + 1} - 2|.$$

If $-1/4 \leq p \leq 0$, then the left side is

$$(1/2 - \sqrt{p + 1/4}) + (3/2 - \sqrt{p + 1/4}) = 2 - \sqrt{4p - 1}$$

and the given equation is satisfied.

Summing up, we have the table:

| Range of p | Is | a | b | c | d | a solution? |
|---|---|---|---|---|---|---|
| $p < -1/4$ | | N | N | N | N | |
| $-1/4 \leq p \leq 0$ | | Y | Y | N | N | |
| $0 < p < 2$ | | Y | N | N | N | |
| $2 \leq p$ | | Y | Y | ? | N | |

When $p = -1/4$, then $a = b$. Hence there is a unique solution if and only if $p = -1/4$ or $0 < p < 2$.

**26.** Let $a$, $b$, $c$ be the zeros of the cubic polynomial. Then, $a$, $b$, $c$ are the lengths of the sides of a triangle if and only if $a$, $b$, $c$ are positive real numbers for which $a + b - c$, $a + c - b$ and $b + c - a$ are all positive.

Suppose that these conditions hold. Then the discriminant $18uvw + u^2v^2 - 27w^2 - 4v^3 - 4u^3w$ is nonnegative and the numbers $u$, $v$, $w$,

$$4uv - u^3 - 8w = (a + b - c)(a + c - b)(b + c - a)$$

are all positive.

On the other hand, let these conditions hold. The discriminant condition guarantees that all the zeros are real. Since the polynomial is negative for nonpositive $x$, $a$, $b$, $c$ must all be positive. Since two of $a + b - c$, $a + c - b$, $b + c - a$ are positive in any case, the condition that their product is positive ensures that all three are positive.

**27.** *First solution.* We look at a number of cases.

(1) $a = b = c = 0$. The system is trivially solvable.

(2) $a = b = 0$, $c \neq 0$. The system is not solvable, since it becomes $z = \sqrt{1 - z^2} = 0$, which is impossible.

(3) $a = 0$, $bc \neq 0$. The system becomes

$$by + cz = b\sqrt{1 - y^2} + c\sqrt{1 - z^2} = 0.$$

If the system is solvable, then $b^2y^2 = c^2z^2$ and $b^2(1 - y^2) = c^2(1 - z^2)$, which implies $b^2 = c^2$. On the other hand, suppose that $b^2 = c^2$. Then

(i) if $b = c$, the system is satisfied by $(x, y, z) = (p, 1, -1)$, where $p$ is arbitrary;

(ii) if $b = -c$, the system is satisfied by $(x, y, z) = (p, q, q)$ where $p, q$ are arbitrary.

Thus, in Case (3), there is a solution iff $|b| = |c|$.

(4) $abc \neq 0$: $a$, $b$, $c$ all positive. Then

$$a\sqrt{1 - x^2} + b\sqrt{1 - y^2} + c\sqrt{1 - z^2} = 0 \Rightarrow x^2 = y^2 = z^2 = 1.$$

Since $ax + by + cz = 0$, $x$, $y$, $z$ cannot have all the same sign, so one of $a$, $b$, $c$ is the sum of the other two. On the other hand, if, say, $a = b + c$, then $(x, y, z) = (-1, 1, 1)$ is a solution.

(5) $abc \neq 0$: $a$, $b$, $c$ all negative. As in (4), it can be shown that there is a solution iff one of $a$, $b$, $c$ is the sum of the other two.

(6) $a$, $b$, $c \neq 0$: not all of $a$, $b$, $c$ have the same sign. Let there be a solution $(x, y, z) = (\cos u, \cos v, \cos w)$ where $0 \leq u, v, w \leq \pi$. Then

$$a \cos u + b \cos v + c \cos w = 0$$

$$a \sin u + b \sin v + c \sin w = 0.$$

Shifting two terms to the right side in each equation, squaring and adding yields

$$\begin{aligned} a^2 &= b^2 + c^2 + 2bc \cos(v - w) \\ b^2 &= a^2 + c^2 + 2ac \cos(u - w) \qquad (*) \\ c^2 &= a^2 + b^2 + 2ab \cos(u - v). \end{aligned}$$

On the other hand, suppose $u$, $v$, $w$ can be chosen to satisfy $(*)$. Then

$$(a \cos u + b \cos v + c \cos w)^2 + (a \sin u + b \sin v + c \sin w)^2$$

$$= a^2 + b^2 + c^2 + 2ab(\cos u \cos v + \sin u \sin v) + \cdots$$

$$= a^2 + b^2 + c^2 + [c^2 - a^2 - b^2] + \cdots = 0.$$

Hence, the given system is solvable iff the system $(*)$ is solvable.

We show that $(*)$ is solvable iff $|a| \leq |b| + |c|$, $|b| \leq |a| + |c|$ and $|c| \leq |a| + |b|$. If $(*)$ is solvable, then $a^2 \leq b^2 + c^2 + 2|b||c| = (|b| + |c|)^2$, etc. On the other hand, suppose the inequality conditions hold. Then we can form a triangle with sides $|a|$, $|b|$, $|c|$ opposite angles $A$, $B$, $C$ respectively. We can select $u$, $v$, $w$ to satisfy

$$|bc| \cos A = -bc \cos(v - w)$$

$$|ac| \cos B = -ac \cos(u - w)$$

$$|ab| \cos C = -ab \cos(u - v).$$

For example, if $bc > 0$, $ab < 0$, $ac < 0$, we can choose $u = B$, $v = B + C = \pi - A$, $w = 0$. [Solution by Gary Baumgartner.]

*Second solution.* Assume that $abc \neq 0$, and let there be a solution $(x, y, z) = (\cos u, \cos v, \cos w)$, where $0 \leq u, v, w \leq \pi$. Then, for each $\theta$, we have that

$$a \cos(u + \theta) + b \cos(v + \theta) + c \cos(w + \theta)$$

$$= \cos \theta (a \cos u + b \cos v + c \cos w)$$

$$+ \sin \theta (a \sin u + b \sin v + c \sin w) = 0.$$

Taking $\theta = -u$ yields $a = -b \cos(v - u) - c \cos(w - u)$, whence $|a| \leq |b| + |c|$. Similarly, $|b| \leq |a| + |c|$ and $|c| \leq |a| + |b|$. The rest can be treated as before.

**28.** Yes. Since there are five odd powers of $x$, the second player can guarantee that after the first player has completed four moves, there is a coefficient of one of these powers which has not been assigned. When he comes to make his fourth move, he finds that the polynomial $f(x)$ has the form $g(x) + ax^r + bx^s$ where $g(x)$ is determined, $r$ is odd and $a, b$ have yet to be selected. If $s$ is even, he notes that, since $f(1) + f(-1) = g(1) + g(-1) + 2b$ does not depend on $a$, he can choose $b$ in such a way that $f(1) + f(-1) = 0$.

Since this guarantees that either 1 is a zero or $f(1)$ and $f(-1)$ have opposite signs, he can win regardless of the last move of the first player. If $s$ is
odd, he needs to take a little more care. In this case, he notes that

$$2^r f(1) + f(-2) = 2^r g(1) + g(-2) + (2^r - 2^s)b.$$

Again, he can choose $b$ to ensure that this quantity vanishes regardless of
the first player's last move and thus, as before, ensure a win.

**29.** For the case $m = n = 4$, the strategy is to raise $AX - BY$ to a
sufficiently high power that $A^4$ or $B^4$ is a factor of each term. Thus,

$$1 = (AX - BY)^7 = A^4(A^3 X^7 - 7A^2 B X^6 Y + 21 A B^2 X^5 Y^2 - 35 B^3 X^4 Y^3)$$

$$- B^4(B^3 Y^7 - 7 A B^2 X Y^6 + 21 A^2 B X^2 Y^5 - 35 A^3 X^3 Y^4).$$

The choice of $u$ and $v$ is clear. For general $m$, $n$, look at $1 = (AX - BY)^{m+n-1}$.

**30.** Let $q(x) = p(x) + p'(x) + \cdots + p^{(n)}(x)$. Then $q(x)$ is a polynomial
of even degree with positive leading coefficient. Suppose $q(x)$ assumes its
minimum value when $x = u$. Then $0 = q'(u) = q(u) - p(u)$. But then
$q(x) \geq q(u) = p(u) \geq 0$ for all $x$ (cf. Problem 5.4.22).

**31.** Suppose $a \leq b \leq c$. Let $f(t) = (t - x)(t - y)(t - z)$ and $g(t) = (t - a)(t - b)(t - c)$. Since $f(c) \geq 0 = g(c)$ and $f(a) \leq 0 = g(a)$, there exists
some point $u$ in the interval $[a, c]$ for which $f(u) = g(u)$. This implies that
$(xy + xz + yz - ab - ac - bc)u = 0$. Since $u \geq a > 0$, it follows that
$xy + xz + yz = ab + ac + bc$ and $f(t) = g(t)$ identically. The result follows.
    For other solutions, see *Amer. Math. Monthly* **72** (1965), 185–186.

**32.** Let $n$ be a given integer. We can write the given polynomial in the
form $p(x) = k \prod(x - r_i)$. Let

$$q(x) = x^n \prod(x^{n-1} + r_i x^{n-2} + \cdots + r_i^{n-2} x + r_i^{n-1}).$$

Then $p(x)q(x) = kx^n \prod(x^n - r_i^n)$. We obtain the result by taking $n = 1\,000\,000$. [Solution by M.S. Klamkin.]

**33.** Suppose that the degree of $p(x)$ is $m$ and that of $q(x)$ is $n$ where
$m \geq n$. If the distinct zeros of $p(x)$ and $q(x)$ are $u_1, u_2, \ldots, u_r$ and the
zeros of $p(x) - 1$ and $q(x) - 1$ are $v_1, v_2, \ldots, v_s$, then the $u_i$ and the $v_j$ are
distinct zeros of the difference $(p - q)(x)$. If we can show that $r + s > m \geq \deg(p - q)(x)$, the result will follow.
    Now, the common derivative of $p(x)$ and $p(x) - 1$ has at least $(m - r) + (m - s)$ zeros counting multiplicity, so that $2m - r - s \leq \deg p'(x) = m - 1$,
whence $r + s \geq m + 1 > m$.

**34.** The system

$$p + q + r = 0$$
$$up + vq + wr = 0 \qquad (*)$$
$$xp + yq + zr = 0$$

in the variables $p$, $q$, $r$ has a nontrivial solution

$$(p, q, r) = (y - z, z - x, x - y).$$

Since $x^3 + u^3 = y^3 + v^3$, it follows that $uy - vx \neq 0$ (otherwise $x = ky$ and $u = kv$). Hence, we can solve the system $bu + cx = bv + cy = 1$ for $b$ and $c$. Multiplying the equations $(*)$ respectively by $1$, $-b$, $-c$ and adding yields $0 = (1 - bw - cz)(x - y)$, whence $bw + cz = 1$.

From $(-cx)^3 = (bu - 1)^3$ and $-x^3 = u^3 - a^3$, etc., we find that the equation

$$(c^3 - b^3)t^3 + 3b^2t^2 - 3bt + (1 - a^3c^3) = 0$$

has (unequal) roots $u$, $v$, $w$, so that $(b^3 - c^3)uvw = 1 - a^3c^3$. Similarly, $(b^3 - c^3)t^3 + 3c^2t^2 - 3ct + (1 - a^3b^3) = 0$ has roots $x$, $y$, $z$, so that $(b^3 - c^3)xyz = a^3b^3 - 1$. Hence

$$(b^3 - c^3)(uvw + xyz) = a^3(b^3 - c^3).$$

Since each of the two cubic equations has three roots and since $b$ and $c$ cannot both be zero, at least one of the cubic equations is nontrivial and $b^3 - c^3 \neq 0$. The result follows.

Suppose $x = y = z$. Then $u = v = w$ and the result is trivial. If $x = y$, then $u = v$ and the linear condition holds automatically. However, the result may fail. Let $(u, v, w) = (12, 12, 10)$, $(x, y, z) = (1, 1, 9)$, so that the cubic condition is satisfied with $a^3 = 1729$. But $uvw + xyz = 1449$. If we put these numbers into the above solution, we find that $(*)$ consists of

$$p + q + r = 12(p + q) + 10r = (p + q) + 9r = 0$$

and that $b = 4/49$, $c = 1/49$. Thus, $u = 12$, $w = 10$ are the roots of the equation

$$0 = 63t^3 - 2352t^2 + 28812t - 115920 = 21(t - 10)(t - 12)(3t - 46)$$

and $x = 1$, $z = 9$ are the roots of

$$0 = 21(3t^3 + 7t^2 - 343t + 333) = 21(t - 1)(t - 9)(3t + 37).$$

Thus, if we take rather $(u, v, w) = (12, 46/3, 10)$, $(x, y, z) = (1, -37/3, 9)$, then the hypotheses and the conclusion of the problem are satisfied.

**35.** The given inequality has no solution when $x < 1$ since all summands on the left side are negative and no solutions for large $x$ when all summands assume very small positive values. The inequality will be satisfied when $x$

slightly exceeds any one of the integers $1, 2, \ldots, 70$ and not satisfied when $x$ is slightly less than one of the integers $2, 3, \ldots, 70$. Let

$$p(x) = 5(x-1)\cdots(x-k)\cdots(x-70) - 4\sum_{k=1}^{70} k(x-1)\cdots(\widehat{x-k})\cdots(x-70).$$

Then $p(r) > 0$ when $r = 1, 3, 5, \ldots, 69$ and $p(r) < 0$ when $r = 2, 4, 6, \ldots, 70$. Since $\deg p(x) = 70$ and $p(x) > 0$ for large $x$, $p(x)$ has exactly one real zero $u_r$ in each of the open intervals $(r, r+1)$ $(1 \le r \le 69)$ and a real zero $u_{70} > 70$. The inequality will be satisfied exactly when $x$ belongs to one of the intervals $(r, u_r]$ $(1 \le r \le 70)$. Let $S = 1 + 2 + \cdots + 70 = (35)(71)$. Since the coefficient of $x^{69}$ in $p(x)$ is $-5S - 4S = -9S$, we find that $\Sigma u_r = 9S/5$. Hence the total length of all the intervals is

$$\sum_{r=1}^{70}(u_r - r) = 9S/5 - S = 4(7)(71) = 1988.$$

**36.** *First solution.* Suppose $a + b \ne 0$. We have that

$$
\begin{aligned}
0 &= (\Sigma a)^3 = a^3 + 3\Sigma a^2 b + 6\Sigma abc \\
&= 0 + 3\Sigma a^2(-a) + 6\Sigma abc = 6\Sigma abc,
\end{aligned}
$$

where the summations are symmetric in the variables $a$, $b$, $c$, $d$, $e$, $f$. The left side multiplied by $a + b$ is

$$(a+b)(a+c)(a+d)(a+e)(a+f)$$

$$
\begin{aligned}
&= a^5 + a^4(\Sigma b) + a^3(\Sigma bc) + a^2(\Sigma bcd) + a(\Sigma bcde) + bcdef \\
&= (a^5 - a^5) + a^2(a\Sigma bc + bcd) + (a\Sigma bcde + bcdef) \\
&= 0 + 0 + (a\Sigma bcde + bcdef),
\end{aligned}
$$

where the summations are symmetric in the variables $b$, $c$, $d$, $e$, $f$. By symmetry, the right side multiplied by $a + b$ is the same, and the result follows. If $a + b = 0$, then the given conditions are $c + d + e + f = c^3 + d^3 + e^3 + f^3 = 0$ and we find that the coefficients of even powers of $(a+c)$ $(a+d)(a+e)(a+f)$ and $(a-c)(a-d)(a-e)(a-f)$ agree while those of odd powers vanish by a similar argument.

*Second solution.* Noting that $(a + c + d) + (b + e + f)$ is a factor of the sum of the corresponding cubes, we have that

$$
\begin{aligned}
0 &= (a+c+d)^3 + (b+e+f)^3 \\
&= 3[(a+c)(a+d)(c+d) + (b+e)(b+f)(e+f)]
\end{aligned}
$$

whence

$$(a+c)(a+d)(c+d) = -(b+e)(b+f)(e+f).$$

Similarly,

$$(a + e)(a + f)(e + f) = -(b + c)(b + d)(c + d).$$

Multiplying the corresponding sides of these two equations yields the result when $(c + d)(e + f) \neq 0$.

Suppose, say, that $c + d = 0$. If $e + f = 0$, too, then $a = -b$ and the result is clear. If $e + f \neq 0$, then $(b + e)(b + f) = (a + e)(a + f) = 0$ and both sides of the required equation vanish.

*Third solution.* For $1 \leq k \leq 6$, let $s_k$ be the sum of all products of $k$ of the six variables and $p_k$ be the sum of their $k$th powers. Define

$$\begin{aligned} f(x) &= (x + a)(x + b)(x + c)(x + d)(x + e)(x + f) \\ &= (x^6 + s_2 x^4 + s_4 x^2 + s_6) + s_5 x. \end{aligned}$$

Then $f(x) - 2s_5 x = f(-x)$, so that $0 = f(a) - 2s_5 a = f(b) - 2s_5 b$. Therefore, $bf(a) = af(b)$.

If $ab(a + b) \neq 0$, this equation leads directly to the result. If $a = b = 0$, the result is obvious. If, say, $a = 0$, $b \neq 0$, then $f(b) = 2s_5 b = 2b^2 cdef$ and the result holds. If $a + b = 0$, $ab \neq 0$, then $f(a) = f(b) = 0$, so that $s_5 = 0$. We can then write

$$f(x) = (x^2 - u^2)(x^2 - v^2)(x^2 - w^2),$$

where $u^2$, $v^2$, $w^2$ are the zeros of $t^3 + s_2 t^2 + s_4 t + s_6$. Hence $a = u$, $b = -u$ (say) and $c$, $d$, $e$, $f$ are, in some order, $v$, $-v$, $w$, $-w$. The result again follows.

**37.** *First solution.* Let $x = \cos u + i \sin u$, $y = \cos v + i \sin v$, $z = \cos w + i \sin w$, and define the symmetric functions $p = x + y + z$, $q = xy + yz + zx$, $r = xyz$. Then

$$a + bi = p$$
$$a - bi = q/r$$
$$c + di = x^2 + y^2 + z^2 = p^2 - 2q$$
$$c - di = x^{-2} + y^{-2} + z^{-2} = (q^2 - 2pr)/r^2.$$

Since $q = r\bar{p}$ and $|x| = |y| = |z| = |r| = 1$, we deduce that

$$|q|^2 = |p|^2 = a^2 + b^2.$$

Since $2q = (a + bi)^2 - (c + di)$, it follows that

$$(a^2 - b^2 - c)^2 + (2ab - d)^2 = 4(a^2 + b^2).$$

*Second solution.* Let $\theta = u + v + w$. Then

$$\begin{aligned} a^2 - b^2 &= c + 2[\cos(\theta - u) + \cos(\theta - v) + \cos(\theta - w)] \\ &= c + 2a \cos \theta + 2b \sin \theta \\ 2ab &= d + 2a \sin \theta - 2b \cos \theta \end{aligned}$$

so that

$$
\begin{aligned}
(a+bi)^2 &= (c+di) + 2(a-bi)(\cos\theta + i\sin\theta) \\
&\Rightarrow |(a+bi)^2 - (c+di)| = 2|a-bi| \\
&\Rightarrow (a^2 - b^2 - c)^2 + (2ab - d)^2 = 4(a^2 + b^2).
\end{aligned}
$$

[Solution by Georges Gonthier.]

*Third solution.* We have

$$(a^2 - b^2 - c)/2 = \cos(v+w) + \cos(u+w) + \cos(u+v) \qquad \dots(1)$$

$$(2ab - d)/2 = \sin(v+w) + \sin(u+w) + \sin(u+v) \qquad \dots(2)$$

$$a^2 + b^2 - 3 = \cos(v-w) + \cos(w-u) + \cos(u-v). \qquad \dots(3)$$

The process that yields (3) from the equations for $a$ and $b$ yields the following equation from (1) and (2):

$$[(a^2 + b^2 - c)/2]^2 + [(2ab - d)/2]^2 - 3$$

$$= \cos((u+w) - (u+v)) + \cos((u+v) - (v+w)) + \cos((v+w) - (u+w))$$

$$= a^2 + b^2 - 3.$$

[Solution by Alexander Pruss.]

**38.**

$$
\begin{aligned}
x^2 - y^2 &= \cot^2\theta + \tan^2\theta - \sec^2\theta - \cos^2\theta + 4 \\
&= -1 + \cos^2\theta(\csc^2\theta - 1) + 4 = 3 + \cos^2\theta\cot^2\theta
\end{aligned}
$$

while

$$
\begin{aligned}
xy &= \csc\theta - \cot\theta\cos\theta + \tan\theta\sec\theta - \sin\theta \\
&= \csc\theta(1 - \cos^2\theta) - \sin\theta + \tan\theta\sec\theta = \tan\theta\sec\theta.
\end{aligned}
$$

Hence $x^2 y^2 (x^2 - y^2 - 3) = 1$.

**39.** The polynomial equals

$$((3/4) - x) + (x^2 + x^4)(1 - x) + x^6 = (x-1)(x^5 + x^3 + x) + (3/4)$$

is clearly positive for $x \le 3/4$ and $x \ge 1$. The derivative of the polynomial is

$$6x^5 - 5x^4 + 4x^3 - 3x^2 + 2x - 1 = (1/2)[(3x^4 + 2x^2 + 1)(4x - 3) + (1 - x^4)]$$

which is positive for $3/4 \le x \le 1$. Hence the given function increases on the closed interval $[3/4, 1]$ and so is positive there.

**40.** The local maximum and minimum values are those values of $k$ for which $y = k$ is tangent to $y = x^3 + 3px^2 + 3qx + r$. This occurs only if the equation $f(x) = 0$ has a double zero, where

$$\begin{aligned} f(x) &= x^3 + 3px^2 + 3qx + (r - k) \\ &= (x + p)^3 + 3(q - p^2)(x + p) + (2p^3 - 3pq + r - k). \end{aligned}$$

Now $f'(x) = 3[(x + p)^2 + (q - p^2)]$. Since $f(x)$ has three real zeros, $f'(x)$ has real zeros and $p^2 - q > 0$. We have, by division, that

$$3f(x) = (x + p)f'(x) + 3[2(q - p^2)(x + p) + (2p^3 - 3pq + r - k)].$$

Since a double zero of $f(x)$ is a zero of $f'(x)$, a zero $u$ of $f(x)$ is a double zero $\Longleftrightarrow$

$$2(q - p^2)(x + p) + (2p^3 - 3pq + r - k) = 0$$

and $(x + p)^2 + (q - p^2) = 0$

$$\begin{aligned} &\Rightarrow \quad 4(p^2 - q)^3 = 4(q - p^2)^2(x + p)^2 = (2p^3 - 3pq + r - k)^2 \\ &\Longleftrightarrow \quad \mp 2(p^2 - q)^{3/2} = 2p^3 - 3pq + r - k \\ &\Longleftrightarrow \quad k = 2p^3 - 3pq + r \pm 2(p^2 - q)^{3/2}. \end{aligned}$$

**41. (a)**

$$\begin{aligned} \frac{2(ab - cd)}{a - b + c - d} &= \frac{[a^2 + b^2 - c^2 - d^2] - [(a - b)^2 - (c - d)^2]}{a - b + c - d} \\ &= (a^2 + b^2 - c^2 - d^2)/(a - b + c - d) - [(a - b) - (c - d)] \\ &= (a^2 - b^2 - c^2 + d^2)/(a + b + c + d) - [(a + d) - (b + c)] \\ &= \frac{[a^2 - b^2 - c^2 + d^2] - [(a + d)^2 - (b + c)^2]}{a + b + c + d} = \frac{2(bc - ad)}{a + b + c + d}. \end{aligned}$$

**(b)** $(a, b, c, d) = (5, 3, 2, 6)$ works.

**42.** Suppose, if possible, that the degree of the polynomial $f(x)$ is a number $n$ exceeding 1. From Lagrange's Formula,

$$f(x) = \sum_{k=0}^{n} \frac{f(k)x(x - 1) \cdots (\widehat{x - k}) \cdots (x - n)}{k(k - 1) \cdots (k - n)}$$

it follows that all the coefficients must be rational. Hence, the polynomial $df(x)$ obtained by multiplying $f(x)$ by the least common multiple $d$ of the denominators of its coefficients is a polynomial with the property of the problem which has integer coefficients. Suppose

$$df(x) = a_n x^n + a_{n-1} x^{n-1} + \cdots + a_1 x + a_0.$$

Let

$$g(x) = a_n^{n-1} df(x/a_n)$$
$$= x^n + a_{n-1}x^{n-1} + a_n a_{n-2}x^{n-2} + \cdots + a_n^{n-2}a_1 x + a_n^{n-1}a_0.$$

Then $g(x)$ is a monic polynomial over $\mathbf{Z}$ which has the property ascribed to $f(x)$.

We now modify $g(x)$ to obtain a polynomial which is sure to have a real zero which is nonrational. To this end, observe that $g(x) - g(0) - x$ has at most finitely many zeros, so that there exists a prime $p$ for which $g(p) - g(0) - p$ is nonzero. Let $h(x) = g(x) - g(0) - p$. Then $h(x)$ is a monic polynomial over $\mathbf{Z}$ of degree $n$ which assumes rational values if and only if $x$ is rational. Since $h(0) < 0$ and $h(x)$ has positive leading coefficient, $h(x)$ must have a positive zero. By hypothesis, this zero must be rational and thus must be an integer dividing the prime $p$, and so must be $p$. But this contradicts $h(p) \neq 0$. The result follows.

**Remark.** Compare this with Problem 5.4.23.

**43.** Let $u = a/(bc - a^2)$, $v = b/(ca - b^2)$, $w = c/(ab - c^2)$. Then $vw - u^2 = s/(bc - a^2)$, $wu - v^2 = s/(ca - b^2)$, $uv - w^2 = s/(ab - c^2)$ where

$$s = \frac{(a^3b^3 + b^3c^3 + c^3a^3 - 3a^2b^2c^2)}{(ab - c^2)(ca - b^2)(bc - a^2)}.$$

Suppose, if possible, that $s = 0$. Then $ab + bc + ca = 0 \Rightarrow ab - c^2 = -c(a + b + c)$, $ca - b^2 = -b(a + b + c)$, $bc - a^2 = -a(a + b + c)$ whence $u = v = w = -1/(a + b + c)$ and so $u + v + w$ is nonzero contrary to hypothesis. Hence $s \neq 0$, and so

$$u/(bc - a^2) + v/(ca - b^2) + w/(ab - c^2) = (3uvw - u^3 - v^3 - w^3)/s$$

$$= -(u + v + w)(u^2 + v^2 + w^2 - uv - uw - vw)/s = 0$$

as required.

**44.** Let $x = 1 + 1/1 + x = (2 + x)/(1 + x)$, $x_{k+1} = 1 + 1/x_k$ for $k = 1, 2, \ldots, n - 1$. By induction, it can be shown that

$$x_k = (F_{k+2} + F_{k+1}x)/(F_{k+1} + F_k x),$$

where $\{F_k\} = \{1, 1, 2, 3, 5, 8, \ldots\}$ is the Fibonacci sequence. Thus, the equation to be solved is $x_n = x$, i.e. $F_{n+2} + F_{n+1}x = (F_{n+1} + F_n x)x$, whence $x = \pm\sqrt{F_{n+2}/F_n}$.

**45.** If $w$ is a zero of $f(t)$, then so also are $w^2 + w + 1$ and $w^2 - w + 1 = (w - 1)^2 + (w - 1) + 1$. Since $2w = (w^2 + w + 1) - (w^2 - w + 1)$, it follows that $2|w| \leq |w^2 + w + 1| + |w^2 - w + 1|$ with equality if and only if

$w^2 + w + 1$ is a positive real multiple of $-(w^2 - w + 1)$. If there is equality, then $|w^2 + w + 1| = |w^2 - w + 1|$ which occurs when

$$2(w^2 + 1) = (w^2 + w + 1) + (w^2 - w + 1) = 0.$$

Hence, except when $w^2 = -1$, it must happen that $|w|$ is strictly less than the maximum of $|w^2 + w + 1|$ and $|w^2 - w + 1|$, so at least one of these three must differ from 1. It follows from this, since $f(t)$ is a polynomial, that the only zeros of $f(t)$ can be $i$ and $-i$, and so $f(t) = c(t^2 + 1)^k$ for $c = 1$ and nonnegative integer $k$. It is readily checked that this works.

**46.** There is no solution. Otherwise, by the AGM inequality, $2\sqrt{3}ux \leq u^2 + 3x^2$, etc., so that $6 < 4\sqrt{3} = 2\sqrt{3}(ux + vy + wz) \leq 6$, which is a contradiction.

**47.** $x^3 + y^3 + z^3 - 3xyz = (x + y + z)(x^2 + y^2 + z^2 - xy - xz - yz) = 0 \Rightarrow$ $x^3 + y^3 + z^3 = 3xyz$. Similarly, $x^{-3} + y^{-3} + z^{-3} = 3x^{-1}y^{-1}z^{-1}$. Thus

$$
\begin{aligned}
x^6 + y^6 + z^6 &= (x^3 + y^3 + z^3)^2 - 2x^3 y^3 z^3 (x^{-3} + y^{-3} + z^{-3}) \\
&= 9x^2 y^2 z^2 - 6x^2 y^2 z^2 = 3x^2 y^2 z^2,
\end{aligned}
$$

and this yields the result.

**48.** Let $w$ be a zero of $z^p + 1$. Then $w$ is simple, so it suffices to show that $z = w$ makes the right side vanish. This is clear for $w = -1$. Otherwise, $1 - w + w^2 - \cdots + w^{p-1} = 0$ and the right side is equal to $(-w^{-1} - 1)(w^{p-1})^k + (w + 1)w^{(p-1)k-1}$ which is 0. See Problem 3.7.17 and its solution for an alternative solution to a special case.

**49.** Suppose that $(x + 1)^3 - x^3 = y^2$. This implies that

$$3(x^2 + x) + 1 = y^2 \Rightarrow 3(2x + 1)^2 = 4y^2 - 1 = (2y - 1)(2y + 1).$$

Since $2y - 1$ and $2y + 1$ are relatively prime, either

(1)  $2y - 1 = a^2$, $2y + 1 = 3b^2$ for some $a$, $b$, or

(2)  $2y - 1 = 3c^2$, $2y + 1 = d^2$ for some $c$, $d$.

Case (2) cannot occur, since it would imply that $d^2 = 3c^2 + 2$, an impossibility. Hence, we must have Case (1) and $4y = a^2 + 3b^2 = 2(a^2 + 1)$. Since $a$ is odd, we can write $a = 2u + 1$, whence $y = u^2 + (u + 1)^2$.

**Remark.** In *Amer. Math. Monthly* **57** (1950), 190, it is noted that solutions of the equation are given by $(x, y) = (x_n, y_n)$ where $(x_0, y_0) = (0, 1)$, $(x_1, y_1) = (7, 13)$ and $(x_{n+1}, y_{n+1}) = (14x_n - x_{n-1} + 6, 14y_n - y_{n-1})$ for $n \geq 1$.

**50.** From the difference of the first two equations, we obtain that

$$x(1 - x^2) + y(y - 1) + z^2(z - 1) = 0. \qquad \qquad \ldots (1)$$

From the difference of the last two equations, we obtain that

$$y(1 - y^2) + z(z - 1) + x^2(x - 1) = 0. \qquad \ldots (2)$$

Subtracting $z$ times (2) from (1) yields

$$x(x - 1)(1 + x + xz) = y(y - 1)(1 + z + yz). \qquad \ldots (3)$$

Similarly

$$y(y - 1)(1 + y + yx) = z(z - 1)(1 + x + zx). \qquad \ldots (4)$$

It is clear from (3) and (4) that, if $x$, $y$, $z$ are positive, then $x$, $y$, $z$ are all equal to 1, all less than 1, or all greater than 1. The last two possibilities contradict the given equations and the result follows.

**51.** Consider the graphs of the equations $y = 6x^2$, $y = 77x - 147$ and $y = 77[x] - 147$. Since $6x^2 - 77x + 147 = (3x - 7)(2x - 21)$, the first two curves cross when $x = 7/3$ and $x = 21/2$. From the graphs, it can be seen that any solution $x$ of the given equation must satisfy $3 \le x < 21/2$. Hence the possible values of $[x]$ are integers between 3 and 10 inclusive. We now consider the following table:

| $[x]$ | $y = 77[x] - 147$ | $y/6$ | $[\sqrt{y/6}]$ |
|---|---|---|---|
| 3 | 84 | 14 | 3 |
| 4 | 161 | 26.8 | 5 |
| 5 | 238 | 39.7 | 6 |
| 6 | 315 | 52.5 | 7 |
| 7 | 392 | 65.3 | 8 |
| 8 | 469 | 78.2 | 8 |
| 9 | 546 | 91 | 9 |
| 10 | 623 | 103.8 | 10 |

The solutions of the equation are those values of $\sqrt{y/6}$ for which $[x] = [\sqrt{y/6}]$, i.e. $\sqrt{14}$, $\sqrt{469/6}$, $\sqrt{91}$, $\sqrt{623/6}$.

**52.** Let

$$\begin{aligned} f(x) &= x^4 + kx^2 - 2k^2(2k + 1)x \\ &= x[x^3 + kx - 2k^2(2k + 1)]. \end{aligned}$$

Then

(1) $f(x)$ has four distinct real zeros $\Longleftrightarrow$ the discriminant of the cubic factor is positive $\Longleftrightarrow k^3(3k + 1)(36k^2 + 24k + 1) < 0$.

(2) $f(k) = -k^3(3k + 1)$.

(3) $f'(x) = (x - k)g(x)$ where
$$g(x) = 4x^2 + 4kx + 2k(2k + 1) = (2x + k)^2 + k(3k + 2).$$

(4) $f''(x) = 2(6x^2 + k)$.

(5) $f(x)$ and $f'(x)$ have a zero in common $\iff k = -1/2, -1/3, 0$ or else $g(x)$ and $x^3 + kx - 2k^2(2k + 1)$ have a zero in common. In the last case we have that
$$(x^3 + kx) + (4kx^2 + 4k^2x) = 0 \Rightarrow x^2 + 4kx + (4k^2 + k) = 0.$$
Since also $x^2 + kx + (k^2 + k/2) = 0$, we obtain $3kx + (3k^2 + k/2) = 0$ or $x = -(k + 1/6)$. Plugging this into the equation $f'(x) = 0$ yields $36k^2 + 24k + 1 = 0$ or $6k = (-2 \pm \sqrt{3})$.

(6) $f'(x)$ has three real zeros $\iff -2/3 \le k \le 0$. Let the zeros of $f'(x)$ be $k$, $r$, $s$ with $r \le s$.

(7) $k$ is a double zero of $f'(x) \iff 0 = g(k) = 2k(6k + 1) \iff k = 0, -1/6$.

We consider various ranges of values for $k$:

(a) $k < -2/3$: $f(x)$ has a single minimum value at $x = k$ and $f''(x) < 0 \iff -\sqrt{-k/6} < x < \sqrt{-k/6}$.

(b) $k = -2/3$: $r = s = 1/3$; $f''(x) < 0 \iff -1/3 < x < 1/3$.

(c) $-2/3 < k < (-2 - \sqrt{3})/6$: Since $r + s = -k > 0$, $rs = k(k + 1/2)$, $r$ and $s$ are both positive.

(d) $k = (-2 - \sqrt{3})/6$: $r = 1/6$, $s = (1 + \sqrt{3})/6$, $f(s) = 0$.

(e) $(-2 - \sqrt{3})/6 < k < -1/2$: $r$ and $s$ are both positive.

(f) $k = -1/2$: $r = 0$, $s = 1/2$.

(g) $-1/2 < k < -1/3$.

(h) $k = -1/3$.

(i) $-1/3 < k < -1/6$.

(j) $k = -1/6$.

(k) $-1/6 < k < (-2 + \sqrt{3})/6$.

(l) $k = (-2 + \sqrt{3})/6$.

(m) $(-2 + \sqrt{3})/6 < k < 0$.

(n) $k = 0$.

(o) $k > 0$.

(a) $k < -\frac{2}{3}$

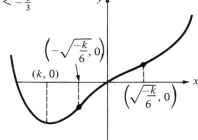

$\left(-\sqrt{\frac{-k}{6}}, 0\right)$

$(k, 0)$

$\left(\sqrt{\frac{-k}{6}}, 0\right)$

(b) $k = -\frac{2}{3}$

$\left(\frac{1}{3}, \cdot\right)$

$\left(-\frac{1}{3}, \cdot\right)$

$\left(-\frac{2}{3}, -\frac{8}{27}\right)$

(c) $-\frac{2}{3} < k < \frac{-2-\sqrt{3}}{6}$

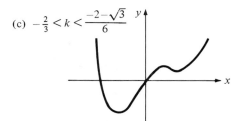

(d) $k = \frac{-2-\sqrt{3}}{6}$

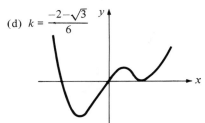

(e) $\frac{-2-\sqrt{3}}{6} < k < -\frac{1}{2}$

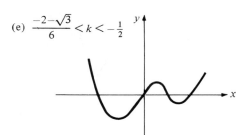

(f) $k = -\frac{1}{2}$

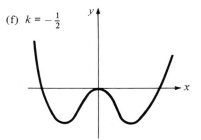

(g) $-\frac{1}{2} < k < -\frac{1}{3}$

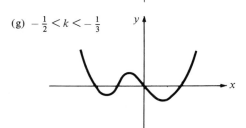

(h) $k = -\frac{1}{3}$

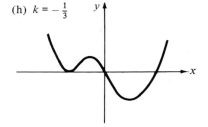

(i) $-\frac{1}{3} < k < -\frac{1}{6}$

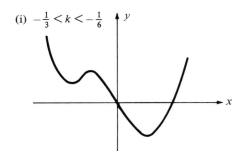

(j) $k = -\frac{1}{6}$

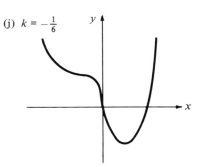

(k) $-\frac{1}{6} < k < (-2 + \sqrt{3})/6$

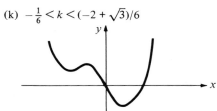

(l) $k = \dfrac{-2 + \sqrt{3}}{6}$

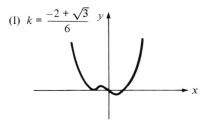

(m)

$\dfrac{-2 + \sqrt{3}}{6} < k < 0$

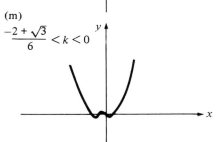

(n) $k = 0$

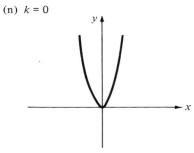

(o) $k > 0$

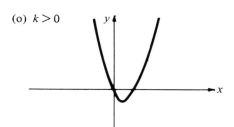

**53.** Let $a, b, c$ be the lengths of the sides. Then $a+b+c = 98$, and, by Heron's formula for the area, $49(49 - a)(49 - b)(49 - c) = 420^2$. Let $u = 49 - a$, $v = 49 - b$, $w = 49 - c$. Then $u + v + w = 49$ and $uvw = 3600$, so that $u$, $v$, $w$ are zeros of a polynomial of the form $f(x) = x^3 - 49x^2 + rx - 3600$.

From the given example, we know that $(u, v, w) = (9, 20, 20)$ works and we can write

$$f(x) = (x - 9)(x - 20)^2 - sx.$$

Now, $f(x)$ will have three real zeros $\Longleftrightarrow$ the line $y = sx$ intersects the cubic curve $y = (x - 9)(x - 20)^2$ in three points (counting multiplicity). This will occur for $0 \le s \le k$, where $k$ is that positive value of $s$ for which $f(x)$ has a double zero, i.e. $f(x) = (x - 9)(x - 20)^2 - sx$ and $f'(x) = (x - 20)[(x - 20) + 2(x - 9)] - s$ have a zero in common.

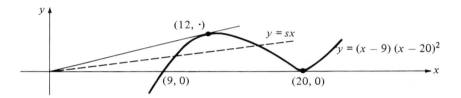

But any common zero of $f(x)$ and $f'(x)$ is a zero of

$$f(x) - x f'(x) = -(x - 20)(x - 12)(2x + 15).$$

Hence, when $s = k$, $f(x)$ has 12 as a double zero. When $0 \le s \le k$, $f(x)$ has a zero between 9 and 12 inclusive. Since such a zero must divide 3600, the possibilities are 9, 10 and 12. Checking these out yields the solutions $(u, v, w) = (9, 20, 20)$, $(10, 15, 24)$, $(12, 12, 25)$ or $(a, b, c) = (40, 29, 29)$, $(39, 34, 25)$, $(37, 37, 24)$.

**54.** $(x - y)^2 + (y - z)^2 + (z - x)^2$.

**55.** Multiply the difference of the two sides by the product of the denominators to obtain

$$x^2 y^2 z^2 (x + y + z) - 2xyz(xy + yz + zx) + (xy^2 + yz^2 + zx^2)$$
$$+ xyz(xy^2 + yz^2 + zx^2) - 2xyz(x + y + z) + (xy + yz + zx)$$
$$= xy(y + 1)(zx - 1)^2 + yz(z + 1)(xy - 1)^2 + zx(x + 1)(yz - 1)^2 \ge 0.$$

**56.** Factoring the denominator, first as a difference of squares, and then completely yields $(a - x)(b - x)(a + x)(b + x)$. The numerator is equal to

$$[a(b - x) + x(a - x)]^2 - [x(a - x) + x(b - x)][a(b - x) + b(a - x)]$$
$$= (a^2 - ax)(b - x)^2 + (x^2 - bx)(a - x)^2 + (ax - bx)(a - x)(b - x)$$
$$= (a - x)(b - x)[a(b - x) - x(a - x) + (a - b)x] = (a - x)^2(b - x)^2.$$

The answer is $(a - x)(b - x)/(a + x)(b + x)$.

# Notes on Explorations

**E.1.** The identity

$$a(ad + bc) + (2bd + c^2) = c(c + ab) + d(2b + a^2)$$

is useful in picking the coefficients of $t^4 + at^3 + bt^2 + ct + d$ in such a way that four coefficients of the nine in the square of this polynomial will vanish. The polynomial

$$t^4 + 2t^3 - 2t^2 + 4t + 4$$

has five terms in its square.

**E.2.** Write $\{a_i\} = (m)\{b_i\}$ if $\Sigma a_i^k = \Sigma b_i^k$ ($k = 0, 1, 2, 3, 4, \ldots, m$). It can be seen that, if $\{a_i\} = (m)\{b_i\}$, then $\{ua_i + b\} = (m)\{ub_i + v\}$ for any $u$ and $v$. To show how to construct pairs of subsets of integers with the first few powers equal, we illustrate with an example how to move from simple sets to more complex sets:

(1) $\{-1, 1\} = (1)\{-2, 2\}$

(2) $\{1, 3\} = (1)\{0, 4\}$

(3) $\{-1, 1, 0, 4\} = (2)\{-2, 2, 1, 3\}$

(4) $\{-1, 0, 4\} = (2)\{-2, 2, 3\}$

(5) $\{-2, -1, 3\} = (2)\{-3, 1, 2\}$

(6) $\{3, 4, 8\} = (2)\{2, 6, 7\}$

(7) $\{-2, -1, 3, 2, 6, 7\} = (3)\{-3, 1, 2, 3, 4, 8\}$

(8) $\{-2, -1, 6, 7\} = (3)\{-3, 1, 4, 8\}$

(9) $\{-9, -7, 7, 9\} = (3)\{-11, -3, 3, 11\}$.

For the problem posed in (d), a solution for $d = 3$, $m = 2$ is

$$\{1, 6, 8, 12, 14, 16, 20, 22, 27\}$$

$$= (2)\{2, 4, 9, 10, 15, 17, 21, 23, 25\}$$

$$= (2)\{3, 5, 7, 11, 13, 18, 19, 24, 26\}.$$

The question of finding distinct sets with equal sums of powers pops up persistently in the literature. The most detailed treatment appears in Albert Gloden, *Mehrgradige Gleichungen,* (Noordhoff, Groningen, 1944) **MR 8,** 441f.

A discussion of the problem, in particular version (d), is found in E.M. Wright, Prouhet's 1851 solution of the Tarry–Escott problem of 1910, *Amer. Math. Monthly* **66** (1959), 199–201. To divide the set of numbers up to $d^{k+1}$ into $d$ sets with the desired property, write each number in the set to base $d$, sum the digits and classify according to the congruence of this sum modulo $d$.

A history of the problem can be found in L.E. Dickson, *History of the Theory of Numbers,* (Washington, 1920; reprint, Chelsea, 1952) Vol. II, Chapt. 24. There is also a section (pages 328–332) on the problem in G.H. Hardy & E.M. Wright, *An Introduction to the Theory of Numbers,* (Oxford, 4th ed., 1960).

A recent reference in which the problem plays a role is Gerald Myerson, How small can a sum of roots be? *Amer. Math. Monthly* **93** (1986), 457–459.

Other references related to the topic are

J. Chernick, Ideal solutions of the Tarry–Escott problem
*Amer. Math. Monthly* **44** (1937), 626–633.

L.E. Dickson, *Introduction to the Theory of Numbers*
(1929), pages 55-58.

H.L. Dorwart & O.E. Brown, The Tarry–Escott problem
*Amer. Math. Monthly* **44** (1937), 613–626.

A. Gloden, Two theorems on multi-degree equalities
*Amer. Math. Monthly* **53** (1946), 205.

A. Gloden, Parametric solutions of two multi-degreed equalities
*Amer. Math. Monthly* **55** (1948), 86–88.

A. Gloden, Normal trigrade and cyclic quadrilaterals with integral sides and diagonals
*Amer. Math. Monthly* **58** (1951), 244–247.

Maurice Kraitchik, *Mathematical Recreations*
(Norton, 1942; Dover), 79

D.H. Lehmer, The Tarry–Escott problem
*Scripta Math.* **13** (1947), 37–41.

Joseph S. Madachy, *Mathematics on Vacation*
(Scribner's, 1966), pages 173–175.

J.B. Roberts, A curious sequence of signs
*Amer. Math. Monthly* **64** (1957), 317–322.

J.B. Roberts, Splitting consecutive integers into classes with equal power sums
  *Amer. Math. Monthly* **71** (1964), 25–37.
J.S. Vidger, Consecutive integers having equal sums of squares
  *Math. Mag.* **38** (1965), 35–42.
T.N. Sinha, On the Tarry–Escott problem
  *Amer. Math. Monthly* **73** (1966), 280–285.
E.M. Wright, On Tarry's problem
  *Quart. J. Math. Oxford* (1) **6** (1935), 261–267.
Number curiosities
  *Crux Mathematicorum* (Eureka) **2** (1976), 62.

Here are some problems references:
  # 963 *Crux Mathematicorum* **11** (1985), 292–296.
  E1504 *Amer. Math. Monthly* **69** (1962), 165, 924.

**E.3.** See for example, pages 111–129 of C.S. Liu, *Introduction to Combinatorial Mathematics,* (McGraw-Hill, 1968), or pages 162–171 of Alan Tucker, *Applied Combinatorics,* (Wiley, 1980).

**E.4.** An account of geometric methods for solving quadratic equations appears on pages 59–62 and pages 69–70 of Howard Eves, *An Introduction to the History of Mathematics,* (5th edition; Saunders, 1983). There is a fairly detailed discussion of the Euclidean technique of application of areas as well as an exercise on the approaches of Carlyle and von Staudt (1798–1867).

For visual methods for equations of higher degree, consult T.R. Running, Graphical solutions of cubic, quartic and quintic, *Amer. Math. Monthly* **50** (1943), 170–173.

**E.5.** It is impossible to find four distinct square integers in arithmetic progression. The problem is discussed in *Crux Mathematicorum* **8** (1982), 281–282 (Problem 677). Proofs appear in W. Sierpinski, *Elementary Theory of Numbers,* (New York, 1964), pp. 74–75; and L.J. Mordell, *Diophantine Equations,* (Academic Press), pp. 20–22. For a history of the problem, consult L.E. Dickson, *History of the Theory of Numbers,* (Washington, 1920; reprint, Chelsea, 1952), Vol. II, p. 440.

The quadratic $60t^2 - 60t + 1$ takes successive square values at $t = -2$, $-1$, $0$, $1$, $2$, $3$. This can be discovered by noting that the second order differences of the sequence $361, 121, 1, 1, 121, 361$ are constant (see Section 2.1, Exploration **E.18**; Exercise 7.1.17).

We can try to generalize this example to have the quadratic take the successive integer values

$$z^2 \ y^2 \ x^2 \ 1 \ 1 \ x^2 \ y^2 \ z^2.$$

The conditions for constant second order differences are

$$3x^2 - y^2 = 2 \tag{1}$$

$$2y^2 - z^2 = 1. \tag{2}$$

The positive solutions to equation (1) are given by

$$(x_k, y_k) = (1,1), \ (3,5), \ (11,19), \ (41,71), \ldots$$

where

$$x_{k+1} = 2x_k + y_k = 4x_k - x_{k-1}$$
$$y_{k+1} = 3x_k + 2y_k = 4y_k - y_{k-1}.$$

The positive solutions to equation (a) are given by

$$(y_j, z_j) = (1,1), \ (5,7), \ (29,41), \ldots$$

where

$$y_{j+1} = 3y_j + 2z_j = 6y_j - y_{j-1}$$
$$z_{j+1} = 4y_j + 3z_j = 6z_j - z_{j-1}.$$

Putting these together, we arrive at two solutions $(x, y, z) = (1,1,1)$, $(3,5,7)$. Both these correspond to quadratics which are identically squares of linear polynomials. The existence of a nontrivial quadratic turns on whether the two "$y$-sequences"

$$1, \ 5, \ 19, \ 71, \ 265, \ 989, \ 3691, \ldots$$

$$1, \ 5, \ 29, \ 169, \ 985, \ 5741, \ldots$$

have a third integer in common. This is a difficult question to deal with. For research into this type of problem, consult

A. Baker & H. Davenport, The equations $3x^2 - 2 = y^2$ and $8x^2 - 7 = z^2$
*Quart. J. Math.* (2) **20** (1969), 129–137.

K. Kubota, On a conjecture of Morgan Ward, I
*Acta Arith.* **33** (1977), 11–48.

R. Loxton, Linear recurrences of order two
*J. Austral. Math. Soc.* **7** (1967), 108–114.

M. Mignotte, Intersection des images de certaines suites récurrentes linéaires
*Theor. Comp. Sci.* **7** (1978), 117–122.

P. Kanagasabapathy & Tharmambikai Ponnudurai, The simultaneous diophantine equations $y^2 - 3x^2 = -2$ and $z^2 - 8x^2 = -7$
*Quart. J. Math.* (2) **26** (1975), 275–278.

Quadratics taking four successive square values $x^2$, $y^2$, $z^2$, $w^2$ when the variable takes the respective values 0, 1, 2, 3 are easy to find. The condition is that

$$x^2 + 3z^2 = w^2 + 3y^2.$$

Taking note of the special case $k = 3$ of the identity

$$
\begin{aligned}
(p^2 + kq^2)(r^2 + ks^2) &= (pr + kqs)^2 + k(ps - qr)^2 \\
&= (pr - kqs)^2 + k(ps + qr)^2,
\end{aligned}
$$

we can take $x = pr - 3qs$, $y = ps - qr$, $z = ps + qr$, $w = pr + 3qs$. For example, $(p, q, r, s) = (4, 1, 2, 1)$ leads to $(x, y, z, w) = (5, 2, 6, 11)$ and the quadratic $(1/2)(53t^2 - 95t + 50)$.

But more can be said. D. Allison gives the example

$$-420t^2 + 2100t + 2809$$

which is square for integers between $-1$ and $6$ inclusive, and the example

$$-4980t^2 + 32100t + 2809$$

which takes distinct square values for integers between $0$ and $6$ inclusive. Consult D. Allison, On square values of quadratics, *Math. Proc. Camb. Phil. Soc.* **99** (1986), 381–383; and Duncan A. Buell, Integer squares with constant second differences, *Mathematics of Computation* **49** (1987), 635–644.

The "1986" problem is due to Andy Liu of the University of Alberta. He bases a solution on the observation that

$$(t - 1)(t - 9)(t - 8)(t - 6) = (t^2 - 12t)^2 + 53t^2 - 606t + 432.$$

The polynomial $(t^2 - 12t)^2 - (t - 1)(t - 9)(t - 8)(t - 6)$ takes a negative value for $t = 1986$ and so provides a suitable example. P. Reiss of Winnipeg considers the polynomial

$$f(t) = k(1 - t)(t - 9) + r^2$$

where $k$ and $r$ are to be chosen to make $f(8)$ and $f(6)$ squares, say $u^2$ and $v^2$ respectively. We need

$$15(u^2 - r^2) = 7(v^2 - r^2).$$

A trial of $u = r + 2$ and $v = r + 4$ leads to $r = 13$ and $k = 8$. Thus $f(t) = -8t^2 + 80t + 97$ works. Another polynomial which works is $2t(9 - t)$.

Is it possible for a polynomial (not necessarily a quadratic) over $\mathbf{Z}$ to assume a square value at every integer, and yet itself not be the square of another polynomial? The answer is no, and we have the following general result due to W.H.J. Fuchs:

**Theorem.** Suppose that $f$ and $g$ are two polynomials and that there is an integer $m$ such that, for each integer $n > m$, there is a number $k$ such that $f(n) = g(k)$. Then there is a polynomial $h$ such that $f = g \circ h$. If $f$

and $g$ have integer coefficients and the leading coefficient of $g$ is 1, then it can be arranged that $h$ has integer coefficients.

The square problem is the special case $g(t) = t^2$. This theorem appears as a solution to problem E869 in the *American Mathematical Monthly* (**56** (1949), 338; **57** (1950), 114). This reference also gives a history of the square problem. See also Problems 114, 190 on pages 132, 143, 325, 341 of G. Pólya & G. Szegö, *Problems and Theorems in Analysis*, (4th ed., Springer).

**E.6.** An *entire* set of polynomials contains at least one polynomial of each positive degree such that any pair commute. An elementary proof that every entire set consists, up to similarity with a linear polynomial, of either the ordinary powers or the Tchebychef polynomials appears in H.D. Block & H.P. Thielman, Commutative polynomials, *Quart. Jour. Math.* (Ser. 2) **1** (1951), 241–243.

This work is related to an attractive conjecture: Let $f$ and $g$ be continuous functions mapping the closed unit interval $\{x : 0 \le x \le 1\}$ into itself for which $f(g(x)) = g(f(x))$ for all $x$; then there exists a point $c$ for which $f(c) = g(c) = c$. Is this conjecture true for polynomials? It is refuted for continuous functions in general by counterexamples given independently in the papers William M. Boyce, Commuting functions with no common fixed point, *Trans. Amer. Math. Soc.* **137** (1969), 77–92, MR 38 # 4267; and John Philip Huneke, On common fixed points of commuting continuous functions on an interval, *Trans. Amer. Math. Soc.* **139** (1969), 371–381, MR 38 # 6005.

**E.7.** The connection between $p_n$ and $T_n$ can be seen by noting that, if $x = \cos\theta + i\sin\theta$, then $x^n = \cos n\theta + i\sin n\theta$ and $t = 2\cos\theta$. Thus, we are essentially interested in expressing $2\cos n\theta$ in terms of $2\cos\theta$. The first few polynomials are as follows:

| $n$ | $p_n(t)$ |
|---|---|
| 0 | 2 |
| 1 | $t$ |
| 2 | $t^2 - 2$ |
| 3 | $t^3 - 3t$ |
| 4 | $t^4 - 4t^2 + 2 = (t^2 - 2)^2 - 2$ |
| 5 | $t^5 - 5t^3 + 5t$ |
| 6 | $t^6 - 6t^4 + 9t^2 - 2 = (t^3 - 3t)^2 - 2$ |
|  | $\qquad\quad = (t^2 - 2)^3 - 3(t^2 - 2)$ |
| 7 | $t^7 - 7t^5 + 14t^3 - 7t$ |
| 8 | $t^8 - 8t^6 + 20t^4 - 16t^2 + 2$ |
|  | $\qquad\quad = (t^4 - 4t^2 + 2)^2 - 2 = (t^2 - 2)^4 - 4(t^2 - 2)^2 + 2$ |

For a study of the role of these functions in determining the algebraic character of certain values of trigonometric functions, see L. Carlitz & J.M.

Thomas, Rational tabulated values of trigonometric functions, *Amer. Math. Monthly* **69** (1962), 789–793.

**E.8.** The problem of showing that $f(x, y)$ is a polynomial if it is so in each variable separately was twice posed in the *American Mathematical Monthly* (# 4897 AMM **67** (1960), 295 & **68** (1961), 187; # E2940 AMM **89** (1982), 273 & **91** (1984), 142). A solution was published in the note, F. W. Carroll, A polynomial in each variable separately is a polynomial, *Amer. Math. Monthly* **68** (1961), 42.

**E.9.** If $n = 1$, then there are essentially three possibilities:

(1) the polynomial is constant and its range is a singleton;

(2) the polynomial is of odd positive degree and its range is all of **R**;

(3) the polynomial is of even degree and its range is a closed semi-infinite interval of the form $[m, \infty)$ for positive leading coefficient or $(\infty, m]$ for negative leading coefficient.

The problem of determining the possible ranges of $f(x, y)$ opened the 1969 Putnam Examination. To the possible surprise of the competitors and perhaps even their supervisors, it turns out that the range can be an open half line. The example given is $(xy - 1)^2 + x^2$. (See G.L. Alexanderson, L.F. Klosinski & L.C. Larson, *The W.L. Putnam Mathematical Competition Problems and Solutions: 1965–1984* (MAA, 1985).) There are no further possibilities when the number of variables exceeds 2. For complex polynomials, the range is either a singleton or all of **C**. For polynomials over **Q** defined on **Q**, the situation is complicated indeed.

**E.10.** (a) (28, 53, 75, 84) and (65, 127, 248, 260) are instances of the solution

$$(x^3 + 1, 2x^3 - 1, x^4 - 2x, x^4 + x).$$

Other examples lead to the polynomial solutions

$$(3x^2, 6x^2 - 3x + 1, 3x(3x^2 - 2x + 1) - 1, 3x(3x^2 - 2x + 1))$$

$$(3x^2, 6x^2 + 3x + 1, 3x(3x^2 + 2x + 1), 3x(3x^2 + 2x + 1) + 1),$$

one of which can be derived from the other by a change of variable $x \longrightarrow -x$.

In his 1761 paper, Solutio generalis quorundam problematum diophanteorum quae vulgo nonnisi solutiones speciales admittere videntur (*Opera Omnia* (Series 1) **2**, 428–458), Leonard Euler presents a number of formulae for three cubes which add up to a fourth cube. One of these is

$$(x(x^3 - y^3), y(x^3 - y^3), y(2x^3 + y^3), x(x^3 + 2y^3)).$$

An equivalent diophantine equation is $X^3 + Y^3 = U^3 + V^3$. One polynomial solution is given by

$$(X, Y, U, V) = (ax^4 + bxy^3, ay^4 + cx^3y, ax^4 + cxy^3, ay^4 + bx^3y),$$

where $a$, $b$, $c$ are constants chosen to satisfy $b(3a^2 - b^2) = c(3a^2 - c^2)$. For example, $(a, b, c) = (7, 11, 2)$ will work. (Cf. Victor Thébault, *El. Math.* **8** (1953), 47.)

For a discussion of other solutions to this equation as well as to $X^4 + Y^4 = U^4 + V^4$ in polynomials of two variables, see G.H. Hardy & E.M. Wright, *An Introduction to the Theory of Numbers,* (Oxford, 4th ed., 1960), pages 199–201.

(b) A simple solution is $(x - 1, x, x + 1, x + 2)$. The solutions (6, 23, 32, 39), (39, 70, 91, 108), (108, 157, 194, 225), etc. are instances of

$$A = 2x^3 - 5x = 2y^3 - 3y^2 - (7/2)y + (9/4)$$

$$B = 2x^3 + 2x^2 - x + 1 = 2y^3 - y^2 - (3/2)y + (7/4)$$

$$C = 2x^3 + 4x^2 + x - 2 = 2y^3 + y^2 - (3/2)y - (7/4)$$

$$D = 2x^3 + 6x^2 + x - 3 = 2y^3 + 3y^2 - (7/2)y - (9/4).$$

The variables are related by $y = x + 1/2$. Note that $D(x) = A(x + 1) = -A(-x - 1)$ and $C(x) = -B(-x - 1)$.

(c) This equation is satisfied by

$$(X, Y, Z) = (2xy + rx^2, y^2 - x^2, x^2 + rxy + y^2).$$

A complex number method of solving the related equation

$$Z^2 = X^2 + Y^2 + 2XY \cos \theta$$

is to let $Z = |(y + x \cos \theta) + ix \sin \theta|$. Then

$$Z^2 = |(y^2 - x^2) + 2x(y + x \cos \theta)(\cos \theta + i \sin \theta)|^2$$

yields the solution

$$(X, Y, Z) = (2x(y + x \cos \theta), y^2 - x^2, x^2 + 2xy \cos \theta + y^2).$$

These equations were the subject of lively correspondence in the Reader Reflections column of the *Mathematics Teacher;* see **78** (1985), 238, 663; **79** (1986), 158, 522; **80** (1987), 343.

(d) Observing that $(12)_3 = 1^2 + 2^2$, $(23)_5 = 2^2 + 3^2$, etc., we find that, for numbers of base $2k + 1$,

$$k(2k + 1) + (k + 1) = k^2 + (k + 1)^2.$$

Similarly, we can discover that

$$k(2k - 1) + k = k^2 + k^2.$$

The number of integers which can be expressed as the sum of its digits to a given base $b$ is given in the solution to Problem E2925 in *Amer. Math. Monthly* **90** (1983), 401.

(e) Observing that $(130)_4 = 1^3 + 3^3 + 0^3$, $(250)_7 = 2^3 + 5^3 + 0^3$, etc., we find that, for numbers of base $3k + 1$,

$$k(3k + 1)^2 + (2k + 1)(3k + 1) = k^3 + (2k + 1)^3,$$

and also that

$$k(3k + 1)^2 + (2k + 1)(3k + 1) + 1 = k^3 + (2k + 1)^3 + 1^3.$$

(f) $\{-5x - 3y, -4x - y, -x - 2y, x + 2y, 4x + y, 5x + 3y\}$ and $\{-4x - 3y, -5x - 2y, x - y, -x + y, 5x + 2y, 4x + 3y\}$ are two sets of polynomials for which the sum of the $k$th powers of the elements of one set are equal to the corresponding power sums of the second set for $0 \leq k \leq 5$. See the notes on Exploration **E.2** for references.

**E.11.** An 1844 result of Gabriel Lamé is that the number of steps in the Euclidean algorithm does not exceed five times the number of digits in the smaller number. If $b$ is the smaller number in a pair $(a, b)$ for which the Euclidean algorithm has $n$ steps, it can be shown readily that $b$ is at least as great as the $n$th Fibonacci number (see Exploration **E.14.**) Consult Ross Honsberger, A theorem of Gabriel Lamé, *Mathematical Gems* II, Dolciani Mathematical Expositions # 2 (MAA, 1976), 54–57; and H. Grossman, On the number of divisions in finding a G.C.D., *Amer. Math. Monthly* **31** (1924), 443.

**E.12.** The congruence $ax \equiv b \pmod{m}$ is soluble iff $\gcd(a, m)$ divides $b$, and there are $\gcd(a, m)$ incongruent solutions modulo $m$. For details, consult G.H. Hardy & E.M. Wright, *An Introduction to the Theory of Numbers*, 4th ed. (Oxford, 1960), Sect. 5.4, 51–52; Sect. 8.1, 94–95; and I. Niven & H.S. Zuckerman, *An Introduction to the Theory of Numbers*, 2nd ed. (Wiley, 1960, 1966), Sect. 2.3, 31–32.

**E.13.** Suppose $p$ is odd and $f(n) = n^2 - n + p$ is prime for $0 \leq n \leq \sqrt{p/3} + 1$. Then $n^2 - n + p$ is prime for $0 \leq n \leq p - 1$. To see this, assume to the contrary that $n^2 - n + p$ is composite for at least one $n$ not exceeding $p - 1$. Let $q$ be the smallest prime that divides any one of the composite values of $f(n)$ $(0 \leq n \leq p - 1)$. Thus $q < \sqrt{f(p)} = p$. (Explain why equality cannot occur.)

Suppose that $u$ is the smallest nonnegative integer for which $q|f(u)$. (Explain why $f(u)$ is composite, i.e. not equal to $q$.) Let $q = 2k - 1$. Since

$f(k - i) \equiv f(k + i) \pmod{q}$ for each integer $i$, it can be argued that $u \leq k$. We have

$$4k(k - 1) + 1 = q^2 \leq f(u) \leq f(k) = k(k - 1) + p$$

whence $3(k - 1)^2 < p$, so that $u \leq \sqrt{p/3} + 1$. But this contradicts the hypothesis.

A polynomial of several variables whose positive values are prime is given in J.P. Jones, D. Sato, H. Wada & D. Wiens, Diophantine representations of the set of prime numbers, *Amer. Math. Monthly* **83** (1976), 449–464; and Problem P. 291, *Canadian Math. Bull.* **24** (1981), 505.

**E.14.** With respect to the assertions in (b), we have the following. If $x \geq y$, then

$$|(y - x)y - x^2| = x^2 + xy - y^2 = x(x + y) - y^2 \geq 2y^2 - y^2 = y^2,$$

so that $(*)$ implies $x = y = 1$. If $y - x \geq x$, then

$$4y^2 - 4xy - 4x^2 = y^2 + 2y(y - 2x) + (y^2 - 4x^2) \geq y^2 \geq 4,$$

so that $(*)$ implies $y = 2$, $x = 1$. In the induction step, if $F_n < x < y \leq F_{n+1}$, then $y - x < F_{n-1}$ and

$$
\begin{aligned}
(y - x)y - x^2 &\leq (F_{n-1} - 1)F_{n+1} - (F_n + 1)^2 \\
&= (F_{n-1}F_{n+1} - F_n^2 - 1) - F_{n+1} - 2F_n \leq -3,
\end{aligned}
$$

which contradicts $(*)$. See James P. Jones, Diaphantine representation of the Fibonacci numbers, *Fibonacci Quart.* **13** (1975), 84–88; and Problem 3, Int. Math. Olympiad 1981, *Math. Mag.* **55** (1982), 55.

The issue raised in this exploration is related to the tenth problem posed by David Hilbert in his famous keynote address to the International Congress of Mathematicians in 1900. He sought to give a prospectus of the main topics requiring the attention of mathematical researchers during the coming century. To provide a focus, he posed thirty-seven problems, and these have tended to become benchmarks for progress in mathematics. For a biographical account, read Constance Reid, *David Hilbert* (Springer, 1970).

The tenth problem is: Specify a procedure which in a finite number of steps enables one to determine whether or not a given diophantine equation with an arbitrary number of indeterminates and with rational integer coefficients has a solution in rational integers. While it was later shown that no such general procedure exists, research into the question has led to significant developments in the foundations of mathematics. One direction has involved diophantine sets, i.e. sets $S$ of natural numbers for which there is a polynomial $f(y, \mathbf{x})$ over $\mathbf{Z}$ for which $y$ belongs to $S$ if and only if there are numbers $\mathbf{x}$ for which $f(y, \mathbf{x}) = 0$. If we define

$$g(y, \mathbf{x}) = (y + 1)(1 - (f(y, \mathbf{x}))^2) - 1,$$

then $y$ belongs to $S$ if and only if $y$ is a positive value assumed by the polynomial $g(y, \mathbf{x})$. A survey article on *Diophantine decision problems* by Julia Robinson appears in W.J. LeVeque (ed.), Studies in number theory, *Studies in Mathematics* 6 Math. Assoc. of America, 1969.

The following references are also relevant: M. Davis, H. Putman & J. Robinson, The decision problem for exponential diophantine equations, *Ann. Math.* (2) **74** (1961), 425–436, MR24 # A3061; and Ju. V. Matijasevic, The Diophantineness of enumerable sets, *Dokl. Akad. Nauk SSSR* **191** (1970), 279–282 = *Soviet Math. Dokl.* **11** (1970), 354–358, MR41 # 3990.

**E.15.** Since a polynomial is irreducible along with every positive multiple of itself, it is enough to look at monic polynomials; the total number of irreducibles will be $p - 1$ times the number of monic irreducibles. Clearly, there are $p$ monic irreducibles of degree 1: $t, t + 1, t + 2, \ldots, t + (p - 1)$.

To determine the number of irreducibles of higher degree, subtract from the total number of polynomials of that degree the number obtainable as products of polynomials of lower degree. For example, there are

$$p^2 - \left( p + \binom{p}{2} \right) = (p^2 - p)/2$$

monic irreducible quadratics. For $p = 2$, the only possibility is $t^2 + t + 1$, while for $p = 3$, there are $t^2 + 1$, $t^2 + t + 2$, $t^2 + 2t + 2$.

There are $(p^3 - p)/3$ monic irreducible cubics. For $p = 2$, they are $t^3 + t + 1$ and $t^3 + t^2 + 1$. There are $p(p - 1)^2(p + 8)/8$ monic irreducible quartics. See Markus Nijmeijer & Mike Staring, A formula that produces all, and nothing but, irreducible polynomials in $\mathbf{Z}_p[x]$, *Mathematics Magazine* **61** (1988), 41–44.

**E.16.** A thorough discussion of this problem can be found in Donald E. Knuth, *The Art of Computer Programming*, Vol. 2: *Seminumerical Algorithms*, (Addison Wesley) pages 441–446.

**E.17.** Horner's table for the expansion of $t^n$ in terms of $t - 1$ is

| 1 | 0 | 0 | 0 | 0 | $\cdots$ | 0 | 0 | 0 | 0 |
|---|---|---|---|---|---|---|---|---|---|
|   | 1 | 1 | 1 | 1 | $\cdots$ | 1 | 1 | 1 | 1 |

| 1 | 1 | 1 | 1 | 1 | $\cdots$ | 1 | 1 | 1 | 1 |
|---|---|---|---|---|---|---|---|---|---|
|   | 1 | 2 | 3 | 4 | $\cdots$ | $n-3$ | $n-2$ | $n-1$ | |

| 1 | 2 | 3 | 4 | 5 | $\cdots$ | $n-2$ | $n-1$ | $\underline{n}$ |
|---|---|---|---|---|---|---|---|---|
|   | 1 | 3 | 6 | 10 | $\cdots$ | | | |

| 1 | 3 | 6 | 10 | 15 | $\cdots$ |
|---|---|---|---|---|---|
| $\cdots$ | | | | | |

Denoting by $\begin{pmatrix} n \\ r \end{pmatrix}$ the coefficient of $x^r$ in the expansion of $(1+x)^n$, from Horner's table we find that

$$\begin{pmatrix} n \\ 0 \end{pmatrix} = 1$$

$$\begin{pmatrix} n \\ 1 \end{pmatrix} = 1 + 1 + 1 + \cdots + 1$$

$$\begin{pmatrix} n \\ 2 \end{pmatrix} = \begin{pmatrix} 1 \\ 1 \end{pmatrix} + \cdots + \begin{pmatrix} n-1 \\ 1 \end{pmatrix} = \frac{n(n-1)}{2}$$

$$\cdots$$

$$\begin{pmatrix} n \\ r \end{pmatrix} = \begin{pmatrix} r-1 \\ r-1 \end{pmatrix} + \begin{pmatrix} r \\ r-1 \end{pmatrix} + \begin{pmatrix} r+1 \\ r-1 \end{pmatrix} + \cdots + \begin{pmatrix} n-1 \\ r-1 \end{pmatrix}$$

$$\cdots$$

It can be shown by induction on $r$ that $\begin{pmatrix} n \\ r \end{pmatrix}$ is equal to

$$\frac{n(n-1)(n-2)\cdots(n-r+1)}{r!}.$$

This can be handled by summation techniques discussed in Exploration **18**.

**E.18.** The standard reference for finite differences is L.M. Milne-Thomson, *The Calculus of Finite Differences,* (Macmillan, London, 1933). For a lighter treatment, consult H. Freeman, *Mathematics for Actuarial Students,* Part II (Cambridge, 1952).

**E.19.** For an introduction to coloring problems with some key references, see W.T. Tutte, Chromials, *Studies in Graph Theory,* Part II (ed. D.R. Fulkerson), (*Studies in Mathematics,* MAA, 1975), p. 361–377. Also, see Chapter 9 of C.L. Liu, *Introduction to Combinatorial Mathematics,* (McGraw-Hill, 1968) and Chapter 8 of Alan Tucker, *Applied Combinatorics,* (Wiley, 1980). The chromatic polynomials for the five Platonic solids are discussed in D.H. Lehmer, Coloring the Platonic solids, *Amer. Math. Monthly* **93** (1986), 288–292.

**E.20.** A survey of the techniques of factoring polynomials is given in Section 4.6.2 (pages 420–441) of Donald E. Knuth, *The Art of Computer Programming,* Vol. 2: *Semi-Numerical Algorithms,* (Addison-Wesley). The greatest common divisor is discussed on pages 434–436.

**E.21.** The remainder for division by $(t-c)^k$ is conveniently provided by Taylor's Theorem which renders the polynomial in the form

$$q(t)(t-c)^k + a_{k-1}(t-c)^{k-1} + \cdots + a_1(t-c) + a_0.$$

In dealing with polynomial divisors with more than one distinct zero, the divided difference technique of Exploration **E.55** may be used. Alternatively, the remainder upon division of a polynomial $p(t)$ by $(t - a_1)$ $(t - a_2) \cdots (t - a_m)$ is the Lagrange polynomial of degree less than $m$ which assumes the value $p(a_i)$ at $a_i$ (see Exercise 7.1.5). In a similar way, the remainder for division by $(t - a)^r (t - b)^s$ can be identified as that polynomial of degree less than $r + s$ for which the $k$th derivative at $a$ (resp. $b$) agrees with that of $p$ at $a$ (resp. $b$) for $0 \leq k \leq r - 1$ (resp. $0 \leq k \leq s - 1$).

**E.22.** The $n$th derivative of $p \circ q$ is a sum of terms of the form

$$c p_k q_1^{a_1} q_2^{a_2} q_3^{a_3} \cdots q_n^{a_n},$$

where $\Sigma i a_i = n$, $\Sigma a_i = k$, and the coefficient is positive. The determination of the coefficients is an interesting combinatorial problem solved by Faà di Bruno in the middle of the last century. For a recent treatment and bibliography, consult Steven Roman, The formula of Faà di Bruno, *Amer. Math. Monthly* **87** (1980), 805–809.

**E.23 & 24.** Partial derivatives are studied in a second calculus course and are discussed in any textbook. The equation $\Sigma x_i \partial f / \partial x_i = kf$ for a homogeneous polynomial of degree $k$ is called Euler's equation.

**E.25.** A natural generalization of polynomials is the class of complex-valued functions $f(z)$ of a complex variable which satisfy the differentiability condition

$$\lim_{h \to 0} \frac{f(z + h) - f(z)}{h} \quad \text{exists} \qquad (*)$$

for each point $z$ in the complex plane. Students who are familiar only with the calculus of real-valued functions will not appreciate the strength of this condition. In contrast to the real case, in which $h$ can tend to 0 from one of two real directions, in the complex plane $h$ is permitted to tend to 0 in any way over a two dimensional neighborhood of 0. As a result, the condition implies that the functions (known as entire) have derivatives of all higher orders and can be represented as the sum of an infinite convergent power series $a_0 + a_1 z + a_2 z^2 + a_3 z^3 + \cdots$ for every complex $z$. These functions share many properties with polynomials.

As for polynomials, we can write $f(z) = u(x, y) + iv(x, y)$ and discover that $(*)$ will be valid if and only if the Cauchy–Riemann conditions

$$\partial u / \partial x = \partial v / \partial y \quad \partial u / \partial y = -\partial v / \partial x$$

hold, where now the partial derivatives are defined through limits.

A pleasant introduction to the theory is George Polya & Gordon Latta, *Complex Variables*, (Wiley, 1974). Another source which will richly reward the patient reader is the set of five short volumes by Konrad Knopp, published by Dover, New York: *Elements of the Theory of Functions; Theory of Functions:* Parts I and II; *Problem Book in the Theory of Functions:* Volumes I and II.

**E.26.** There are many relations which are satisfied by Legendre polynomials. One of the most striking is this formula

$$2^n n! \, P_n(x) = D^n(x^2 - 1)^n$$

where $D$ is the differentiation operator. The polynomials turn up as coefficients in a generating function expansion

$$(1 - 2xt + t^2)^{-1/2} = \sum_{n=0}^{\infty} P_n(x)t^n.$$

A list of properties of Legendre polynomials occurs on pages 50–53 of W. Magnus & F. Oberhettinger, *Formulas and Theorems for the Functions of Mathematical Physics*, (Chelsea, 1949).

To get some idea of the richness of the area of mathematics which includes the study of these functions, consult Gabor Szegö, *Orthogonal Polynomials*, (AMS Colloquium, 1939).

A discussion of Legendre's use of these polynomials in 1782 in dealing with a problem of potential theory can be found on pages 525–528 of Morris Kline, *Mathematical Thought from Ancient to Modern Times*, (Oxford, New York, 1972). Legendre's own paper (referred to in a book review in *Bull. A.M.S.* (NS) **19** (1988), 346–348) is A.M. Legendre, Recherches sur l'attraction des sphéroides homogènes, *Mém. Math. Phys. Prés. à l'Acad. Roy. Sci.* (Paris) *par divers savants* **10** (1785), 411–434.

For other work, see Mary L. Boas, A formula for the derivatives of Legendre polynomials, *Amer. Math. Monthly* **70** (1963), 643–644.

**E.28.** Although Rolle's Theorem is a standard topic of a first calculus course, Rolle himself was interested in using it to locate zeros of polynomials. An English translation of an excerpt of his work with commentary can be found in D.E. Smith, *A Source Book In Mathematics*, Volume One (Dover, 1929, 1959).

For a discussion of generalizing Rolle's Theorem to the complex plane and an interesting open problem, consult I.J. Schoenberg, A conjectured analogue of Rolle's Theorem to polynomials with real or complex coefficients, *Amer. Math. Monthly* **93** (1986), 8–13.

**E.29.** Modulo the polynomial $t^2 - t + a$ we have that

$$t^2 \equiv t - a$$

$$t^3 \equiv (1 - a)t - a$$

and, in general,

$$t^k \equiv f_k(a)t - af_{k-1}(a)$$

(i.e. $t^k - f_k(a)t + af_{k-1}(a)$ is a multiple of $t^2 - t + a$) where $f_1(a) = 1$, $f_2(a) = 1$ and $f_{k+1}(a) = f_k(a) - af_{k-1}(a)$ for $k \geq 2$. Then $t^n + t + b$ is divisible by $t^2 - t + a$ if and only if $f_n(a) + 1 = 0$ and $b = af_{n-1}(a)$. Hence we need to examine the values of $n$ and $a$ for which $f_n(a) = -1$. When $a = 1$, the situation is straightforward and $f_n(1) = -1 \Longleftrightarrow n = 6k + 4$ or $6k + 5$ for some $k$. For $n = 6k + 4$, we have that

$$t^2 - t + 1 \,|\, t^{6k+4} + t = t(t^3 + 1)(t^{6k} - \cdots + 1).$$

For $n = 6k + 5$, we have that

$$t^2 - t + 1 \,|\, t^{6k+5} + t - 1.$$

This is easily checked since

$$t^5 + t - 1 = (t^2 - t + 1)(t^3 + t^2 - 1)$$

and

$$t^{6k+5} + t - 1 = t^{6(k-1)+5}(t^6 - 1) + t^{6(k-1)+5} + t - 1$$

for $k \geq 1$.

For higher values of $a$, the problem becomes more interesting. $f_n(2) = -1$ at least for $n = 3, 5, 13$, and we obtain

$$t^3 + t + 2 = (t^2 - t + 2)(t + 1)$$

$$t^5 + t - 6 = (t^2 - t + 2)(t^3 + t^2 - t - 3)$$

$$t^{13} + t + 90 = (t^2 - t + 2)(t^{11} + t^{10} - t^9 - 3t^8 - t^7 + 5t^6$$
$$+ 7t^5 - 3t^4 - 17t^3 - 11t^2 + 23t + 45).$$

The question of the existence of other solutions to the equation $f_n(2) = -1$ is apparently quite difficult. It is known that a second order recurrence like $f_n(a)$ for $a \neq 1$ visits a given number at most finitely often. See the following papers:

R. Alter & K. Kubota, Multiplicities of second order linear recurrences
   *Trans. Amer. Math. Soc.* **178** (1973), 271–284.

K. Kubota, On a conjecture of Morgan Ward, I
   *Acta Arithmetica* **33** (1977), 11–48.

R. Loxton, Linear recurrences of order two
   *J. Austral. Math. Soc.* **7** (1967), 108–114.

M. Mignotte, A note on recursive sequences
   *J. Austral. Math. Soc.* **20** (A) (1975), 242–244.

For the special case $n = 5$, see Stanley Rabinowitz, The factorization of $x^5 \pm x + n$, *Math. Mag.* **61** (1988), 191–193.

**E.30.** The sequences $u_n(4)$ and $u_n(6)$ arise in the following context. The number 5040 is not only 1 less than a perfect square, but differs from each of the next three larger perfect squares by a perfect square:

$$5040 = 71^2 - 1^2 = 72^2 - 12^2 = 73^2 - 17^2.$$

The existence of other such numbers turns on making $8k^2 + 1$ and $3k^2 + 1$ both squares for a fixed value of $k$. Now $8k^2 + 1$ is square $\Longleftrightarrow k = u_n(6)$ for some $n$, and $3k^2 + 1$ is square $\Longleftrightarrow k = u_n(4)$ for some $n$. Hence the question arises as to what numbers $u_n(4)$ and $u_n(6)$ have in common.

In the paper, M. Mignotte, Intersection des images de certaines suites récurrentes linéaires, *Theor. Comp. Sci.* **7** (1978), 117–122 it is shown that there are only finitely many such common numbers. Remarkably, in A. Baker & H. Davenport, The equations $3x^2 - 2 = y^2$ and $8x^2 - 7 = z^2$, *Quart. J. Math.* (2) **20** (1969), 129–137 a similar problem is solved using deep results in diophantine approximation. See also E.J. Barbeau, Numbers differing from consecutive squares by squares, *Canad. Math. Bull.* **28** (1985), 337–342. See Notes on Explorations **E.5** and **E.29**.

A similar problem involving sums instead of differences is E3080 found in *Amer. Math. Monthly* **92** (1985), 215; **95** (1988), 141.

The paper by Carlitz and Thomas cited in the notes for Exploration **E.7** indicates how the sequence $\{u_n\}$ is tied in with the sequence involved in the reciprocal substitution.

**E.31.** The polynomial has rational zeros if and only if its discriminant $5(n + 1)^2 - 4$ is a perfect square. Since $5(n + 1)^2 - 4$ has the same value for $n = m$ and $n = -2 - m$, it suffices to determine the situation for nonnegative $n$. A little experimentation reveals that the discriminant is square when $n + 1$ takes alternate values of the Fibonacci sequence, i.e. when $n + 1$ is one of 1, 2, 5, 13, 34, 89, . . .. Indeed

$$0t^2 + t - 2 = t - 2$$

$$t^2 + 2t - 3 = (t - 1)(t + 3)$$

$$4t^2 + 5t - 6 = (4t - 3)(t + 2)$$

$$12t^2 + 13t - 14 = (3t - 2)(4t + 7)$$

$$33t^2 + 34t - 35 = (11t - 7)(3t + 5)$$

$$88t^2 + 89t - 90 = (8t - 5)(11t + 18).$$

Finding a complete set of $n$ amounts to solving the diophantine equation $x^2 - 5y^2 = -4$ (and then taking $n = y - 1$). A complete set of solutions is given by the recursion $(x_1, y_1) = (1, 1)$ and

$$(x_{n+1}, y_{n+1}) = ((3x_n + 5y_n)/2, \ (x_n + 3y_n)/2).$$

The reader may wish to explore further the intervention in the factorization of the Fibonacci sequence 1, 1, 2, 3, 5, 8, ... and the related Lucas sequence 1, 3, 4, 7, 11, 18, 29, 47, .... For a reference, see Steven Schwartzman, Factoring polynomials and Fibonacci, *Math. Teacher* **79** (1986), 54–56, 65.

Equations of the type $x^2 - dy^2 = k$ come under the general heading of Pell's equation. Since they arise in many number theoretic problems, their theory is covered in most elementary number theory texts. For a brief, insightful introduction to this area and its significance, see Chapter 7 of H. Rademacher, *Higher Mathematics from an Elementary Point of View*, (Birkhäuser, 1983).

**E.33.** For an appreciation of the role of $p$-adic numbers in the solution of diophantine equations, see the article *Diophantine equations: p-adic methods* in W.J. LeVeque (ed.), Studies in number theory, *Studies in Mathematics* 6 (Math. Assoc. of Amer., 1969).

For other references on $p$-adic numbers, see G. Bachman, *Introduction to p-Adic Numbers and Valuation Theory*, (Academic, 1964); and Kurt Mahler, *p-Adic Numbers and Their Functions*, (Cambridge, 1981).

**E.35.** The proof of the irreducibility of $Q_n(t)$ requires more advanced theory. See, for example, Section 53 of B.L. van der Waerden, *Modern Algebra*, Volume I (Ungar, New York, revised edition, 1953); and Theorem 41, Chapter 12 of Jean-Pierre Tignol, *Galois' Theory of Algebraic Equations*, (Longman, 1988). In Solomon W. Golomb, Cyclotomic polynomials and factorization theorems, *Amer. Math. Monthly* **85** (1978), 734–737; **88** (1981), 338–339 criteria for reducibility of $Q_n(t^r)$ and factorizability over **Z** of $Q_n(m)$ are discussed.

**E.36.** For

$$Q_{pq} = \sum_{n=0}^{(p-1)(q-1)} c_n t^n,$$

it is shown in Sr. Marion Beiter, The midterm coefficient of the cyclotomic polynomial $F_{pq}(x)$, *Amer. Math. Monthly* **71** (1964), 769–700 that

$$c_n = \begin{cases} (-1)^k & \text{if } n = aq + bp + k \text{ in exactly one way} \\ 0 & \text{otherwise.} \end{cases}$$

Let $m$ be the smallest value of $n$ for which $Q_n(t)$ has coefficients other than 0, 1, −1. By Exercise 3.5.12, it is clear that $m$ must be odd. It can be shown that $m$ is not a prime power or a product of two distinct primes (see Exploration **E.35** for a reference). The smallest possibilities for $m$ are 45, 63, 75, 99, 105. It turns out that $m = 105$ and that

$$Q_{105}(t) = 1 + t + t^2 - t^5 - t^6 - 2t^7 - \cdots .$$

In fact, the coefficients of cyclotomic polynomials can be arbitrarily large. For a discussion, consult Section 6 of R.C. Vaughan, Adventures in Arithmetick, or: How to make good use of a Fourier transform, *Mathematical*

*Intelligencer* **9** (1987), no. 2, 53–60. Also see P. Erdös & R.C. Vaughan, Bounds for the *r*th coefficients of cyclotomic polynomials, *J. Lond. Math. Soc.* (2) **8** (1974), 393–400; MR 50, # 9835.

**E.37.** A recent and somewhat advanced approach to Fermat's Little Theorem which obtains the result in the general form

$$\sum_{d|n} a^d \mu(n/d) \equiv 0 \pmod{n}$$

where $\mu$ is the Möbius function (Exploration **E.34**) is found in C.J. Smyth, A coloring proof of a generalization of Fermat's Little Theorem, *Amer. Math. Monthly* **93** (1986), 469–470.

**E.38.** The theory of functions of complex variables treats functions which can be regarded as generalizations of polynomials and rational functions. These are assumed to possess derivatives (see the note on Exploration **E.25**) except possibly at a discrete set of points on a region of the complex plane. The principal parts and residues of such functions become significant in the evaluation of integrals; indeed, some definite integrals of real-valued functions of a real variable can be evaluated by applying a "calculus of residues" for the determination of a corresponding complex integral. The understanding of this theory depends on a background of a second college calculus course.

For a clear treatment, consult George Polya & Gordon Latta, *Complex Variables,* (Wiley, 1974). An older reference which will reward careful study is Konrad Knopp, *Theory of Functions*, Part I (Dover, 1952).

**E.39 & 40.** Probably the best elementary account of the treatment of solvability of equations and ruler and compasses constructions is to be found in Charles R. Hadlock, *Field theory and classical problems,* (MAA, 1978: Carus Monograph # 19). These problems are also treated in D.E. Littlewood, *The Skeleton Key of Mathematics,* (Hutchinson University Library, London, 1949, 1957). An excellent historically sensitive account appears in Jean-Pierre Tignol, *Galois' Theory of Algebraic Equations,* (Longman, 1988). A more advanced treatment of Galois theory is contained in Chapter 4 of Nathan Jacobson, *Basic Algebra I.* 2nd ed. (Freeman, 1985).

**E.41.** The theorem that the zeros of the derivative of a polynomial are contained within the smallest polygon containing the zeros of the polynomial is called the Gauss–Lucas Theorem. A thorough treatment can be found in Morris, Marden, *Geometry of Polynomials,* (AMS, 1949, 1966). In Chapter 1, the result is interpreted physically and geometrically, while in Chapter 2, it is established and extended.

A section of problems on this result is found in Part III, Chapter 1, Section 3 of G. Polya & G. Szegö, *Problems and Theorems in Analysis,* (Springer-Verlag, 1972). In W.H. Echols, Note of the roots of the derivative of a polynomial, *Amer. Math. Monthly* **27** (1920), 299–300 it is shown that, for a polynomial $f(z)$ with real coefficients, the nonreal roots of $f'(z)$ are in the closed discs whose diameters are segments joining the pairs of conjugate nonreal roots of $f(z)$ (Jensen's theorem). For further results, see J.L. Walsh, A new generalization of Jensen's theorem on the zeros of the derivative of a polynomial, *Amer. Math. Monthly* **68** (1961), 978–983.

In the cubic case, the result is especially interesting. Let $T$ be a non-degenerate triangle in the complex plane whose vertices are the zeros of a cubic $f(t)$ and let $E$ be the (Steiner) ellipse inscribed in $T$ which touches the sides of $T$ at their midpoints. Then the zeros of $f'(t)$ are the foci of $E$. For an application of this result, see I.J. Schoenberg, A conjectured analogue of Rolle's theorem for polynomials with real or complex coefficients, *Amer. Math. Monthly* **93** (1986), 8–13.

**E.42.** Newton's Method inspired A. Cayley, a nineteenth century British mathematician, to study the sets of starting points which would yield a sequence of approximants to a given zero of the polynomial. This has been recently taken up and integrated with the study of fractals. See, for example, the article H.-O. Peitgen, D. Saupe & F. v. Haeseler, Cayley's problem and Julia sets, *Math. Intelligencer* **6** (no. 2) (1984), 11–20.

**E.43.** The extract is taken from Newton's tract, *Analysis of equations of an infinite number of variables* (page 320). This has been reprinted in Volume 1 of *The Mathematical Works of Isaac Newton*. Assembled with an introduction by Dr. Derek T. Whiteside, (Johnson Reprint, NY, 1964, 1967).

Readers may also be interested in the facsimile of a 1728 English translation of another Newton work, *Universal Arithmetick,* originally written in Latin in 1684, and reproduced in Volume 2. The last part of the paper treats solution of equations and location of roots.

**E.45 & 46.** Continued fractions are of use, not only for approximating the solutions of equations, but also in the treatment of diophantine equations and the close approximation of nonrationals by rationals. Irrationals which are roots of quadratic equations over **Z** can be characterized by the periodicity of the numbers occurring in their continued fraction expansion.

For a rich high school level introduction to the topic, consult C.D. Olds, *Continued fractions,* (MAA, 1963; New Mathematical Library). A recent book which provides an historical perspective on continued fractions through a study of the Greek theory of ratio is D.H. Fowler, *The Mathematics of Plato's Academy: A New Reconstruction,* (Oxford, 1987).

An excerpt of Lagrange's work along with a brief history of continued fractions appears in Chapter II, Article 12 (p. 111–115) of D.J. Struik (ed.), *A Source Book in Mathematics, 1200–1800,* (Harvard, 1969). See also the excerpts of work of Bombelli (c. 1526–1573) and Cataldi (1548–1626) reproduced in translation on pages 80–84 of D.E. Smith, *A Source Book in Mathematics,* Volume One (Dover, 1959).

For a general introduction on the nature of the real number field, which includes a chapter on continued fractions, see Ivan Niven, Irrational numbers (Carus monographs 11), (Math. Assoc. of Amer., 1956, 1967).

A definitive work is H.S. Wall, *Analytic Theory of Continued Fractions,* (D. van Nostrand, 1948).

**E.49.** Interest in the iterations of the function $f(x) = ax(1-x)$ has greatly increased during the last decade with the study of chaotic behaviour and the advent of high speed computers capable of dealing with complex problems. For a gentle introduction, see A.K. Dewdney, Probing the strange attractions of chaos (Computer Recreations), *Scientific American* **257** (#1) (1987), 108–111.

A recent book which explores the visual beauty of this area of mathematics is H.-O. Peitgen & P.H. Richter, *The Beauty of Fractals: Images of Complex Dynamical Systems,* (Springer-Verlag, 1986).

A captivating layman's introduction to this new branch of mathematics is the book James Gleick, *Chaos: Making a New Science,* (Viking Penguin, 1987). This book also recounts the story of the Mandelbrot set, introduced in Exploration **E.67**.

**E.54.** The formula for the sum of the first $n$ $k$th powers for small values of $k$ are given by the following formulae:

| $k$ | *sum of first $n$ $k$th powers* |
|---|---|
| 1 | $n(n+1)/2 = n^2/n + n/2$ |
| 2 | $n(n+1)(2n+1)/6 = n^3/3 + n^2/2 + n/6$ |
| 3 | $n^2(n+1)^2/4 = n^4/4 + n^3/2 + n^2/4$ |
| 4 | $n^5/5 + n^4/2 + n^3/3 - n/30$ |
| 5 | $n^6/6 + n^5/2 + 5n^4/12 - n^2/12$ |
| 6 | $n^7/7 + n^6/2 + n^5/2 - n^3/6 + n/42$ |
| 7 | $n^8/8 + n^7/2 + 7n^6/12 - 7n^4/24 + n^2/12$ |

The coefficients involve a special sequence of numbers called the Bernoulli numbers. For some exercises on this topic, consult M. Spivak, *Calculus* (2nd ed., Publish or Perish, Washington, 1980), Exercise 7, p. 29–30; Exercises 16, 17, p. 538–541.

These sum formulae were derived by Jakob Bernoulli in his book, *Ars conjectandi,* published in 1713. For an English translation of the relevant excerpt, consult pages 316–320 of D.J. Struik (ed.), *A Source Book in Mathematics, 1200–1800,* (Harvard, 1969); or pages 85–90 of D.E. Smith, *A Source Book in Mathematics,* Volume One, (Dover, 1959).

An elementary derivation can be found in John G. Christiano, On the sum of powers of natural numbers, *Amer. Math. Monthly* **68** (1961), 149–151; and in Dumitru Acu, Some algorithms for the sums of integer powers, *Math. Mag.* **61** (1988), 189–191 are obtained other identities involving these sums. The paper C. Kelly, An algorithm for sums of integer powers, *Math. Mag.* **57** (1984), 296–297 gives an elementary derivation of

$$1 + \sum_{k=0}^{m} \binom{m+1}{k} \left(1^k + 2^k + \cdots + n^k\right) = (n+1)^{m+1}.$$

A simple recursive technique based on the lemma:

$$\text{if } \sum_{r=1}^{n} a_r = s_n, \text{ then } \sum_{r=1}^{n} r a_r = (n+1)s_n - \sum_{r=1}^{n} s_r$$

can be found in D. Sullivan, The sums of powers of integers (note 71.23), *Math. Gaz.* **71** (1987), 144–146.

An application of matrices to the evaluation of power sums appears in A.W.F. Edwards, Sums of powers of integers: a little of the history, *Math. Gazette* **66** (1982), 22–28; and in A.W.F. Edwards, A quick route to sums of powers, *Amer. Math. Monthly* **93** (1986), 451–455. A discussion of approximations for the sums is found in B.L. Burrows & R.F. Talbot, Sums of powers of integers, *Amer. Math. Monthly* **91** (1984), 394–403.

Let $S_1(n)$ be the sum of the first $n$ squares, and for $k \geq 1$, let $S_{k+1}(n) = S_k(1) + S_k(2) + \cdots + S_k(n)$. A straightforward induction argument shows that

$$(k+2)! S_k(n) = n(n+1)(n+2) \cdots (n+k)(2n+k).$$

This result is generalized in Problem 4380 in *Amer. Math. Monthly* **57** (1950), 119; **58** (1951), 429.

**E.58.** Even though $f_n(t)$ interpolates more and more values of $|t|$, in fact $\lim_{n \to \infty} f_n(t) = |t|$ is true only for $t = -1, 0, 1$. For a reference to this fact, see page 37 of G.G. Lorentz, *Approximation of Functions*, (Holt, Rinehart & Winston, 1966).

This phenomenon occurs for other functions as well. For example, if $p_n(t)$ interpolates $(1+t^2)^{-1}$ on $[-5, 5]$ at $n+1$ equally spaced points, then $\lim_{n \to \infty} p_n(t) = (1+t^2)^{-1}$ when $|t|$ does not exceed approximately 3.63. Otherwise, the sequence diverges. A paper which analyses this rather subtle issue is James F. Epperson, On the Runge example, *Amer. Math. Monthly* **94** (1987), 329–341.

**E.61.** In general

$$x_1^n + x_2^n + \cdots + x_n^n - n x_1 x_2 \cdots x_n$$

is the sum of positive polynomials. For this and related results, consult Sections 2.18–2.23 of G.H. Hardy, J.E. Littlewood & G. Polya, *Inequalities*, (Cambridge, 1964).

David Hilbert (1862–1942) considered the following problem: Let $f(\mathbf{x})$ be a polynomial of degree $n$ with $m$ variables for which $f(\mathbf{x}) \geq 0$ for all real vectors $\mathbf{x}$. Is it true that $f(\mathbf{x})$ is the sum of squares of finitely many real polynomials? He showed that the answer is affirmative in the cases:

(i)  $m = 2$, $n$ is even;

(ii)  $n = 2$, $m$ is arbitrary;

(iii)  $m = 3$, $n = 4$.

See Section 2.23 of the above reference for a brief discussion.

Research on expressing polynomials as sums of squares has recently become active. See, for example, the paper M.-D. Choi & T.-Y. Lam, An old question of Hilbert, Conference on quadratic forms, 1976 (G. Orzech, ed.), *Queen's Papers in Pure and Applied Mathematics*, **46** (1977), 385–405 (Queen's University, Kingston, Ontario).

**E.62.** It is readily verified that

$$q_3(z) = 3(1 + z)$$

$$q_4(z) = 4(2 + 3z + 2z^2)$$

$$q_5(z) = 5(1 + z)(1 + z + z^2)$$

$$q_7(z) = 7(1 + z)(1 + z + z^2)^2$$

all have zeros on the unit circle. When $n \geq 8$, $q_n'(z)$ has a zero whose absolute values exceeds 1, so that, by the Gauss–Lucas Theorem, the same is true of $q_n(z)$ itself. See Problem E3078 in *Amer. Math. Monthly* **92** (1985), 215; **95** (1988), 140. As for $q_n(6)$, it can be written in the form $(a + bz + az^2)(c + dz + cz^2)$ where $ad$ and $bc$ are the zeros of the quadratic $t^2 - 15t + 48$. From this it follows that $q_6(z)$ has two real quadratic factors and it can be further shown that each has nonreal zeros, so that all the zeros of $q_6(z)$ lies on the unit circle.

**E.63.** For an example of a polynomial mapping in two variables with polynomial inverse, consult page 694 of Gary H. Meisters, Jacobian problems in differential equations and algebraic geometry, *Rocky Mountain J. Math.* **12** (1982), 679–705. For an indication of the significance of this question in the study of differential equations and for further references, see Hyman Bass & Gary Meisters, Polynomial flows in the plane, *Advances in Math.* **55** (1985), 173–208.

**E.67.** For an elementary introduction to the Mandelbrot set, see the following Computer Recreations columns by A.K. Dewdney in *Scientific American:* **253** (no. 2), 198, 16–24; **257** (no. 5), 1987, 140–145. See the note on Exploration **E.49** for related references.

# Glossary

**abscissa** the first coordinate of a point in the cartesian plane

**arithmetic progression** a sequence in which the difference between adjacent entries is always the same

**closed interval** a set of reals of the form $\{t : a \leq t \leq b\}$ denoted by $[a, b]$

**closed unit disc** the set $\{z : |z| \leq 1\}$ in the complex plane

**composite** not prime

**coprime** with greatest common divisor 1

**dilatation** central similarity, homothety; a transformation which reduces the scale about a fixed center

**even function** a function $f$ satisfying $f(-x) = f(x)$

**factored** written as a product

**fraction in lowest terms** fraction for which the numerator and denominator are coprime

**geometric progression** a sequence in which the ratio of adjacent entries is always the same

**harmonic progression** a sequence of numbers whose reciprocals form an arithmetic progression

**identity** equation valid for all values of the variable

**iff** if and only if

**locus** of an equation: the set of points in the plane or more generally in $n$-dimensional space whose coordinates satisfy the equation

**lower bound** of a set of numbers: a number which does not exceed any number in the set

**negative** strictly less than 0

**nonnegative** greater than or equal to 0

**nonpositive** less than or equal to 0

**odd function** a function $f$ satisfying $f(-x) = -f(x)$

**one-one correspondence** function from one set onto another which is invertible, i.e. distinct elements of the domain have distinct images; a pairing of elements of one set with those of another

**open interval** a set of reals of the form $\{t : a < t < b\}$ denoted by $(a, b)$

**open unit disc** the set $\{z : |z| < 1\}$ in the complex plane

**ordinate** the second coordinate of a point in the cartesian plane

**over D** with coefficients in **D**

**parity** the property of being even or odd

**positive** strictly greater than 0

**pure imaginary** a complex number equal to $i$ times a nonzero real

**sign** the property of being positive, negative or zero

**supplementary angles** angles whose sum is 180°

**surd conjugate** of $a + \sqrt{b}$, $a - \sqrt{b}$

**trivial** inconsequential, nonessential, extremely straightforward (according to context)

**unique** only one, exactly one

**unit circle** the circle in the complex plane with center 0 and radius 1, the set $\{z : |z| = 1\}$

**unity** one

**upper bound** of a set of numbers: a number not less than any number in the set

**vanishes** equals zero

$[x]$ the greatest integer not exceeding $x$

$\Rightarrow$ implies, is sufficient for

$\Leftarrow$ is implied by, is necessary for

$\Longleftrightarrow$ if and only if, is logically equivalent to

**N** the set of natural numbers 1, 2, 3, 4, . . .

**Z** ("Zahlen") the ring of integers

**Q** the field of rational numbers

**R** the field of real numbers

**C** the field of complex numbers

**Heron's Formula** The area of a triangle with sides $a$, $b$, $c$ and semi-perimeter $s = \frac{1}{2}(a + b + c)$ is $\sqrt{s(s-a)(s-b)(s-c)}$.

**Intermediate Value Theorem** Let $f$ be a continuous, real-valued function (in particular, a polynomial) defined on a closed interval $[a, b]$. Then $f$ assumes on $[a, b]$ every value between $f(a)$ and $f(b)$. If $f(a)$ and $f(b)$ have opposite signs, then $f$ has a zero in $[a, b]$.

# Further Reading

A gentle first reader is the Mir tract by **Kurosh**. Designed for a one-semester course is the book by MacDuffee (**MacD**), who treats selected topics. The text of Dobbs & Hanks (**DH**), written for prospective teachers, is an admirable reference. More comprehensive texts are those of Borofsky (**Bo**) and Uspensky (**Us**), who cover most of the topics handled in this book and provide proofs. Older texts which may be consulted are those of Burnside (**Bu**) and Dickson (**Di**). Chapter II of D.J. Struik's *Source Book* contains excerpts from a number of historical papers on the theory of equations. For a modern abstract approach to polynomials, the experienced college student can have recourse to **Lausch & Nöbauer**. An overview is provided by the essay on the theory of algebraic equations by B.N. Delone, which is Chapter 4 of Volume 1 of the collection edited by Aleksandrov, Kolmgorov and Lavrent'ev.

A recent and highly recommended book by J.-P. **Tignol** deals with the theory of equations from an historical perspective, recapturing the methodology of the pioneers in the field. This treats many of the topics of the first four and sixth chapters, in particular the question of solvability by radicals.

**Chapter 1.** Solution of cubic and quartic equations: **Bo**, Ch. 8; **Us**, Ch. V and **Bu**, Ch. VI, **Cajori** (Hist), p. 133–139, **DH**, Ch. 3.
Complex numbers: **Bo**, Ch. 1.

**Chapter 2.** Horner's method: **Bo**, Ch. 7; **Bu**, Ch. X, where it is used as a tool in approximately solving equations.
Graphing: **Bu**, Ch. I; **MacD**, Ch. 4.
Multiple roots and derivatives: **Bu**, Ch. VII; **MacD**, Ch. 3; **DH**, Sect. 2.3.
Factor and remainder theorem: **DH**, Sect. 2.1.

**Chapter 3.** Greatest common divisor, factorization: **Bo**, Ch. 2; **MacD**, Ch. 7; **Us**, Ch. I.
Factoring and irreducibility: **DH**, Ch. 4.
Partial fractions: **MacD**, Ch. 3.

A survey article with many references on the current state of the art in factoring polynomials is Susan Laudau, Factoring polynomials quickly, *Notices A.M.S.* **34** (1987), Issue 253, 3–8.

**Chapter 4.** The Fundamental Theorem: **Di** (1914), Ch. V; **Bu**, Ch. X. Uspensky (**Us**, Ch. X) discusses Lagrange's method for solving cubic and biquadratic equations, and provides a different insight into Galois theory.

See also **DH**, Ch. 3, for a general discussion of solvability. An historical survey of the theory of equations and group theory appears in **Cajori** (Hist), p. 349–363.

Ruler and compasses constructions: **Bo**, Ch. 9; **Moise**, Ch. 19.

For a more comprehensive, but elementary, introduction to Galois theory, see the book by **Hadlock**. An elementary approach to the topological ideas behind the fundamental theorem is found in the book by **Chinn & Steenrod**. Those who would like to see the Fundamental Theorem in the hands of its discoverers should consult **Struik**.

**Chapter 5.** General: **Di** (1914), Ch. X; **Cajori** (Hist), 363–366.

Descartes rule: **Bu**, Ch. II; **Di** (1939), Ch. VII; **DH**, Sect. 5.3.

Bounds: **Bu**, Ch. VIII; **Us**, Ch. IV.

Fourier–Budan: **Bo**, Ch. 6; **Bu**, Ch. IX.

Approximation of roots: **Bo**, Ch. 7; **Us**, Ch. VIII; **Bu**, Ch. X; **Di** (1939), Ch. VIII; **Vilenkin**.

Separation of roots: Rolle's Theorem: **Us**, Ch. VI.

Sturm's Theorem: **Bo**, Ch. 6; **Bu**, Ch. IX; **Us**, Ch. VII; **MacD**, Ch. 4; **Di** (1939), Ch. VII; **DH**, Sect. 5.4.

Continued fractions: **Us**, App. 2; **Olds**

Location of zeros: **Polya–Szegö**, Part III, Ch. 1, Sect. 2.

A discussion of criteria for stability of polynomials appears in Chapter 7 of **Kaplan**.

**Chapter 6.** Symmetric functions of roots: **Bo**, Ch. 11; **Us**, Ch. XI; **MacD**, Ch. 6; **DH**, Sect. 2.4, 2.5.

Newton's formula for power sums: **Us**, Ch. XI.

**Chapter 7.** Finite differences and interpolation: **Milne–Thomson**, Ch. I-IV; **Scarborough**; **Ralston**, Ch. 3.

Lagrange interpolation: **DH**, Sect. 2.2.

A succinct, but advanced, introduction to approximation theory is given in **Lorentz**. This includes a treatment of Bernstein and Tchebychev polynomials. The alternation property is discussed in Section 7.3.

# Books

A.D. Aleksandrov, A.N. Kolmogorov & M.A. Lavrent'ev (eds.), *Mathematics: Its Content, Method and Meaning* (3 volumes)
Moscow, 1956 (Russian); MIT, 1963 (English)
Vol. 1, Ch. 4: B.N. Delone, Algebra: theory of algebraic equations.

George A. Baker, *Essentials of Padé Approximants*
Academic, 1975.

S. Barnett, *Polynomials and Linear Control Systems*
Marcel Dekker, NY, 1983.

Maxime Bocher, *Introduction to Higher Algebra*
Dover, 1964.

Samuel Borofsky, *Elementary Theory of Equations*
Macmillan, NY, 1950

William Snow Burnside, *The Theory of Equations, With an Introduction to the Theory of Binary Algebraic Forms*
Dover, 1960.

William Snow Burnside & Arthur William Panton, *The Theory of Equations,* 2nd ed.
Dublin, London, 1886

Florian Cajori, *A History of Mathematics,* 3rd ed.
Chelsea, 1980.

Girolamo Cardano, *The Great Art, or the Rules of Algebra (Artis Magnae, Sive de Regulis Algebraicis;* 1545)
Translated by T.R. Witmer
MIT, 1968.

W.G. Chinn & N.E. Steenrod, *First Concepts of Topology*
New Mathematical Library #18
Math. Assoc. of America, 1966.

George Chrystal, *Algebra, An Elementary Textbook for the Higher Classes of Secondary Schools and for Colleges*
2nd ed., Black, 1926; 7th ed., Chelsea, 1964.

Allan Clark, *Elements of Abstract Algebra*
Wadsworth, 1971.

Nelson Bush Conkwright, *Introduction to the Theory of Equations*
Ginn, 1957.

Richard J. Crouse & Clifford W. Sloyer, *Mathematical Questions From the Classroom*
Prindle, Weber & Schmidt, 1977.

L.E. Dickson, *Introduction to the Theory of Algebraic Equations*
Wiley, 1903.

ibid, *Elementary Theory of Equations*
Wiley, 1914.

ibid, *First Course in the Theory of Equations*
Wiley, 1922.

ibid, *New First Course in the Theory of Equations*
Wiley, 1939.

P. Dienes, *The Taylor Series*
Dover, 1957, pages 66–70.

David E. Dobbs & Robert Hanks, *A Modern Course in the Theory of Equations*
Polygonal Pub. House, 80 Passaic Ave., Passaic, NJ 07055, 1980.

H.D. Ebbinghaus, et al., *Zahlen*
(Grundwissen Mathematik I)
Springer-Verlag, 1983.

Howard Eves, *An Introduction to the History of Mathematics*
5th ed., Saunders, 1983
Exercise 3.10, p. 69; p. 59.

Robert P. Feinerman & Donald J. Newman, *Polynomial Approximation*
Williams & Wilkins, Baltimore, 1974.

Harry Freeman, *Mathematics for Actuarial Students*, Part II: *Finite Differences, Probability and Statistics*
Cambridge, 1952.

Joseph A. Gallian, *Contemporary Abstract Algebra*
Heath, 1986
Chapter 20: Factorization of polynomials.

Albert Gloden, *Mehrgradige Gleichungen*
Noordhoff, 1944 (2nd ed.).

Lois Wilfred Griffiths, *Introduction to Theory of Equations*
Wiley, 1945.

Charles R. Hadlock, *Field Theory and Its Classical Problems*
Carus Monograph No. 19, MAA, 1978.

H.S. Hall & S.R. Knight, *Higher Algebra*
MacMillan, 1887.

G.H. Hardy, J.E. Littlewood & G. Pólya, *Inequalities*
Cambridge, 1964
Chapter 2.

V.E. Hoggatt, Jr., *Fibonacci and Lucas Numbers*
Mathematics Enrichment Series; Houghton Mifflin, Boston, 1969.

Nathan Jacobson, *Basic Algebra I* 2nd ed.
W.H. Freeman, NY, 1985.

Wilfrid Kaplan, *Operational Methods for Linear Systems*
Addison-Wesley, 1962
Chapter 7: Stability of polynomials, pp. 403–435.

Felix Klein, *Elementary Mathematics from an Advanced Standpoint*
2 volumes; Dover, 1939.

Donald E. Knuth, *The Art of Computer Programming*
Addison-Wesley, 1969
Section 4.6, pp. 360–444 (Vol. 2).

Kaiser S. Kunz, *Numerical Analysis*
New York, 1957, pages 34–37.

A.G. Kurosh, *Algebraic Equations of Arbitrary Degree*
Little Mathematical Library
Mir, Moscow, 1977.

Hans Lausch & Wilfried Nöbauer, *Algebra of Polynomials*
North Holland/American Elsevier, 1973.

Howard Levi, *Polynomials, Power Series and Calculus*
Van Nostrand, Princeton, 1968.

D.E. Littlewood, *The Skeleton Key of Mathematics: A Simple Account of Complex Algebraic Theories*
Hutchinson University Library, London, 1949, 1957.

C.S. Liu, *Introduction to Combinatorial Mathematics*
McGraw-Hill, 1968
p. 111 seq.; p. 248 seq.

G.G. Lorentz, *Approximation of Functions*
Holt, Rinehart & Winston, Athena Series, 1966.

Cyrus Colton MacDuffee, *Theory of Equations*
Wiley, 1954

Morris Marden, *Geometry of Polynomials*
Math. Surveys, No. 3; AMS, 1966.

W.E. Milne, *Numerical Analysis*
Princeton, 1949, pages 53–57.

L.M. Milne-Thomson, *The Calculus of Finite Differences*
Macmillan, London, 1933

A.P. Mishina & I.V. Proskuryakov, *Higher Algebra: Linear Algebra, Polynomials and General Algebra*
Pergamon, 1965.

Edwin Moise, *Elementary Geometry From an Advanced Standpoint*
Addison-Wesley, 1963.

G. Pólya & G. Szegö, *Problems and Theorems in Analysis* Vol. I
4th ed., Springer-Verlag, 1972, pages 105–109, 300–305.

G. Pólya & G. Szegö, *Problems and Theorems in Analysis* Vol. II
4th ed., Springer-Verlag, 1976
pages 36–91, 129–134, 212–278, 319–330
Part Six: Polynomials and trigonometric functions
Part Eight, Chapter 2: Polynomials with integer coefficients and integer-valued functions.

H. Rademacher, *Higher Mathematics From an Elementary Point of View*
Birkhäuser, 1983.

A. Ralston, *A First Course in Numerical Analysis*
McGraw-Hill, 1965, 1978, Chapter 3.

Hans Reisel, *Prime Numbers and Computer Methods for Factorization*
Birkhäuser, 1985.

Theodore J. Rivlin, *An Introduction to the Approximation of Functions*
Blaisdell, 1969.

James B. Scarborough, *Numerical Mathematical Analysis,* 2nd ed.
Johns Hopkins, Baltimore, 1950.

Arthur Schultze, *Graphic Algebra*
Macmillan, NY, 1922
Sections 58, 59, 65.

David E. Smith (ed.), *A Source Book in Mathematics* (2 vols.)
McGraw-Hill, 1929; Dover, 1959.

D.J. Struik (ed.), *A Source Book in Mathematics, 1200–1800*
Harvard, 1969.

Jean-Pierre Tignol, *Galois' Theory of Algebraic Equations*
Longman, 1988.

Joseph Miller Thomas, *Theory of Equations*
McGraw-Hill, 1938.

John Todd, *Basic Numerical Mathematics*
Vol. 1: *Numerical Mathematics*
Birkhäuser, 1979
p. 86: Aitken's algorithm for computing the Lagrange polynomial by a succession of linear interpolations.

Herbert Westren Turnbull, *Theory of Equations,* 5th ed.
Interscience, 1957.

James Victor Uspensky, *Theory of Equations*
McGraw-Hill, 1948.

Bartel Leender van der Waerden, *Modern Algebra*
  Ungar, 1949.

Richard S. Varga, *Topics in Polynomial and Rational Interpolation and Approximation*
  University of Montreal, 1982.

N.Ya. Vilenkin, *Method of Successive Approximations*
  Little Mathematical Library
  Mir, Moscow, 1979.

N.N. Vorobyov, *The Fibonacci Numbers*
  Topics in Mathematics; D.C. Heath, Boston, 1963.

Louis Weisner, *Introduction to the Theory of Equations*
  Macmillan, NY, 1938.

Edmund Taylor Whittaker, *The Calculus of Observations; a Treatise on Numerical Mathematics*
  Blackie, London, 1960.

James H. Wilkinson, *The Perfidious Polynomial*
  Studies in Numercial Analysis
  ed. Gene H. Golub (MAA, 1984).

I.M. Yaglom, *Felix Klein and Sophus Lie: Evolution of the Idea of Symmetry in the Nineteenth Century*
  Birkhäuser, 1988, Chapter 1.

# Selected Papers

Other papers which the reader may find interesting are listed below. After the title is the indication **E**, **M**, **A** often followed by a number. **E** indicates that the paper is elementary and accessible to a high school student, **M** that it is of moderate difficulty and accessible to an undergraduate, **A** that it is advanced for a reader with mathematical maturity and considerable background. In some cases, the classification is difficult to make and so it should be regarded only as a first approximation. The number indicates the chapter in this text for the topic.

Most of the abbreviations for journals are straightforward. North American readers may not be familiar with the British journal, *Mathematical Gazette (Math. Gaz.)*. Other abbreviations which require some explanation are

| | |
|---|---|
| J. Math. & Phys. | Journal of Mathematics and Physics |
| Math. Comp. | Mathematics of Computation |
| Math. Ann. | Mathematische Annalen |
| Math. Intell. | Mathematical Intelligencer |

N.C. Ankeny, One more proof of the fundamental theorem of algebra
*Amer. Math. Monthly* **54** (1947), 464.

I.N. Baker, Fixpoints of polynomials and rational functions
*J. Lond. Math. Soc.* (1) **39** (1964), 615–622. [**A**]

J.P. Ballantine, Complex roots of real polynomials
*Amer. Math. Monthly* **66** (1959), 411–414. [**M** 5]

Ed Bergdal, Complex graphs
*Math. Mag.* **24** (1950), 195–202. [**E** 4]

E.R. Berlekamp, Factoring polynomials over large finite fields
*Math. Comp.* **24** (1970), 713–735 [**A** 3]

W.A. Blankinship, A new version of the Euclidean algorithm
*Amer. Math. Monthly* **70** (1963), 742–745. [**E** 1, 3]

R.P. Boas, Yet another proof of the fundamental theorem of algebra
*Amer. Math. Monthly* **71** (1964), 180. [**M** 4]

R.P. Boas, Extremal problems for polynomials
*Amer. Math. Monthly* **85** (1978), 473–475. [**A** 7]

David W. Boyd, The diophantine equation $x^2 + y^m = z^{2n}$
*Amer. Math. Monthly* **95** (1988), 544–547. [M 1]

Louis Brand, The roots of a quaternion
*Amer. Math. Monthly* **49** (1942), 519–520.

Rolf Brandl, Integer polynomials that are reducible modulo all primes
*Amer. Math. Monthly* **93** (1986), 286–288. [A 3]

J.L. Brenner & R.C. Lyndon, Proof of the fundamental theorem of algebra
*Amer. Math. Monthly* **88** (1981), 253–256. [M 4]

Louis Brickman, On nonnegative polynomials
*Amer. Math. Monthly* **69** (1962), 218–221. [M 1]

J.R. Britton, A note on polynomial curves
*Amer. Math. Monthly* **42** (1935), 306–310. [M 2]

Duane Broline, Renumbering the faces of dice
*Math. Magazine* **52** (1979), 312–315. [E 3]

W.S. Brown, Reducibility properties of polynomials over the rationals
*Amer. Math. Monthly* **70** (1963), 965–969. [A 3]

R. Creighton Buck, Sherlock Holmes in Babylon
*Amer. Math. Monthly* **87** (1980), 335–345. [E 1]
polynomial identities to generate pythagorean triples

F.J. Budden, Functions which permute the roots of an equation
*Math. Gaz.* **60** (1976), 24–38. [M 6]

Sylvan Burgstahler, An algorithm for solving polynomial equations
*Amer. Math. Monthly* **93** (1986), 421–430. [M 5]

R. Butler, The rapid factorization to high accuracy of quartic expressions
*Amer. Math. Monthly* **69** (1962), 138–141. [M 3]

D.G. Cantor, Irreducible polynomials with integral coefficients have succinct certificates
*J. Algorithms* **2** (1981), 385–392, MR 83a:12003. [A 3]

D.G. Cantor & H. Zassenhaus, A new algorithm for factoring polynomials over finite fields
*Math. Comp.* **36** (1981), 587–592, MR 82e:12020. [A 3]

L. Carlitz & J.M. Thomas, Rational tabulated values of trigonometric functions
*Amer. Math. Monthly* **69** (1962), 789–793. [M 1]

F.W. Carroll, A polynomial in each variable separately is a polynomial
*Amer. Math. Monthly* **68** (1961), 42. [M 1,2]

Roger Chalkley, Cardan's formula and biquadratic equations
*Math. Mag.* **47** (1974), 8–14. [M 1]

Geng-Zhe Chang, Bernstein polynomials via the shifting operator
*Amer. Math. Monthly* **91** (1984), 634–638. [M 7]

John G. Christiano, On the sum of powers of natural numbers
  *Amer. Math. Monthly* **68** (1961), 149–151. [**E** 6]

G.E. Collins, Computer algebra of polynomials and rationals
  *Amer. Math. Monthly* **80** (1973), 725–755. [**A** 3,5]

Brian Conrey & Amit Ghosh, On the zeros of the Taylor polynomials associated with the exponential function
  *Amer. Math. Monthly* **95** (1988), 528–533. [**A** 5]

Carl H. Denbow, Some types of elementary equations
  *Math. Mag.* **23** (1949), 137–141. [**E** 1]

Emeric Deutsch, Bounds for the zeros of polynomials
  *Amer. Math. Monthly* **88** (1981), 205–206. [**E** 5]

H.L. Dorwart, Irreducibility of polynomials
  *Amer. Math. Monthly* **42** (1935), 369–381. [**M** 3]

J.E. Eaton, The Fundamental Theorem of Algebra
  *Amer. Math. Monthly* **67** (1960), 578–579. [**A** 4]

A.W.F. Edwards, A quick route to the sums of powers
  *Amer. Math. Monthly* **93** (1986), 451–455. [**M** 6]

Evelyn Frank, On the calculation of the roots of equations
  *J. Math. & Phys.* **34** (1955), 187–197. [**M** 5]
  Generalization of Newton's method to $f(x, y) = g(x, y) = 0$
  Determination of complex roots of complex polynomials

Kenneth W. Frank, Bent wire—an application of quadratic equations and inequalities
  *Math. Teacher* **79** (1986), 57–58. [**E** 1]

J.A. Gallian & D.J. Rusin, Cyclotomic polynomials and nonstandard dice
  *Discrete Math.* **27** (1979), 245–259. [**M** 3]

Edward F. Gardner, Some little discussed facts about cubic polynomials (Sharing teaching ideas)
  *Math. Teacher* **81** (1988), 112–113.

J. von zur Gathen & E. Kaltofen, Factorization of multivariate polynomials over finite fields
  *Math. Comp.* **45** (1985), 251–261. [**A** 3]

H.M. Gehman, Complex roots of a polynomial equation
  *Amer. Math. Monthly* **48** (1941), 237–239. [**M** 2,4]

H.W. Gould, Euler's formula for $n$th differences of powers
  *Amer. Math. Monthly* **85** (1978), 450–467. [**M** 7]

J.H. Grace, The zeros of a polynomial
  *Proc. Cambridge Phil. Soc.* **11** (1902), 352–357.

D.R. Green, The historical development of complex numbers
  *Math. Gaz.* **60** (1976), 99–107. [**E** 1]

Louisa S. Grinstein, Upper limits to the real roots of polynomial equations
*Amer. Math. Monthly* **60** (1953), 608–615. [**M** 5]

C.W. Groetsch, J.T. King, The Bernstein polynomials and finite differences
*Math. Mag.* **46** (1973), 280–282. [**M** 7]
Bernstein polynomials appear as a finite difference approximant to a first order partial differential equation

Emil Grosswald, Recent applications of some old work of Laguerre
*Amer. Math. Monthly* **86** (1979), 648–658. [**M** 5]

H. Guggenheimer, Bounds for roots of algebraic equations
*Amer. Math. Monthly* **69** (1962), 915–916. [**E** 5]

J.E. Hacke, Jr., A simple solution of the general quartic
*Amer. Math. Monthly* **48** (1941), 327–328. [**E** 1]

Morton J. Hellman, A unifying technique for the solution of the quadratic, cubic and quartic
*Amer. Math. Monthly* **65** (1958), 274–276. [**E** 1,6]

Morton J. Hellman, The insolvability of the quintic re-examined
*Amer. Math. Monthly* **66** (1959), 410. [**A**]

Garcia Henriquez, The graphical interpretation of the complex roots of cubic equations
*Amer. Math. Monthly* **42** (1935), 383–384. [**E** 1,2]

Irwin Hoffman & Larry Kauvar, Computer oriented mathematics: polynomial synthetic division
*Math. Teacher* **63** (1970), 429–431. [**E** 1]

F.E. Hohn, The number of terms in a polynomial
*Amer. Math. Monthly* **48** (1941), 686. [**E** 1]

J.S. Huang, Is a sequence of polynomials complete?
*Amer. Math. Monthly* **85** (1978), 107–108. [**A** 7]

C.A. Hutchinson, On Graeffe's method for the numerical solution of algebraic equations
*Amer. Math. Monthly* **42** (1935), 149–161. [**M** 5]

Margaret Wiscomb Hutchinson, Using synthetic division by quadratics to find rational roots
*Math. Teacher* **64** (1971), 349–352 = Crouse & Sloyer, 286–289. [**E** 1,3]

I.M. Isaacs, Solution of polynomials by real radicals
*Amer. Math. Monthly* **92** (1985), 571–575. [**A** 4]

Elbert Johnson & C.R. Wylie, Jr., A nomographic solution of the quartic
*Amer. Math. Monthly* **68** (1961), 461–464. [**M** 1]

E.C. Kennedy, Bounds for the roots of a trinomial equation
*Amer. Math. Monthly* **47** (1940), 468–470 [**E** 5]

E.C. Kennedy, Concerning nearly equal roots
*Amer. Math. Monthly* **48** (1941), 42–43. [**E** 5]

Clark Kimberling, Microcomputer-assisted mathematics: Roots: half-interval search
  *Math. Teacher* **78** (1985), 120–123. [**E** 5]

Clark Kimberling, Microcomputer-assisted mathematics: Roots: Newton's method
  *Math. Teacher* **78** (1985), 626–629. [**E** 5]

Clark Kimberling, Factoring and unfactoring
  *Math. Teacher* **79** (1986), 48–53. [**E** 1,3]

Murray S. Klamkin, A polynomial functional equation
  *Eureka* (*Crux Mathematicorum*) **4** (1978), 32–33.

P.G. Laird & R. McCann, On some characterizations of polynomials
  *Amer. Math. Monthly* **91** (1984), 114–116. [**M** 1]

Susan Landau, Factoring polynomials quickly
  *Notices A.M.S.* **34** (1987), 3–8 (No. 1, Issue 253).

M.A. Lee, Some irreducible polynomials which are reducible mod $p$ for all $p$
  *Amer. Math. Monthly* **76** (1969), 1125. [**A** 3]

A.K. Lenstra, H.W. Lenstra, L. Lovàsz, Factoring polynomials with rational coefficients
  *Math. Ann.* **261** (1982), 515–534. [**A** 3]

John S. Lew, Polynomials in two variables taking distinct integer values at lattice points
  *Amer. Math. Monthly* **88** (1981), 344–346. [**A** 1]

Tien-Yi Li, Solving polynomial systems
  *Math. Intell.* **9** (1987), no. 3, 33–39. [**M** 4]

W.B.R. Lickorish & K.C. Millett, The new polynomial invariants of knots and links
  *Math. Magazine* **61** (1988), 3–23. [**M**]

Shih-Nge Lin, A method of successive approximations of evaluating the real and complex roots of cubic and higher-order equations
  *J. Math. Phys.* **20** (1941), 231–242. [**M** 5]

Daniel B. Lloyd, Factorization of the general polynomials by means of its homomorphic congruential functions
  *Amer. Math. Monthly* **71** (1964), 863–870. [**M** 3]

David London, On a connection between the permanent function and polynomials
  *Linear & Multilinear Alg.* **1** (1973), 231–240. [**M**]

Calvin T. Long, Gregory interpolation: a trick for problem solvers from out of the past
  *Math. Teacher* **76** (1983), 323–325. [**E** 7]

C.C. MacDuffee, Some applications of matrices in the theory of equations
*Amer. Math. Monthly* **57** (1950), 154–161. [**M** 2]

D. Mackie & T. Scott, Pitfalls in the use of computers for the Newton–Raphson method
*Math. Gaz.* **69** (1985), 252–257.

Pavan K. Malhotra, A new method of solving a quartic
*Amer. Math. Monthly* **65** (1958), 280–282. [**A** 1]

Morris Marden, Location of zeros of infrapolynomials
*Amer. Math. Monthly* **70** (1963), 361–371. [**A** 5,7]

Morris Marden, Conjectures on the critical points of a polynomial
*Amer. Math. Monthly* **90** (1983), 267–276. [**A** 5]

Kenneth O. May & Henry S. Tropp, Some algebraic equations do not have exactly $n$ roots
*Math. Teacher* **66** (1973), 179–182 = Crouse & Sloyer, 279–282. [**E** 4]

John McKay, On computing discriminants
*Amer. Math. Monthly* **94** (1987), 523–527. [**M** 6]

Katherine E. McLain & Hugh M. Edgar, A note on Golomb's "Cyclotomic polynomials and factorization theorems"
*Amer. Math. Monthly* **88** (1981), 753. [**A** 3]

M. Mignotte, An inequality about factors of polynomials
*Math. Comp.* **28** (1974), 1153–1157. [**A** 3,5]

Harold Willis Milnes, Conditions that the zeros of a polynomial lie in the interval $[-1, 1]$ when all zeros are real
*Amer. Math. Monthly* **70** (1963), 746–750. [**M** 5]

L. Mirsky, A note on cyclotomic polynomials
*Amer. Math. Monthly* **69** (1962), 772–775. [**A** 3]

Q.G. Mohammad, On the zeros of polynomials
*Amer. Math. Monthly* **69** (1962), 901–904. [**M** 5]

Q.G. Mohammad, On the zeros of polynomials
*Amer. Math. Monthly* **72** (1965), 631–633. [**A** 3]

J.C. Molluzzo, A representation theorem for polynomials of two variables
*Amer. Math. Monthly* **82** (1975), 385–387. [**M** 6]

Gerald Myerson, How small can a sum of roots of unity be? (unsolved problem)
*Amer. Math. Monthly* **93** (1986), 457–459. [**A** 1,3]

Goro Nagase, Using the discriminant for problems involving extrema
*Math. Teacher* **79** (1986), 145–146. [**E** 1,2]

Goro Nagase, Existence of real roots of a radical equation
*Math. Teacher* **80** (1987), 369–370.

D.J. Newman, Norms of polynomials
*Amer. Math. Monthly* **67** (1960), 778–779. [**A**]

Ivan Niven, The roots of a quaternion
*Amer. Math. Monthly* **49** (1942), 386–388. [**M** 1]

William J. O'Donnell, Computers and the rational root system
*Math. Teacher* **81** (1988), 142–145.

Patrick J. O'Hara & Rene S. Rodriguez, Polynomials with zeros uniformly
distributed on the unit circle
*Amer. Math. Monthly* **85** (1978), 814–817. [**A** 5]

E.L. Ortiz & T.J. Rivlin, Another look at the Chebyshev polynomials
*Amer. Math. Monthly* **90** (1983), 3–10. [**M** 1]

George H. Palagi, A conversation on factoring
*Math. Teacher* **66** (1973), 671–672 = Crouse & Sloyer, 86–89. [**A** 3]

R.V. Parker, Multiplication and division by binomial factors
*Amer. Math. Monthly* **65** (1958), 39–42. [**A** 1]

Ricarda A. Perez, Solving a polynomial equation of degree greater than
two
*Math. Teacher* **80** (1987), 207–208. [**E** 1]

Gregg N. Petruno, Sums of irrational square roots are irrational
*Math. Magazine* **61** (1988), 44-45. [**E** 1]

John W. Pratt, Finding how many roots a polynomial has in $(0, 1)$ or $(0, \infty)$
*Amer. Math. Monthly* **86** (1979), 630–637. [**M** 5]

S. Ramanujan, Note on a set of simultaneous equations
*J. Ind. Math. Soc.* **4** (1912), 94–96 = Collected papers (# 3), 18–19
(Chelsea, 1927, 1962).

R.M. Redheffer, What! Another note just on the fundamental theorem of
algebra?
*Amer. Math. Monthly* **71** (1964), 180–185 [**M** 4]

R.E. Rice, B. Schweizer & A. Sklar, When is $f(f(z)) = az^2 + bz + c$?
*Amer. Math. Monthly* **87** (1980), 252–263. [**A** 1]

T.J. Rivlin, On the maximum modulus of polynomials
*Amer. Math. Monthly* **67** (1960), 251–253. [**A**]

D.W. Robinson, A matrix application of Newton's identities
*Amer. Math. Monthly* **68** (1961), 367–369. [**M** 6]

Raphael M. Robinson, Three old problems about polynomials with real
roots (unsolved problems)
*Amer. Math. Monthly* **95** (1988), 329–330.

Steven Roman, The formula of Faà di Bruno
*Amer. Math. Monthly* **87** (1980), 805–809. [**M** 2]

David Ruelle, Is our mathematics natural? The case of equilibrium statis-
tical mechanics
*Bull. Amer. Math. Soc.* (NS) **19** (1988), 259–268.

T.R. Running, Graphical solutions of cubic, quartic and quintic
*Amer. Math. Monthly* **50** (1943), 170–173. [M 1,4]

Herbert E. Salzer, Polynomials for best approximation over semi-infinite and infinite intervals
*Math. Mag.* **23** (1949), 59–69. [M 7]

John Savage, Factoring quadratics
*Math. Teacher* **82** (1989), 35–36. [E 1,3]

I.J. Schoenberg & G. Szegö, An extremum problem for polynomials
*Compositio Math.* **14** (fasc. 3), 260–268. [A 7]

E.J. Scott, A formula for the derivatives of Tchebychef polynomials of the second kind
*Amer. Math. Monthly* **71** (1964), 524–525. [A 1,2]

Henry S. Sharp, A comparison of methods for evaluating the complex roots of a quartic equation
*J. Math. Phys.* **20** (1941), 243–258. [M 1,5]

Allen Shields, Polynomial approximation (Years Ago)
*Math. Intell.* **9** (1987), no. 3, 5–7.

W.M. Snyder, Factoring repunits
*Amer. Math. Monthly* **89** (1982), 462–466. [A 3]

D.D. Stancu, A method of obtaining polynomials of Bernstein type of two variables
*Amer. Math. Monthly* **70** (1963), 260–264. [M 7]

W.J. Sternberg, On polynomials with multiple roots
*Amer. Math. Monthly* **52** (1945), 440. [M 2]

H. Joseph Straight & Richard Dowds, An alternate method for finding the partial fraction decomposition of a rational function
*Amer. Math. Monthly* **91** (1984), 365–367. [E 3]

R.J. Stroeker, How to solve a diophantine equation
*Amer. Math. Monthly* **91** (1984), 385–392. [A 8]

J.M. Thomas, The linear diophantine equation in two unknowns
*Math. Mag.* **24** (1950), 59–64. [E 1]

Jan Turk, The fixed divisor of a polynomial
*Amer. Math. Monthly* **93** (1986), 282–286. [A 3]

C.E. van der Ploeg, Duality in nonreal quartic fields
*Amer. Math. Monthly* **94** (1987), 279–284. [A 4]

R.C. Vaughan, Adventures in Arithmetick, or: How to make good use of a Fourier transform
*Math. Intelligencer* **9** (no. 2) (1987), 53–60.

J.H. Wahab, Irreducibility of polynomials
*Amer. Math. Monthly* **68** (1961), 366–367. [M 3]

H.S. Wall, Polynomials whose zeros have negative real parts
*Amer. Math. Monthly* **52** (1945), 308–322. [**M** 5]

J.L. Walsh, On the location of the roots of certain polynomials
*Trans. Amer. Math. Soc.* **24** (1922), 163–180. [**M** 5]

J.L. Walsh, An inequality for the roots of an algebraic equation
*Annals Math.* **25** (1924), 285–286. [**E** 5]

P.S. Wang & L.P. Rothschild, Factoring multivariate polynomials over the integers
*Math. Comp.* **29** (1975), 935–950. [**A** 3]

Paul S. Wang, An improved multivariate factoring algorithm
*Math. Comp.* **32** (1978), 1215–1231. [**M** 3]

L.E. Ward, Jr., Linear programming and approximation problems
*Amer. Math. Monthly* **68** (1961), 46–53. [**M** 7]

William C. Waterhouse, A neglected note showing Gauss at work
*Historia mathematica* **13** (1986), 147–156 [**M** 3]

E.E. Watson, A test for the nature of the roots of the cubic equation
*Amer. Math. Monthly* **48** (1941), 687. [**E** 1]

John J. Wavrik, Computers and the multiplicity of polynomial roots
*Amer. Math. Monthly* **89** (1982), 34–56. [**A** 5]

Kenneth W. Wegner, Trigonometric excursions and side trips
*Math. Teacher* **60** (1967), 33–37 = Mathematics in the secondary school classroom: Selected Readings (eds. Rising & Wiesen; publ. T.Y. Crowell, 1972), 376–382. [**E** 1,3]

P.J. Weinberger, Finding the number of factors of a polynomial
*J. Algorithms* **5** (1984), 180–186, MR 86h:11110. [**A** 3]

C.R. White, Definitive solutions of general quartic and cubic equations
*Amer. Math. Monthly* **69** (1962), 285–287. [**M** 1]

H.S. Wilf, Curve fitting matrices
*Amer. Math. Monthly* **65** (1958), 272–274. [**M** 7]

Wm. Douglas Withers, Folding polynomials and their dynamics
*Amer. Math. Monthly* **95** (1988), 399–413. [**A**]

H. Zassenhaus, On Hensel factorization I
*J. Number Theor.* **1** (1969), 291–311. [**A** 3]

H. Zassenhaus, A remark on the Hensel factorization method
*Math. Comp.* **32** (1978), 287–292. [**A** 3]

# Index